The Stille Reaction

The Stille Reaction

VITTORIO FARINA AND VENKAT KRISHNAMURTHY
Department of Process Research
Boehringer Ingelheim Pharmaceuticals
Ridgefield, Connecticut

WILLIAM J. SCOTT
Department of Medicinal Chemistry
Institute of Chemistry
Bayer Corporation
West Haven, Connecticut

Published in association with
Organic Reactions, Inc.

JOHN WILEY & SONS, INC.
New York · Chichester · Weinheim · Brisbane · Singapore · Toronto

Originally published as Volume 50 of *Organic Reactions*.

This book is printed on acid-free paper.

Library of Congress Cataloging-in-Publication Data

Farina, Vittorio.
    The Stille reaction / Vittorio Farina and Venkat Krishnamurthy,
William J. Scott.
        p. cm.
    Includes index.
    ISBN 0-471-31273-8 (pbk. : alk. paper)
    1. Stille reaction. I. Krishnamurthy, Venkat. II. Scott,
William Johnston, 1951–   . III. Title.
QD262.F37   1998
547'.2—dc21   98-39154
                                                                    CIP

FOREWORD

A historic meeting occurred in December, 1939 at the Eighth National Organic Chemistry Symposium being held in St. Louis. At that time, agreement was reached between a group of academicians headed by Roger Adams and representatives of John Wiley and Sons, Inc. concerning the establishment of a corporation to be known as *Organic Reactions*. These actions, formalized in 1942, were taken for the explicit purpose of advancing the educational and research objectives of the worldwide organic chemical community. The sound planning established in the early days has made it possible to publish 52 volumes of *Organic Reactions* by mid-1998. Indeed, the special format of the many extensive chapters contained therein and the very considerable amount of volunteer service provided by a relatively small group of Editorial Board members and authors have played a significant role in serving the research needs of organic chemists, even as our field of endeavor evolves so very rapidly.

During the past forty years, *Organic Reactions* has given increased attention to its educational mission. Since December, 1956, an outreach program has been in place wherein undergraduates, graduate students, postdoctorals, and research fellows have found it possible to purchase individual volumes of *Organic Reactions* at half the list price. The cost differential, paid by the corporation, has helped young chemists add approximately 28,000 volumes to their personal libraries during this period. Since 1958, *Organic Reactions* has also co-sponsored the Roger Adams Award with *Organic Synthesis*. For the past decade, an Organic Reactions Lecturer has been named on a biennial basis with the stipulation that the outstanding young non-U.S. investigator lecture primarily at those institutions represented by the Editorial Board members. More recently, *Organic Reactions* has sponsored an ACS Division of Organic Chemistry Graduate Fellowship on an annual basis.

Despite the above efforts, we believe that more can be done to foster our educational mission. The present paperback volume comprises a further step in this direction. The growth of organic chemistry has led to longer chapters covering topics of more widespread relevance. The coverage of "The Stille Reaction" by Vittorio Farina, Venkat Krishnamurthy, and William J. Scott is illustrative of this trend. For the purpose of disseminating the contents of this chapter as widely as possible, *Organic Reactions* has authorized its incorporation into a single lower-priced paperback edition. In so doing, we hope that individual copies of this valuable book would find their way into a significantly greater number of personal libraries and individual laboratories as a ready reference.

The response to this experiment will dictate whether future *Organic Reactions* chapters will be handled in a similar manner. Obviously, the Editorial Board of *Organic Reactions* is highly optimistic as to its success.

Leo Paquette
Editor-in-Chief
Columbus, Ohio

PREFACE

The Stille reaction is just one member of a large family of transition metal-catalyzed cross-coupling reactions. However, its generality and high degree of success have made it most popular among synthetic chemists as attested to by the 570 pages of examples offered in tabular form in this review.

One obvious reason for such popularity is the reaction's versatility. As opposed to most cross-coupling reactions which operate in more or less strongly basic media, the Stille reaction works under neutral conditions, and is therefore compatible with virtually all functional groups without requiring protection. It is one of very few reactions that can form carbon-carbon bonds under such mild conditions, and it does so often with complete stereospecificity. It is therefore often used in the total syntheses of complex molecules as a way of coupling together complex subunits. It is an excellent tool for medicinal chemists, allowing the rapid and systematic modification of lead structures, again simply because it can be predictably performed regardless of other functional groups present in the molecule.

One of the major limitations usually associated with the Stille reaction is the presence of toxic and hard-to-remove tin-containing by-products. This has indeed limited the industrial applications of the reaction, because, in such setting, the tolerance for tin-containing impurities is low (in the ppm range). For the academic lab, however, this limitation is not a serious one because there are several methods that can be used for the removal of tin, as discussed and exemplified in the Experimental Conditions section. The acute toxicity of tin-containing compounds is usually low: tributyltin chloride, the most common tin-containing by-product, for example, has an LD_{50} in rats that is over 100 mg/kg. For industrial applications, work is still needed to render the work-up more practical. Preliminary results using water-soluble organostannates are promising and may point to a general solution. Polymer-bound tin has also been used to facilitate removal of tin from the products.

The book begins with a brief mechanistic discussion; there is little known in detail about the mechanism, but the basic concepts need to be understood in order to apply the reaction successfully in non-trivial cases.

The section on scope and limitations surveys the impressive range of electrophiles that have been employed successfully. This is then followed by a survey of the types of stannanes that can be used. Since any cross-coupling can be approached in two ways (i.e. either fragment can be an electrophile or a nucleophile), it is important to understand the scope of the reaction to make an intelligent choice. We have added a brief section that illustrates how the Stille reaction can be successfully incorporated in reaction cascades, some totally palladium-catalyzed, some not. Since there are many insertion reactions (olefins, carbon monoxide etc.) quite common in palladium chemistry, it is possible to devise sequences that alternate one or more insertions with a transmetallation involving tin. Several examples are already known, but this area is open to further creative applications. We have added a section on side reactions, because, in spite of the generality of the method, failures are not uncommon. We already understand enough about some of these side reactions to be able to adjust conditions (solvent, catalyst, co-catalyst) in order to minimize or avoid these processes. The experimental conditions are then given, with a short introduction on how to select the experimental parameters. There is no "general best protocol" for the Stille reaction, and it is likely that many conditions will all work. Unavoidably, fine-tuning of the experimental conditions has taken place after Stille's pioneering work, and it is now possible to carry out a huge range of cross-couplings without significant problems, sometimes using special ligands, additives or even special stannanes. The survey of procedures presents a few that originate with Stille's work and are still quite useful, as well some new protocols that have been successful in more challenging transformations. The tables are designed to inform the reader of precedents that may be related to the synthetic operation he intends to carry out: the reactions are organized by electrophile class, and in each class the electrophiles are listed in ascending order of carbon atoms. Because all the electrophiles are organized in 33 different tables, each table has an average of 17 pages and browsing should be easy. The table lists reaction conditions, yields and structure of the major product. Well-documented failures are also listed, as this may interest the reader attempting a similar transformation.

I think this volume should be of considerable utility to all practicing synthetic organic chemists. It offers easy-to-follow recipes for the casual user of the reaction, but also underlines areas where more work is needed for the chemist who may be interested in developing the reaction further.

ACKNOWLEDGMENTS

We thank Dr. Gregory P. Roth for help with parts of the manuscript.

Vittorio Farina
February 1998

CONTENTS HIGHLIGHTS

For complete contents see page 1.

ACKNOWLEDGMENTS 3
INTRODUCTION 3
MECHANISTIC CONSIDERATIONS, REGIOCHEMISTRY AND
    STEREOCHEMISTRY 4
SCOPE AND LIMITATIONS: THE ELECTROPHILE 9
SCOPE AND LIMITATIONS: THE STANNANE 25
CARBONYLATIVE COUPLINGS 36
COMPLEX SYNTHETIC SEQUENCES INVOLVING TIN-TO-
    PALLADIUM(II) METATHESIS STEPS 42
SIDE REACTIONS 46
COMPARISON WITH OTHER METHODS 51
EXPERIMENTAL CONDITIONS 52
EXPERIMENTAL PROCEDURES 55
TABULAR SURVEY 59
REFERENCES 633

INDEX 653

THE STILLE REACTION

Vittorio Farina and Venkat Krishnamurthy

*Department of Process Research, Boehringer Ingelheim Pharmaceuticals,
Ridgefield, Connecticut*

William J. Scott

*Department of Medicinal Chemistry, Institute for Chemistry, Bayer Corp.,
West Haven, Connecticut*

CONTENTS

	Page
Acknowledgments	3
Introduction	3
Mechanistic Considerations, Regiochemistry and Stereochemistry	4
Scope and Limitations: The Electrophile.	9
Alkenyl Halides	9
Aryl and Heterocyclic Halides	12
Acyl Chlorides	16
Allylic, Benzylic, and Propargylic Electrophiles	17
Alkenyl Sulfonates and Other Electrophiles	19
Aryl and Heterocyclic Sulfonates and Other Derivatives	21
Miscellaneous Electrophiles.	23
Scope and Limitations: The Stannane	25
Alkylstannanes	25
Alkenylstannanes	27
Aryl and Heterocyclic Stannanes	30
Alkynylstannanes	32
Allylstannanes	32
Other Stannanes	34
Carbonylative Couplings	36
Alkenyl Halides	36
Aryl and Heterocyclic Halides	37
Allylic and Benzylic Halides	39
Alkenyl Sulfonates	40
Aryl and Heterocyclic Sulfonates	40
Miscellaneous Substrates	41

COMPLEX SYNTHETIC SEQUENCES INVOLVING TIN-TO-PALLADIUM(II) METATHESIS STEPS   .   .   42
SIDE REACTIONS   46
   Homocoupling Reactions   46
   Transfer of "Nontransferable" Ligands   47
   Destannylation   48
   Cine Substitution   48
   Phosphorus-to-Palladium Aryl Migration   49
   Electrophile Reduction   49
   Product Isomerization   49
   Miscellaneous Side Reactions   50
COMPARISON WITH OTHER METHODS   51
EXPERIMENTAL CONDITIONS   52
   The Stannane: Preparation and Handling   52
   Alkenyl and Aryl Triflates   52
   Choice of Nontransferable Ligands   52
   Choice of Catalyst and Ligands   53
   Choice of Solvent   53
   Additives.   54
   Workup: Removal of Tin Halides   54
EXPERIMENTAL PROCEDURES   55
   Trimethyl([3-(cyclohexen-1-yl)-2-propynyl]oxy)silane [Cross-Coupling of a Vinyl
      Halide with an Alkynylstannane Using Pd(PPh$_3$)$_2$Cl$_2$]   55
   4-tert-Butyl-1-vinylcyclohexene [Cross-Coupling of a Vinyl Triflate with a
      Vinylstannane Using Pd(PPh$_3$)$_4$ and LiCl]   55
   1-(4-Methoxyphenyl)-4-tert-butylcyclohexene [Cross-Coupling of a Vinyl Triflate
      with an Arylstannane Using Pd$_2$(dba)$_3$ and AsPh$_3$]   56
   3-Methyl-2-(4-tolyl)-2-cyclopentenone [Cross-Coupling of an Unreactive Alkenyl
      Halide under "Modified" Conditions Using Pd(PhCN)$_2$Cl$_2$, AsPh$_3$, and CuI as
      Cocatalyst]   56
   1-(4-Nitrophenyl)-2-propenone (Cross-Coupling of an Acid Chloride with an
      Arylstannane)   57
   4-Allylacetophenone [Cross-Coupling of an Aryl Triflate under Mild Conditions
      using Tri(2-furyl)phosphine as Ligand]   57
   8-(Trimethylstannyl)quinoline (Preparation of an Arylstannane by Cross-Coupling
      an Aryl Triflate with Hexamethyldistannane)   58
   4-(tert-Butyl-1-vinylcyclohexen-1-yl)-2-propenone [Carbonylative Cross-Coupling
      of an Alkenyl Triflate with an Alkenylstannane using Pd(PPh$_3$)$_4$ and LiCl]   .   58
   (E)-1-(4-Methoxyphenyl)-3-phenyl-2-propenone [Carbonylative Cross-Coupling
      of an Aryl Triflate with an Alkenylstannane using Pd(dppf)Cl$_2$ and LiCl]   .   59
TABULAR SURVEY   59
   Table I. Direct Cross-Coupling of Alkenyl Electrophiles   62
   Table II. Intramolecular Cross-Coupling of Alkenyl Electrophiles   144
   Table III. Direct Cross-Coupling of Aryl Electrophiles   153
   Table IV. Intramolecular Cross-Coupling of Aryl Electrophiles   283
   Table V. Direct Cross-Coupling of Furan and Benzofuran Electrophiles   .   .   289
   Table VI. Direct Cross-Coupling of Pyrrole and Indole Electrophiles   .   .   .   293
   Table VII. Direct Cross-Coupling of Thiophene and Benzothiophene Electrophiles   .   300
   Table VIII. Direct Cross-Coupling of Pyran and Benzopyran Electrophiles .   .   .   311
   Table IX. Direct Cross-Coupling of Pyridine Electrophiles   319
   Table X. Direct Cross-Coupling of Pyrimidine Electrophiles   333
   Table XI. Direct Cross-Coupling of Quinoline and Isoquinoline Electrophiles   .   .   349
   Table XII. Direct Cross-Coupling of Miscellaneous Heterocyclic Electrophiles   .   .   359
   Table XIII. Direct Cross-Coupling of Acyl Chlorides: Alkyl Systems   .   .   .   406
   Table XIV. Direct Cross-Coupling of Acyl Chlorides: Aryl Systems   .   .   .   428

Table XV. Direct Cross-Coupling of Acyl Chlorides: Benzyl Systems 454
Table XVI. Direct Cross-Coupling of Acyl Chlorides: Alkenyl Systems . . . 456
Table XVII. Direct Cross-Coupling of Acyl Chlorides: Heterocyclic Systems . . 462
Table XVIII. Direct Cross-Coupling of Chloroformates and Carbamoyl Chlorides . 467
Table XIX. Intramolecular Cross-Coupling of Acyl Chlorides and Chloroformates . 471
Table XX. Direct Cross-Coupling of Allyl and Propargyl Electrophiles . . . 474
Table XXI. Direct Cross-Coupling of Benzyl Electrophiles 512
Table XXII. Intramolecular Cross-Coupling of Allyl and Benzyl Electrophiles . . 517
Table XXIII. Direct Cross-Coupling of Organometallic Electrophiles 521
Table XXIV. Direct Cross-Coupling of Miscellaneous Electrophiles 534
Table XXV. Carbonylative Cross-Coupling of Alkenyl Electrophiles 550
Table XXVI. Carbonylative Cross-Coupling of Aryl Electrophiles. 559
Table XXVII. Carbonylative Cross-Coupling of Heterocyclic Electrophiles . . . 575
Table XXVIII. Carbonylative Cross-Coupling of Allyl and Benzyl Electrophiles . 578
Table XXIX. Carbonylative Cross-Coupling of Miscellaneous Electrophiles . . 584
Table XXX. Intramolecular Carbonylative Cross-Coupling Reactions 585
Table XXXI. Cross-Coupling Reactions that Form Polymers 587
Table XXXII. Multi-Step Transformations Involving Direct
    Cross-Coupling Reactions. 596
Table XXXIII. Multi-Step Transformations Involving Carbonylative Cross-Coupling. 626
REFERENCES 633

ACKNOWLEDGMENTS

We thank Dr. Gregory P. Roth for help with parts of the manuscript.

INTRODUCTION

Examples of the palladium-catalyzed coupling of organotin compounds with carbon electrophiles were first reported in 1977 by Kosugi, Shimizu, and Migita.[1-3] The first study by Stille appeared in 1978.[4] The early work of Beletskaya, using "ligandless" catalysts in cross-coupling reactions, also often employed organostannanes.[5] In recognition of Stille's comprehensive synthetic and mechanistic studies, this coupling is now referred to as the Stille reaction.[6] The Stille reaction is schematically defined in Eq. 1.

$$R^1Sn(R^2)_3 + R^3X \xrightarrow{Pd(0)L_n} R^1\text{-}R^3 + (R^2)_3SnX \qquad \text{(Eq. 1)}$$

In Eq. 1, R^1 is typically an unsaturated moiety (e.g., vinyl, aryl, heteroaryl, alkynyl, allyl) or less often an alkyl group, and R^2, the nontransferable ligand, is almost always butyl or methyl. Electrophiles participating in the coupling include halides (almost always bromides or iodides) and sulfonates (most often used are the triflates). Other leaving groups have been used in special cases.

The Stille reaction belongs to the larger family of palladium- and nickel-catalyzed cross-coupling reactions which features, e.g., organomagnesium,[7] organozinc,[8] organoboron,[9] and organosilicon reagents.[10]

Organotin reagents are air- and moisture-stable organometallics, and can be conveniently purified and stored. Since they do not react with most common functional groups, the use of protecting groups is almost always unnecessary in conjunction with the Stille reaction. This is a very unusual and attractive feature for an organometallic process. Also, the reaction is often neither air nor moisture sensitive. In some cases, water and oxygen have actually been shown to promote the coupling. Although the reaction as initially described by Stille is often carried out under rather drastic conditions (temperatures of $\geq 100°$ are not uncommon), newly developed ligands[11] and the addition of copper(I) salts[12] have solved some of the problems associated with low reactivity. The utility and mildness of the Stille reaction are demonstrated by its frequent use in the final stages of complex natural product syntheses.

This review attempts a critical and comprehensive coverage of the reaction scope. Our mechanistic description of the reaction is rather brief, and we refer the reader to the pertinent literature for a more detailed analysis. All of the relevant literature is covered up to the end of 1994. The reaction was reviewed by Stille in 1986,[6] and by Mitchell in 1992;[13] a rather comprehensive account by Farina and Roth has appeared more recently.[14] Developments that occurred in 1995, as this work was in progress, and that were deemed important were incorporated as much as possible in this review.

MECHANISTIC CONSIDERATIONS, REGIOCHEMISTRY, AND STEREOCHEMISTRY

The three-step catalytic cycle proposed for the Stille reaction follows the general principles of transition metal-mediated cross-coupling reactions and is shown in Scheme 1.[6]

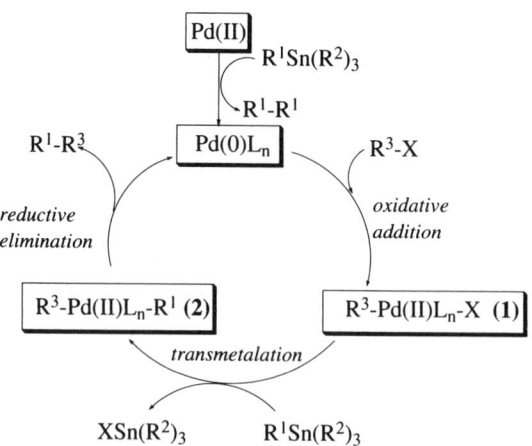

Scheme 1. Catalytic cycle of the Stille reaction.

When the catalyst is introduced as Pd(II), fast reduction by the stannane to a Pd(0) complex ensues, and the resulting Pd(0) species enters the cycle. Alternatively, the catalyst can be introduced directly as Pd(0). The rate or yield differences sometimes observed between Pd(II) and Pd(0) catalysts are not likely to be due to the initial difference in oxidation state, but rather to the stoichiometric ratio of palladium to ligand or other factors.[11]

The first step of the cycle is termed *oxidative addition* and is a quite general process for low-valent transition metal complexes.[15] The reaction is represented as a simple process in Scheme 1, but is likely to be a rather complex one. There is substantial evidence that a coordinatively unsaturated Pd(0) species, for example Pd(PPh$_3$)$_2$, is responsible for the oxidative process.[16] When the substrate is an aryl iodide, the reaction is accelerated by electron-withdrawing substituents on the ring ($\rho = +2$).[17] Oxidative additions are also accelerated by electron-rich phosphorus ligands on the palladium center.[18] In the coupling of aryl bromides with tetramethylstannane, the overall rate is strongly enhanced by electron-withdrawing groups on the aryl moiety ($\rho = +3.38$), suggesting that in this case the oxidative addition is rate limiting.[19]

At least with alkenyl halides, the oxidative addition may be a reversible process. Such a reaction generally proceeds with retention of olefin geometry.[20] Benzylic bromides undergo oxidative addition with partial or total racemization;[21] this has been explained by invoking a one-electron transfer process for this oxidative addition,[22] and CIDNP studies have supported the suggestion.[23] In these cases, the oxidative addition may be accelerated by the presence of oxygen in solution.[19] Intermediate **1** (Scheme 1) is generally formed as a *trans* square-planar complex, i.e., the two phosphine moieties are *trans* to each other, although the intermediacy of the less stable *cis* complex is assumed.[6]

In allylic systems, i.e., allylic chlorides, the oxidative addition was initially shown to proceed with complete inversion of configuration, through the intermediacy of η^3-complexes,[24] but subsequent studies have revealed a more complex situation (Eq. 2).[25]

Solvent	4:5
Benzene	100:0
THF	95:5
Acetone	75:25
DMF	29:71
DMSO	3:97

(Eq. 2)

Specifically, it was shown that, in the absence of strong coordinating ligands, the stereochemistry depends on the solvent, nonpolar solvents favoring retention and polar ones leading to inversion. Furthermore, olefin ligands promote *syn* oxidative addition, and phosphines favor the *anti* pathway.[26]

Although it is known that the transmetallation is very often the rate-determining step of the Stille reaction, much less is known mechanistically about this metathesis reaction.

In early studies, Stille et al. showed that, in the coupling of benzylic stannanes with acid chlorides, electron-releasing substituents slightly increased the transmetallation rate ($\rho = +1.2$), suggesting that carbon-tin bond breaking precedes palladium-carbon bond formation. The stereochemical outcome with benzylic stannanes is predominantly inversion at the tin-bearing carbon, suggesting an "open" S_E2 mechanism.[27]

More recently, it has been shown that the transmetallation of **1** to **2** proceeds via prior ligand dissociation and that ligands with lower donicity toward Pd(II) than PPh$_3$ [i.e., tri(2-furyl)phosphine and triphenylarsine] can lead to major (up to 1,000-fold) rate enhancements in the transmetallation.[11] With these ligands, many Stille couplings previously requiring vigorous conditions can be performed at room temperature.

In studies of the synthetically important coupling of organic triflates,[28,29] LiCl is necessary to induce coupling of organic triflates in THF as solvent. This has been rationalized by postulating that the initial oxidative addition product (**6**, Scheme 2), which was isolated in one case, is catalytically incompetent, whereas ligand substitution with chloride ion leads to the reactive species **7**.[28]

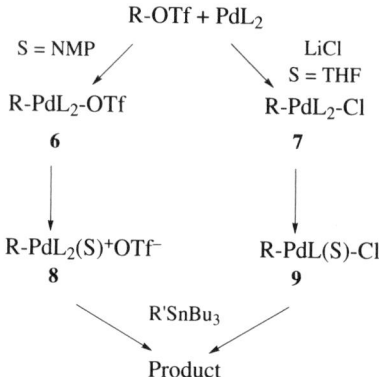

Scheme 2. Two possible pathways in the Stille coupling with organic triflates.

More recently, it has been found that addition of LiCl is often not necessary when operating in highly polar solvents like NMP, and in many cases LiCl is actually an inhibitor of the coupling. This was explained by invoking two pathways in the transmetallation, i.e.. a faster one proceeding via cationic species **8** and a slower one (with L = PPh$_3$) proceeding via ligand dissociation (through **9**). Hammett studies confirmed that there are two pathways with opposite electronic demands. Thus, in the absence of chloride the reaction is faster when the arylstannane contains electron-releasing groups ($\rho = -0.89$), whereas in the presence of LiCl, electron-withdrawing substituents also enhance the rate. The transmetallation is affected in a complex way by the combination of LiCl, ligands, and solvent, and the highest rates are obtained with AsPh$_3$ as ligand. With

this superior ligand, the effect of halide additives on the rate of the transmetallation is minimal.[30]

Intramolecular couplings of triflates with stannanes do not require LiCl even
in THF.[31] The recently reported ability of Ag(I) salts to improve some Stille couplings may also be explained by a switch of the transmetallation pathway via **8**
and away from **9** (Scheme 2).[32]

The cocatalytic effect of Cu(I) in the Stille coupling was first reported by
Liebeskind and Fengl.[12] Later studies have shown that Cu(I) performs a dual role:
In ethereal solvents (THF, dioxane) and in conjunction with highly coordinating
ligands (PPh$_3$), Cu(I) acts as a ligand scavenger to facilitate formation of the coordinatively unsaturated Pd(II) intermediate (**9** in Scheme 2) needed to effect
transmetallation, whereas in highly dipolar solvents (NMP) in the presence of
"soft" ligands (AsPh$_3$) formation of an organocopper species is likely.[33] Thus, it
seems simply that in the presence of inorganic Cu(I) salts, an organostannane
may be in equilibrium with an organocopper species (Eq. 3). Another important
role of Cu(I), enhancing the selectivity of group transfer in the Stille reaction, is
discussed in a later section.

$$RSnBu_3 + CuI \underset{}{\overset{NMP}{\rightleftharpoons}} RCu + ISnBu_3 \qquad (Eq.\ 3)$$

Similar transmetallations have been postulated in order to explain the beneficial effect of stoichiometric Zn(II) salts on certain Stille couplings, but no experimental evidence is available.[28]

From the standpoint of the stereochemistry at Pd(II), the transmetallation usually proceeds with retention of configuration and is probably followed by *cis-
trans* isomerization. The reductive elimination that follows probably proceeds
through a T-shaped intermediate via prior ligand dissociation at Pd(II).[15] Pd(IV)
species have been implicated as intermediates in the reductive elimination,[34] but
factors that influence this step are not discussed further since reductive elimination is not rate determining in the Stille coupling. In the coupling of allylic electrophiles, however, reductive elimination will determine the regiochemistry of
coupling, and in this case detailed understanding of this step is very important.

Allylic halides, typically chlorides, couple smoothly with organostannanes
under normal conditions, and the regiochemistry of the coupling is usually the
one resulting from attack of the organostannane at the less hindered terminus of
the allylic moiety (Eqs. 4 and 5).[24]

(87%)

(Eq. 5)

When the organostannane is also allylic, the situation is more complicated. Apparently, the coupling is somewhat regiospecific, and the C-C bond is formed between the more substituted end of the allylic stannane and the less substituted one in the allylic halide.[35,36] To explain the predominant allylic transposition of the stannane, both Stille and Trost postulated a direct attack of the stannane at the carbon terminus of an intermediate π-allyl complex, but there is no proof for such a mechanism. Indeed, indicator substrates for nucleophilic attack at π-allyl complexes classify allylstannanes as reacting directly at Pd(II) and not at carbon.[37] This mechanistic issue is still unresolved, even though a simple stereochemical probe could resolve the issue. On the other hand, in the presence of maleic anhydride the coupling takes place in a preferred head-to-head mode, and the stereochemistry indicates attack of the stannane at the Pd center of the π-allyl complex, followed by reductive elimination with retention of configuration.[38,39]

Exceptions to these regiochemical trends, however, can be found in the literature. One is shown in Eq. 6 and is mechanistically difficult to explain. One must also note that the two regiochemistries are interconvertible by Cope rearrangement.[40]

(Eq. 6)

An important mechanistic issue that has recently begun to be addressed by several investigators concerns the effect of nucleophilic assistance at tin(IV) during the transmetallation. Two studies[41,42] have independently shown that a nucleophilic moiety placed within the stannane considerably enhances transmetallation rates, whereas other studies in related systems have failed to detect such enhancements.[30] The increased reactivity of stannanes **10** has been explained by invoking internal N-Sn coordination in the transition state,[41] and a similar rationalization has been applied to the increased reactivity of systems such as **11**.[42]

These stannanes are able to effect transfers of alkyl moieties, which occur sometimes with difficulty or not at all using traditional Stille chemistry. The mechanistic and synthetic significance of these intriguing observations should be further explored.

SCOPE AND LIMITATIONS: THE ELECTROPHILE

In this section, the range of electrophiles used in the Stille coupling is surveyed. Details of experimental conditions and side reactions are more fully described in separate sections. The examples discussed are a select few. A complete survey is found in the tables. Limitations are discussed whenever carefully documented in the literature. Occasionally, low yields are reported in a number of isolated Stille couplings. These may be due to incomplete optimization of the reaction. Therefore, these examples are considered a real limitation only if the authors reported a thorough study exploring a comprehensive list of catalysts and conditions.

Alkenyl Halides

Alkenyl chlorides have been used very little in Stille couplings, presumably because of their lack of reactivity in the oxidative addition with Pd(0). Scattered examples of successful coupling exist, but appear limited to activated systems.[43,44]

Alkenyl bromides and iodides are generally useful partners. Their coupling is often stereospecific. Since bromides undergo oxidative addition only at elevated temperatures, E/Z isomerizations are sometimes observed. More consistently stereospecific is the coupling with vinyl iodides, which takes place at room temperature or slightly above. The higher reactivity of the iodides vs. the bromides is nicely illustrated in Eq. 7, where under the mild conditions employed the bromide moiety is left unreacted.[45]

$$\text{(Eq. 7)}$$

Two general studies on the cross-coupling between simple alkenyl iodides with both alkenyl[46] and alkynyl[47] stannanes are reported. Bromides also couple, but in lower yield. In each case, the preferred catalyst is the "ligandless" species Pd(CH$_3$CN)$_2$Cl$_2$. The reaction proceeds in DMF or THF at room temperature, and E/Z isomerization is negligible (Eq. 8).

$$\text{(Eq. 8)}$$

The palladium-catalyzed reduction of vinyl iodides with tributyltin hydride or other hydride reagents can be loosely classified as a Stille coupling. The reaction is highly stereospecific, in contrast with the radical-induced reduction, which leads to geometrical isomerization.[48]

Very few limitations of this coupling reaction have been clearly documented. Even tetrasubstituted vinyl iodides couple in good yields.[49,50] However, β-silyl vinyl bromide **12** couples with stannane **13** to yield only a completely isomerized product even under the mildest conditions.[51] This lack of stereospecificity is attributed to the bulky silyl group (Eq. 9).

$$\text{(Eq. 9)}$$

Special classes of alkenyl halides that have been made the objects of specific studies include β-halo-α,β-unsaturated ketones and esters, which couple smoothly with a variety of stannanes,[52–54] quinone halides, which also couple well (preferentially using CuBr as cocatalyst),[55–58] and β-halo-α,β-unsaturated sulfoxides, which couple with alkenyl-[59] and alkynylstannanes[60] without E/Z isomerization and without epimerization at the chiral sulfur center.

Certain systems, on the other hand, appear difficult to couple and require carefully optimized conditions. For example, α-iodo-α,β-unsaturated ketones must be coupled using the "soft" ligand AsPh$_3$ and cocatalytic Cu(I). Even under these conditions, high temperatures are required, but the reaction is general and gives very good yields (Eq. 10).[61]

(85%)

(Eq. 10)

On the other hand, α-bromo-α,β-unsaturated ketones can be coupled with aryl stannanes using P(o-Tol)$_3$ as ligand in the absence of Cu(I) additives.[62]

Halocyclobutenediones couple with stannanes, and CuI cocatalyst is necessary to obtain good yields (Eq. 11).[63,64]

(91%)

(Eq. 11)

Cyclooctatetraenyl bromide couples with stannanes at room temperature, and P(2-furyl)$_3$ or AsPh$_3$ are the ligands of choice.[65,66]

Bromotropolones can be coupled with a variety of arylstannanes, to yield analogs of the antimitotic agent colchicine.[67]

Intramolecular versions of this coupling reaction yield a variety of ring sizes, from four[68] and five[69] to medium-size rings,[70–74] and even macrocycles.[75] Equation 12 illustrates the key step in the total synthesis of leinamycin.[76] The

(37%)

(Eq. 12)

mildness and generality of this method is demonstrated by its frequent application to the late stages of complex natural product syntheses. Thus, the alkenyl halide/organostannane coupling has been applied in recent years to the total syntheses of neooxazolomycin,[77] onnamide A,[78] 22,23-dihydroavermectin,[79] calyculin A,[80–82] lankacidin C,[83] lepicidin A,[84] and rapamycin.[85]

Probably the most spectacular application of this reaction is represented by the final step of Nicolaou's total synthesis of rapamycin, in which a tandem Stille coupling is carried out on the fully functionalized skeleton.[86] The yield is modest, but an intermediate iodostannane could be isolated and resubjected to the reaction conditions, affording more cyclized product and thereby increasing the overall yield to 46% (Eq. 13).

(Eq. 13)

Aryl and Heterocyclic Halides

An early study reports that in the coupling of aryl halides with organostannanes, aryl bromides are the optimal electrophiles in the coupling reaction with allyltributylstannane. Aryl chlorides react only if strongly activated toward oxidative addition (e.g., p-nitrochlorobenzene), whereas aryl iodides couple only in low yields.[3]

In independent studies of the scope and utility of the reaction, it was found that both aryl bromides and iodides couple with a number of stannanes in high yield.[19,87] The coupling of aryl bromides requires more vigorous conditions and is facilitated by electron-withdrawing substituents in the *para* position of the halide derivative, indicating that oxidative addition is the rate-determining step. A specific study deals with the preparation of styrene derivatives.[88] The method was applied to the synthesis of indole derivatives (Eq. 14).[89]

(Eq. 14)

A synthetically useful variant of the Stille reaction is the coupling of aryl halides with aminostannanes.[90-92] The reaction so far is limited to aryl bromides. Secondary amines can generally be coupled, whereas among primary amines, only anilines have been reported to couple. The aminostannanes can be conveniently generated in situ from the corresponding amines and (diethylamino)tributylstannane. This is obviously a reaction with much potential, and it is likely that its scope will grow after further scrutiny. An example is shown in Eq. 15.[91] Other carbon-heteroatom bonds can be made through the intermediacy of organostannanes, as detailed later in the section describing the scope and limitation with respect to the types of stannanes that can be used.

(Eq. 15)

Heteroaryl halides also couple with organostannanes. Although the scope of these reactions has generally not been studied in detail, many examples in the literature exist to support some generalizations. For example, 2-, 3-, or 4-bromopyridines couple well with aryl and heteroaryl stannanes,[93-95] whereas 3-iodopyridines couple in only fair yields.[96] 2-Chloro-3-fluoropyridine derivatives couple specifically at the 2 position with a variety of alkenyl stannanes.[97] Even 4-chloropyridine can be coupled. 3-Bromoquinolines also couple with stannanes.[93,98] Equation 16 illustrates the key step in the synthesis of a lavendamycin analog.[99]

(Eq. 16)

2- and 3-Furyl[100] and thienyl[96,101-106] halides are easily coupled with stannanes. 2-Halothiazoles couple smoothly, as illustrated by a key step in a recent synthesis of micrococcinic acid (Eq. 17).[107] 2,5-Dibromothiazole couples first at the 2 position, then at C-5 (Eq. 18).[108]

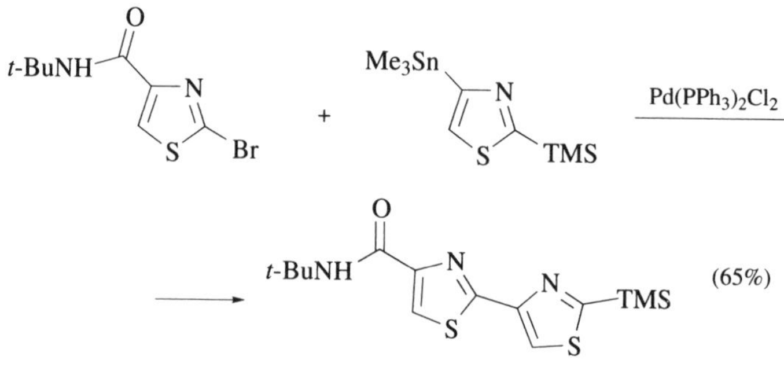

(Eq. 17)

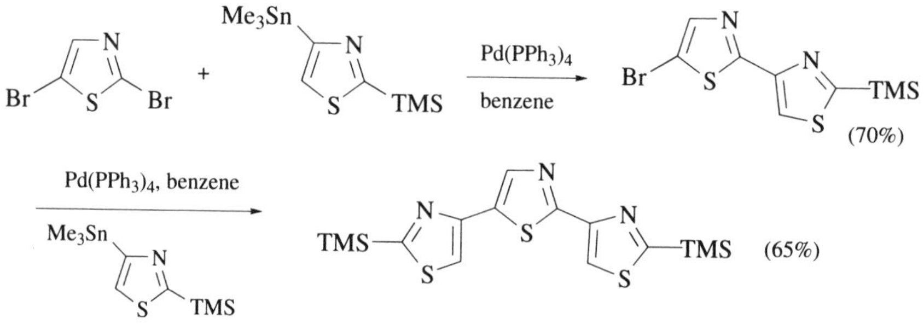

(Eq. 18)

Both 2-[109] and 3-indolyl[110] halides have been coupled with stannanes. Interestingly, 5-bromo-3-iodotosylindole couples specifically at C-3 (Eq. 19).[110]

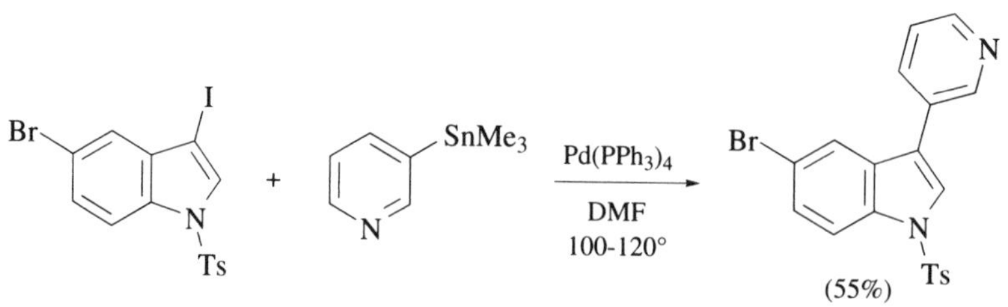

(Eq. 19)

2-Imidazolyl bromides couple with phenyltrimethylstannane, and 2,4-imidazolyl dibromides couple selectively at the 2 position with aryl stannanes, contrary to the corresponding arylboronates, which couple at both positions without selectivity.[111]

4(5)-Imidazolyl iodides, however, can be successfully coupled.[112,113] 4-Iodoisoxazoles can be coupled with a large number of stannanes.[114] 2,5-Dibromosiloles couple with alkynylstannanes,[115] and 2-bromo- and 2,4,6-tribromophosphinines couple with stannanes in an interesting selectivity pattern.[116]

Many applications of the Stille reaction to nucleoside chemistry have been made since the first application of the reaction to 2-iodopurines (Eq. 20).[117-120]

$$\text{(Eq. 20)}$$

Similar chemistry has been reported for 5-iodouridines,[121-125] and 5-bromo- or 5-iodouracil[126-128] derivatives. 5-Arylcytosines have been prepared from the corresponding 5-iodo derivatives by Stille coupling.[129] Stannane coupling in purine chemistry has been extended to 8-bromoadenosines,[130] 8-iodoadenosines,[131] 6-iodouridines,[132] and 6-chloropurines.[133,134] A number of 4- and 5-halopyrimidines (halo = Cl, Br, I) have been coupled with stannanes.[135-140] In polyhalogenated pyrimidines the order of reactivity in the coupling is C-4 > C-5 > C-2, regardless of the halide (Eq. 21).[141]

$$\text{(Eq. 21)}$$

2-Chloropyrazines can be coupled with stannanes,[142] and even bromo-substituted porphyrins have been subjected to the Stille coupling.[143,144] Finally, aryl iodides attached to a polymer have been subjected to Stille couplings in relation to the building of combinatorial libraries.[145]

Acyl Chlorides

It was reported in 1977 that stannanes can be coupled with acyl chlorides under palladium[1] or rhodium catalysis.[2] Stille subsequently explored the scope of the reaction and showed that it is general for a wide variety of acyl chlorides (Eq. 22).[146]

$$R^1COCl \ + \ R^2SnBu_3 \quad \xrightarrow[\text{CHCl}_3, \text{ reflux}]{\text{BnPd(PPh}_3)_2\text{Cl}} \quad R^1COR^2$$

(Eq. 22)

R^1= Aryl, alkyl, alkenyl; R^2= Alkyl, alkenyl, alkynyl, aryl

Few limitations are encountered in this reaction. Allylstannanes may react further with the ketone products in a nonpalladium catalyzed nucleophilic carbonyl addition. Decarbonylation is seen in some cases, but can be avoided by running the reaction under a CO atmosphere. Product isomerization is a complication when allyl- and alkenylstannanes are employed. This reaction can be run under milder conditions (room temperature) by using tri(2-furyl)phosphine or AsPh$_3$ as ligands.[11] Use of the former often prevents the unwanted geometric isomerization. Oxalyl chloride is not a good substrate for this reaction.[147] Coupling with β-stannyl enones yields butene-1,4-diones, which are directly reduced to 1,4-diketones under the reaction conditions.[148]

The coupling of acyl chlorides and alkynylstannanes is quite general and affords good yields of α,β-acetylenic ketones.[149]

Examples of this reaction in the absence of palladium are well known,[150] and, although the uncatalyzed reaction is outside the scope of this chapter, in some cases it is claimed to be higher yielding than its palladium-promoted counterpart.[151] Acyl chlorides from dicarboxylic acids also participate in the coupling. If a distannane is used, an annulation reaction results (Eq. 23).[152]

(Eq. 23)

Intramolecular couplings are also quite useful synthetically.[153,154] An example is shown in Eq. 24.[155]

(Eq. 24)

When the stannane used is tributyltin hydride, a general synthesis of aldehydes results.[156]

Chloroformates and carbamoyl chlorides also couple with stannanes[157] to yield esters and amides, respectively, in good yields (Eq. 25).[158] Intramolecular examples have been reported.[159]

(Eq. 25)

Allylic, Benzylic, and Propargylic Electrophiles

The coupling of allylic electrophiles with organostannanes is a reaction of general utility. Stille studied the scope of the reaction of allylic chlorides and bromides with organostannanes. With allylic electrophiles, a regiochemical issue exists: Since these couplings probably proceed via η^3-allylpalladium intermediates, coupling at either the α or the γ position is possible. Stille reports that coupling generally occurs at the less substituted terminus of the allyl moiety. An example is shown in Eq. 26.[24]

(Eq. 26)

Aryl- and alkenylstannanes couple in good yields. Allylic stannanes react to yield mixtures in which coupling at the more substituted terminus of the stannane is favored.[36,37] Among the applications to compounds of biological interest, the coupling of chloromethylcephems with stannanes constitutes a versatile approach to novel semisynthetic cephalosporins.[41]

Allylic acetates[36,160,161] and allylic phosphates[162] also couple with stannanes under special conditions. A study on the cross-coupling of allylic acetates showed that the reaction is quite general and is best carried out in the absence of phosphine but in the presence of LiCl. Again, coupling takes place at the less substituted allyl terminus, and both alkenyl- and arylstannanes couple in high yields. An example is given in Eq. 27.[163]

(Eq. 27)

Alkenyl epoxides can be considered allylic electrophiles. They also undergo coupling with aryl- and alkenyl- (but not allyl-, benzyl-, alkyl-, and alkynyl-) stannanes to yield mixtures of 1,2 and 1,4 coupling products.

As with allylic acetates, the less substituted terminus is the more reactive. Added water increases the yield and the regioselectivity, but further work aimed at better control of the regiochemistry is necessary to make this reaction synthetically useful. Equation 28 shows a typical example.[164]

Propargylic acetates do not couple with organostannanes,[165] and alkynylstannanes may undergo anomalous coupling with allyl halides.[41] Allenyl acetates have been coupled with stannanes to yield polysubstituted 1,3-dienes (Eq. 29).[166]

(Eq. 29)

Intramolecular examples of the coupling of organostannanes with allylic electrophiles have also been reported. Under optimized conditions, large rings can be constructed in fair yields (Eq. 30).[167]

(Eq. 30)

Allyl esters and carbamates are important in the protection of carboxy and amine functional groups. Deprotection conditions sometimes involve use of Pd(0) catalysts in conjunction with tributyltin hydride.[168] Specific examples are not discussed, since they are outside the scope of this review.

Few studies on the coupling of benzyl halides with stannanes have appeared. Benzyl bromide itself couples with tetramethylstannane, vinyltributylstannane, and tetraphenylstannane in good yields under the catalysis of $BnPd(PPh_3)_2Cl$ in HMPA.[19] Reaction with hexaalkyldistannanes yields benzylic stannanes in fair to good yields.[169] Propargyl halides have not generally been used as substrates in the Stille reaction. Propargyl bromide couples to some stannanes to yield allene derivatives.[170] The coupling of benzylic bromides containing β hydrogens takes place smoothly, without substantial β elimination, in the presence of the catalyst (2,2′-bipyridine)fumaronitrile palladium(0) (Eq. 31)[171]. Further applications of

$$(Eq. 31)$$

this interesting catalyst to other cross-coupling chemistry have not been reported. Finally, a nice application of this coupling to natural product synthesis is found in an approach to furanocembranolides (Eq. 32).[172]

$$(Eq. 32)$$

Alkenyl Sulfonates and Other Electrophiles

The coupling of vinyl sulfonates is, in general, limited to triflates. In a few special cases where extra activation is present, mesylates[173] and tosylates[174] can be used, but these substrates have limited utility and are not discussed further. The coupling of vinyl triflates with organostannanes is a truly general reaction of paramount importance in organic synthesis, owing in part to the ready availability of isomerically pure alkenyl triflates.[175] An initial study shows that the coupling takes place in high yield in THF with alkenyl-, alkynyl-, and allylstannanes, but arylstannanes do not react.[28] The reaction requires addition of excess LiCl (Eq. 33).

(90%)

(Eq. 33)

The reaction of alkenyl triflates with hexamethyldistannane constitutes an important approach to alkenylstannanes (Eq. 34).[176]

(80%)   (Eq. 34)

A more recent study has shown that even arylstannanes couple smoothly under optimized conditions, using the "soft" ligand $AsPh_3$ and highly polar solvents such as NMP.[30] A careful reexamination of the LiCl effect has shown that this additive is often unnecessary for the reaction to proceed if one operates in NMP as solvent. LiCl is generally an inhibitor of the reaction in NMP when strong ligands (PPh_3) are used , but has little effect on the rate when "soft" ligands ($AsPh_3$) are employed. For a discussion of this complex behavior, the reader is referred to the mechanistic section. E/Z isomerization of the product can be a problem with these couplings (Eq. 35).[30] Use of CuI as a cocatalyst often reduces such isomerization.[177]

(72%)

(Eq. 35)

The intramolecular version of this reaction has been developed. The cyclization precursors were assembled using an array of tin-containing bifunctional synthons developed for this purpose. A variety of small- and medium-size rings was assembled, and applications to the total synthesis of terpenoids were reported.[31,69,178-183] Once again, LiCl behaved as an inhibitor of the coupling. An example of this powerful methodology is shown in Eq. 36.[184] An extension to macrocyclizations is reported.[185,186]

(Eq. 36)

Alkenyl phenyliodonium salts also couple with alkenylstannanes under mild conditions, as shown in Eq. 37.[187,188]

(Eq. 37)

Aryl and Heterocyclic Sulfonates and Other Derivatives

The Stille coupling of aryl triflates has been extensively studied. In the presence of LiCl, these substrates couple with alkyl-, alkenyl-, allyl-, alkynyl-, and arylstannanes in high yields under relatively harsh conditions (ca. 100°). Dioxane and DMF are the solvents of choice. Equation 38 shows a typical example.[189]

(Eq. 38)

Aryl triflates are less reactive than aryl iodides, but their reactivity is comparable to that of aryl bromides. A direct competition experiment showed that product distribution depends strongly on the coordinative level of the catalyst used (Eq. 39). Unfortunately, no firm conclusions can be drawn about the mechanistic

Catalyst	
Pd(PPh$_3$)$_4$, dioxane	1 : 6
Pd(PPh$_3$)$_2$Cl$_2$, DMF	5 : 1

(Eq. 39)

basis for this dichotomy, since the two catalysts were used in different solvents, and it is likely that the solvent is also a key factor in the ease of oxidative addition.[30] Ether, nitro, amido, and carbonyl groups (even aldehydes) are tolerated on the aryl triflate. Because of the harsh conditions employed, double bond migrations and isomerizations are recurring problems. As for vinyl triflates, a reexamination of the reaction showed that the coupling of aryl triflates is best carried out in NMP with AsPh$_3$ as ligand. In this solvent, LiCl reduces the coupling rate, but is sometimes beneficial to catalyst stability. An *ortho* methyl group on the aryl triflate slows the coupling by a factor of 3.[30]

Separate studies have shown that electron-rich aryl triflates also couple in good yields, especially with Cu(I) cocatalysts.[190,191] Both 1- and 2-naphthyl triflates couple as expected,[192] as do indolyl,[193] quinolyl, and isoquinolyl triflates.[194,195] Pyrimidyl triflates couple with organostannanes in good yields.[196] Among the derivatives of medicinal interest as targets, one must note the utility of the coupling of cephem,[40] carbacephem,[197] and carbapenem[198] triflates with stannanes for the synthesis of antibacterial β-lactams, the coupling of uridine triflates with stannanes,[199] and an application to the synthesis of anthramycin (Eq. 40).[200]

(70%)

(Eq. 40)

In addition to triflates, other sulfonates can be used, including long-chain polyfluorinated sulfonates,[29,201] *p*-fluorophenyl sulfonates,[202] and fluorosulfonates.[203] The last appears to be of practical utility, considering the low cost of fluorosulfonic acid vs. the expense of triflic acid (Eq. 41).

(92%)   (Eq. 41)

Among the aryl electrophiles, diazonium salts participate in the Stille coupling with alkenyl-, alkyl-, and arylstannanes, and an example is shown in Eq. 42.[204] Given their ready availability, the under-utilization of these substrates is hard to understand.

(66%)   (Eq. 42)

Even some ether derivatives, notably some *pseudo*-saccharyl *O*-ethers, couple with stannanes in low to fair yield, especially under Ni(0) catalysis, but this reaction is restricted to tetramethylstannane so far, and therefore its scope is still to be fully explored.[205] Diaryliodonium salts also participate in the Stille reaction.[206]

Miscellaneous Electrophiles

Alkyl halides do not normally cross-couple with organostannanes, but some α-activated substrates do undergo the Stille coupling. Among them, the α-halo ethers and α-halo thioethers couple smoothly, even if β hydrogens are present (Eq. 43),[207] whereas α-halolactones couple with allylic and acetonyl stannanes.[208]

(Eq. 43)

α-Halocarbonyl compounds react with allyl and acetonyl stannanes in an anomalous fashion, i.e., by attack at the carbonyl followed by oxirane formation (Eq. 44).[209]

(Eq. 44)

Perfluorinated alkyl iodides, in which β-hydride elimination after oxidative addition is impossible, couple with stannanes in good yields, although the reaction is proposed to be radical mediated.[210] Imidoyl chlorides couple with stannanes in low to fair yields, thus providing a route to imines from amides. An example is shown in Eq. 45.[211] Alkynylstannanes react in particularly good yields.[212]

(Eq. 45)

Although no general study has appeared on the use of alkynyl halides in the Stille reaction, sporadic but useful applications of these electrophiles have been recorded.[213-215] A remarkable result is reported in a dynemicin total synthesis (Eq. 46).[216]

(Eq. 46)

Many examples of arene or polyene metallocarbonyls in the Stille cross-coupling have been reported.[217-226] The purpose of the metallocarbonyl moiety is often to activate the aryl electrophile toward oxidative addition, as in Eq. 47.[227]

(Eq. 47)

Several heteroatom-halogen bonds can be activated toward coupling by Pd(0) catalysts, including P-Cl,[228] S-Cl,[229] and Fe-I bonds.[230] The last appears to be the first example of the formation of a transition metal-carbon bond under the catalysis of a Pd(0) complex. An example is shown in Eq. 48.[231]

R = H, Pr, Bu, Ph

(Eq. 48)

Bifunctional electrophiles and stannanes, when coupled, usually give rise to polymeric materials. Many examples of this strategy have been reported, as is evident from Table XXXI. A typical example is shown in Eq. 49.[232]

(Eq. 49)

SCOPE AND LIMITATIONS: THE STANNANE

Unfortunately, most studies on the Stille reaction emphasize a specific type of electrophile, and very few studies examine a particular class of stannanes. General studies of stannane reactivity are therefore lacking. It is impossible to discuss all examples in which a particular type of stannane has been used. In this section we attempt to focus on a limited number of more general papers in an effort to delineate the current scope and limitations in the use of stannanes for the Stille reaction.

Alkylstannanes

It is generally accepted that transfer of alkyl groups from tin is much slower than that of unsaturated substituents.[6] Indeed, it is this property that makes the methyl and especially the butyl group such excellent "dummy," i.e., "nontransferable," ligands. Nevertheless, in many cases coupling of tetraalkylstannanes occurs in high yields at elevated temperatures. Among the tetraalkylstannanes, tetramethylstannane and tetrabutylstannane are most often used, the former being more reactive. The coupling of these stannanes with aryl and benzyl halides is carried out in HMPA and proceeds in good yields.[19] Use of triphenylarsine as ligand facilitates the coupling of these sluggish nucleophiles with aryl triflates.[30]

One of the problems associated with the coupling of symmetrical tetraalkylstannanes is that only the first alkyl group is transferred at a sufficient rate to be of synthetic utility,[6] successive transfer becoming more and more difficult with increasing halogen substitution at tin. The need therefore arises for the use of "dummy" ligands; selectivity in the transfer of alkyl groups, however, is quite poor. In special cases, when the alkyl group is activated by particular substituents, some selectivity may be observed. Thus, benzyl trialkylstannanes selectively transfer the benzyl group[27] with inversion of configuration at carbon.

The reaction is facilitated by electron-withdrawing substituents on the aryl ring of the stannane.

Other activated stannanes have been coupled successfully, including transfer of hydroxymethyl,[233] methoxymethyl,[234] and cyanomethyl[235] groups onto a number of aryl bromides (Eqs. 50 and 51).

$$Bu_3SnCH_2OMe, Pd(PPh_3)_2Cl_2$$
HMPA, 80°
(70%)

(Eq. 50)

$$Bu_3SnCH_2CN, Pd[P(o\text{-}Tol_3)]_2Cl_2$$
xylene, 120°
(66%)

(Eq. 51)

The successful coupling of ethyl α-(tributylstannyl)acetate is reported; the addition of Zn(II) salts is needed for optimum results (Eq. 52).[236] Unfortunately, in none of these studies was a quantitative assessment carried out regarding the transfer selectivity of the activated alkyl vs. the "dummy" butyl group.

$$Bu_3SnCH_2CO_2Et, Pd[P(o\text{-}Tol_3)]_2Cl_2$$
ZnBr_2, DMF, 80°
(93%)

(Eq. 52)

Acetonylation is also possible using acetonyltributylstannane,[237] but in general these α-stannyl ketones are unstable, and their coupling is best carried out by generating them in situ from enol acetates[238-240] or enol silanes.[241] This reaction amounts to a net α-arylation (or alkenylation) of enolates, a rather difficult operation. The above methodology, however, is limited: Only methylene enolates are arylated in good yields, whereas more substituted derivatives couple poorly (Eqs. 53[240] and 54[241]). Further synthetic studies in this important area are warranted.

$$Pd[P(o\text{-}Tol_3)]_2Cl_2$$
Bu_3SnOMe,
toluene, 100°

R^1	R^2	
H	H	(62%)
H	Me	(35%)
Me	Me	(8%)

(Eq. 53)

(Eq. 54)

Cyclopropyltributylstannane transfers the cyclopropyl group in low yield.[126] The coupling of α-amino- and α-alkoxystannanes[242] with acyl chlorides takes place in good yields and with retention of configuration at the sp^3 carbon of the stannane, provided Cu(I) salts are added as cocatalysts (Eq. 55).[243] The intermediacy of an organocopper species has been implicated. 4-(Tributylstannyl)-2-azetidinones also couple with acid chlorides.[244]

(Eq. 55)

An important advance in the selective transfer of alkyl groups from tin has been reported.[41] Using alkylstannanes **10**, selective transfer of alkyl groups, including sec-butyl and α-trimethylsilylmethyl, is achieved under rather mild conditions. Further research is needed to expand the synthetic utility of systems containing a substituent capable of triggering pentacoordination at tin.

Alkenylstannanes

The coupling of alkenylstannanes with a variety of electrophiles is a quite general reaction, and it is difficult to find specific limitations in the literature. Some failures, however, have been reported. Most studies on the cross-coupling of alkenylstannanes are limited to readily accessible 1,2-disubstituted substrates. These couple efficiently and often with good stereospecificity.[47] More heavily substituted or more complex stannanes couple sometimes with difficulty or not at all. In particular, alkenylstannanes that bear another substituent α to tin appear difficult to couple. For example, stannane **14** does not couple with internal alkenyl iodide **15**, but couples normally with its terminal isomer **16**.[244a] This difference is most likely due to steric hindrance.

(14) (15) (16)

Methyl α-(tributylstannyl)acrylates couple abnormally with iodobenzene, owing to their tendency to yield cine-substitution products (vide infra).[245] Normal *ipso* reactivity is restored by the addition of Cu(I) salts.[246] β-Substituted α-(tributylstannyl)acrylates, however, couple normally with both acyl chlorides[247] and allylic halides (Eq. 56).[248] Evidently, the β substitution dramatically slows the cine-substitution process.

$$(56\%)$$

(Eq. 56)

α-Styrylstannanes yield cine substitution when coupled with aryldiazonium compounds (vide infra),[249] but can be coupled with acyl chlorides without side reactions.[250] Again, β substitution restores normal Stille reactivity, although in poor yield.[251] In general, densely substituted stannanes react poorly, and their coupling must be carefully optimized. An example from the total synthesis of lacrimin A is shown in Eq. 57.[252]

$$(43\%)$$

(Eq. 57)

Examples where every attempt to induce coupling fails include stannanes **17**[253] and **18**.[51] Other stannanes with seemingly comparable steric hindrance, however,

(17)   (18)

couple under standard conditions. For example, α-trialkylsilyl substitution in alkenyltrimethylstannanes prevents Stille coupling with allyl halides because the methyl groups on tin transfer more rapidly.[254] However, 1-triethylsilyl-2-trialkyl-stannyl-1-alkenes similar to **18** can be coupled with acyl halides (Eq. 58).[255]

$$(Eq. 58)$$

α-Phenyl and α-methyl substitution of olefinic stannanes does not seem to hinder Stille coupling in some cases (Eqs. 59[49] and 60[256]). The latter coupling, however, is successful only in the presence of cocatalytic copper. This may represent a general solution to the problem of coupling hindered alkenylstannanes.

$$(Eq. 59)$$

$$(Eq. 60)$$

Another example of this trend is shown by the difficult coupling of cyclohexenylstannanes with aryl triflates. Butyl transfer is an important side reaction here, unless one employs cocatalytic copper (Eq. 61).[33]

Additives	19:20
P(2-furyl)$_3$	36:64
AsPh$_3$	10:90
AsPh$_3$ + CuI	0:100

$$(Eq. 61)$$

In general, 1-tributylstannylcycloalkenes couple very sluggishly under Stille conditions,[257,258] and the reason must be attributed to some type of steric hindrance. β-Stannyl enones,[259] β-sulfonyl alkenylstannanes,[260] and 3- (or 4-) tributylstannyl-2-(5H)-furanones[261] have been made the objects of special investigations. In each case coupling with electrophiles is successful. Other types of alkenylstannanes that have been separately investigated include a variety of fluorinated alkenyl stannanes,[262-266] cyclobutenone,[267] and cyclobutenedione[12,64,268] stannanes.

α-Alkoxy-substituted alkenylstannanes seem to be especially reactive partners in the Stille reaction.[269-271] β-Alkoxyalkenylstannanes have also been coupled successfully.[272-274] Polyunsaturated alkenylstannanes have been studied in a few sporadic cases. Thus, allenylstannanes couple with aryl iodides[275] and triflates in modest yields (Eq. 62).[276] With allylic electrophiles, these stannanes

(Eq. 62)

yield propargylic derivatives, the result of allylic inversion.[165] A variety of dienyl-[277] and ynenyl-[278] stannanes have also been coupled with a number of electrophiles. 1,1-Distannylalkenes have been coupled with allylic halides, double substitution being the result.[254] With 1,2-bis(stannyl)ethylenes, on the other hand, monocoupling can be controlled to produce substituted alkenylstannanes. A large excess of the bis(stannane) is not necessary, because the first cross-coupling is faster than the second one. The second coupling can be carried out under more forcing conditions (Eqs. 13 and 63[279]).

(Eq. 63)

Aryl and Heterocyclic Stannanes

Arylstannanes couple readily with a variety of electrophiles. Both electron-withdrawing and electron-releasing substituents on the aryl ring can accelerate coupling, an indication of a dual mechanism for the transmetallation (see mechanistic section).[30] In general, however, electronic effects in the transmetallation are minor. On the other hand, steric effects can be important. An alkyl group *ortho* to the tin residue can slow the coupling by a factor of ca. 20. An *ortho* methoxy group, which is sterically much smaller, leads to only a 2-fold rate reduction.[30] In general, therefore, coupling with *ortho*-substituted arylstannanes can be difficult, and substantial transfer of the dummy ligand can take place (see section on side reactions). This problem has been tackled successfully by using Cu(I) salts. Under these conditions aryl group transfer is exclusive.[30,280]

Aryl trichlorostannanes have been used as coupling partners in aqueous media employing vigorous conditions,[281] under which the tin-chlorine bond is probably hydrolyzed to a tin-hydroxy species, because coupling does not take place in organic media (Eq. 64).[282] This protocol obviates the use of organic solvents, but

(Eq. 64)

appears limited to water-soluble electrophiles. In a similar vein, tetrabutylammonium difluorotriphenylstannate can be used to transfer a phenyl group onto vinyl triflates.[283]

Pyridyl-, quinolyl-, and isoquinolylstannanes have been the objects of separate studies. They couple smoothly with acyl chlorides.[284,285] Electron-rich heterocyclic stannanes, such as the 2-furyl-, 2-thienyl-, 2-pyrrolyl-, and 2-thiazolylstannanes, couple with aryl halides under rather mild conditions. An example is shown in Eq. 65.[286]

(Eq. 65)

3,4-Distannylfurans have been studied in great detail as bifunctional reagents,[287] and 3-stannylfurans have been used as substrates with acyl chlorides.[288] 2-Stannyl-[289,290] and 3-stannylindoles[291] have also been coupled with a variety of electrophiles. 5-Isoxazolylstannanes have been coupled with aryl iodides.[292,293]

2-Tributylstannylfuran couples with a number of α-chlorocyclobutenones in low yields, and it is postulated that this is due to further attack of the electrophile on the 5 position of the heterocycle, which is very electron-rich. These electrophilic palladations of electron-rich heteroaromatics are indeed precedented.[294] However, 5-trimethylsilyl-substituted stannylfurans couple in excellent yields.[295]

Equation 66 shows the application of the Stille reaction to the synthesis of 5-substituted furanones.[296]

(Eq. 66)

Couplings of nonaromatic, heterocyclic stannanes are often found in the literature. A popular target has been α-substituted glycals.[297-300] One example is shown in Eq. 67.[301]

(Eq. 67)

Alkynylstannanes

Alkynylstannanes couple smoothly with a variety of electrophiles, including alkenyl halides.[47] This class of stannanes is the most reactive of all, according to Stille,[6] and few limitations exist. Alkoxy-substituted alkynylstannanes have been used in an interesting approach to α-aryl and heteroaryl acetates (Eq. 68).[302]

(Eq. 68)

In general, although these stannanes are quite reactive, their use in cross-coupling chemistry is often unnecessary, since terminal alkynes couple directly with organic electrophiles using a palladium catalyst, cocatalytic copper, and amines as bases (Sonogashira coupling).[303]

Allylstannanes

Allylstannanes have been underutilized in the Stille coupling, presumably because of the difficulties with the synthesis of regiochemically defined substrates and their tendency to undergo allylic isomerization, thus making it hard to predict the regiochemistry of the coupling. Simple allylic stannanes couple more slowly than alkenylstannanes,[6] but at acceptable rates in most cases. One problem that has been documented with allylstannanes is the tendency of the double bond to move into conjugation after coupling, especially in reactions with acyl halides[146] and aryl triflates.[189] This can sometimes be prevented by operating at lower temperatures using tri(2-furyl)phosphine as the palladium ligand (Eq. 69).[11]

Conditions	
Pd(PPh$_3$)$_4$, LiCl, dioxane, 98°	I (18%) + II (54%)
Pd$_2$(dba)$_3$, LiCl, P(2-furyl)$_3$, NMP, rt	I (78%)

(Eq. 69)

Allylstannanes may couple at the α or the γ position, and not enough data are presented in the literature to draw firm conclusions.[2] Thus, crotyltrimethylstannane couples with acyl chlorides to yield a 1:1 mixture of α and γ products, but the product resulting from γ attack predominates at lower temperatures.[146]

Terpenic allylstannanes undergo regioselective Rh-catalyzed acylation at the α or γ position, depending on the structure of the substrate (Eqs. 70 and 71).[150,304]

(Eq. 70)

(54%) (Eq. 71)

A few special classes of allylstannanes have been described as substrates for the Stille reaction. An interesting one is shown in Eq. 72.[305] Thus, α-alkoxyallyl-

(72%), $E{:}Z = 75{:}25$

(Eq. 72)

stannanes couple with acyl chlorides to yield the allylically inverted β,γ-unsaturated ketones, which can be further converted to 1,4-dicarbonyl compounds by acid hydrolysis.

On the other hand, γ-carbalkoxy-substituted allylstannanes undergo selective coupling at the α position with alkenyl, aryl, and acyl halides (Eq. 73), but only at

$$\text{(Eq. 73)}$$

the γ position with allylic electrophiles.[306] This confirms early results, in which allylstannanes were coupled with allylic electrophiles with predominant allylic inversion.[35,36] Further aspects of this reaction are discussed in the mechanistic section.

The use of an allylic bis(stannane) as an annulation reagent has already been discussed (Eq. 23).

In conclusion, although allylstannanes are useful partners in the Stille reaction, they have been used infrequently, probably because the regiochemistry of the coupling is still unpredictable. This area certainly deserves further in-depth research.

Other Stannanes

Acylstannanes have been coupled in a few cases with acyl chlorides to provide unsymmetrical α-diketones (Eq. 74).[307] A CO atmosphere may help to prevent decarbonylation.

$$\text{(Eq. 74)}$$

Distannane derivatives are useful reagents in conjunction with a variety of electrophiles. Upon reaction with acyl halides, they yield mixtures of symmetrical ketones and α-diketones. Diketones predominate under a CO atmosphere.[308] Under suitable conditions, the reaction stops at the acylstannane stage, and this is preparatively useful (Eq. 75).[309]

$$\text{(Eq. 75)}$$

The couplings of hexamethyl- and hexabutyldistannanes with aryl bromides and iodides, and also with benzylic bromides, are high yielding, homocoupling of the electrophile being the only detectable side reaction (Eq. 76). Most substituents on the aryl ring are tolerated except *p*-amino and *p*-nitro. Under these conditions, allyl and alkenyl halides give the corresponding stannanes in low yields.[169]

$$\text{MeO—C}_6\text{H}_4\text{—I} \xrightarrow[\text{toluene, 115°}]{\text{Pd(PPh}_3)_2\text{Br}_2,\ \text{Me}_3\text{SnSnMe}_3} \text{MeO—C}_6\text{H}_4\text{—SnMe}_3 \quad (96\%)$$

(Eq. 76)

The coupling of distannanes with aryl halides has been studied independently,[310,311] and another investigator found that some of the above limitations can be overcome by using "ligandless" conditions.[312,313] A problem with this protocol is, however, disproportionation of the distannane, and an excess of the reagent must be used. A typical example of this protocol as it applies to allylic acetates, bromides, and chlorides is shown in Eq. 77.[314] Nickel catalysis has also been used in this reaction.[315]

$$\text{(cyclohexenyl)—Br} \xrightarrow[\text{Me}_3\text{SnSnMe}_3,\ \text{HMPA, rt}]{[(\eta^3\text{-C}_3\text{H}_5)\text{PdCl}]_2} \text{(cyclohexenyl)—SnMe}_3 \quad (83\%) \quad \text{(Eq. 77)}$$

The reaction of distannanes with vinyl triflates is an important route to regiochemically and geometrically defined vinylstannanes, as previously shown (Eq. 34).[176] Even some activated vinylic chlorides couple with hexamethyldistannane.[260]

Aminostannanes react with electrophiles, such as aryl and alkenyl bromides, in variable yields (Eq. 78).[90,316] This process was recently reinvestigated and improved,[91,92] as already illustrated (Eq. 15).

$$\text{Ph—CH=CH—Br} \xrightarrow[\text{Pd[P(}o\text{-Tol)}_3]_2\text{Cl}_2,\ 100\text{-}120°]{\text{Bu}_3\text{SnNEt}_2,\ \text{xylene}} \text{Ph—CH=CH—NEt}_2 \quad (50\%) \quad \text{(Eq. 78)}$$

The formation of C-S bonds via organotin sulfides is also well precedented. Alkenyl,[317] aryl,[318] and heteroaryl halides[319] participate. An example is shown in Eq. 79.[320]

$$\text{O}_2\text{N—C}_6\text{H}_4\text{—I} \xrightarrow[\text{DMSO, 100°}]{(\text{Et}_3\text{Sn})_2\text{S},\ \text{PhPd(PPh}_3)_2\text{I}} \text{O}_2\text{N—C}_6\text{H}_4\text{—S—C}_6\text{H}_4\text{—NO}_2 \quad (100\%)$$

(Eq. 79)

Among related reactions that have received only scant attention, (trimethylstannyl)diphenylphosphine couples with iodoaromatics to provide substituted triarylphosphines,[321] and tin alkoxides have been coupled with allylic electrophiles.[322] These methods have not been further applied to organic synthesis.

CARBONYLATIVE COUPLINGS

When a Stille coupling is carried out under a CO atmosphere, carbonyl incorporation under catalytic conditions is possible. The reaction is general for alkenyl, aryl, heteroaryl, and allyl electrophiles (Eq. 80).

$$R^1\text{-}X \; + \; CO \; + \; R^2SnR^3_3 \; \xrightarrow{[Pd(0)]} \; R^1(CO)R^2 \; + \; R^3_3SnX \quad \text{(Eq. 80)}$$

The earliest report of a successful carbonylative coupling between a stannane and an organic halide showed that several simple aryl, alkenyl, and benzyl halides could be coupled with simple stannanes under rather vigorous conditions (Eq. 81).[323] A considerable body of research has been reported as this procedure has been refined and its scope defined.

$$R^1\text{-}X \; + \; CO \; + \; R^2_4Sn \; \xrightarrow[\substack{450 \text{ psi CO, } 120° \\ HMPA}]{PhPd(PPh_3)_2I} \; R^1COR^2 \quad \text{(Eq. 81)}$$

$R^1 = Ph, PhCH_2, PhCH=CH, EtO_2CCH_2; R^2 = Me, Bu, Ph; X = Cl, Br, I$

Alkenyl Halides

The palladium-catalyzed carbonylative coupling of alkenyl iodides with alkenylstannanes affords the corresponding dialkenyl ketones in good yield (Eq. 82).[324] The reaction takes place under neutral, mild conditions (40–50°,

$$R^4 = \text{alkenyl} \quad \text{(Eq. 82)}$$

THF) and low CO pressure (1–3 atm). One may assume that all of the functional groups compatible with the standard, noncarbonylative cross-coupling reactions are also compatible with the carbonylative conditions, although no comprehensive study has been reported.

The outcome of the reaction can be sensitive to CO pressure, and slightly elevated pressures (45 psi) typically eliminate the competing direct coupling. An example can been seen in Eq. 83. β-Iodostyrene requires 45 psi CO for exclusive

$$\text{Ph}\diagdown\diagup\text{I} + \text{Ph}\diagdown\diagup\text{SnBu}_3 \xrightarrow[\text{45 psi CO}]{\text{Pd(PPh}_3)_2\text{Cl}_2\text{, THF, rt}} \text{Ph}\diagdown\diagup\overset{\displaystyle O}{\underset{(70\%)}{\diagdown}}\diagup\diagdown\text{Ph}$$

(Eq. 83)

carbonylative coupling, because under 15 psi CO a 1:1 mixture of direct and carbonylative coupling products is formed.[324] Double bond isomerization can be a problem. Alkenes with Z geometry have a propensity to isomerize, especially under harsh reaction conditions.

Alkenyl iodides can also be transformed into the corresponding α,β-unsaturated aldehydes through carbonylative cross-coupling using tributyltin hydride as a partner. As with ketone formation, partial Z/E isomerization is a problem (Eq. 84).[325]

$$n\text{-Bu}\diagup\diagdown\text{I} + \text{Bu}_3\text{SnH} \xrightarrow[\text{45 psi CO, 50°}]{\text{Pd(PPh}_3)_4\text{, THF}} n\text{-Bu}\diagup\diagdown\text{CHO} \qquad \text{(Eq. 84)}$$

$Z{:}E = 85{:}15$

Aryl and Heterocyclic Halides

Aryl iodides and bromides, but not chlorides, can be carbonylatively coupled with organostannanes to furnish ketones. The number of examples in the literature for aryl iodides and bromides is limited, and although bromides couple, the yields are low. The moderate interest in aryl halides is due to the extensive versatility of aryl triflates in this coupling strategy. The protocol using "ligandless" conditions is illustrated in Eq. 85.[326,327]

$$R^1\text{—}\underset{}{\underbrace{\bigcirc}}\text{—X} + \text{Me}_3\text{SnR}^2 \xrightarrow[\text{45 psi CO, HMPA, 20°}]{[(\eta^3\text{-C}_3\text{H}_5)\text{PdCl}]_2} R^1\text{—}\underset{}{\underbrace{\bigcirc}}\text{—}\overset{\displaystyle O}{\text{C}}\text{—R}^2$$

X= I, Br

(Eq. 85)

A recent example, which uses more vigorous conditions but employs a nonpolar solvent, is shown in the coupling of aryl and heteroaryl iodides with cyclobutenedionestannanes (Eq. 86).[268]

$$R^1\text{I} + \underset{}{\text{Bu}_3\text{Sn}}\overset{O}{\underset{O}{\diagup}}\text{—O} \xrightarrow[\text{30 psi CO, C}_6\text{H}_6\text{, 80°}]{\text{BnPd(PPh}_3)_2\text{Cl}} R^1\text{CO}\overset{O}{\underset{O}{\diagup}}\text{—O} \qquad \text{(Eq. 86)}$$

R^1= Ph, 2-thienyl

The role of additives, as well as potential ligand effects, has not been experimentally determined for the carbonylation reaction. There is a report on the beneficial effect of $AsPh_3$ in the context of a key step in a total synthesis of strychnine (Eq. 87).[328]

(Eq. 87)

A variety of heterostannanes (R_3Sn-OR', $-SR'$, $-NR'_2$) can also be used as nucleophilic partners in the carbonylative Stille reaction (Eq. 88).[329,330] Esters and

$$XR^2 = NEt_2, SPh, OMe$$

(Eq. 88)

amides are formed under mild conditions using HMPA as solvent. Electron-withdrawing groups on the aromatic ring appear to slow down CO insertion, and when such functional groups are present, there is competing direct coupling between the aryl moiety and the heterostannane.

The formylation of aryl iodides appears to be a general process. Aryl bromides furnish the desired aldehydes in moderate to low yield. A competing side reaction is direct reduction of the halide. Aryl iodides containing electron-releasing groups are formylated under 15 psi CO, whereas those containing electron-withdrawing groups need at least 45 psi CO to minimize reduction. Slow addition of tributyltin hydride to the reaction mixture under CO pressure is necessary in order to optimize the ratio of aldehyde to reduced product. A single example using 3-iodofuran demonstrates that heterocycles can also be formylated in this manner (Eq. 89).[325,331]

(Eq. 89)

Ortho substituents adversely affect the yield, and those containing a heteroatom also present a unique problem: the potential for competitive alkoxycarbonylation or amidation (Eq. 90).[332]

(Eq. 90)

Allylic and Benzylic Halides

Allyl and benzyl chlorides insert CO when reacted with stannanes, forming the corresponding ketones.[24] Diallylic ketones have been prepared under very mild conditions.[333] Higher pressures of CO favor ketone formation over direct coupling. The major side reaction is the carbonylative homocoupling of the organostannane. Carbonylative couplings occur with inversion of stereochemistry at the halide-bearing carbon, at least under the conditions specified in Eq. 91.[24]

R^1= Ph, alkenyl, allyl, H

(Eq. 91)

Allyl and benzyl chlorides are also formylated readily. Double bond migration to the α,β-unsaturated aldehyde is a common problem with allylic chlorides, as is competing reduction.[331]

Alkenyl Sulfonates

Alkenyl triflates are popular substrates for carbonylative coupling, which leads to α,β-unsaturated ketones and aldehydes. Many coupling examples can be found in the literature, and the scope of the reaction is broad. This strategy has been used in the total synthesis of natural products such as $\Delta^{9(12)}$-capnellene (Eq. 92)[334] and jatrophone.[335]

$$(87\%)$$

$$(Eq. 92)$$

Aryl-, alkynyl-, and alkenylstannanes all couple well, but double bond migration is a problem with allylstannanes. It has been reported that lithium chloride is a required additive for successful reaction. In several examples, the addition of zinc chloride improves the yields.[335] Macrocycles can be effectively prepared through intramolecular carbonylative ketone formation using a polymer-supported palladium catalyst.[186]

Aryl and Heterocyclic Sulfonates

The palladium-catalyzed carbonylative coupling of aryl triflates to give aryl ketones takes place under mild conditions.[336] Alkenyl-, alkynyl-, and arylstannanes all work well as coupling partners, but the presence of electron-withdrawing groups (e.g., NO_2) in these stannanes adversely affects the reaction because the aryl triflate is cleaved at the oxygen-sulfur bond. Allylstannanes are ineffective, resulting in high proportions of directly coupled products. As with alkenyl triflates, the presence of lithium chloride is required, but here the catalyst dichloro[1,1'-bis(diphenylphosphino)ferrocene]palladium gives superior yields (Eq. 93). If a competitive coupling site such as bromide is present on the

$$(Eq. 93)$$

aryl triflate, carbonylative cross-coupling takes place selectively at the triflate moiety even in the absence of lithium chloride (Eq. 94).

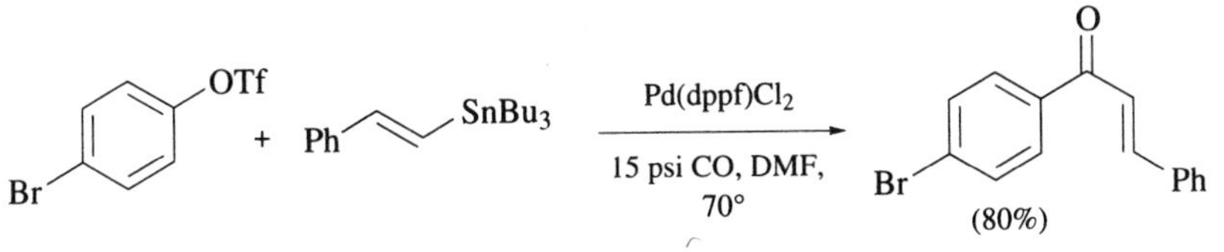

(Eq. 94)

Miscellaneous Substrates

Some activated organic halides containing β hydrogens can be carbonylatively cross-coupled under high CO pressures, and the ligand of choice for this reaction is triphenylarsine (Eq. 95).[337] The reported scope of this reaction is limited to the

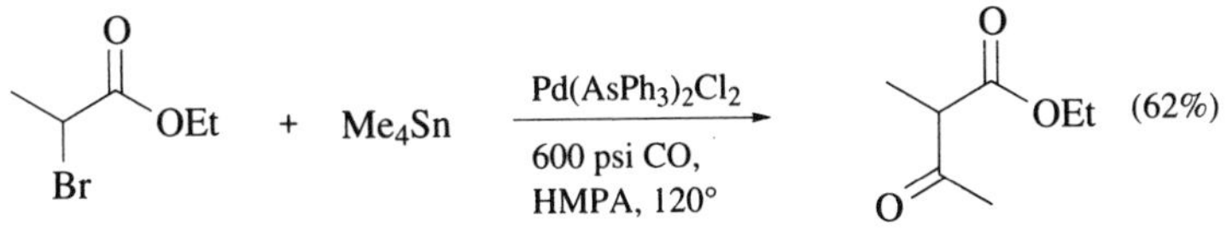

(Eq. 95)

use of α-phenethyl bromide, ethyl α-bromopropionate, and α-phenylpropyl bromide as substrates for the formation of methyl ketones, and the major side product is the result of elimination to the corresponding alkene. In a single example tetraphenylstannane has also been coupled.[323]

An interesting example of carbonylation has been applied to the synthesis of (+)-negamycin and (−)-5-*epi*-negamycin (Eq. 96).[338] The intermediate from the

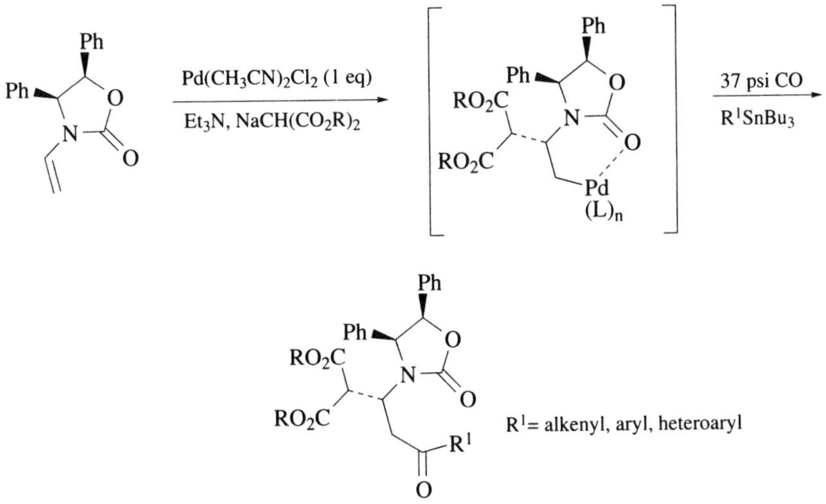

(Eq. 96)

palladium-assisted alkylation of an optically active enecarbamate is effectively carbonylated in the presence of an alkenylstannane to furnish the desired optically active ketone. Although this transformation requires a stoichiometric amount of palladium, it appears to be quite general and works well with a variety of alkenyl, aryl, and heteroarylstannanes.[339]

Aryl diazonium salts are also effective substrates for ketone formation (Eq. 97).[340] Diaryl and arylalkyl ketones can be prepared under very mild conditions. The presence of electron-withdrawing and electron-releasing groups on the ring is tolerated, and products from direct coupling are not observed.

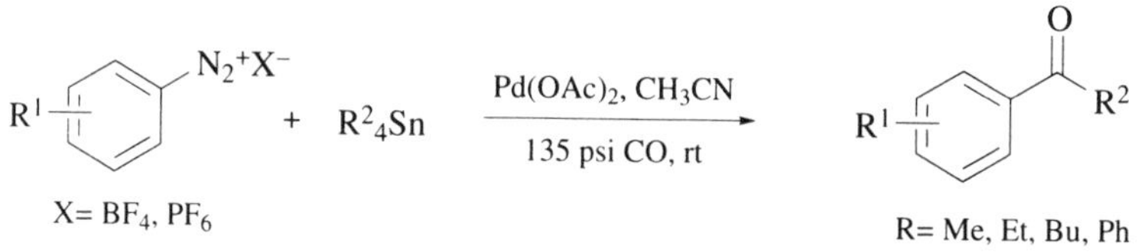
N2+X-
R1
+ R24Sn
Pd(OAc)2, CH3CN
135 psi CO, rt
X= BF4, PF6
O
R1
R2
R= Me, Et, Bu, Ph

(Eq. 97)

COMPLEX SYNTHETIC SEQUENCES INVOLVING TIN-TO-PALLADIUM(II) METATHESIS STEPS

A strategy that is receiving considerable attention in palladium chemistry is the tandem Heck-Stille sequence. Under suitable conditions, the organopalladium(II) intermediate resulting from a Heck insertion can be trapped by an organostannane, resulting in the formation of two C-C bonds at once. This strategy works best when the Heck adduct cannot undergo palladium hydride β elimination. The norbornyl system is used often in this sequence because the initially formed adduct **21** (Scheme 3) has no easily accessible *syn* β hydrogens, which are needed for a stereocontrolled elimination, and it is stable enough to be intercepted by the stannane to yield **22**.

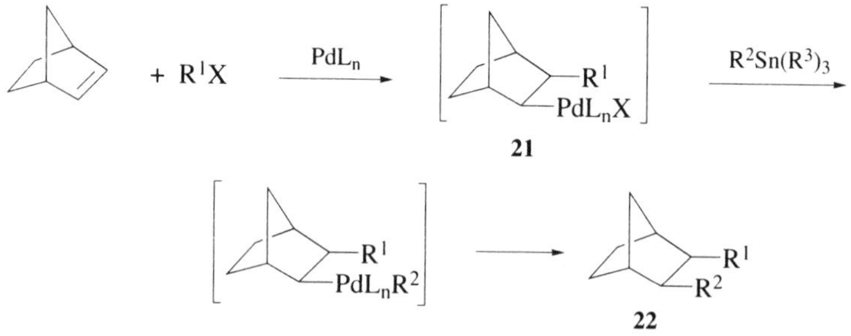
+ R1X
PdLn
R1
PdLnX
21
R2Sn(R3)3
R1
PdLnR2
R1
R2
22

Scheme 3. The Tandem Heck/Stille Strategy.

This strategy can be used in conjunction with $Pd(PPh_3)_4$ as catalyst, alkenyl or aryl bromides as electrophiles, and alkenyl-, alkynyl-, aryl- or allylstannanes as traps. The yields are low to fair, and direct coupling is the major side process.[341]

Allyl, benzyl, and acyl halides do not participate in this reaction. Among the stannanes that do not participate are the activated alkylstannanes, aminostannanes, alkoxystannanes, and thioalkoxystannanes.[342] For the analogous reaction with norbornadiene as substrate, the best ligand is (o-tolyl)diphenylphosphine. The additive tetraethylammonium chloride is needed for best results.[343]

More generally useful is the analogous sequence in which the initial Heck insertion is intramolecular. An elegant application to the synthesis of benzoprostacyclins is shown in Eq. 98.[344]

(Eq. 98)

This method can be extended to situations in which the initially formed organopalladium(II) intermediate is, in principle, capable of undergoing ready β-hydride elimination. Nevertheless, fine-tuning of the process with the help of tri(2-furyl)phosphine to accelerate the metathesis, in conjunction with zinc chloride, affords the Heck-Stille coupling product in high yield. The generality of these observations remains to be verified (Eq. 99).[345]

(Eq. 99)

When C-C triple bonds are used as intramolecular traps in this strategy, competing β elimination is not possible, and the tandem process is often successful, the only competition originating from the direct coupling (intermolecular) process. The initial 5-*exo* and 6-*exo* cyclizations are faster than direct coupling, and the tandem process succeeds, even though Al, Zr, and Zn derivatives often yield better results.[346-350] An application of this strategy to a neocarzinostatin synthesis is shown in Eq. 100.[351-353]

(Eq. 100)

Similar applications to the synthesis of vitamin D are reported.[354] Carbon monoxide insertion can be included in this sequence. An example of this interesting intramolecular Heck-CO insertion-transmetallation strategy is shown in Eq. 101.[355]

(Eq. 101)

In special cases, even the intermolecular insertion of alkynes can be carried out. When the electrophile is an allylic halide, apparently the direct coupling with stannanes is slow enough that the alkyne is first to react with the intermediate allylpalladium complex. Aryl-, alkenyl-, and alkynylstannanes can be used as traps. The yields, however, are quite modest (10–53%). An example is shown in Eq. 102.[356] A Ni(0)-catalyzed version of this reaction proceeds in higher yields, at least with alkynylstannanes as traps.[357]

$$TMSC\equiv CH \quad + \quad \diagup\!\!\!\!\diagdown\!\!\!\!\diagup^{Br} \quad + \quad PhSnBu_3 \quad \xrightarrow[\substack{P(2\text{-furyl})_3, \\ Et_4NCl, HMPA}]{Pd(CH_3CN)_2Cl_2}$$

$$TMS\diagup\!\!\!\!\!\!\overset{\overset{Ph}{|}}{\diagup}\!\!\!\!\diagdown\!\!\!\!\diagup\!\!\!\!\diagdown \quad (23\%)$$

(Eq. 102)

An interesting variant of the tandem Heck-Stille protocol is the reverse strategy. A bis(electrophile) can undergo monocoupling with an alkenylstannane, and this is followed by a fast intramolecular Heck reaction (Eq. 103).[358] This interest-

(Eq. 103)

ing strategy deserves further investigation. There are a number of interesting strategies for the construction of aromatic rings based on the ring opening of complex cyclobutenones, which on thermolysis rearrange to arenes via dienylketenes, as exemplified in Eq. 104.[359] Both alkenyl- and arylstannanes can be

(Eq. 104)

used in this coupling, leading to benzene and naphthalene derivatives, respectively, after electrocyclic ring opening/reclosure.

Variants of this technique are the synthesis of benzofurans and benzothio-phenes,[295] an approach to naphthoquinones and anthraquinones,[360] and new routes to benzocyclobutenedione derivatives,[361] azaheteroaromatics,[362] and 2-pyrones, the last involving a carbonylative step.[363]

Finally, the oxidative addition of Pd(0) onto silicon halides can be incorpo-rated in a three-component condensation involving 1-alkynes, TMSI, and alkenyl-, alkynyl-, or allylstannanes. An example of this powerful protocol is shown in Eq. 105.[364]

$$
PhC \equiv CH \xrightarrow[\substack{TMSI, \text{ dioxane} \\ 60°}]{Pd(PPh_3)_4} \left[\begin{array}{c} Ph \\ \diagup \!\!\!= \!\!\!\diagdown \\ IL_nPd \qquad TMS \end{array} \right] \xrightarrow{\diagup\!\!\!\!\diagdown SnBu_3}
$$

$$
\begin{array}{c} Ph \\ \diagup \!\!\!= \!\!\!\diagdown \\ =\!\!\!\diagdown \qquad TMS \end{array} \quad (80\%)
$$

(Eq. 105)

The use of complex strategies centered on, or terminated by, cross-coupling chemistry is an important and expanding synthetic tool that allows the formation of two or more C-C bonds, usually in a regioselective and stereoselective manner.

SIDE REACTIONS

Homocoupling reactions

Homocoupling of stannanes is apparently the most common side reaction ob-served when attempting Stille couplings.[30,106,204,286,297,299,365] The reaction may even be synthetically useful when symmetrical dienes[366] or biaryls[30] are desired. An obvious source of small amounts of homocoupled product is the reaction of the stannane with the Pd(II) precatalyst when this is employed. Each molar equivalent of Pd(II) reacts with two equivalents of the stannane to afford a sym-metrical product. In many cases, however, larger amounts of homocoupling prod-ucts are observed than can be accounted for in this way, and homocoupling takes place even when employing preformed Pd(0) catalysts. The reaction involves a catalytic cycle that has a radical component and requires atmospheric oxygen. In-sertion of Pd(0) in the carbon-tin bond of the stannane is postulated as the first step of the cycle.[30]

Homocoupling of the electrophile is often observed in transition metal-catalyzed cross-coupling reactions,[367] and there is evidence for a mechanism in-volving the exchange of organic groups between palladium and tin.[368] These au-thors used bidentate nitrogen-based ligands, and it is not clear whether this exchange occurs in reactions that use phosphorus-based ligands. A similar phe-nomenon with PPh₃ as ligand, on the other hand, has been documented.[34]

Transfer of "Nontransferable" Ligands

The Stille reaction usually employs three groups on tin that are not meant to be transferred in the coupling. Overwhelmingly, trialkyl derivatives are used because alkyl groups transfer slowly. Typically, trimethyl- or tributylstannane derivatives are used because of the ready availability of the corresponding trialkyltin halides. Selectivity is not, however, always complete.

For example, phenyltrimethylstannane couples with aryl triflates to yield products resulting from both aryl and methyl group transfer.[189] The selectivity is solvent dependent, dioxane yielding more aryl transfer than DMF or NMP. The phenyl group transfers 37 times more readily than *n*-butyl in NMP, using an aryl triflate as the electrophile. This ratio shows little dependence on the type of ligand. The ratio of the transfer rates of phenyl vs. methyl, on the other hand, is only 5.[30] These data strongly suggest that *n*-butyl groups are preferable to methyl groups as nontransferable moieties. The use of Cu(I) salts as cocatalysts improves this selectivity to >50:1,[33] and this may represent a potentially general solution to the selectivity problem (see also Eq. 61).

An interesting selectivity switch occurs in a hindered Stille coupling using stannane **24**. Whereas exclusive methyl transfer is observed under traditional conditions, use of Cu(I) salts leads to the aryl transfer product **26** in moderate yields (Eq. 106).[280]

Conditions: (i) PdCl$_2$(dppf), DMF, **25** (80%); (ii) Pd(PPh$_3$)$_4$, LiCl, dioxane, CuBr, 90°, **26** (60-64%).

(Eq. 106)

Other reports of alkyl group transfer in competition with the intended transfer of an aryl group are rather widespread,[55,191,369,370] and alkyl group transfer can sometimes be competitive even with alkynyl[219] and alkenyl coupling.[40,259] Once again, use of Cu(I) has resulted in substantial selectivity improvement in a butyl vs. alkenyl transfer competition.[33]

Further studies aimed at more careful quantification of alkyl group transfer as a side process and at discovering new tools to increase selectivity are definitely warranted.

Destannylation

Hydrolytic destannylation, probably brought about by traces of water and/or acids in the reaction medium, has been reported in very few cases, perhaps only because such a process in structurally simple stannanes yields volatile products that are difficult to detect. Organostannanes are quite stable hydrolytically, but when electron-rich aryl- or heteroarylstannanes are employed, destannylation may be a serious side reaction.[371,372]

Cine Substitution

Cine substitution can be a side process in a cross-coupling reaction, and Scheme 4 illustrates an example, together with a proposed mechanism.[204]

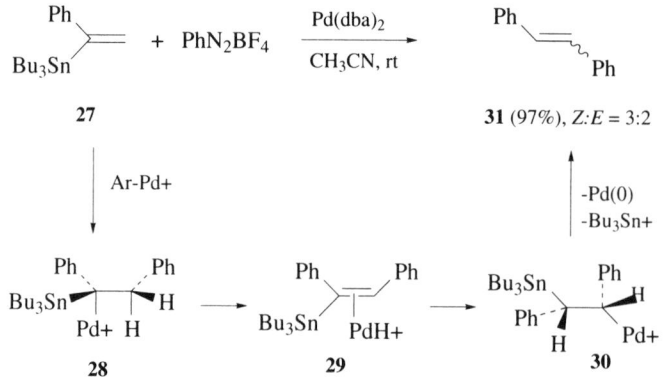

Scheme 4. Mechanistic interpretation of cine-substitution.

The first step is obviously an insertion of the arylpalladium intermediate across the double bond of the olefin. Evidently, a direct transmetallation is hindered by the α-phenyl substituent on the stannane. The following steps of β elimination and protodestannylation are reasonable and precedented. Another example of cine substitution requires an *anti* β elimination of palladium and hydrogen, which is a stereoelectronically disfavored pathway.[373]

It has been proposed that species like **28** may be able to undergo an unprecedented α elimination of Bu$_3$SnX to yield a Pd(0)-carbene species. A study of cine substitution with α-(tributylstannyl)acrylate showed that nonpolar solvents favor cine substitution, whereas ligands of different donicity have remarkably little effect on the product distribution.[245] Other authors have independently observed similar cine substitutions,[374-376] and high-yielding Stille coupling can be restored, once again, by using cocatalytic Cu(I).[246]

Cine substitution is a rare event in the coupling of organostannanes and is so far limited to 1-substituted 1-stannylethylenes, but it is a mechanistically intriguing process. From the mechanistic point of view, use of Cu(I) salts presumably yields intermediate organocopper species,[33] which undergo transmetallation with the "correct" regiochemistry. Silver carbonate has been used in one reaction to avoid cine substitution.[375] The generality of these observations remains to be verified.

Phosphorus-to-Palladium Aryl Migration

Arylpalladium(II) complexes like **32** (Eq. 107) undergo exchange of substituents between phosphorus and palladium at temperatures as low as 50° to yield **33-35**.[377] Thus, it is remarkable that this scrambling has not been detected

(Eq. 107)

in most of the classical Stille couplings. Recently, however, some examples of side products originating from aryl transfer by the phosphine were reported.[375,378] Triphenylarsine and tri(2-furyl)phosphine also lead to this side reaction. An obvious way to limit this unwanted process is to run the coupling at as low a temperature as possible.

Electrophile Reduction

Electrophile reduction is often a side reaction in Stille couplings, especially at high temperatures. It has been observed in the coupling of aryl triflates,[189,379] heteroaryl iodides,[126,128] alkenyl halides,[380] and allylic electrophiles.[163] The origin of this side process is uncertain. Alkyl transfer with β elimination prior to reductive elimination may be involved, although a radical mechanism is also possible.

Product Isomerization

In the coupling of acyl chlorides with alkenylstannanes, E/Z isomerization is observed under the coupling conditions.[146] Allylic stannanes, on the other hand, may yield mixtures of α,β- and β,γ-unsaturated ketones.[146] Geometric isomerization of olefins has often been reported as a side reaction.[46,51,153,157,269,289,381,382] Double bond migration has also been observed quite frequently.[56,135,383] It is likely that isomerization occurs at the product stage, but it is not clear whether it is catalyzed by palladium. Mild thermal conditions are believed to prevent or reduce isomerization. In addition, tri(2-furyl)phosphine-based catalysts prevent E/Z isomerization in the coupling of acyl chlorides and (Z)-alkenylstannanes.[11] The generality of this observation must be verified.

Miscellaneous Side Reactions

When using aryl triflates, hydrolytic cleavage to the corresponding phenols is a side reaction, especially at high temperatures.[55] Replacement of triflate with chloride owing to the presence of LiCl is a rare event, but it must be kept in mind as a possibility, especially for activated substrates.[40,173,195]

When carrying out Stille reactions on substrates containing isolated double bonds, the intermediate organopalladium species may undergo insertion across the double bond (Heck reaction), as discussed in the section on complex strategies.[336]

Reduction of enones has also been observed. The reducing agent is the tributyltin halide produced in the coupling.[148]

In one example, attempted coupling of an acyl chloride with vinyltributylstannane has led to dehydrodecarbonylation. Thus, proline derivative **36** gives **37** in unreported yield (Eq. 108). Use of the catalyst Pd(dppf)Cl$_2$ obviates the problem.[381]

$$\text{(Eq. 108)}$$

In reactions where the electrophile contains a quinone system, reduction to a dihydroquinone is a serious side reaction.[58,384]

1,1-Dibromoolefins couple with stannanes only once, whereas the second bromine moiety is eliminated (Eq. 109).[385] This side reaction may not be palladium catalyzed.

$Z:E = 74:26, (39\%)$

$$\text{(Eq. 109)}$$

The large variety of side reactions described for the Stille coupling does not reflect serious weaknesses in this cross-coupling method, but rather the careful scrutiny given to this important synthetic method in recent years. The side reactions can often be minimized or eliminated by using simple modifications of the traditional conditions, such as the use of appropriate ligands, solvents, additives, and temperatures, as described in this section.

COMPARISON WITH OTHER METHODS

A direct comparison between the Stille reaction and other cross-coupling protocols has been made in only a few cases, and these studies must be regarded with skepticism, since often each particular coupling was not separately optimized, as it should for the comparison to be legitimate. Thus, in a study of several alkenyl-alkenyl couplings in an approach to vitamin A,[386] it was concluded that the Stille coupling was unsatisfactory because of extensive homocoupling and that the reaction of alkenyl iodides with organozinc reagents gave better results. However, a limited set of conditions was explored.

Similarly, it has been concluded that zinc acetylides are better partners than alkynylstannanes in the coupling with certain alkenyl iodides.[387] In the coupling of an iodoglucal with arylmetals, the yields using arylzinc and arylboron compounds were quite superior to the ones obtained with the corresponding stannanes, but only under one set of conditions.[388] Similar conclusions were reached in a related system.[389] The synthesis of polyphenylenes by the Suzuki coupling appears to be superior to the corresponding Stille approach.[390]

Conversely, in other reactions, the Stille protocol outperforms the competition. In the 2-arylation of benzofuran derivatives, the use of organostannanes gives better results than the corresponding zinc derivatives.[391,392] In the synthesis of tamoxifen analogs, coupling of an alkenyl bromide with organotin, organozinc, and organoboron derivatives gives excellent results in each case.[50] Coupling of tetraalkylstannanes is reported to be superior to alkylaluminum and alkylzinc derivatives.[43] The Stille coupling is also the preferred route to substituted nucleosides.[132,374] A commonly given reason for preferring the use of organozinc and organoboron reagents over organostannanes is the toxicity of the latter. Conversely, the stannanes are often preferred because of the unusually mild and absolutely neutral conditions their coupling involves.

Bifunctional derivatives bearing a 9-BBN moiety and a tributylstannane residue couple selectively at the boron end under basic conditions (Eq. 110).[393]

(Eq. 110)

In general, the Stille reaction will continue to be a favorite method for carbon-carbon bond formation, owing to the lack of cross-reactivity displayed by the organostannanes with most functional groups. Its general utility is demonstrated by the many diverse applications reported in the tables.

EXPERIMENTAL CONDITIONS

The Stannane: Preparation and Handling

Caution! Many organotin compounds are toxic, especially the lower alkyl derivatives. Their acute toxicity decreases dramatically with increasing alkyl group length.[394,395] As a precaution, the preparation and use of all stannanes should only be carried out in a well-ventilated hood. After use, all glassware should be thoroughly washed, preferably after soaking in a KOH/alcohol bath to remove surface-bound tin alkoxides and/or halides.

Organostannanes are typically synthesized by reaction of organolithium or organomagnesium derivatives with trialkyltin halides. Another important method is the radical-induced or Pd-promoted addition of tin hydrides to unsaturated systems (e.g., alkynes, alkenes). Very important also is the transition metal-catalyzed cross-coupling of hexaalkyldistannanes with organic electrophiles, as discussed in the section on scope and limitations. Tin acetylides are best formed by the reaction of trialkyltin diethylamide with an alkyne.[396] A thorough treatment of the synthesis of organostannanes is outside the scope of this review, and the reader is referred to reviews on organostannanes.[6,395]

Most organostannanes are stable to air and moisture and can therefore be distilled and/or chromatographed. Stannanes are often too nonpolar to be efficiently purified on silica gel, but C-18 flash chromatography appears to be useful.[397] Given their ease of purification, for best results stannanes should not be used as crude preparations in Stille couplings.

Alkenyl and Aryl Triflates

Alkenyl triflates are typically synthesized by the reaction of triflic anhydride with a ketone or aldehyde in the presence of a hindered base, such as 2,6-di-*tert*-butylpyridine.[398,399] Enolates can be trapped with *N*-aryltriflimides, such as *N*-phenyltriflimide.[400,401] Vinyl triflates are also available from the addition of triflic acid to alkynes, though regio- and stereochemical considerations may be a problem.[402,403]

Aryl triflates are readily prepared by the reaction of triflic anhydride with a phenol in the presence of a base such as triethylamine or pyridine.[189] *N*-Phenyltriflimide can also be used for this transformation.[404] A thorough treatment of the synthesis of vinyl and aryl triflates is beyond the scope of this review, and the reader is referred to reviews on the formation and reactions of triflates.[405,406]

Choice of Nontransferable Ligands

Using nontransferable ligands is an area of the Stille reaction that needs further improvement. As discussed above, tributylstannane derivatives are usually preferred because of the low cost and low toxicity of tributyltin chloride, as well as the fact that competitive transfer of the butyl groups is a rare event. On the other hand, removal of traces of tributylstannane derivatives from the product can be problematic. Trimethylstannane derivatives have the disadvantage that methyl group transfer can often compete with the desired transfer of the unsaturated group, but the trimethylstannane derivatives produced in the coupling can usually be removed from the product by simple aqueous wash. Nontransferable ligands that speed up the transmetallation have been described in recent years, but have not yet found general acceptance.[41] Trichlorostannates have recently been used and can be employed to carry out Stille reactions in aqueous systems.[282,283]

Choice of Catalyst and Ligands

As discussed earlier, both Pd(0) and Pd(II) catalysts may be used to promote the cross-coupling reaction. Pd(II) catalysts have the advantage of being air stable, but must be reduced before entering the catalytic cycle. Typically, reduction is achieved in situ through the homocoupling of two equivalents of stannane, or with some reductant such as carbon monoxide. In rare instances, Pd(II) catalysts are pre-reduced by the addition of a Grignard or hydride reagent (often L-Selectride or DIBAL).[43] Pd(0) catalysts can enter the catalytic cycle directly, but can suffer from air and/or light stability problems.

Most catalysts are commercially available. Some of the most commonly used are: tetrakis(triphenylphosphine)palladium(0),[407] bis(dibenzylideneacetone)-palladium(0),[408] bis(acetonitrile)palladium(II) dichloride,[409] bis(triphenylphos-phine)palladium(II) chloride,[410,411] benzyl[bis(triphenylphosphine)]palladium(II) chloride,[21,412] 1,1'-bis(diphenylphosphino)ferrocenepalladium(II) dichloride,[413] and allylpalladium(II) chloride dimer.[414] Catalysts that do not incorporate strong ligands are often used in conjunction with added phosphines. Particularly useful among them are the Pd-dibenzylideneacetone complexes, which are commercially available and air stable. They can be used in conjunction with a variety of ligands. In addition to the traditional triphenylphosphine, ligands of reduced donicity, such as tri(2-furyl)phosphine and triphenylarsine, or increased steric bulk, such as tri(o-tolyl)phosphine, usually lead to much faster coupling.[11] These ligands are all commercially available. Nitrogen-based ligands have been used in a few cases, but their scope and utility have not been well established.[169,171,415] In some instances, it is advantageous to completely omit the ligand from the Stille reaction.[5] Ligandless catalysts usually afford high coupling rates but also premature interruption of the catalytic cycle.

Choice of Solvent

Solvents used include benzene, toluene, xylene, mesitylene, chloroform, 1,2-dichloroethane, THF, DME, dioxane, DMF, DMA, NMP, DMSO, HMPA, and water. Given the stable nature of the stannane organometallic species, it is fair to say that almost any conceivable solvent is likely to be compatible with the Stille protocol. Most couplings are carried out either in an ethereal solvent like THF or dioxane, or in highly dipolar solvents, such as DMF or NMP. Any of these four solvents represents a reasonable first choice when studying a new Stille coupling. The solvents are typically of anhydrous quality, but there does not seem to be a compelling reason to avoid traces of moisture. In many cases the literature specifically mentions that moisture accelerates the reaction. The same can be said about air: Whereas many Pd(0) complexes are air sensitive, during the Stille coupling the active catalyst is normally in the air-stable Pd(II) oxidation state (owing to rapid oxidative addition), and oxygen has no deleterious effect on the reaction. Many Stille reactions have been run in the presence of oxygen: Under these conditions a black precipitate of Pd metal signals the end of the reaction, where air-sensitive Pd(0) species accumulate. However, atmospheric oxygen can sometimes induce efficient homocoupling of the stannane (as discussed in the section on side reactions). In this event, careful deoxygenation by multiple freeze-thaw cycles is recommended.

Additives

The use of copper salts to facilitate the Stille cross-coupling is one the more significant recent developments in this area; the "copper effect" was discussed in the mechanistic section. The use of silver salts was also mentioned. Zinc chloride has often been used as additive. Yields are often better in the presence of stoichiometric amounts of Zn(II) salts, although the origin and the generality of the effect are not understood. The use of a stabilizing halide source, such as LiCl, and its complex effect on reaction rates in conjunction with the coupling of triflates have been discussed in the mechanistic section. When coupling triflates in ethereal solvents, LiCl appears to be necessary to induce coupling; in DMF or NMP (and presumably other dipolar solvents), LiCl is often unnecessary when coupling alkenyl triflates, whereas it sometimes appears to be necessary when coupling the less reactive aryl triflates. The experimentalist is urged to try the reaction both with and without LiCl. Bases such as triethylamine,[54,416] diisopropylethylamine,[80] lithium carbonate,[417] sodium carbonate,[298,418] pyridine,[419] and 2,6-di-*tert*-butyl-4-methylpyridine,[417] have also been employed as additives, presumably to minimize degradation of stannanes by adventitious acid.

Antioxidants, such as BHT, di-*tert*-butylphenol, or *tert*-butylcatechol are sometimes added to minimize side product formation via radical pathways.

Some reactions proceed more rapidly or in higher yield when run under dry air.[19] Palladium compounds catalyze the oxidation of triphenylphosphine to triphenylphosphine oxide by atmospheric oxygen. The rate enhancement found when running reactions under air may simply be due to the depletion of excess phosphine (see the "Mechanistic Considerations" section).

Workup: Removal of Tin Halides

A major consideration in working up reaction mixtures from the Stille cross-coupling is the removal of tin byproducts. Trimethyltin chloride is water soluble and rather volatile and is therefore readily removed on normal aqueous workup. Tributyltin chloride has low volatility (bp 171–173° at 25 mm Hg) and is soluble in most common organic solvents. Separation by chromatography on silica gel is made difficult by the tendency for tributyltin chloride to elute under relatively nonpolar conditions and to streak. A variety of methods have been devised to remove bulk tributyltin chloride prior to final purification. Aqueous KF solutions react with tributyltin halides under biphasic conditions to form polymeric tributyltin fluoride, which may be removed by filtration. Ammonia complexes with tributyltin halides, making them somewhat water soluble. Thus, washing of organic solutions with dilute ammonium hydroxide can remove the stannane.[88] Tributyltin chloride is insoluble in acetonitrile. Thus, dissolving crude or partially purified reaction mixtures in acetonitrile followed by washing with hexanes (in which tributyltin chloride is soluble) will remove most of the tin.[420] DBU in wet diethyl ether, followed by filtration through silica, has also been used to remove tributyltin residues.[420a] Scott and Stille proposed that CsF as a coupling additive might cause the formation of tributyltin fluoride in situ, thus facilitating workup.[28]

EXPERIMENTAL PROCEDURES

Trimethyl([3-(cyclohexen-1-yl)-2-propynyl]oxy)silane [Cross-Coupling of a Vinyl Halide with an Alkynylstannane Using Pd(PPh₃)₂Cl₂].[47] To a solution of 1-iodocyclohexene (0.424 g, 2.04 mmol), and trimethyl[3-(trimethylstannyl)-2-propynyl)oxy]silane (0.592 g, 2.04 mmol) in dry THF (25 mL) was added Pd(PPh₃)₂Cl₂ (0.0215 g, 0.031 mmol). The resulting mixture was stirred at 22–25° for 2 hours. The progress of the reaction was followed by TLC. The reaction mixture was diluted with CH_2Cl_2, coated onto alumina (10 g), and eluted with pentane. The resulting pentane solution was washed with water (3 × 25 mL) and a saturated NaCl solution (25 mL), dried (K_2CO_3), and concentrated under reduced pressure to give a pale yellow liquid (0.388 g, 92%): ^{1}H NMR (CDCl$_3$) δ 0.14 (s, 9 H), 1.48–1.68 (m, 4 H), 2.00–2.15 (m, 4 H), 4.36 (s, 2 H), 6.04–6.12 (m, 1 H); ^{13}C NMR (CDCl$_3$) δ -0.3, 21.5, 22.3., 25.6, 29.1, 51.5, 84.9, 86.8, 120.5, 134.5; IR (neat) 3040, 2218, 1442, 1322, 1258 cm^{-1}; Anal. Calcd for $C_{12}H_{20}OSi$: C, 69.17; H, 9.67. Found: C, 68.93; H, 9.70.

4-*tert*-Butyl-1-vinylcyclohexene [Cross-Coupling of a Vinyl Triflate with a Vinylstannane Using Pd(PPh₃)₄ and LiCl)][421] A slurry of Pd(PPh₃)₄ (1.18 g, 1.02 mmol) and LiCl (12.9 g, 0.305 mol) in dry THF (500 mL) was stirred for 15 minutes under a static Ar atmosphere, then a solution of 4-*tert*-butylcyclohexenyl triflate (28.0 g, 0.0979 mol) and trimethylvinylstannane (19.0 g, 0.0997 mol) in dry THF (250 mL) was added, followed by an additional 250 mL of THF. The resulting solution was heated under gentle reflux for 48 hours, then was cooled to room temperature and partitioned between water (500 mL) and pentane (250 mL). The aqueous layer was back-extracted with pentane (2 × 250 mL), and the combined organics were washed with a saturated NaHCO$_3$ solution (2 × 250 mL), water (2 × 250 mL), and a saturated NaCl solution (2 × 250 mL). The organic extracts were dried (MgSO$_4$), filtered through a pad of silica gel (4 cm × 4 cm), and concentrated by distillation using a 10-cm Vigreux column. Bulb-to-bulb distillation (Kugelrohr; oven temperature 65–68° at 0.55 mm Hg) gave the desired product (12.6–12.8 g, 78–79%): bp 45° (0.1 mm Hg); ^{1}H NMR (CDCl$_3$) δ 0.87 (s, 9 H), 1.08–1.34 (m, 3 H), 1.84–2.36 (m, 4 H), 4.88 (d, J = 10.7 Hz, 1 H), 5.04 (d, J = 17.5 Hz, 1 H), 5.73–5.75 (m, 1 H), 6.35 (dd, J = 17.5, 10.7 Hz, 1 H); ^{13}C NMR (CDCl$_3$) δ 23.8, 25.3, 27.2 (3C), 27.4, 32.2, 44.4, 109.7, 129.8, 136.0, 139.7; IR (neat) 3100, 3020, 1650, 1610, 1395, 1365, 985, 890 cm^{-1}.

1-(4-Methoxyphenyl)-4-*tert*-butylcyclohexene [Cross-Coupling of a Vinyl Triflate with an Arylstannane Using Pd$_2$(dba)$_3$ and AsPh$_3$].[30] A solution of Pd$_2$(dba)$_3$ (0.0083 g, 0.0184 mmol), AsPh$_3$ (0.023 g, 0.0734 mmol), and 4-*tert*-butylcyclohexenyl triflate (0.263 g, 0.918 mmol) in anhydrous degassed NMP (5 mL) was allowed to stand until the purple color was discharged (5 minutes), and (4-methoxyphenyl)tributylstannane (0.430 g, 1.083 mmol) in dry NMP (2 mL) was added. The resulting solution was stirred at room temperature for 16 hours, then stirred with a 1 M aqueous KF solution (1 mL) for 30 minutes, diluted with EtOAc, and filtered. The filtrate was washed extensively with water, dried, and concentrated to give a crude oil. The oil was purified by reverse phase flash chromatography (C-18, 10% CH$_2$Cl$_2$, 90% CH$_3$CN) to give a white solid which was recrystallized (MeOH), (0.201 g, 89%): mp 78–79°; ^{1}H NMR (CDCl$_3$) δ 0.91 (s, 9 H), 1.22–1.39 (m, 2 H), 1.89–2.02 (m, 2 H), 2.19–2.54 (m, 3 H), 3.80 (s, 3 H), 6.04 (m, 1 H), 6.84 (d, J = 9.0 Hz, 2 H), 7.32 (d, J = 9.0 Hz, 2 H); Anal. Calcd. for C$_{17}$H$_{24}$O: C, 83.55; H, 9.90. Found: C, 83.58; H, 9.85.

3-Methyl-2-(4-tolyl)-2-cyclopentenone [Cross-Coupling of an Unreactive Alkenyl Halide Under "Modified" Conditions Using Pd(PhCN)$_2$Cl$_2$, AsPh$_3$, and CuI as Cocatalyst].[61] A solution of 2-iodo-3-methyl-2-cyclopentenone (0.222 g, 1.00 mmol), CuI (0.019 g, 0.10 mmol), AsPh$_3$ (0.031 g, 0.10 mmol), and Pd(PhCN)$_2$Cl$_2$ (0.019 g, 0.05 mmol) in NMP (1 mL) was treated under Ar with *p*-tolyltributylstannane (0.37 mL, 1.20 mmol), and the mixture was heated in an oil bath at 100° for 30 minutes. After cooling, the solution was diluted with EtOAc (100 mL) and washed with aqueous KF (0.67 satd., 3 × 30 mL) and water (2 × 20 mL). The combined aqueous layers were back-extracted with EtOAc (60 mL). The combined organics were dried (MgSO$_4$), filtered, and evaporated to dryness. The resulting oil was purified by silica gel chromatography (gradient 2–10% EtOAc in pet. ether) to yield a white solid (0.165 g, 89%): mp 102–103° (EtOAc/pet. ether); ^{1}H NMR (CDCl$_3$) δ 7.20 (m, 4 H), 2.61 (m, 2 H), 2.51 (m, 2 H), 2.35 (s, 3 H), 2.15 (s, 3 H); ^{13}C NMR (CDCl$_3$) δ 207.6, 171.2, 140.1, 137.2, 128.9, 34.7, 31.7, 21.2, 18.2. IR (CHCl$_3$) 1685 cm^{-1}; Anal. Calcd for C$_{13}$H$_{14}$O: C, 83.87; H, 7.54. Found: C, 84.06; H, 7.42.

1-(4-Nitrophenyl)-2-propenone (Cross-Coupling of an Acid Chloride with an Arylstannane).[146] To a solution of 4-nitrobenzoyl chloride (5.00 mmol) and BnPd(PPh$_3$)$_2$Cl (0.015–0.020 g, 0.020–0.026 mmol) in chloroform (1 mL) was added a solution of tributylvinylstannane (5.20 mmol) in chloroform (4 mL). The resulting yellow solution was heated at 65° under dry air until palladium metal precipitated (20 minutes). The reaction mixture was diluted with Et$_2$O (30 mL) and washed with water (30 mL). The organic phase was shaken with an aqueous KF solution (15 mL of saturated KF solution/15 mL of water) and allowed to stand for 15–30 minutes. The resulting white precipitate (Bu$_3$SnF) was removed by filtration. The organic layer was separated and again treated with an aqueous KF solution. After decantation from the resulting white precipitate, the organic phase was washed with concentrated NaCl solution, dried (MgSO$_4$), and concentrated under reduced pressure. Treatment of the residue with EtOAc afforded an additional crop of white precipitate, which was removed by filtration through a Celite pad. Following concentration under reduced pressure, recrystallization from chloroform/hexanes gave the product as a yellow solid (0.780 g 88%): mp 87–89°; ^{1}H NMR (CDCl$_3$) δ 6.0 (dd, J = 10.2 Hz, 1 H), 6.4 (dd, J = 18.2 Hz, 1 H), 7.1 (dd, J = 18.1 Hz, 1 H), 8.0 (d, J = 9 Hz, 2 H), 8.3 (d, J = 9 Hz, 2 H); IR (KBr) 1670 cm^{-1}; Anal. Calcd. for C$_9$H$_7$NO$_3$: C, 61.02; H, 3.93. Found: C, 61.23; H, 4.11.

4-Allylacetophenone [Cross-Coupling of an Aryl Triflate Under Mild Conditions Using Tri(2-furyl)phosphine as Ligand].[11] A solution of 4-

(triflyloxy)acetophenone (0.566 g, 2.11 mmol) in NMP (3 mL) was treated with anhydrous LiCl (0.268 g, 6.30 mmol), tri(2-furyl)phosphine (0.0392 g, 0.168 mmol), and Pd$_2$(dba)$_3$ (0.0193 g, 0.042 mmol Pd). After 10 minutes at room temperature, the solution was treated with allyltributylstannane (0.72 mL, 2.464 mmol) and the mixture was stirred at room temperature for 24 hours. The solution was stirred with a saturated aqueous KF solution, diluted with EtOAc, and filtered. Washing the organics with water, drying (anhydrous Na$_2$SO$_4$), and evaporation of the solvent gave a crude oil which was purified by flash chromatography (silica gel, 5% EtOAc in hexanes) to yield a colorless liquid (0.264 g, 78.5%); bp (Kugelrohr) 90–95° (0.2 mmHg); ^{1}H NMR (CDCl$_3$) δ 7.89 (d, J = 8.3 Hz, 2 H), 7.27 (d, J = 8.2 Hz, 2 H), 5.94 (m, 1 H), 5.13–5.06 (m, 2 H), 3.43 (d, J = 6.7 Hz, 2 H), 2.57 (s, 3 H); Anal. Calcd for C$_{11}$H$_{12}$O: C, 82.46; H, 7.55. Found: C, 82.11; H, 7.56.

8-(Trimethylstannyl)quinoline (Preparation of an Arylstannane by Cross-Coupling of an Aryl Triflate with Hexamethyldistannane).[189] To a solution of 8-(triflyloxy)quinoline (1.98 mmol) in dioxane (9 mL) were added hexamethyldistannane (2.05 mmol), LiCl (0.252 g, 5.94 mmol) Pd(PPh$_3$)$_4$ (0.046 g, 0.040 mmol), and a few crystals of BHT. The mixture was heated to reflux for 75 hours, cooled, and treated with pyridine (1 mL) and pyridinium fluoride (1.4 M in THF, 2 mL) for 16 hours at room temperature. The mixture was diluted with Et$_2$O, filtered through Celite, and washed with water, 10% HCl, water, and brine. Drying (MgSO$_4$) and concentration afforded an oil. Silica gel chromatography and bulb-to-bulb distillation (bp: 103–104° at 0.4 mm Hg) gave a colorless oil in 67% yield; ^{1}H NMR (CDCl$_3$) δ 8.86 (dd, J = 4.2, 1.7 Hz, 1 H), 8.07 (dd, J = 8.2, 1.8 Hz, 1 H), 7.88 (d, J = 6.5, 1.3 Hz, 1 H), 7.75 (dd, J = 8.1, 1.3 Hz, 1 H), 7.49 (dd, J = 8.1, 6.6 Hz, 1 H), 7.31 (dd, J = 8.2, 4.2 Hz, 1 H), 0.30 (s, 9 H); ^{13}C NMR (CDCl$_3$) δ 153.17, 153.06, 149.35, 147.56, 136.94, 127.97, 126.21, 125.83, -8.32; IR (neat) 3050, 2970, 2905, 1485, 810, 785 cm^{-1}; Anal. Calcd. for C$_{12}$H$_{15}$NSn: C, 49.37; H, 5.18. Found: C, 49.50; H, 5.25.

4-(tert-Butyl-1-vinylcyclohexen-1-yl)-2-propenone [Carbonylative Cross-coupling of an Alkenyl Triflate with an Alkenylstannane Using Pd(PPh$_3$)$_4$ and LiCl].[421] A slurry of Pd(PPh$_3$)$_4$ (1.12 g, 0.968 mmol) and LiCl (13.2 g,

0.312 mol) in dry THF (500 mL) was stirred for 15 minutes under a static Ar atmosphere, then a solution of 4-*tert*-butylcyclohexenyl triflate (28.6 g, 0.100 mol) and trimethylvinylstannane (19.1 g, 0.100 mol) in dry THF (250 mL) was added, followed by an additional 250 mL of THF. The reaction mixture was flushed with carbon monoxide and maintained under a carbon monoxide atmosphere (15–20 psi) while heating to 55°. After 40 hours the reaction mixture darkened and was cooled to room temperature. The resulting solution was diluted with pentane (500 mL), washed with water (2 × 200 mL), saturated NaHCO₃ solution (2 × 200 mL), and brine (2 × 200 mL), then was dried (MgSO₄), filtered through a 4-cm × 4-cm pad of silica gel, and concentrated under reduced pressure. Bulb-to-bulb distillation (Kugelrohr) at 85–95° (0.35 mm Hg) gave the desired product (14.3–14.5 g, 74–75%): bp 75° (0.1 mm Hg); ^{1}H NMR (CDCl₃) δ 0.81 (s, 9 H), 1.21–2.65 (m, 7 H), 5.58 (d, J = 9.0 Hz, 1 H), 6.14 (d, J = 17.2 Hz, 1 H), 6.75–7.00 (m, 2 H); ^{13}C NMR (CDCl₃) δ 23.3, 24.6, 26.9 (3C), 27.8, 32.0, 43.4, 127.1, 131.5, 141.1, 190.8; IR (neat) 1665, 1645, 1612 cm^{-1}.

(E)-1-(4-Methoxyphenyl)-3-phenyl-2-propenone [Carbonylative Cross-Coupling of an Aryl Triflate With an Alkenylstannane Using Pd(dppf)Cl₂ and LiCl].[336] To a solution of 4-methoxyphenyl triflate (0.390 g, 1.52 mmol) in DMF (7 mL) was added (*E*)-(β-tributylstannyl)styrene (0.645 g, 1.64 mmol), LiCl (0.200 g, 4.72 mmol), Pd(dppf)Cl₂ (0.045 g, 0.060 mmol), a few crystals of BHT, and 4 Å molecular sieves (0.10 g). The resulting mixture was heated at 70° under 15 psi of CO. After 23 hours the reaction was cooled to room temperature, diluted with Et₂O, and filtered. The filtrate was washed with water (3 times) and saturated NaCl solution, dried (MgSO₄), and concentrated. The resulting material was purified by chromatography (silica gel, 10:1 hexanes/EtOAc) to give the product as a white solid (0.250 g, 68%), which was recrystallized from 20:1 hexanes/EtOAc: mp 105–106°; ^{1}H NMR (CDCl₃) δ 3.82 (s, 3 H); 6.94, (d, J = 8.8 Hz, 2 H), 7.36–7.39 (m, 3 H), 7.53 (d, J = 15.7 Hz, 1 H), 7.59–7.63 (m, 2 H), 7.79 (d, J = 15.7 Hz, 1 H), 8.03 (d, J = 8.9 Hz, 1 H).

TABULAR SURVEY

The literature was searched to the end of 1994 by Chemical Abstracts, extensive citation searches and browsing. A few of the papers which describe Stille couplings but are missing a vital piece of information (i.e., clear structure of substrates and/or products) were not abstracted. No attempts were made to cover the patent literature. A dash indicates lack of reported yield. When only GLC, NMR, or HPLC yields were reported, these were simply incorporated in the tables without specific notation. When both isolated and "estimated" yields were given, the isolated yields are shown in the tables. If experimental conditions were not given, the appropriate column usually contains the generic statement "Pd(0)". Reactions that appear well documented but afford none of the anticipated product are still reported, and 0% yield is shown next to the structure of the expected product. We think failed reactions may stimulate further research and new thinking. In some papers, the attempt to optimize a reaction led to many experiments done on the same substrate under slightly different catalytic conditions. In most cases, for the sake of simplicity, we report only the highest yielding of all these experiments. However, in some cases the comparison of two or more sets of conditions on the same substrate proves a point which, in our opinion, was important enough to warrant a separate entry.

Some of the 1995 papers were incorporated in the tables as they appeared in the literature, but only those which, in our opinion, reported new catalytic systems or new classes of substrates.

The substrates are broken down into specific classes according to electrophile type, to reflect the classification made in the "Scope and Limitations" section. Some classes (heterocyclic or acyl electrophiles) are further broken down into subclasses to facilitate target finding. The electrophiles are listed in order of increasing carbon count for the moiety that is being transferred (the leaving group is not included in the carbon or heteroatom count). Within a given C count, they are listed in order of increasing numbers of heteroatoms, the priority being assigned alphabetically except for H, which has *the lowest* priority. For example $C_6H_5\underline{C}lO$ has priority over C_6H_5O and/or C_6H_6ClO. This ranking was the simplest and visually the most pleasing of a number of alternatives that we examined.

Electrophiles where the halide moiety is attached to a heterocyclic system or an aryl ring fused to a heterocyclic system (be it aromatic or partially saturated) are considered heterocyclic electrophiles. If the heterocyclic portion is *isolated* from the electrophilic moiety, then it is not considered.

The stannanes are similarly arranged according to the moiety that is being transferred. Tin hydrides are listed first, then all the C-based nucleophiles in the order explained above (in addition, trimethylstannanes have priority over tributylstannanes and bis[stannanes] are listed after all the monostannanes within a given electrophile), then the heterostannanes are listed (priority is assigned based on the alphabetical rank of the atom whose bond to tin is being broken). Intramolecular Stille couplings are listed in separate tables. A special case is the coupling of bis(stannanes) with bis(electrophiles), ultimately yielding a cyclic product. These reactions are listed twice: once in the appropriate table for the Stille coupling which our mechanistic knowledge tells us is taking place first, the second time in the intramolecular table. We realize this is cumbersome and causes duplication, but it seems the only logical way of dealing with the problem in an informative way. Other, more complex strategies in which the Stille reaction is coupled to other reactions are listed separately in Tables XXXII (no CO involved) and XXXIII (CO involved). The structures of stannanes that were formed in situ are enclosed in brackets.

The following abbreviations are used in the tables:

BINAP	2,2′-bis(diphenylphosphino)-1,1′-binaphthyl
Bn	benzyl
Boc	*tert*-butoxycarbonyl
BOM	benzyloxymethyl
Bz	benzoyl
Cbz	benzyloxycarbonyl
d	day(s)
dba	dibenzylideneacetonyl
DIOP	2,3-*O*-isopropylidene-2,3-dihydroxy-1,4-bis-(diphenylphosphino)butane
DME	1,2-dimethoxyethane, glyme
DMF	dimethylformamide
DMSO	dimethyl sulfoxide
dppb	1,3-bis(diphenylphosphino)butane
dppf	1,1′-bis(diphenylphosphino)ferrocene
dppp	1,3-bis(diphenylphosphino)propane
EE	(1-ethoxy)ethyl
FMOC	fluorenylmethyloxycarbonyl
HMPA	hexamethylphosphoric triamide
MEM	methoxyethoxymethyl
MOP	2-(diphenylphosphino)-2′-methoxy-1,1-binaphthyl
MOM	methoxymethyl
Ms	methanesulfonyl
NMP	*N*-methylpyrrolidinone
Ph-BIAN	bis(phenylimino)acenaphthene
PMB	*p*-methoxybenzyl
PNB	*p*-nitrobenzyl
rt	room temperature
SEM	(2-trimethylsilylethoxy)methyl
TBDMS	*tert*-butyldimethylsilyl
TBDPS	*tert*-butyldiphenylsilyl
Tf	trifluoromethanesulfonyl
Thexyl	1-(1,1,2-trimethyl)propyl
TIPS	tri(isopropyl)silyl
THF	tetrahydrofuran
THP	tetrahydropyranyl
TMS	trimethylsilyl
p-Tol	*p*-tolyl
Ts	*p*-toluenesulfonyl

TABLE I. DIRECT CROSS-COUPLING OF ALKENYL ELECTROPHILES

Substrate	Stannane	Conditions	Product(s) and Yield(s) (%)	Refs.
C₂ $CH_2=CHBr$	Me_3Sn, OEt	Pd(PPh₃)₄ (2%), LiCl, neat, 80°, 12 h	OEt diene (80)	270
	Bu_3Sn cyclopropane, TMS	Pd(PPh₃)₄ (5%), THF, 66°	TMS (0)	422
	Me_3Sn, OEt, TBDMS	BnPd(PPh₃)₂Cl (1%), CuI, DMF, 50°	OEt TBDMS (66)	49
	indole, SEM, H_2N, Bu₃Sn	Pd(PPh₃)₄ (10%), DMF, 90°, 1 h	(84)	74
	uracil–SnBu₃ (HN, O, O, O, MOMO)	Pd(PPh₃)₄ (10%), CuI (20%), DMF, 80°, 100 min	(80)	170
	MOMO sugar, OBn, OBn, OBn	Pd(0)	OBn, OBn, OBn (22-30)	423, 424
$CH_2=CHI$	Bu₃Sn cyclopropane, TMS	Pd(PPh₃)₄ (5%), THF, 66°	TMS (0)	422
Cl–CH=CH–Cl	Bu₃Sn, OH, C₅H₁₁-n, OH	Pd(PPh₃)₄, DMF	OH, C₅H₁₁-n, OH, Cl (≥46)	44
	Bu₃Sn, OH, C₅H₁₁-n, OH	Pd(PPh₃)₄, DMF	OH, C₅H₁₁-n, OH, Cl (61)	44

62

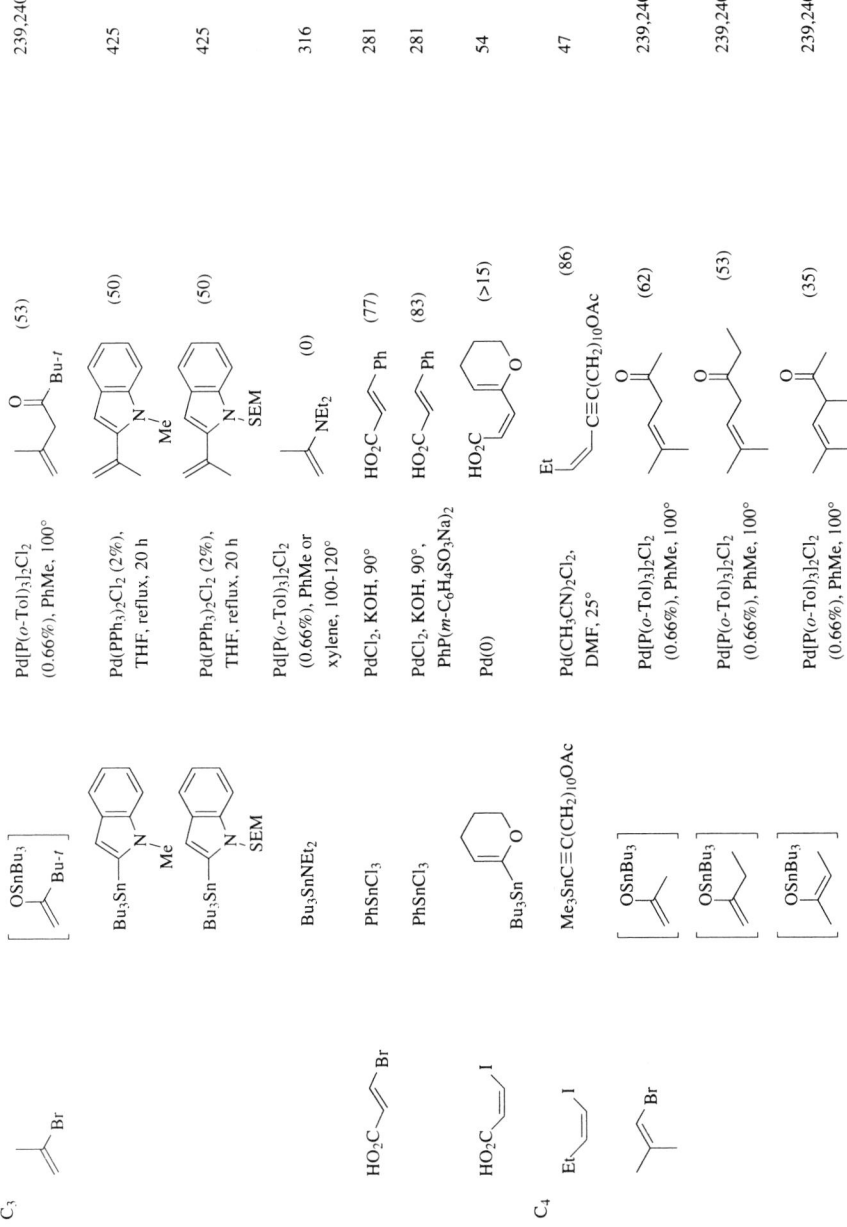

C₃				
		Pd[P(o-Tol)₃]₂Cl₂ (0.66%), PhMe, 100°	(53)	239,240
		Pd(PPh₃)₂Cl₂ (2%), THF, reflux, 20 h	(50)	425
		Pd(PPh₃)₂Cl₂ (2%), THF, reflux, 20 h	(50)	425
		Pd[P(o-Tol)₃]₂Cl₂ (0.66%), PhMe or xylene, 100-120°	(0)	316
		PdCl₂, KOH, 90°	(77)	281
		PdCl₂, KOH, 90°, PhP(m-C₆H₄SO₃Na)₂	(83)	281
		Pd(0)	(>15)	54
C₄		Pd(CH₃CN)₂Cl₂, DMF, 25°	(86)	47
		Pd[P(o-Tol)₃]₂Cl₂ (0.66%), PhMe, 100°	(62)	239,240
		Pd[P(o-Tol)₃]₂Cl₂ (0.66%), PhMe, 100°	(53)	239,240
		Pd[P(o-Tol)₃]₂Cl₂ (0.66%), PhMe, 100°	(35)	239,240

TABLE I. DIRECT CROSS-COUPLING OF ALKENYL ELECTROPHILES (*Continued*)

Substrate: Bu₃Sn— (alkenyl oxazoline)

Stannane	Conditions	Product(s) and Yield(s) (%)	Refs.
(oxazoline)	Pd(PPh₃)₂Cl₂ (1%)	(51)	426
OSnBu₃	Pd[P(o-Tol)₃]₂Cl₂ (0.66%), PhMe, 100°	(74)	239, 240
OSnBu₃	Pd[P(o-Tol)₃]₂Cl₂ (0.66%), PhMe, 100°	(8)	239, 240
OSnBu₃ Bu-t	Pd[P(o-Tol)₃]₂Cl₂ (0.66%), PhMe, 100°	Bu-t (81)	239, 240
OSnBu₃ (cyclohexenyl)	Pd[P(o-Tol)₃]₂Cl₂ (0.66%), PhMe, 100°	(32)	239, 240
CO₂Et, SnBu₃ E:Z = 1:7	Pd(PPh₃)₄ (5%), C₆H₆, reflux, 21 h	CO₂Et E:Z = 1:3 (20)	306, 427
OSnBu₃ Ph	Pd[P(o-Tol)₃]₂Cl₂ (0.66%), PhMe, 100°	Ph (74)	239, 240
Bu₃SnNEt₂	Pd[P(o-Tol)₃]₂Cl₂ (0.66%), PhMe, 100-120°	NEt₂ (50)	316
(Bu₃Sn)₂S	Pd(PPh₃)₄, (1%), PhMe, 120°	S (25)	426

Conditions	Yield	Ref.
BnPd(PPh₃)₂Cl (5%), CuI (8%), DMF, rt, 30 min	(70)	188
Pd[P(o-Tol)₃]₂Cl₂ (0.66%), PhMe, 100°	(76)	239, 240
Pd[P(o-Tol)₃]₂Cl₂ (0.66%), PhMe, 100°	(90)	239, 240
Pd(PPh₃)₂Cl₂, DMF	(73)	80
Pd(PPh₃)₂Cl₂, DMF	(>59)	80
Pd(CH₃CN)₂Cl₂, NMP, rt	(64)	80
Pd(PPh₃)₂Cl₂, THF, 60°	(55)	81
Pd(PPh₃)₄, Cl(CH₂)₂Cl, 20°, 2 h	(51)	428

TABLE I. DIRECT CROSS-COUPLING OF ALKENYL ELECTROPHILES (Continued)

Substrate	Stannane	Conditions	Product(s) and Yield(s) (%)	Refs.
	Bu₃SnC≡CTMS	Pd(PPh₃)₄, Cl(CH₂)₂Cl, 20°, 2 h	(TMSC≡C / C≡CTMS squarate) (30)	428
	Bu₃SnC≡CPh	Pd(PPh₃)₄, Cl(CH₂)₂Cl, 20°, 2 h	(PhC≡C / C≡CPh squarate) (70)	428
	Bu₃SnC≡C–C≡CPh	Pd(PPh₃)₄, Cl(CH₂)₂Cl, 20°, 6 h	(PhC≡C–C≡C / C≡C–C≡CPh squarate) (11)	428
MeO₂C—Br, Z:E = 7:1	(uracil-SnBu₃ derivative, MOMO, HN, O=, SEM)	Pd(PPh₃)₄ (10%), CuI (20%), DMF, 80°, 10 min	(uracil–CH=CH–CO₂Me product) (81) Z:E = 3:1	170
	(2-Bu₃Sn-indole, N-SEM)	Pd(CH₃CN)₂Cl₂ (10%), DMF, rt, 1 h	(indolyl–CH=CH–CO₂Me, SEM) (68) Z:E = 1.5:1	289
	"	Pd(PPh₃)₄ (10%), DMF, 90°, 2 h	" (87)	289
MeO₂C—I	(2-Bu₃Sn-dihydrofuran)	Pd(OAc)₂, P(o-Tol)₃, NEt₃, CH₃CN, reflux, 2 h	(dihydrofuranyl–CH=CH–CO₂Me) (64)	429

66

Substrate	Reagent	Conditions	Product (yield)	Refs.
Bu₃Sn–(dihydropyranyl)	HO₂C–C(=CH₂)CH₂–I	Pd(OAc)₂, P(o-Tol)₃, NEt₃, CH₃CN, reflux, 2 h	(dihydropyranyl)CH=CH–CO₂Me (60), Z:E = 4:1	429
Bu₃Sn–CH=CH₂		Pd(CH₃CN)₂Cl₂ (5%), DMF, 45°, 12 h	HO₂C–CH₂–C(=CH₂)–CH=CH₂ (60)	430
Bu₃Sn–CH=CH–CH(OEt)₂		Pd(CH₃CN)₂Cl₂ (5%), DMF, 45°, 12 h	HO₂C–CH₂–C(=CH₂)–CH=CH–CH(OEt)₂ (50)	430
Bu₃Sn–CH=CH–Ph		Pd(CH₃CN)₂Cl₂ (5%), DMF, 45°, 12 h	HO₂C–CH₂–C(=CH₂)–CH=CH–Ph (82)	430

C₅

Substrate	Reagent	Conditions	Product (yield)	Refs.
cyclopentenyl–OTf	Ph₃SnF₂⁻ Bu₄N⁺	Pd(PPh₃)₄, THF, reflux, 30 min	Ph–cyclopentene (81)	283
n-Pr–CH=CH–I	Me₃SnC≡CTMS	Pd(PPh₃)₄ (5%), THF, 50°	n-Pr–CH=CH–C≡C–TMS (—)	431
n-Pr–CH=CH–I (Z)	Me₃SnC≡C(CH₂)₉OTHP	Pd(CH₃CN)₂Cl₂, DMF, 25°	n-Pr–CH=CH–C≡C(CH₂)₉OTHP (83)	47
(CH₃)₂C=C(CH₃)–Br	Bu₃Sn–(oxazoline)	Pd(PPh₃)₂Cl₂ (1%)	(oxazoline) (65)	426
	[OSnBu₃ / Bu-t]	Pd[P(o-Tol)₃]₂Cl₂ (0.66%), PhMe, 100°	O=C–Bu-t ketone (86)	239, 240
	Bu₃SnNEt₂	Pd[P(o-Tol)₃]₂Cl₂ (0.66%), PhMe, 100–120°	NEt₂ (0)	316
	(Bu₃Sn)₂S	Pd(PPh₃)₄ (1%), PhMe, 120°	S (0)	318

TABLE I. DIRECT CROSS-COUPLING OF ALKENYL ELECTROPHILES (*Continued*)

Substrate	Stannane	Conditions	Product(s) and Yield(s) (%)	Refs.
1-Ph⁺OTf⁻	$Bu_3SnC{\equiv}CPh$	BnPd(PPh₃)₂Cl (5%), CuI (8%), DMF, rt, 30 min	[structure] C≡CPh (64)	188
1-Ph⁺OTf⁻	$Bu_3SnC{\equiv}CPh$	"	[structure] C≡CPh (66)	188
[bromocyclopentenone]	[indole, SEM, Bu₃Sn]	Pd(PPh₃)₄ (10%), DMF, 110°, 6 h	[SEM indole-cyclopentenone] (92)	289
	[TsHN, indole, SEM, Bu₃Sn]	Pd(PPh₃)₄ (10%), DMF, 110°, 3 h	[NHTs SEM indole-cyclopentenone] (78)	74
[Br vinyl ketone]	[furan, Bu₃Sn]	Pd(PPh₃)₂Cl₂ (4%), DMF, 70°, 6 h	[furan enone] (70)	287
[Br vinyl ketone]	"	"	[furan enone] (77)	287
	[Bu₃Sn furan, OH]	[(η³-C₃H₅)PdCl]₂, DMF, rt, 24 h	[furan enone, OH] (66)	287, 432
	[Bu₃Sn furan SnBu₃]	[(η³-C₃H₅)PdCl]₂, DMF, 70°, 1 h	[difuran dienone] (79)	287

68

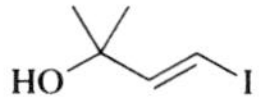 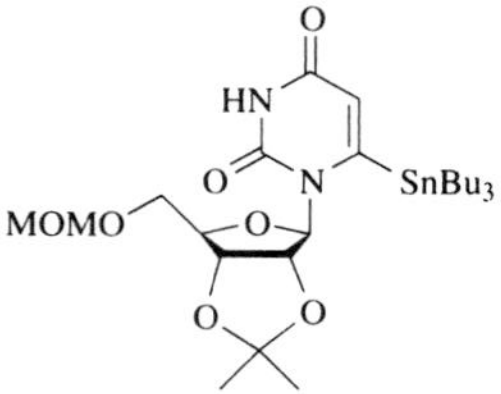 Pd(PPh$_3$)$_4$ (10%), CuI (20%), DMF, 80°, 1 h	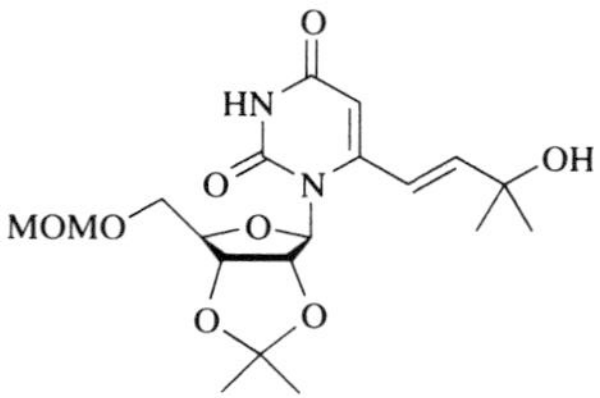 (70) 170
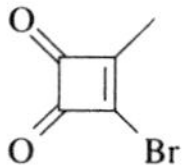 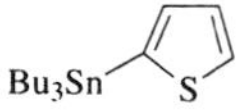 BnPd(PPh$_3$)$_2$Cl (2%), THF, 50°, 13 h	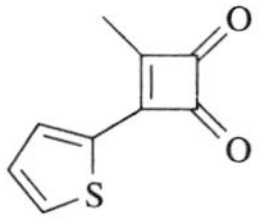 (73) 63
BnPd(PPh$_3$)$_2$Cl (2%), THF, 50°, 20 h	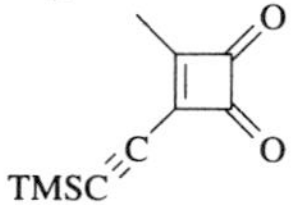 (57) 63
BnPd(PPh$_3$)$_2$Cl (2%), THF, 50°, 6 h	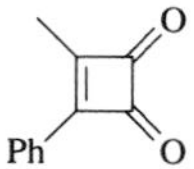 (85) 63
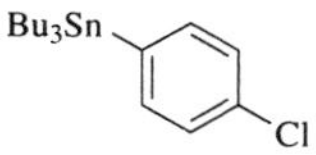 BnPd(PPh$_3$)$_2$Cl (2%), THF, 50°, 15 h	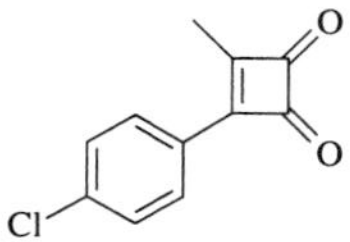 (81) 63
BnPd(PPh$_3$)$_2$Cl (2%), THF, 50°, 20 h	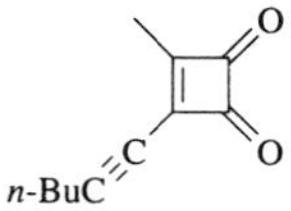 (55) 63
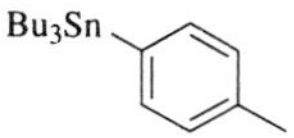 BnPd(PPh$_3$)$_2$Cl (2%), THF, 50°, 10 h	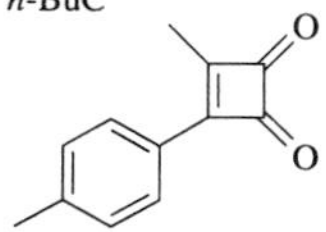 (83) 63

Bu$_3$SnC≡CTMS

Bu$_3$SnPh

Bu$_3$SnC≡CBu-n

TMSC≡C

n-BuC≡C

TABLE I. DIRECT CROSS-COUPLING OF ALKENYL ELECTROPHILES (Continued)

Substrate	Stannane	Conditions	Product(s) and Yield(s) (%)	Refs.
	Bu₃SnSPh	BnPd(PPh₃)₂Cl (2%), THF, 50°, 3 h	(88)	63
	Bu₃Sn (OEt vinyl)	BnPd(PPh₃)₂Cl (2%), THF, 50°, 10 h	(62)	63
	Bu₃Sn (thiophene)	BnPd(PPh₃)₂Cl (2%), THF, 50°, 15 h	(65)	63
	Bu₃SnC≡CTMS	BnPd(PPh₃)₂Cl (2%), THF, 50°, 20 h	(56)	63
	Bu₃SnPh	BnPd(PPh₃)₂Cl (2%), THF, 50°, 20 h	(70)	63
	Bu₃SnC≡CBu-n	BnPd(PPh₃)₂Cl (2%), THF, 50°, 20 h	(50)	63
	Bu₃Sn (tolyl)	BnPd(PPh₃)₂Cl (2%), THF, 50°, 20 h	(81)	63
	Bu₃Sn—OTHP	BnPd(PPh₃)₂Cl (2%), THF, 50°, 20 h	(61)	63

Substrate	Organometallic	Conditions	Product (Yield %)	Refs.
CO_2Et / Br	Bu_3SnSPh	$BnPd(PPh_3)_2Cl$ (2%), THF, 50°, 8 h	(80)	63
	Bu_3Sn — (indole, SEM)	$Pd(PPh_3)_4$ (10%), DMF, 110°, 30 min	(55) EtO_2C, SEM	289
	MOMO…SnBu₃ (uracil derivative)	$Pd(PPh_3)_4$ (10%), CuI (20%), DMF, 80°, 75 min	(56) CO_2Et	170
EtO_2C / Br	Me_3Sn (imidazole, Bu-n, SEM)	$Pd(PPh_3)_4$ (2%), m-xylene, 120°, 20 h	(40)	433
EtO_2C / I	Me_3Sn (thiophene, Bu-n)	$Pd(PPh_3)_2Cl_2$, THF, reflux	(74) EtO_2C	434
	EtO_2C…SnBu₃ (Bu-n)	$Pd(PPh_3)_4$ (10%), CuI (75%), DMF, rt, 7 h	(66) EtO_2C	435
EtO_2C / I	Bu_3Sn (CO_2Et, NHAc)	$Pd_2(dba)_3$ (5%), $AsPh_3$ (40%), THF, 65°	(53) $Z{:}E = 87{:}13\ +$ (17) $Z,E{:}E,E = 72{:}28$	375

TABLE I. DIRECT CROSS-COUPLING OF ALKENYL ELECTROPHILES (Continued)

Substrate	Stannane	Conditions	Product(s) and Yield(s) (%)	Refs.
AcO—CH=CH—I	Bu$_3$Sn—CH=CH—CH(CO$_2$Et)NHAc	Pd$_2$(dba)$_3$ (5%), AsPh$_3$ (40%), THF, 65°	EtO$_2$C···/=\···CO$_2$Et, NHAc (84)	375
	Me$_3$SnC≡CTMS	Pd(PPh$_3$)$_4$ (5%), THF, 50°	AcO—CH=CH—C≡CTMS (—)	431
(OTf, CO$_2$Me, TMS structure)	Bu$_3$SnC≡CPh	Pd(OAc)$_2$ (7%), PPh$_3$ (14%), THF, 55°, 30 min	PhC≡C—C=C(CO$_2$Me) (56)	436
TMS—CH=CH—Br, $E:Z$ = 87:13	Me$_3$SnSPh	Pd(PPh$_3$)$_4$ (3%), C$_6$H$_6$, 80°, 40 h	(TMS, SPh) **I** (67) + (TMS, SPh) **II** (21)	317
(isopropenyl, TMS, Br)	Me$_3$SnSPh	Pd(PPh$_3$)$_4$ (3-4%), C$_6$H$_6$, 40°, 2 h	**II** (91)	317
$E:Z$ = 9:1 (Bu$_3$Sn indole, SEM)	Bu$_3$Sn-indole-SEM	Pd(PPh$_3$)$_4$ (10%), DMF, 110°, 1 h	TMS—CH=CH—indole-SEM (93)	289
$E:Z$ = 100:0 (H$_2$N, Bu$_3$Sn indole, SEM)	H$_2$N···Bu$_3$Sn-indole-SEM	Pd(PPh$_3$)$_4$ (10%), DMF, 90°, 1.5 h	TMS—CH=CH—indole(CH$_2$CH$_2$NH$_2$)-SEM (98)	74
C$_6$ cyclohexenyl-OTf	Me$_3$SnSnMe$_3$	Pd(PPh$_3$)$_4$ (0.45%), LiCl, THF, 60°, 4 h	cyclohexadienyl-SnMe$_3$ (72)	176
methylcyclopentenyl-OTf	Me$_3$SnSnMe$_3$	Pd(PPh$_3$)$_4$ (5%), LiCl, Li$_2$CO$_3$, THF, 60°, 96 h	methylcyclopentenyl-SnMe$_3$ (80)	176

72

Reactant: cyclohexenyl iodide

Organostannane	Conditions	Product (% yield)	Refs.
Me₃SnC≡CH	Pd(PPh₃)₄, THF, 22-25°, 24 h	(cyclohexenyl–C≡CH) (90)	47
Bu₃Sn–CH=CH–CH₂OH	Pd(CH₃CN)₂Cl₂ (1-2%), DMF, 25°, 8 h	(cyclohexenyl–CH=CH–CH₂OH) (61)	46
"	Pd(PPh₃)₄ (1.5-2%), DMF, 25°, 23 h	" (80)	46
"	Pd(PPh₃)₂Cl₂ (1.5-2%), DMF, 25°, 23 h	" (90)	46
Me₃SnC≡CTMS	Pd(PPh₃)₄, THF, 22-25°, 3 h	(cyclohexenyl–C≡CTMS) (96)	47
(pyrrolidine amide Me₃Sn)	Pd(PPh₃)₄ (5%), THF, 50°	" (—)	431
(pyrrolidine amide Bu₃Sn)	Pd(PPh₃)₄ (5%), PhMe, reflux	(pyrrolidine amide) (62)	437
(pyrrolidine amide Bu₃Sn)	Pd(PPh₃)₄ (5%), THF, 65°	" (48)	437
(pyrrolidine amide Ph₃Sn)	Pd(PPh₃)₄ (5%), THF, 65°	" (53)	437
(polymer-bound Bu₂Sn reagent)	Pd(CH₃CN)₂Cl₂ (2%), DMF, 80°, 68 h	(cyclohexenyl–CH=CH–CO₂Me) (68)	376
Bu₃Sn–CH=CH–CO₂Et	Pd(CH₃CN)₂Cl₂, DMF, 25°, 23.5 h	(cyclohexenyl–CH=CH–CO₂Et) (69)	46

TABLE I. DIRECT CROSS-COUPLING OF ALKENYL ELECTROPHILES (*Continued*)

Substrate	Stannane	Conditions	Product(s) and Yield(s) (%)	Refs.
	Bu₃Sn⌒CO₂Et	Pd(PPh₃)₄ (1-2%), DMF, 25°, 12 h	(59) CO₂Et	46
	Bu₃Sn (thiophene dioxide)	Pd(PPh₃)₄ (5%), THF, rt, 24 h	(33)	438
	Me₃SnC≡C⌒OTMS	Pd(PPh₃)₄, THF, 22-25°, 3 h	(76) C≡C⌒OTMS	47
	"	Pd(PPh₃)₂Cl₂, THF, 22-25°, 2 h	" (92)	47
	Me₃SnC≡C—N(piperidinyl)C=O	Pd(PPh₃)₄ (5%), PhMe, reflux	(58)	437
	Bu₃Sn—N(piperidinyl)C=O	Pd(PPh₃)₄ (5%), PhMe, 111°	" (63)	437
	Me₃SnC≡CPh	Pd(PPh₃)₂Cl₂, THF, 22-25°, 10 h	(90) C≡CPh	47
	Bu₃SnC≡CPh	Pd(PPh₃)₂Cl₂, THF, 22-25°, 50 h	" (92)	47
	Bu₂Sn(CH₂CH₂—C₆H₄—C(Me)(Et))ₙ—C≡CPh	Pd(PPh₃)₄, PhMe, 55°, 50 h	" (51)	376
	Me₃SnC≡C⌒OBn	Pd(PPh₃)₂Cl₂, THF, 22-25°, 18 h	(91) C≡C⌒OBn	47

74

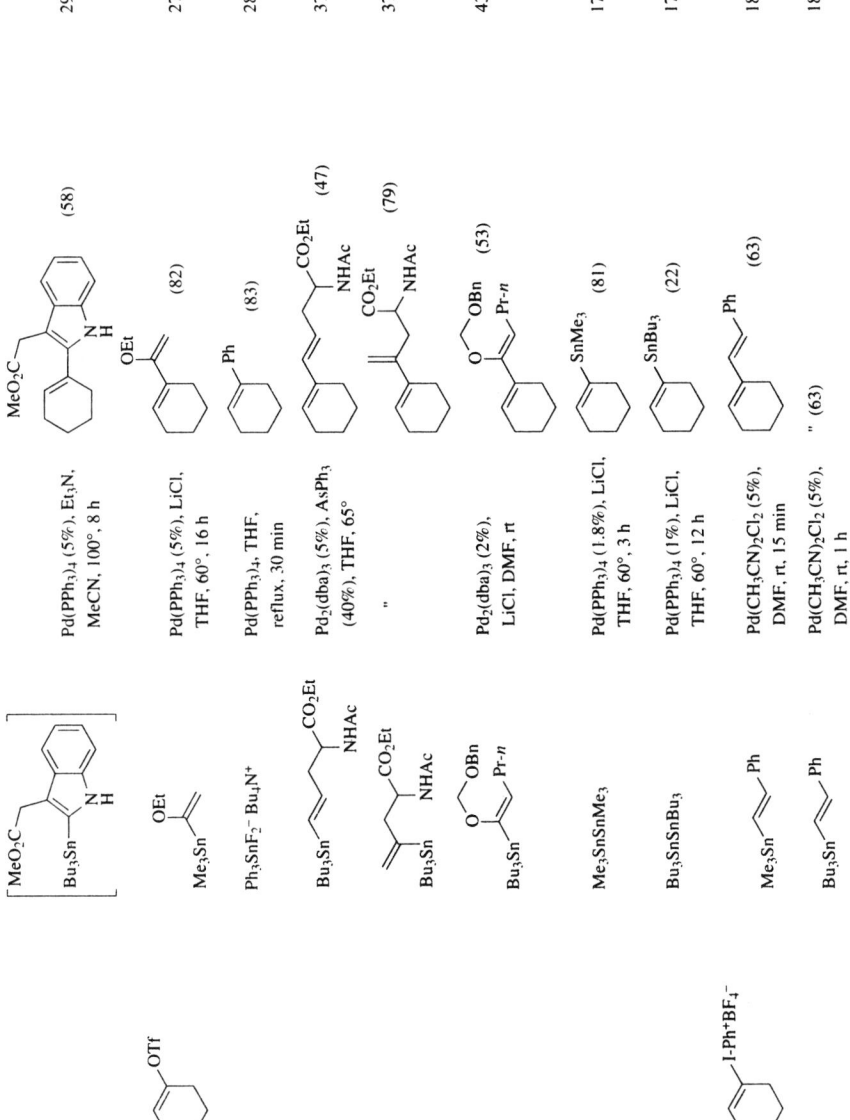

Pd(PPh₃)₄ (5%), Et₃N. MeCN, 100°, 8 h	(58)	290
Pd(PPh₃)₄ (5%), LiCl, THF, 60°, 16 h	(82)	270
Pd(PPh₃)₄, THF, reflux, 30 min	(83)	283
Pd₂(dba)₃ (5%), AsPh₃ (40%), THF, 65°	(47)	375
"	(79)	375
Pd₂(dba)₃ (2%), LiCl, DMF, rt	(53)	439
Pd(PPh₃)₄ (1.8%), LiCl, THF, 60°, 3 h	(81)	176
Pd(PPh₃)₄ (1%), LiCl, THF, 60°, 12 h	(22)	176
Pd(CH₃CN)₂Cl₂ (5%), DMF, rt, 15 min	(63)	187
Pd(CH₃CN)₂Cl₂ (5%), DMF, rt, 1 h	" (63)	187

TABLE I. DIRECT CROSS-COUPLING OF ALKENYL ELECTROPHILES (Continued)

Substrate	Stannane	Conditions	Product(s) and Yield(s) (%)	Refs.
n-Bu (alkenyl iodide)	Me$_3$SnC≡CH	Pd(PPh$_3$)$_4$, Et$_2$O, 22-25°, 22 h	(50)	47
	Bu$_3$Sn⟶OH	Pd(CH$_3$CN)$_2$Cl$_2$ (1-2%), DMF, 25°, 8.5 h	(74)	46
	HO⟶Bu$_3$Sn	Pd(PPh$_3$)$_4$ (3%), DMF, 70°, 5 h	HO⟶ (41)	440
	Bu$_3$Sn⟶CO$_2$Et	Pd(CH$_3$CN)$_2$Cl$_2$ (1-2%), DMF, 25°, 4 h	n-Bu⟶CO$_2$Et (83) E,E:Z,E = 94:6	46
	Bu$_3$Sn⟶CO$_2$Et	Pd(CH$_3$CN)$_2$Cl$_2$ (1-2%), DMF, 25°, 4 h	n-Bu⟶CO$_2$Et (78)	46
	MeO$_2$C, Bu$_3$Sn, indole N-H	Pd(PPh$_3$)$_4$ (5%), Et$_3$N, CuI, CH$_3$CN, 100°, 8h	(71)	290
	TMS⟶O⟶CO$_2$Et, Bu$_3$Sn⟶, H	Pd(CH$_3$CN)$_2$Cl$_2$ (2-5%), DMF, 50°, 70 h	(10) / (30) + (13) / (5)	441
	"	Pd(CH$_3$CN)$_2$Cl$_2$ (2-5%), HMPA, 80°, 20 h		441
	Bu$_3$Sn⟶ sulfone structure	Pd(PPh$_3$)$_4$ (5%), THF, rt, 24 h	(41)	438

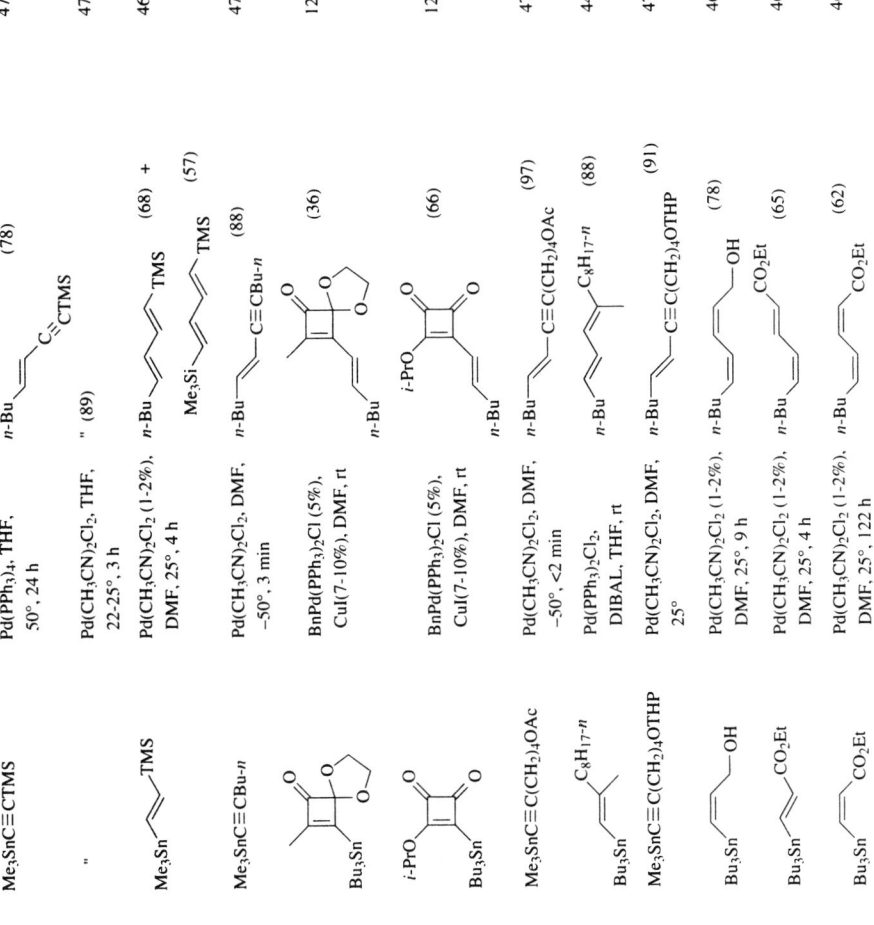

Me₃SnC≡CTMS	Pd(PPh₃)₄, THF, 50°, 24 h	(78) *n*-Bu C≡CTMS	47
"	Pd(CH₃CN)₂Cl₂, THF, 22-25°, 3 h	" (89)	47
Me₃Sn TMS	Pd(CH₃CN)₂Cl₂ (1-2%), DMF, 25°, 4 h	(68) + *n*-Bu TMS (57) Me₃Si TMS	46
Me₃SnC≡CBu-*n*	Pd(CH₃CN)₂Cl₂, DMF, –50°, 3 min	(88) *n*-Bu C≡CBu-*n*	47
(structure, Bu₃Sn)	BnPd(PPh₃)₂Cl (5%), CuI(7-10%), DMF, rt	(36)	12
i-PrO, Bu₃Sn (structure)	BnPd(PPh₃)₂Cl (5%), CuI(7-10%), DMF, rt	(66) *i*-PrO *n*-Bu	12
Me₃SnC≡C(CH₂)₄OAc	Pd(CH₃CN)₂Cl₂, DMF, –50°, <2 min	(97) *n*-Bu C≡C(CH₂)₄OAc	47
C₈H₁₇-*n* Bu₃Sn	Pd(PPh₃)₂Cl₂, DIBAL, THF, rt	(88) *n*-Bu C₈H₁₇-*n*	442
Me₃SnC≡C(CH₂)₄OTHP	Pd(CH₃CN)₂Cl₂, DMF, 25°	(91) *n*-Bu C≡C(CH₂)₄OTHP	47
Bu₃Sn OH	Pd(CH₃CN)₂Cl₂ (1-2%), DMF, 25°, 9 h	(78) *n*-Bu OH	46
Bu₃Sn CO₂Et	Pd(CH₃CN)₂Cl₂ (1-2%), DMF, 25°, 4 h	(65) *n*-Bu CO₂Et	46
Bu₃Sn CO₂Et	Pd(CH₃CN)₂Cl₂ (1-2%), DMF, 25°, 122 h	(62) *n*-Bu CO₂Et	46

n-Bu I

77

TABLE I. DIRECT CROSS-COUPLING OF ALKENYL ELECTROPHILES (*Continued*)

Substrate	Stannane	Conditions	Product(s) and Yield(s) (%)	Refs.
(methyl sulfolene, Bu$_3$Sn)	(methyl sulfolene, Bu$_3$Sn)	Pd(PPh$_3$)$_4$ (5%), THF, rt, 24 h	(40)	438
	Me$_3$Sn—TMS	Pd(CH$_3$CN)$_2$Cl$_2$ (1-2%), DMF, 25°, 6 h	*n*-Bu—TMS (25) + TMS—TMS (53)	46
	Me$_3$SnC≡C(CH$_2$)$_4$OAc	Pd(CH$_3$CN)$_2$Cl$_2$, DMF, 22-25°, 1 h	*n*-Bu—C≡C(CH$_2$)$_4$OAc (91)	47
	Me$_3$Sn—Cl	Pd(CH$_3$CN)$_2$Cl$_2$ or Pd(0)/P(2-furyl)$_3$, DMF	*t*-Bu—Cl (64)	443
	Me$_3$Sn—SnMe$_3$	"	*t*-Bu—Bu-*t* (42)	443
t-Bu—I	Me$_3$SnC≡CTMS	Pd(PPh$_3$)$_4$ (5%), THF, 50°	Bu-*n*—C≡CTMS (—)	431
Bu-*n*—I	Me$_3$Sn—TMS	Pd(PPh$_3$)$_4$ (1.6%), LiCl, THF, reflux, 17 h	*n*-Bu—TMS (90)	28, 444
Bu-*n*—OTf	(OEt stannane, Me$_3$Sn)	Pd(PPh$_3$)$_4$ (2%), LiCl, THF, 60°, 48 h	*n*-Bu—OEt (94)	270
(CN cyclopentenyl OTf)	(CHO aryl, Me$_3$Sn)	Pd(PPh$_3$)$_4$ (3%), LiCl, dioxane, reflux, 16 h	(CN cyclopentenyl–C$_6$H$_4$CHO) (76)	445
(cyclohexenone, Br)	(indole, SEM, Bu$_3$Sn)	Pd(PPh$_3$)$_4$ (10%), DMF, 110°, 4 h	(cyclohexenone–indole, SEM) (87)	289

78

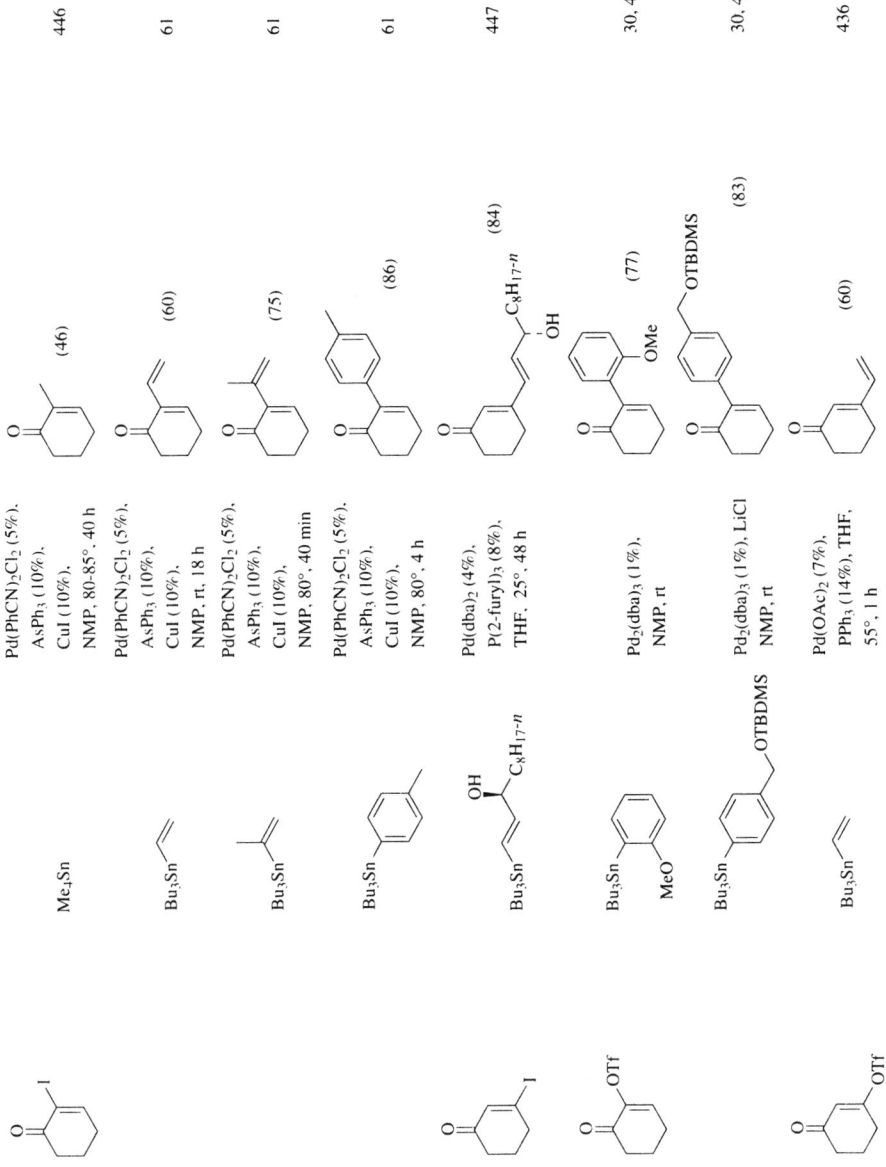

446

61

61

61

447

30, 448

30, 448

436

79

TABLE I. DIRECT CROSS-COUPLING OF ALKENYL ELECTROPHILES (Continued)

Substrate	Stannane	Conditions	Product(s) and Yield(s) (%)	Refs.
(cyclohexenone–OMs)	Bu$_3$Sn–CH=CH$_2$	Pd(PPh$_3$)$_4$, LiCl, THF	(—) + (vinylcyclohexenone / 3-chlorocyclohexenone)	173
	Bu$_3$SnC≡CPh	Pd(OAc)$_2$ (7%), PPh$_3$ (14%), THF, 55°, 40 min	(91)	436
	Bu$_3$Sn~~~Bu-n	Pd(PPh$_3$)$_4$ (5%), LiBr, THF, reflux, 36 h	(50) $E{:}Z = 91{:}9$	173
HO(CH$_2$)$_4$—〜—I	Me$_3$Sn~~~Bu-n	Pd(CH$_3$CN)$_2$Cl$_2$ (1-2%), DMF, 25°	n-Bu ... (CH$_2$)$_4$OH (73)	46
(methylcyclopentenone–I)	Me$_4$Sn	Pd(PhCN)$_2$Cl$_2$ (5%), AsPh$_3$ (10%), CuI (10%), NMP, 80-85°	(84)	446
	Bu$_3$Sn–CH=CH$_2$	Pd(PhCN)$_2$Cl$_2$ (5%), AsPh$_3$ (10%), CuI (10%), NMP, 90°, 20 min	(81)	61
	Bu$_3$Sn–C(CH$_3$)=CH$_2$	Pd(PhCN)$_2$Cl$_2$ (5%), AsPh$_3$ (10%), CuI (10%), NMP, 100°, 30 min	(93)	61
	Bu$_3$Sn–C$_6$H$_4$CH$_3$	"	(89)	61

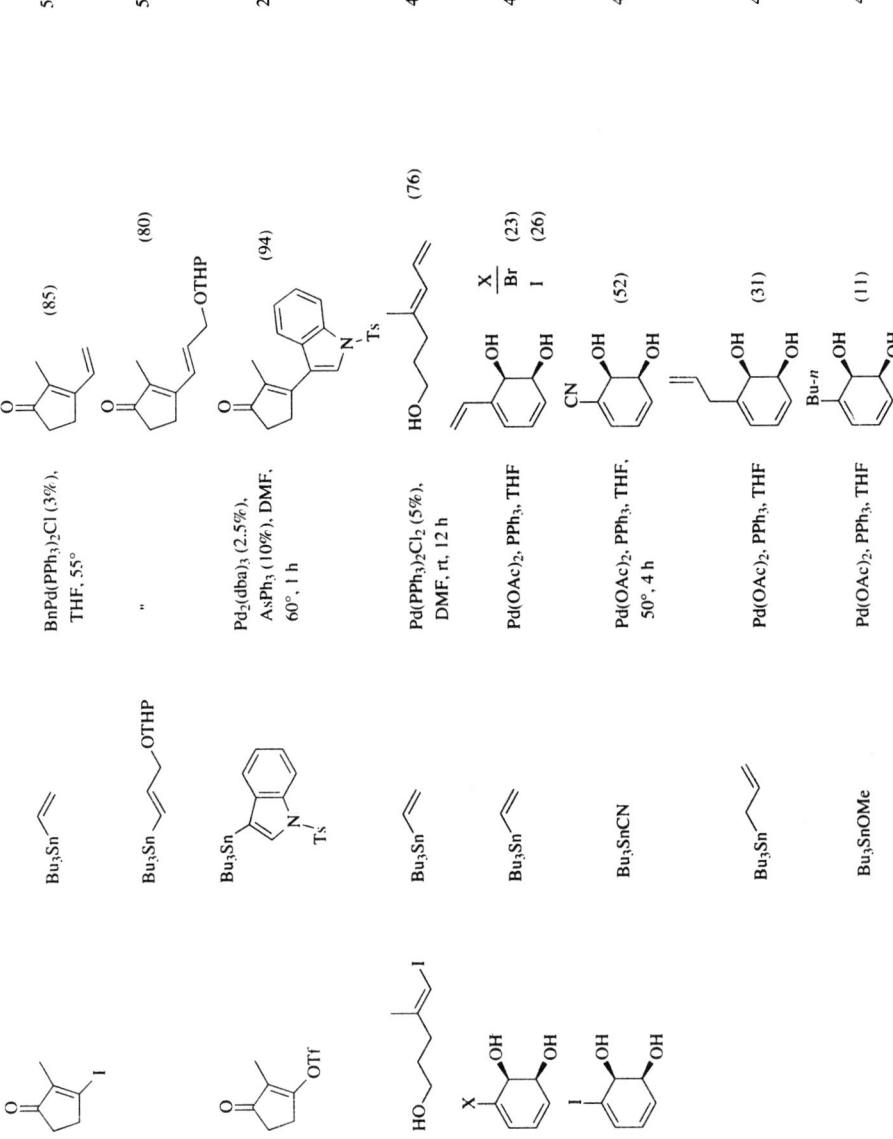

	(85)	BnPd(PPh₃)₂Cl (3%), THF, 55°	52
	(80)	"	52
	(94)	Pd₂(dba)₃ (2.5%), AsPh₃ (10%), DMF, 60°, 1 h	291
	(76)	Pd(PPh₃)₂Cl₂ (5%), DMF, rt, 12 h	449
	X (23) Br (26) I	Pd(OAc)₂, PPh₃, THF	450
	(52)	Pd(OAc)₂, PPh₃, THF, 50°, 4 h	450
	(31)	Pd(OAc)₂, PPh₃, THF	450
	(11)	Pd(OAc)₂, PPh₃, THF	450

TABLE I. DIRECT CROSS-COUPLING OF ALKENYL ELECTROPHILES (*Continued*)

Substrate	Stannane	Conditions	Product(s) and Yield(s) (%)	Refs.
OTf / CO$_2$Et	Bu$_3$Sn—⟨⟩—OMe	Pd$_2$(dba)$_3$ (1%), AsPh$_3$ (8%), NMP, rt	(72) *E:Z* = 95:5	30, 448
E:Z = 68:32	Bu$_3$Sn / C≡CTMS	Pd$_2$(dba)$_3$ (3%), AsPh$_3$ (6%), NMP, 35–40°	(92)	278
⟨S⟩—Br	Bu$_3$SnSPh	Pd(PPh$_3$)$_4$ (3–4%), PhMe, 60°, 3 h	(85) *E:Z* = 98:2	317
C$_7$ OTf (cycloheptenyl)	Me$_3$SnSnMe$_3$	Pd(PPh$_3$)$_4$ (2%), LiCl, THF, reflux, 3.5 h	SnMe$_3$ (71)	451
OTf (cyclohexenyl, Me)	Bu$_3$Sn /	Pd(PPh$_3$)$_4$, LiCl, THF, reflux, 12 h	(97)	452
	Bu$_3$Sn /	Pd(OAc)$_2$ (10%), CH$_2$Cl$_2$, rt, 30 min	(>90)	453
	Me$_3$Sn / TMS	Pd(PPh$_3$)$_4$ (1.6%), LiCl, THF, reflux, 6 h	TMS (100)	28, 444
	Ph$_3$SnF$_2^-$ Bu$_4$N$^+$	Pd(PPh$_3$)$_4$, THF, reflux, 30 min	Ph (85)	283

OTf	Me$_3$SnSnMe$_3$	Pd(PPh$_3$)$_4$ (1.8%), LiCl, THF, 60°, 9 h	SnMe$_3$ (84)	176
	Me$_3$Sn–TMS	Pd(PPh$_3$)$_4$ (1.6%), LiCl, THF, reflux 100 h	TMS (90)	28, 444
	Ph$_3$SnF$_2^-$ Bu$_4$N$^+$	Pd(PPh$_3$)$_4$, THF, reflux, 30 min	Ph (84)	283
	Me$_3$SnSnMe$_3$	Pd(PPh$_3$)$_4$ (5%), LiCl, Li$_2$CO$_3$, THF, 60°, 168 h	SnMe$_3$ (80)	176
	Me$_3$SnSnMe$_3$	Pd(PPh$_3$)$_4$ (1.8%), LiCl, THF, 60°, 0.75 h	SnMe$_3$ (73)	176
OTf + OTf Z,Z:Z,E = 4:1	Bu$_3$Sn–OH	Pd(PPh$_3$)$_4$, LiCl, THF	OH (47) + OH (11)	454
n-Bu I	Me$_3$Sn– OTBDMS	Pd(PPh$_3$)$_4$, DMF, rt, 20 h	OTBDMS (58) + OTBDMS (34)	455
n-C$_5$H$_{11}$ I	N–Sn Me	[(η^3-C$_3$H$_5$)PdCl]$_2$, CH$_3$CN, THF, rt, 5 min	n-C$_5$H$_{11}$ (85)	41

TABLE I. DIRECT CROSS-COUPLING OF ALKENYL ELECTROPHILES (Continued)

Substrate	Stannane	Conditions	Product(s) and Yield(s) (%)	Refs.
	(azabicyclic–Sn–CH2OMOM)	Pd(CH3CN)2Cl2, DMF, rt, 3 h	n-C5H11 —OMOM (81)	41
	Bu3Sn–(furan)	Pd(PPh3)2Cl2 (3%), CuI (8%), DMF, 60°, 24 h	n-C5H11–(furan) (71)	287
	Bu3Sn–(sulfolene, S O2)	Pd(PPh3)4 (5%), THF, rt, 24 h	n-C5H11–(sulfolene) (34)	438
	Me3Sn–C(O)N(pyrrolidine)	Pd(PPh3)4 (5%), PhMe, reflux	n-C5H11–C(O)N(pyrrolidine) **I** (79)	437
	Bu3Sn–C(O)N(pyrrolidine)	Pd(PPh3)4 (5%), PhMe, 111°	**I** (67)	437
	Ph3Sn–C(O)N(pyrrolidine)	Pd(PPh3)4 (5%), PhMe, reflux	**I** (46) + n-C5H11 Ph (24)	437
	Bu3Sn–(methyl sulfolene, S O2)	Pd(PPh3)4 (5%), THF, rt, 24 h	n-C5H11–(methyl sulfolene) (36)	438
	Me3Sn–C(O)N(piperidine)	Pd(PPh3)4 (5%), PhMe, reflux	n-C5H11–C(O)N(piperidine) **I** (64)	437
	Bu3Sn–C(O)N(piperidine)	Pd(PPh3)4 (5%), THF, 65°	**I** (43)	437

84

Substrate	Reagent	Conditions	Product	Refs.
n-C$_5$H$_{11}$ I $E:Z = 83:17$	Ph$_3$Sn—C(O)—N(piperidine)	Pd(PPh$_3$)$_4$ (5%), PhMe, reflux	I (52) + n-C$_5$H$_{11}$ Ph (19)	437
	Me$_3$Sn—C(O)—N(Pr-i)$_2$	Pd(PPh$_3$)$_4$ (5%), PhMe, reflux	n-C$_5$H$_{11}$ C(O)—N(Pr-i)$_2$ (71)	437
	Bu$_3$SnSPh	Pd(PPh$_3$)$_4$ (3–4%), PhMe, 65°, 6 h	n-C$_5$H$_{11}$ SPh (70)	317
n-C$_5$H$_{11}$ I	Me—Sn(N-ligand)	[(η^3-C$_3$H$_5$)PdCl]$_2$, CH$_3$CN, THF, rt, 5 min	n-C$_5$H$_{11}$ (85)	41
	OMOM—Sn(N-ligand)	Pd(CH$_3$CN)$_2$Cl$_2$, DMF, rt, 3 h	n-C$_5$H$_{11}$ OMOM (85)	41
	Bu$_3$Sn (sulfolene)	Pd(PPh$_3$)$_4$ (5%), THF, rt, 24 h	n-C$_5$H$_{11}$ (32)	438
	Bu$_3$Sn (methyl sulfolene)	Pd(PPh$_3$)$_4$ (5%), THF, rt, 24 h	n-C$_5$H$_{11}$ (39)	438
	Bu$_3$SnSPh	Pd(PPh$_3$)$_4$ (3%), PhMe, 110°, 4.5 h	C$_5$H$_{11}$-n SPh (77) + SPh n-C$_5$H$_{11}$ (22)	317
C$_5$H$_{11}$-n Br	Bu$_3$Sn—CH$_2$—C(CO$_2$Et)(NHAc)	Pd$_2$(dba)$_3$ (5%), AsPh$_3$ (40%), THF, 65°	CO$_2$Et NHAc n-C$_5$H$_{11}$ (91)	375
C$_5$H$_{11}$-n OTf	Bu$_3$Sn—CH=CH—CH$_2$—C(CO$_2$Et)(NHAc)	Pd$_2$(dba)$_3$ (5%), AsPh$_3$ (40%), THF, 65°	CO$_2$Et NHAc C$_5$H$_{11}$-n (58)	375

TABLE I. DIRECT CROSS-COUPLING OF ALKENYL ELECTROPHILES (Continued)

Substrate	Stannane	Conditions	Product(s) and Yield(s) (%)	Refs.
(CN, OTf cyclohexene)	Me$_3$Sn–C$_6$H$_4$–CHO	Pd(PPh$_3$)$_4$ (3%), LiCl, dioxane, reflux, 16h	(aryl–CN, CHO) (80)	445
(I, methyl cyclohexenone)	Bu$_3$Sn–CH=CH$_2$	Pd(PhCN)$_2$Cl$_2$ (5%), AsPh$_3$ (10%), CuI (10%), NMP, 80°, 30 min	(95)	61
	Bu$_3$Sn–C(=CH$_2$)CH$_3$	Pd(PhCN)$_2$Cl$_2$ (5%), AsPh$_3$ (10%), CuI (10%), NMP, 110°, 5 h	(83)	61
	Bu$_3$Sn–C$_6$H$_4$–CH$_3$	Pd(PhCN)$_2$Cl$_2$ (5%), AsPh$_3$ (10%), CuI (10%), NMP, 110°, 7 h	(89)	61
(methyl cyclohexenone OTf)	Bu$_3$Sn–CH=CH–CO$_2$Me	Pd(OAc)$_2$ (7%), PPh$_3$ (14%), THF, 65°	(CO$_2$Me product) (68)	259
	Bu$_3$Sn (cyclohexenyl, EtO$_2$C)	Pd(OAc)$_2$ (7%), PPh$_3$ (14%), THF, 65°	(10) + (Bu-n product) (86)	259
(tropolone, OH, Br)	Me$_3$Sn–C(=CH$_2$)CH$_3$	Pd(PPh$_3$)$_2$Cl$_2$ (10%), dioxane, reflux, 80 min	(≥54)	67, 456

Organostannane	Conditions	Product (yield, %)	Ref.
Me₃SnPh	Pd(PPh₃)₂Cl₂ (10%), dioxane, reflux, 16 h	(≥80)	456
Me₃Sn-C₆H₂(OMe)₃	Pd(PPh₃)₂Cl₂ (10%), dioxane, reflux	(≥36) + (≥24)	67
Me₄Sn	Pd(PPh₃)₄ (5%), dioxane, reflux, 48 h	(≥80)	67, 456
Me₃SnPh	Pd(PPh₃)₂Cl₂ (10%), dioxane, reflux, 16 h	(≥98)	67, 456
Me₃Sn-C₆H₄(OMe)	Pd(PPh₃)₂Cl₂ (10%), dioxane, reflux	(≥31)	67
Me₃Sn-C₆H₄(OMe)	Pd(PPh₃)₂Cl₂ (10%), dioxane, reflux	(≥39)	67

TABLE I. DIRECT CROSS-COUPLING OF ALKENYL ELECTROPHILES (*Continued*)

Substrate	Stannane	Conditions	Product(s) and Yield(s) (%)	Refs.
Br / CO₂Me alkene	Me₃Sn-aryl (R¹, R², R³, R⁴)	Pd(PPh₃)₂Cl₂ (10%), dioxane, reflux	tropolone product (R¹, R², R³, R⁴)	67
	Bu₃Sn—	Pd(CH₃CN)₂Cl₂ (1%), P(2-furyl)₃ (2%), NMP, 80°, 24 h	CO₂Me diene (99)	356
	Bu₃SnPh	Pd(CH₃CN)₂Cl₂ (1%), P(2-furyl)₃ (2%), HMPA, 80°, 24 h	Ph / CO₂Me (65)	356
	Bu₃Sn—C≡CPh	Pd(CH₃CN)₂Cl₂ (1%), P(2-furyl)₃ (2%), PhMe, 80°, 24 h	MeO₂C / C≡CPh (88)	356
OMs O OMe cyclopentene	Bu₃Sn—	Pd(PPh₃)₄ (5%), LiBr, THF, reflux, 3 6h	O OMe cyclopentene (>82)	173

Stannane Me_3Sn-substituted arene with substituents R^1, R^2, R^3, R^4:

R^1	R^2	R^3	R^4	
H	H	OMe	H	(≥53)
H	OCH₂O		H	(≥60)
OMe	OMe	H	H	(≥80)
H	OMe	OMe	H	(≥57)
H	OMe	H	OMe	(≥80)
OMe	H	OMe	OMe	(≥61)
OMe	H	H	OMe	(≥79)
H	OMe	OMe	OMe	(≥59)
H	OMe	OTBDMS	OMe	(≥55)

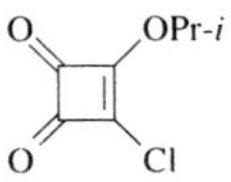

Bu₃SnPh

Bu₃SnC≡CBu-n

(p-tolyl) Bu₃Sn

R¹–N–R²
Bu₃Sn

R¹	R²	Time (h)	
H	H	4	(50)
H	Me	3	(47)
H	n-Pr	4	(64)
H	i-Pr	3	(91)
(CH₂)₅		4	(60)

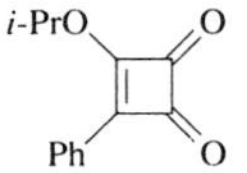

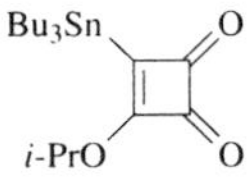

Bu₃Sn i-PrO

OEt
Bu₃Sn

OPr-i
Cl

BnPd(PPh₃)₂Cl (5%), CuI (5%), CH₃CN, 70°, 20 h

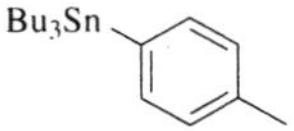

BnPd(PPh₃)₂Cl (5%), CuI (5%), CH₃CN, 70°, 10 h

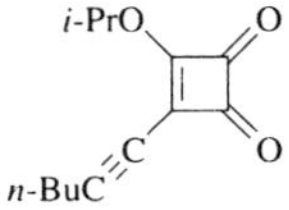

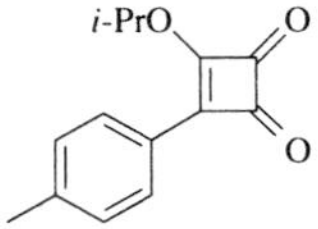

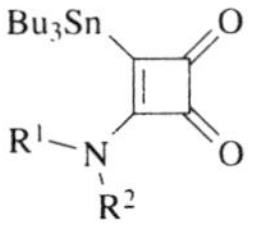

BnPd(PPh₃)₂Cl (5%), CuI (5%), CH₃CN, 70°, 20 h

BnPd(PPh₃)₂Cl (5%), CuI (5%), CH₃CN, 50°

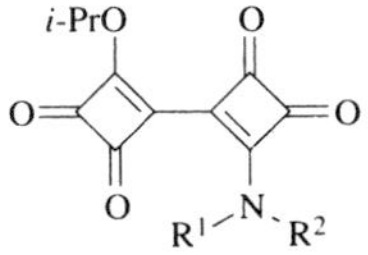

BnPd(PPh₃)₂Cl (5%), CuI (5%), CH₃CN, 50°, 4 h

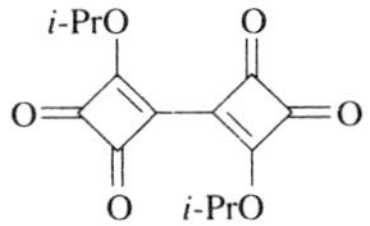

BnPd(PPh₃)₂Cl (5%), CuI (5%), CH₃CN, 70°, 20 h

---

i-PrO Ph   (87)

i-PrO C≡C-Bu-n (58)

i-PrO (p-tolyl)   (76)

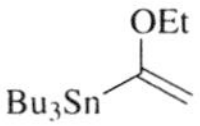

R¹–N–R²   (76)

i-PrO i-PrO   (84)

i-PrO OEt   (70)

63

63

63

64

64

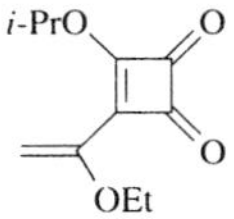

63

89

TABLE I. DIRECT CROSS-COUPLING OF ALKENYL ELECTROPHILES (*Continued*)

Substrate	Stannane	Conditions	Product(s) and Yield(s) (%)	Refs.
	Bu$_3$Sn⁀OTHP	BnPd(PPh$_3$)$_2$Cl (5%), CuI (5%), CH$_3$CN, 70°, 20 h	*i*-PrO (49)	63
	Bu$_3$SnC≡CTMS	BnPd(PPh$_3$)$_2$Cl (5%), CuI (5%), CH$_3$CN, 70°, 10 h	THPO, *i*-PrO, TMSC≡C (57)	63
	Bu$_3$SnO (2,4-dichlorophenyl)	BnPd(PPh$_3$)$_2$Cl (5%), CuI (5%), CH$_3$CN, 70°, 24 h	*i*-PrO, Cl, Cl (68)	63
	Bu$_3$SnSPh	BnPd(PPh$_3$)$_2$Cl (5%), CuI (5%), CH$_3$CN, 70°, 3 h	*i*-PrO, PhS (87)	63
	Bu$_3$Sn (vinyl)	Pd(PhCN)$_2$Cl$_2$ (5%), AsPh$_3$ (10%), CuI (10%), NMP, rt, 12 h	AcO (80)	61
	Bu$_3$Sn (isopropenyl)	Pd(PhCN)$_2$Cl$_2$ (5%), AsPh$_3$ (10%), CuI (10%), NMP, 80°, 1 h	AcO (76)	61
	Bu$_3$Sn (p-tolyl)	Pd(PhCN)$_2$Cl$_2$ (5%), AsPh$_3$ (10%), CuI (10%), NMP, 50°, 18 h	AcO (66)	61

90

C_8	Reagent	Conditions	Product (% yield)	Refs.
cyclooctatetraenyl–Br	$Bu_3SnC≡CR$	$Pd_2(dba)_3$ (2.5%), P(2-furyl)$_3$ (10%), THF, rt	cyclooctatetraenyl–C≡CR, R = TMS (63), R = Ph (64)	65, 66
	$Bu_3SnC≡CSnBu_3$	$Pd_2(dba)_3$ (2.5%), PPh$_3$ (10%), THF, rt, 24 h	bis(cyclooctatetraenyl)–C≡C– (45)	65, 66
	$Bu_3SnC≡CSnBu_3$	$Pd_2(dba)_3$ (2.5%), P(2-furyl)$_3$ (10%), THF, rt, 3 h	" (55)	65, 66
	$Bu_3SnC≡CSnBu_3$	$Pd_2(dba)_3$ (2.5%), AsPh$_3$ (10%), THF, rt, 1 h	" (45)	65, 66
	$Bu_3Sn\!-\!CH\!=\!CH\!-\!SnBu_3$	$Pd_2(dba)_3$ (2.5%), PPh$_3$ (10%), THF, rt, 24 h	(E)-1,2-bis(cyclooctatetraenyl)ethene (45)	65, 66
	$Bu_3Sn\!-\!CH\!=\!CH\!-\!SnBu_3$	$Pd_2(dba)_3$ (2.5%), P(2-furyl)$_3$ (10%), THF, rt, 5 h	" (60)	65, 66
	$Bu_3Sn\!-\!CH\!=\!CH\!-\!SnBu_3$	$Pd_2(dba)_3$ (2.5%), AsPh$_3$ (10%), THF, rt, 2 h	" (40)	65, 66
	$Bu_3Sn\!-\!C_6H_4\!-\!SnBu_3$	$Pd_2(dba)_3$ (2.5%), PPh$_3$ (10%), THF, rt, 48 h	1,4-bis(cyclooctatetraenyl)benzene (0)	65, 66
	$Bu_3Sn\!-\!C_6H_4\!-\!SnBu_3$	$Pd_2(dba)_3$ (2.5%), P(2-furyl)$_3$ (10%), THF, rt, 48 h	" (23)	65, 66
	$Bu_3Sn\!-\!C_6H_4\!-\!SnBu_3$	$Pd_2(dba)_3$ (2.5%), AsPh$_3$ (10%), THF, rt, 24 h	" (30)	65, 66

TABLE I. DIRECT CROSS-COUPLING OF ALKENYL ELECTROPHILES (*Continued*)

Substrate	Stannane	Conditions	Product(s) and Yield(s) (%)	Refs.
Ph⌒⌒Br	Bu₃Sn—C(OTMS)=CH₂	Pd(PPh₃)₄	Ph⌒⌒C(O)CH₃ (53)	457
	2-(Bu₃Sn)pyridine	Pd(PPh₃)₂Cl₂ (3%), DMF, 80°	pyridin-2-yl–CH=CH–Ph (58)	458
	3-(Bu₃Sn)-5-OMe-1-Ts-indole	Pd₂(dba)₃ (2.5%), AsPh₃ (10%), CuI (10%), DMF, 60°, 3 h	indolyl–CH=CH–Ph (62)	291
	5-(Bu₃Sn)-1-Bn-pyrimidin-2(1H)-one	Pd(PPh₃)₂Cl₂ (5%), Cl(CH₂)₂Cl, reflux, 12 h	pyrimidinonyl–CH=CH–Ph (81)	459
	Bu₃SnSnBu₃	Pd(PPh₃)₂Br₂ (1.3%), PhMe, 115°, 15 h	Ph–CH=CH–SnBu₃ (21)	460
Ph⌒⌒Br *E:Z* = 93:7	Me₃Sn—CH=CH–TMS	Pd(CH₃CN)₂Cl₂ (1-2%), DMF, rt, 23 h	Ph⌒⌒CH=CH–TMS (56) *E:E:E:Z* = 95:5 + Ph⌒⌒CH=CH–Ph (34) *E:E:E:Z* = 95:5	46
Ph⌒⌒Br *E:Z* = 89:11	Bu₃SnSPh	Pd(PPh₃)₄ (3-4%), PhMe, 110°, 2 h	TMS–CH=CH–CH=CH–SPh (22) ... Ph⌒⌒SPh (91) *E:Z* = 98.5:1.5	317
Ph⌒⌒Br *E:Z* = 85:15	2-(Bu₃Sn)-1-Boc-indole	Pd/C (0.5%), AsPh₃ (20%), CuI (10%), NMP, 80°, 12 h	indolyl–CH=CH–Ph (85) *E:Z* = 75:25	461
Ph⌒⌒Br *E:Z* = 81:19 *Z:E* = 86:14	Bu₃Sn–CH=CH₂	Pd(CH₃CN)₂Cl₂, CuI, DMF, 100°	Ph⌒⌒CH=CH₂ (30) *E,Z:E,E* = 85:15	385
Ph⌒⌒Br	Bu₃Sn–CH₂CH=CH–NH₂	Pd(PPh₃)₄ (2%), PhMe, reflux	Ph⌒⌒CH=CH–NH₂ (51)	462

92

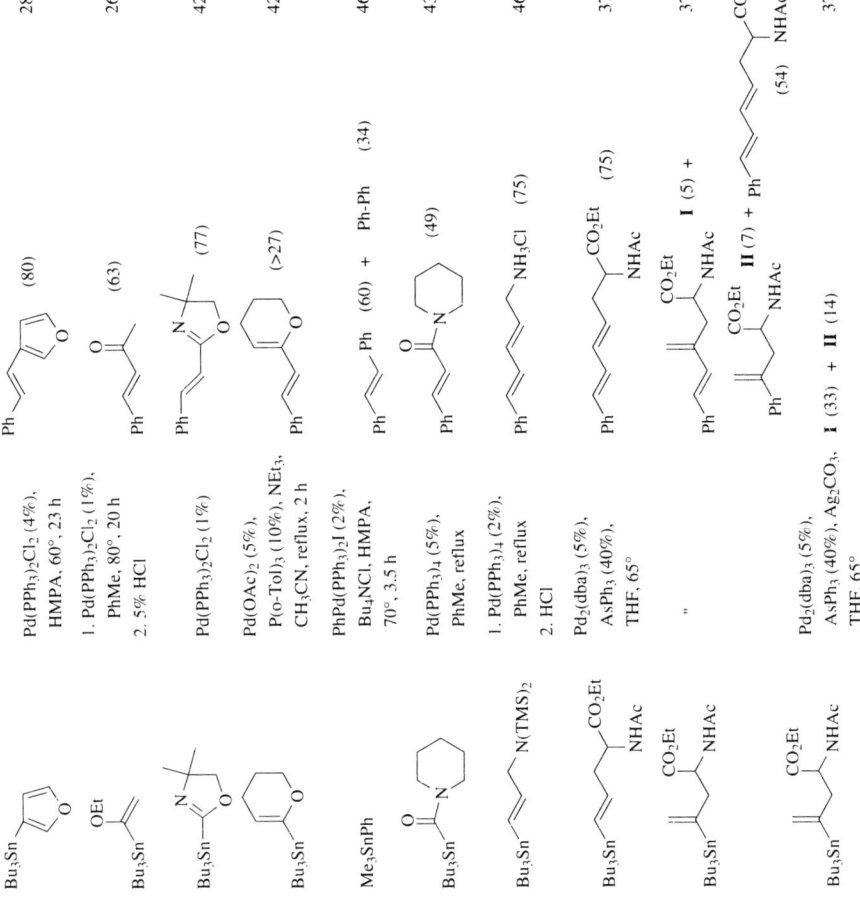

TABLE I. DIRECT CROSS-COUPLING OF ALKENYL ELECTROPHILES (*Continued*)

Substrate	Stannane	Conditions	Product(s) and Yield(s) (%)	Refs.
	Bu₃Sn—[CO₂Et, NHAc]	Pd(CH₃CN)₂Cl₂ (5%), THF, 65°	**I** (91)	375
	Bu₃Sn—[furan-COPh]	[η³-C₃H₅)PdCl]₂, HMPA, 60°, 6 h	Ph—[furan-COPh] (82)	287, 432
	HO—[cyclohexene-furan]	[η³-C₃H₅)PdCl]₂, HMPA, DMF, rt, 24 h	HO—[cyclohexene-furan] (76)	287, 432
	Bu₃Sn—[furan-SnBu₃]	[η³-C₃H₅)PdCl]₂, HMPA, rt, 1 h	Ph—[furan]—Ph (69)	287, 432
	Bu₃SnNEt₂	Pd[P(o-Tol)₃]₂Cl₂ (0.66%), PhMe, 100-120°	Ph—NEt₂ (50)	316
	(Et₃Sn)₂S	PhPd(PPh₃)₂I (5%), DMSO, 100°	Ph—S—Ph (99)	320
	(Bu₃Sn)₂S	Pd(PPh₃)₄ (1%), PhMe, 120°	" (77)	318
Ph—[=]—Br	Bu₃Sn—[OEt]	1. Pd(PPh₃)₂Cl₂ (1%), PhMe, 80°, 20 h 2. HCl (5%)	Ph—[=]—COCH₃ (73)	269
	Bu₃Sn—[oxazoline]	Pd(PPh₃)₂Cl₂ (1%)	Ph—[oxazoline] (95) *Z:E* = 1:1	426
	Bu₃SnNEt₂	Pd[P(o-Tol)₃]₂Cl₂ (0.66%), PhMe, 100-120°	Ph—NEt₂ (43)	316
Ph—[=]—I	Bu₃Sn—[=]	Pd(CH₃CN)₂Cl₂ (1-2%), DMF, 25°, 6 min	Ph—[=]—[=] (85)	46

94

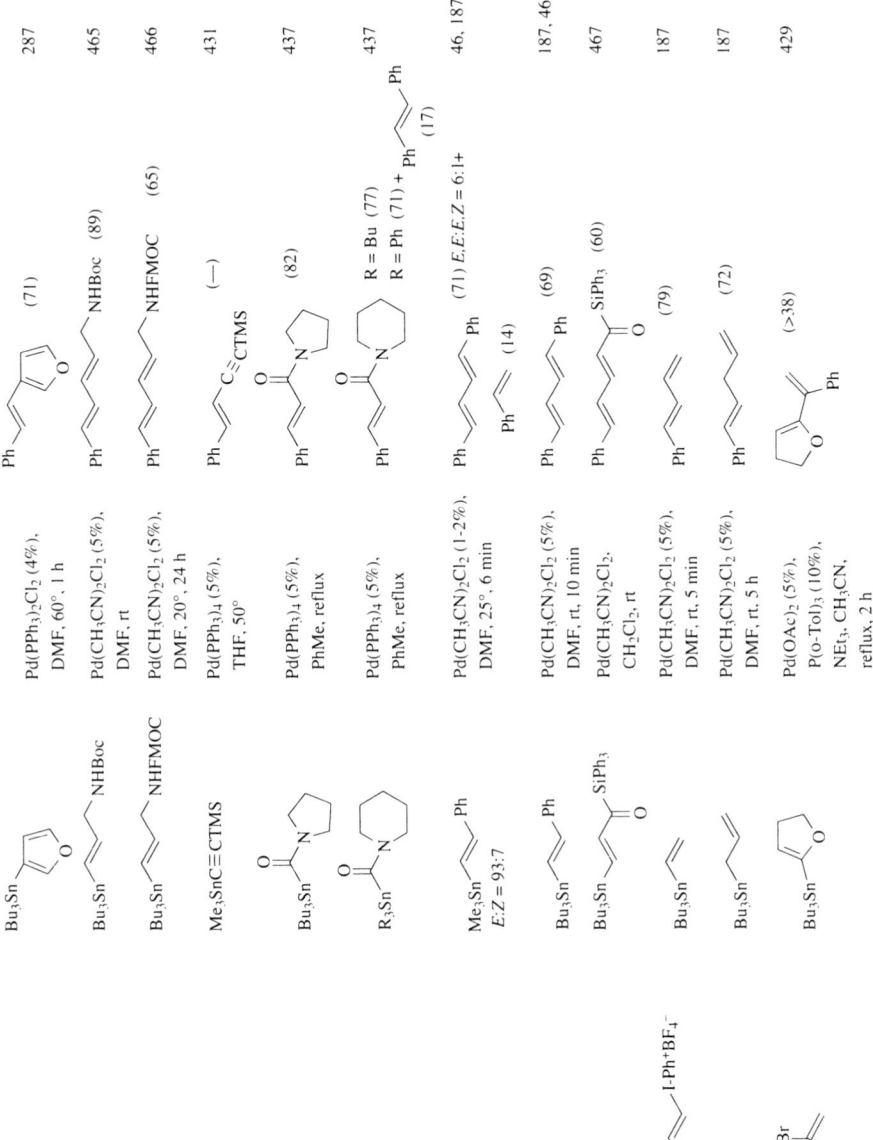

Bu₃Sn [furan]	[furan]CH=CHPh (71)	Pd(PPh₃)₂Cl₂ (4%), DMF, 60°, 1 h	287
Bu₃Sn NHBoc	Ph NHBoc (89)	Pd(CH₃CN)₂Cl₂ (5%), DMF, rt	465
Bu₃Sn NHFMOC	Ph NHFMOC (65)	Pd(CH₃CN)₂Cl₂ (5%), DMF, 20°, 24 h	466
Me₃SnC≡CTMS	Ph C≡CTMS (—)	Pd(PPh₃)₄ (5%), THF, 50°	431
Bu₃Sn [pyrrolidine]	Ph [pyrrolidine] (82)	Pd(PPh₃)₄ (5%), PhMe, reflux	437
R₃Sn [piperidine]	Ph [piperidine] R = Bu (77) R = Ph (71) + Ph Ph (17)	Pd(PPh₃)₄ (5%), PhMe, reflux	437
Me₃Sn Ph E:Z = 93:7	Ph Ph (71) E:E:E:Z = 6:1+ Ph Ph (14)	Pd(CH₃CN)₂Cl₂ (1-2%), DMF, 25°, 6 min	46, 187
Bu₃Sn Ph	Ph Ph (69)	Pd(CH₃CN)₂Cl₂ (5%), DMF, rt, 10 min	187, 46
Bu₃Sn SiPh₃ O	Ph SiPh₃ O (60)	Pd(CH₃CN)₂Cl₂. CH₂Cl₂, rt	467
Bu₃Sn	Ph (79)	Pd(CH₃CN)₂Cl₂ (5%), DMF, rt, 5 min	187
Bu₃Sn	Ph (72)	Pd(CH₃CN)₂Cl₂ (5%), DMF, rt, 5 h	187
Bu₃Sn [dihydrofuran]	[dihydrofuran]Ph (>38)	Pd(OAc)₂ (5%), P(o-Tol)₃ (10%), NEt₃, CH₃CN, reflux, 2 h	429

Ph⌒⌒1-Ph⁺BF₄⁻

Br ⌐ Ph

TABLE I. DIRECT CROSS-COUPLING OF ALKENYL ELECTROPHILES (Continued)

Substrate	Stannane	Conditions	Product(s) and Yield(s) (%)	Refs.
(structure with OTf, Ph)	Bu₃Sn (pyran structure)	Pd(OAc)₂ (5%), P(o-Tol)₃ (10%), NEt₃, CH₃CN, reflux, 2 h	(>46)	429
	Bu₃SnPh	Pd₂(dba)₃ (2%), NMP, 24°	(72)	30
(OTf cyclopentene structure)	Bu₃Sn (isoprenyl)	Pd(PPh₃)₄ (1.6%), LiCl, THF, reflux, 133 h	(80)	28, 443
	Me₃Sn (OEt vinyl)	Pd(PPh₃)₄ (2%), LiCl, THF, 60°, 48 h	(94)	270
	Me₃Sn—TMS	Pd(PPh₃)₄ (1.6%), LiCl, THF, reflux, 36 h	(100)	28, 444
(OTf cyclohexene)	Me₃SnSnMe₃	Pd(PPh₃)₄ (2%), THF, Li₂CO₃, 60°, 120 h	(62)	176
(OTf cyclohexene)	Me₃SnSnMe₃	Pd(PPh₃)₄ (2%), THF, Li₂CO₃, 60°, 120 h	(67)	176
(OTf cyclohexene)	Bu₃Sn (furan structure)	Pd(PPh₃)₄ (1.6%), LiCl, THF, reflux	(75)	28, 444

96

Substrate	Organotin reagent	Conditions	Product (% yield)	Refs.
(cyclohexyl-CH=CH–I) $E:Z = 89:11$	Bu$_3$Sn–CH=CH–NHFMOC	Pd(CH$_3$CN)$_2$Cl$_2$ (5%), DMF, 20°, 24 h	(cyclohexyl)–CH=CH–CH=CH–CH$_2$NHFMOC (63) $E:Z = 90:10$	466
n-C$_6$H$_{13}$–CH=CH–Br $E:Z = 64:36$	Bu$_3$SnSPh	Pd(PPh$_3$)$_4$ (3-4%), PhMe, 50°, 7.5 h	n-C$_6$H$_{13}$–CH=CH–SPh (84) $E:Z = 96:4$	317
n-C$_6$H$_{13}$–CH=CH–I	Me$_3$SnC≡CTMS	Pd(PPh$_3$)$_4$ (5%), THF, 50°	n-C$_6$H$_{13}$–CH=CH–C≡CTMS (—)	431
n-C$_6$H$_{13}$–CH=CH–I (Z)	Me$_3$SnC≡CTMS	Pd(PPh$_3$)$_4$ (5%), THF, 50°	n-C$_6$H$_{13}$–CH=CH–C≡CTMS (—)	431
(vinyl triflate, OTf, C$_6$H$_{13}$-n)	Me$_3$SnSnMe$_3$	Pd(PPh$_3$)$_4$ (1.8%), LiCl, THF, 60°, 5 h	(SnMe$_3$ vinyl, C$_6$H$_{13}$-n) (74)	176
n-Bu ... I-Ph$^+$OTf$^-$	i-PrO squarate–Bu$_3$Sn	BnPd(PPh$_3$)$_2$Cl (5%), CuI (8%), DMF, rt, 30 min	i-PrO squarate, n-Bu (60)	188
	Bu$_3$SnC≡C ...	"	pyrrolidine amide (67)	188
n-Bu ... I-Ph$^+$OTf$^-$ $Z:E = 86:14$	i-PrO squarate–Bu$_3$Sn	"	i-PrO squarate, n-Bu (76)	188
Ph–CH=CBr$_2$	Bu$_3$Sn–...	Pd(CH$_3$CN)$_2$Cl$_2$, CuI, DMF, rt	PhC≡C–CH=CH–... (39) $Z:E = 76:24$	385
(cycloheptene, CN, OTf)	Me$_3$Sn–C$_6$H$_4$–CHO	Pd(PPh$_3$)$_4$ (3%), LiCl, dioxane, reflux, 16 h	(cycloheptene, CN, C$_6$H$_4$–CHO) (88)	445
BocHN–CH=CH–CH=CH–I	Bu$_3$Sn–CH=CH–C(O)–SiPh$_3$	Pd(CH$_3$CN)$_2$Cl$_2$, CH$_2$Cl$_2$, rt, 12 h	BocHN–...–C(O)–SiPh$_3$ (86)	467

TABLE I. DIRECT CROSS-COUPLING OF ALKENYL ELECTROPHILES (Continued)

Substrate	Stannane	Conditions	Product(s) and Yield(s) (%)	Refs.
(4-nitrophenyl alkenyl sulfoxide, =CHCl)	Me₃SnSnMe₃	Pd₂(dba)₃ (1%), PPh₃ (8%), NMP, rt, 40 min	(72)	260
(4-nitrophenyl alkenyl sulfone, =CHCl)	Me₃SnSnMe₃	Pd₂(dba)₃ (1%), PPh₃ (8%), NMP, rt, 85 min	(47)	260
(Br-dienyl ether, OEt)	Bu₃Sn–allyl	Pd(CH₃CN)₂Cl₂ (1%), P(2-furyl)₃, PhMe, 80°	(8)	356
(same)	Bu₃SnPh	Pd(CH₃CN)₂Cl₂ (1%), Et₄NCl, HMPA, 80°	(55)	356
(dimethyl cyclohexenone OTf)	Bu₃Sn—C≡C—CTMS	Pd₂(dba)₃ (3%), AsPh₃ (6%), NMP, 35–40°	(80) E:Z = 80:20	278
(Br-cyclohexenone, OEt)	(2-SEM-indol-3-yl)stannane	Pd(PPh₃)₄ (10%), DMF, 110°, 1 h	(42)	289
(OMs cyclohexenyl methyl ketone)	Me₃Sn–allyl	Pd(PPh₃)₄ (10%), LiBr, THF, reflux, 36 h	(73)	173
(same)	Bu₃Sn–allyl	"	" (69)	173
(same)	Bu₃Sn (propenyl)	"	**I** (24) E:Z = 93:7	173

98

Substrate	Reagent	Conditions	Product(s)	Refs.
![allyl cyclohexenyl ketone structures]	Bu₃Sn (allyl)	"	**I** (69) + (3) + (28)	173
	Bu₄Sn	Pd(PPh₃)₄	(—)	173
	Bu₃SnPh	Pd(PPh₃)₄ (5%), LiBr, THF, reflux, 36 h	Ph (33)	173
	Bu₃Sn–R	Pd(PPh₃)₄ (5%), LiBr, THF, reflux, 36 h	R (structure) R = Bu-*n* (79) R = Ph (57)	173
	Bu₃Sn␣CO₂Et	Pd(PPh₃)₄ (1.25%), CsF, THF, reflux, 6 h	CO₂Et (58)	468, 469
	Me₃SnSnMe₃	Pd₂(dba)₃ (1%), PPh₃ (8%), NMP, rt, 4 h	Me₃Sn␣S(O)Ph (78)	260
	Bu₃Sn	Pd(PPh₃)₄ (5%), THF	X = Br (<5) X = I (80)	452
	Bu₃Sn	Pd(PPh₃)₂Cl₂ (3%), LiCl, THF	" X = OTf (85)	452
	Bu₃Sn	Pd(OAc)₂ (7%), PPh₃ (14%), THF, 55°, 6 h	CO₂Et (61)	436

TABLE I. DIRECT CROSS-COUPLING OF ALKENYL ELECTROPHILES (*Continued*)

Substrate	Stannane	Conditions	Product(s) and Yield(s) (%)	Refs.
	Bu₃Sn⌇	Pd(OAc)₂ (7%), PPh₃ (14%), THF, 55°, 12 h	[cyclopentene with CO₂Et and allyl groups] (56)	436
	Bu₃Sn⌇⌇CO₂Me	Pd(OAc)₂ (7%), PPh₃ (14%), THF, 65°	[cyclopentene with CO₂Et and CO₂Me groups] (59)	259
	Bu₃SnPh	Pd(OAc)₂ (7%), PPh₃ (14%), THF, 55°, 20 min	[cyclopentene with CO₂Et and Ph groups] (0)	436
	Bu₃SnC≡CPh	Pd(OAc)₂ (7%), PPh₃ (14%), THF, 55°	[cyclopentene with CO₂Et and C≡CPh groups] (83)	436
	Bu₃Sn / CO₂Et / NHAc	Pd₂(dba)₃ (5%), AsPh₃ (40%), THF, 65°	EtO₂C [product with CO₂Et, NHAc] (91)	375
	Bu₃Sn / CO₂Et / NHAc	Pd₂(dba)₃ (5%), AsPh₃ (40%), THF, 65°	[cyclopentene with CO₂Et, CO₂Et, NHAc] (93)	375
	Bu₃Sn C≡CTMS	Pd₂(dba)₃ (3%), AsPh₃ (6%), NMP, 35-40°	[cyclopentene with CO₂Et, C≡CTMS] (90)	278
	Bu₃Sn C≡CTMS	Pd₂(dba)₃ (3%), AsPh₃ (6%), NMP, 35-40°	[cyclopentene with CO₂Et, C≡CTMS] (79)	278

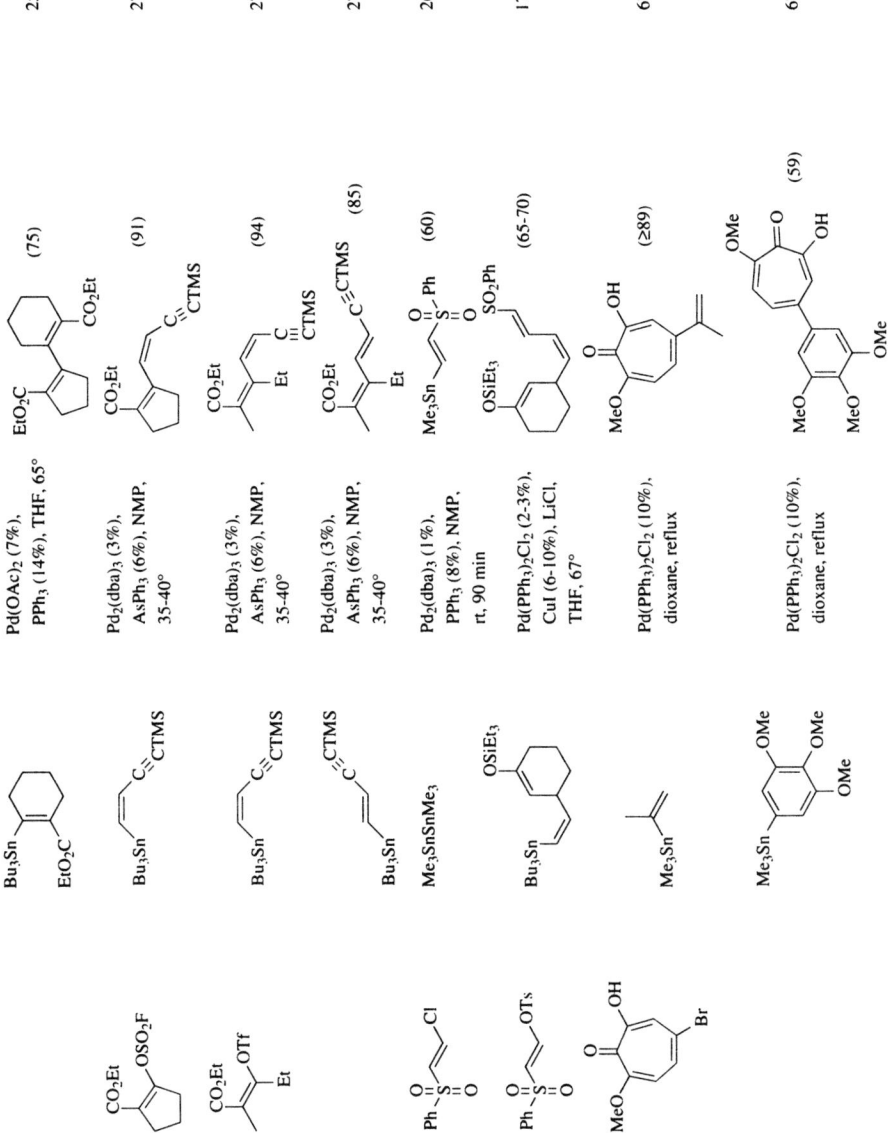

Bu₃Sn, EtO₂C (ring with CO₂Et)	Pd(OAc)₂ (7%), PPh₃ (14%), THF, 65°	EtO₂C, EtO₂C (bicyclic with CO₂Et) (75)	259
CO₂Et, OSO₂F (cyclopentene) + Bu₃Sn—CH=CH—C≡CTMS	Pd₂(dba)₃ (3%), AsPh₃ (6%), NMP, 35-40°	CO₂Et (cyclopentene)—CH=CH—C≡CTMS (91)	278
CO₂Et, OTf, Et (structure) + Bu₃Sn—CH=CH—C≡CTMS	Pd₂(dba)₃ (3%), AsPh₃ (6%), NMP, 35-40°	CO₂Et, Et—C≡CTMS (94)	278
(same, E isomer) + Bu₃Sn—CH=CH—C≡CTMS	Pd₂(dba)₃ (3%), AsPh₃ (6%), NMP, 35-40°	CO₂Et, Et—C≡CTMS (85)	278
PhO₂S—CH=CH—Cl + Me₃SnSnMe₃	Pd₂(dba)₃ (1%), PPh₃ (8%), NMP, rt, 90 min	Me₃Sn—CH=CH—S(=O)Ph (60)	260
PhO₂S—CH=CH—OTs + Bu₃Sn (OSiEt₃ cyclohexene)	Pd(PPh₃)₂Cl₂ (2-3%), CuI (6-10%), LiCl, THF, 67°	SO₂Ph, OSiEt₃ (65-70)	174
(tropone Br structure: HO, O, MeO, Br) + Me₃Sn—C(=CH₂)CH₃	Pd(PPh₃)₂Cl₂ (10%), dioxane, reflux	(tropone: OH, O, MeO, isopropenyl) (≥89)	67, 456
(tropone Br structure) + Me₃Sn (trimethoxyphenyl)	Pd(PPh₃)₂Cl₂ (10%), dioxane, reflux	(tropone with trimethoxyphenyl, OMe, OH, O) (59)	67

101

TABLE I. DIRECT CROSS-COUPLING OF ALKENYL ELECTROPHILES (Continued)

Substrate	Stannane	Conditions	Product(s) and Yield(s) (%)	Refs.
MeO_2C, OH, OTf (methylcyclopentene)	Bu_3Sn (vinyl)	$Pd(PPh_3)_4$ (1%), LiCl, THF, reflux, 6 h	MeO_2C, OH (86)	470
(quinone, I, OMe)	Bu_3Sn isoxazole (OMe trimethoxyphenyl)	$Pd(PPh_3)_4$ (10%), CuI (5%), dioxane, 101°	(55)	471
	''	$Pd(PPh_3)_4$ (10%), dioxane, 101°	(80)	471
Br, TMS (diene)	Bu_3Sn (allyl)	$Pd(CH_3CN)_2Cl_2$ (1%), P(2-furyl)$_3$ (2%), PhMe, 80°, 24 h	TMS (40)	356
	Bu_3SnPh	$Pd(CH_3CN)_2Cl_2$ (1%), Et$_4$NCl, HMPA, 80°, 24 h	Ph, TMS (52)	356
	$Bu_3SnC{\equiv}CPh$	$Pd(CH_3CN)_2Cl_2$ (1%), HMPA, 80°, 24 h	CPh, C, TMS (26)	356
C$_9$ (indene OTf)	$Bu_3SnC_6H_4Cl\text{-}p$	$Pd_2(dba)_3$ (2%), NMP, 24°	$C_6H_4Cl\text{-}p$ (68)	30
(trimethylcyclohexenyl OTf)	$Me_3SnSnMe_3$	$Pd(PPh_3)_4$ (1.8%), LiCl, Li$_2$CO$_3$, THF, 60°, 4 h	$SnMe_3$ (80)	176

Organic Electrophile	Organotin Reagent	Conditions	Product (%)	Refs.
(cyclohexenyl) OTf	Bu₃Sn–C₆H₄–CF₃	Pd₂(dba)₃ (2%), NMP, 24°	(60)	30, 448
Br dienyl (Bu-n)	Me₃SnSnMe₃	Pd(PPh₃)₄ (8%), LiCl, Li₂CO₃, THF, 60°, 240 h	SnMe₃ (39)	176
	Bu₃Sn (vinyl)	Pd(CH₃CN)₂Cl₂ (1%), P(2-furyl)₃ (2%), HMPA, 80°, 24 h	Bu-n (35)	356
	Bu₃SnPh	Pd(CH₃CN)₂Cl₂ (1%), Et₄NCl, HMPA, 80°, 24 h	Ph···Bu-n (52)	356
	Bu₃SnC≡CPh	Pd(CH₃CN)₂Cl₂ (1%), HMPA, 80°, 24 h	CPh C Bu-n (0)	356
n-Pr, OTf, Bu-n	Ph₃SnF₂⁻ Bu₄N⁺	Pd(PPh₃)₄, THF, reflux, 30 min	Ph n-Bu Pr-n (83)	283
CHO Br Ph	HSnBu₃	Pd(CH₃CN)₂Cl₂	CHO Br (—)	472
EtO₂C I Br NHAcI	Me₃Sn (isopropenyl)	Pd(CH₃CN)₂Cl₂, or Pd(AsPh₃)₄, DMF	CO₂Et NHAc (0)	473
	Bu₃Sn (allyl)	Pd(CH₃CN)₂Cl₂, DMF, 90°, 2 h	CO₂Et NHAc (23)	473
	Bu₃Sn TMS	Pd(CH₃CN)₂Cl₂, DMF, 90°, 1 h	CO₂Et NHAc TMS (86)	473

TABLE I. DIRECT CROSS-COUPLING OF ALKENYL ELECTROPHILES (*Continued*)

Substrate	Stannane	Conditions	Product(s) and Yield(s) (%)	Refs.
	Bu₃SnPh	Pd(CH₃CN)₂Cl₂, DMF, 90°, 1 h	(56)	473
	Me₃Sn—Ph	Pd(CH₃CN)₂Cl₂, DMF, 80°, 1 h	(0)	473
	Me₃Sn	Pd(CH₃CN)₂Cl₂, DMF, 40°, 18 h	(65)	473
	Bu₃Sn	Pd(CH₃CN)₂Cl₂, DMF, 80°, 4 h	(41)	473
	Bu₃Sn—TMS	Pd(CH₃CN)₂Cl₂, DMF, 80°, 4 h	(23)	473
	Bu₃SnPh	Pd(CH₃CN)₂Cl₂, DMF, 15°, 15 h	(31)	473
	Me₃Sn—Ph Z:E = 86:14	Pd(CH₃CN)₂Cl₂, DMF, 40°, 17 h	(21)	473
	Bu₃Sn	Pd(CH₃CN)₂Cl₂, CuI, DMF or PhMe, 100° to reflux	(19-21)	385
	Bu₃Sn	Pd(PPh₃)₄ (5%), LiBr, THF, reflux, 48 h	(51)	474
		Pd₂(dba)₃ (5%), AsPh₃ (40%), THF, 65°	(44)	375

Substrate	Reagent	Conditions	Product (Yield)	Ref.
I–C(COMe)=CH–n-C₅H₁₁ ($n\text{-}C_5H_{11}$, COMe, I)	Bu₃Sn–CH=CH₂	Pd(PhCN)₂Cl₂ (5%), AsPh₃ (10%), CuI (10%), NMP, 45°, 19 h	(COMe, $n\text{-}C_5H_{11}$) (57)	446
	Me₃SnPh	Pd(PhCN)₂Cl₂ (5%), AsPh₃ (10%), CuI (10%), NMP, 55°, 4.5 h	(Ph, COMe, $n\text{-}C_5H_{11}$) (84)	446
O=S(p-Tol)–CH=CH–Br	Bu₃Sn–CH=CH₂	Pd₂(dba)₃•CHCl₃ (2%), PPh₃ (8%), THF, reflux, 30 min	(87)	59
	Bu₃Sn–CH=C(CH₃)₂	"	(80)	59
	Bu₃Sn–CH=CH–Ph	"	(Ph) (87)	59
	Bu₃Sn–CH=CH–CH(OEt)₂	"	(OEt, OEt) (85)	59
	Bu₃Sn–CH=CH–(CH₂)₃–dithiolane	"	(S, S, 3) (83)	59
O=S(p-Tol)–CH=CH–I (E)	Bu₃Sn–C(=CH₂)CH₃	Pd(CH₃CN)₂Cl₂ (2%), DMF, rt	(83)	59
	Bu₃Sn–C(OEt)=CH₂	"	(OEt) (90)	59
	Bu₃Sn–CH=CH₂	"	(91)	59
O=S(p-Tol)–CH=CH–I (Z)	Bu₃Sn–C(=CH₂)CH₃	"	(80)	59
	Bu₃Sn–CH=C(CH₃)₂	"	(81)	59

TABLE I. DIRECT CROSS-COUPLING OF ALKENYL ELECTROPHILES (Continued)

Substrate	Stannane	Conditions	Product(s) and Yield(s) (%)	Refs.
OTf / CO$_2$Et (cyclohexene)	Bu$_3$Sn–C(OEt)=CH$_2$	"	O=S, p-Tol, OEt (83)	59
	Bu$_3$SnC≡CBu-n	Pd(CH$_3$CN)$_2$Cl$_2$ (2%), DMF/THF, rt	O=S–p-Tol, C≡CBu-n (77)	60
	Bu$_3$SnC≡CCH(OEt)$_2$	Pd(CH$_3$CN)$_2$Cl$_2$ (2%), DMF, rt	O=S–p-Tol, C≡CCH(OEt)$_2$ (82)	60
	OEt / OEt stannane (Bu$_3$Sn–CH=CH–CH(OEt))	"	O=S–p-Tol, OEt, OEt (76)	59
	Bu$_3$SnC≡CSiEt$_3$	"	O=S–p-Tol, C≡CSiEt$_3$ (87)	60
	Bu$_3$SnC≡C–OTBDMS	"	O=S–p-Tol, C≡C–OTBDMS (75)	60
	Bu$_3$SnC≡C–(CH$_2$)$_3$–OBu-t	Pd(CH$_3$CN)$_2$Cl$_2$ (2%), DMF/THF, rt	O=S–p-Tol, C≡C–(CH$_2$)$_3$–OBu-t (74)	60
	Bu$_3$Sn—CH=CH$_2$	Pd(OAc)$_2$ (7%), PPh$_3$ (14%), THF, 55°, 4 h	(vinyl)/CO$_2$Et (81)	436
	Bu$_3$Sn—CH$_2$CH=CH$_2$	Pd(OAc)$_2$ (7%), PPh$_3$ (14%), THF, 55°, 6 h	(allyl)/CO$_2$Et (66)	436
	Bu$_3$SnPh	Pd(OAc)$_2$ (7%), PPh$_3$ (14%), THF, 55°	Ph/CO$_2$Et (0)	436
	Bu$_3$SnC≡CPh	Pd(PPh$_3$)$_4$ (7%), THF, 55°	C≡CPh/CO$_2$Et (89)	436

106

Substrate	Reagent	Conditions	Product (%)	Ref.
$n\text{-}C_5H_{11}$ / CO_2Me vinyl iodide	Bu_3Sn–cyclohexenyl, EtO_2C	$Pd(OAc)_2$ (7%), PPh_3 (14%), THF, 65°	biscyclohexenyl, CO_2Et, EtO_2C (72)	259
	Me_4Sn	$Pd(PhCN)_2Cl_2$ (5%), $AsPh_3$ (10%), CuI (10%), DMF, 80°, 24 h	$n\text{-}C_5H_{11}$, CO_2Me (38)	435
	Bu_3Sn vinyl	$Pd(PhCN)_2Cl_2$ (5%), $AsPh_3$ (10%), CuI (10%), DMF, 20°, 3 h	$n\text{-}C_5H_{11}$, CO_2Me (69)	435
	Bu_3Sn OAc	$Pd(PhCN)_2Cl_2$ (5%), $AsPh_3$ (10%), CuI (10%), DMF, 20°, 64.5 h	$n\text{-}C_5H_{11}$, CO_2Me, AcO (40)	435
$n\text{-}Bu$, CO_2Et vinyl iodide	Bu_3Sn vinyl	$Pd(PhCN)_2Cl_2$ (5%), $AsPh_3$ (10%), CuI (10%), DMF, 20°, 2.5 h	EtO_2C, $n\text{-}Bu$ (83)	435
	Bu_3Sn OEt	$Pd(PhCN)_2Cl_2$ (5%), $AsPh_3$ (10%), CuI (10%), DMF, 20–40°	EtO_2C, $n\text{-}Bu$, OEt (47)	435
$n\text{-}Bu$, CO_2Et, OTf	Bu_3Sn vinyl	$Pd(PhCN)_2Cl_2$ (5%), $AsPh_3$ (10%), CuI (10%), DMF, 20°, 23 h	$n\text{-}Bu$, CO_2Et (80)	435
EtO_2C, OTf, $Bu\text{-}n$	Bu_3Sn imidazole (N–SEM, $Bu\text{-}n$)	$Pd(PPh_3)_4$ (2%), LiCl, THF, reflux, 22 h	imidazole (N, SEM, $Bu\text{-}n$), EtO_2C (74)	433
aryl sulfinyl chlorovinyl (F_3C)	$Me_3SnSnMe_3$	$Pd_2(dba)_3$ (1%), PPh_3 (8%), NMP, rt, 4 h	Me_3Sn vinyl–S(O)–C6H4–CF_3 (84)	260
aryl sulfonyl chlorovinyl (F_3C)	"	"	Me_3Sn vinyl–S(O)(O)–C6H4–CF_3 (81)	260

TABLE I. DIRECT CROSS-COUPLING OF ALKENYL ELECTROPHILES (Continued)

Substrate	Stannane	Conditions	Product(s) and Yield(s) (%)	Refs.
	$Bu_3SnC{\equiv}CH$	Pd(PPh$_3$)$_4$, C$_6$H$_6$, reflux, 30 min	(74)	475
	Bu_3Sn	Pd(PPh$_3$)$_4$, C$_6$H$_6$, reflux, 30 min	(52)	475
	$Bu_3SnSnBu_3$	Pd(PPh$_3$)$_3$, C$_6$H$_6$, reflux, 15 min	(34)	475
1-Ph$^+$BF$_4^-$	Bu_3Sn	Pd(CH$_3$CN)$_2$Cl$_2$ (5%), DMF, rt, 5 h	(72)	187
	Bu_3Sn	Pd(PPh$_3$)$_4$, LiCl, THF, reflux	(84-95)	452, 476
	Bu_3Sn	Pd(PPh$_3$)$_4$, LiCl, THF, reflux	(84-95)	452, 476
	$Bu_3SnC{\equiv}CH$	Pd(CH$_3$CN)$_2$Cl$_2$, DMF, 25°	(91)	47
		Pd$_2$(dba)$_3$•CHCl$_3$ (1.5%), P(o-Tol)$_3$ (12%), NMP, 70°, 2 h	R = H (78), R = OTBDMS (69)	62

108

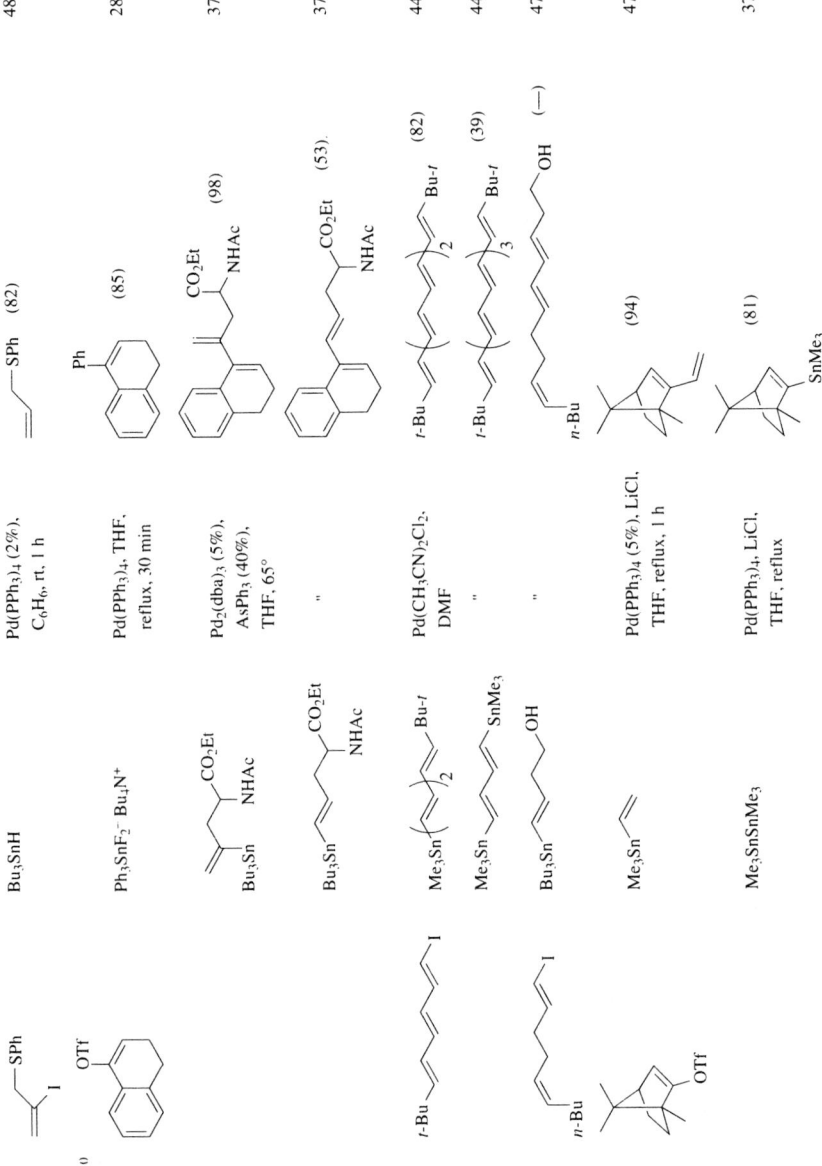

TABLE I. DIRECT CROSS-COUPLING OF ALKENYL ELECTROPHILES (*Continued*)

Substrate	Stannane	Conditions	Product(s) and Yield(s) (%)	Refs.
(3,5,5-trimethylcyclohexenyl iodide)	(5-OMe indole, Bu₃Sn, N-Ts)	Pd₂(dba)₃ (2.5%), AsPh₃ (40%), CuI (10%), DMF, 60°, 1 h	(indole-OMe product) (74)	291
(4-t-Bu cyclohexenyl iodide)	Me₃Sn⁀	Pd(PPh₃)₄ (1-2%), DMF, 25°, 6.5 h	(80)	46
(4-t-Bu cyclohexenyl OTf)	Bu₃SnH	Pd(PPh₃)₄ (1.6%), LiCl, THF, reflux, 0.5 h	(78)	444
	Bu₃Sn⁀	Pd(PPh₃)₄ (1.6%), LiCl, THF, reflux, 17 h	(91)	28, 444
	Bu₃Sn⁀	Pd₂(dba)₃ (1%), P(2-furyl)₃ (8%), NMP, 35°	" (>95)	11
	Bu₃Sn⁀	Pd₂(dba)₃ (1%), LiCl, AsPh₃ (8%), NMP, 35°	" (>95)	11
	Bu₃Sn⁀⁀	Pd(PPh₃)₄ (1.6%), LiCl, THF, reflux, 31 h	(96)	28, 444
	Bu₄Sn	Pd(PPh₃)₄ (1.6%), LiCl, THF, reflux, 41 h	Bu-n (80)	28, 444
	Me₃Sn (OEt)	Pd(PPh₃)₄ (2%), LiCl, THF, 60°, 18 h	OEt (93)	270
	"	Pd(PPh₃)₄ (2%), LiCl, THF, reflux, 15 h	" (97)	480

Me₃SnC≡CTMS	(structure, 90)	Pd(PPh₃)₄ (2%), LiCl, THF, reflux, 15 h	28, 444
Me₃SnPh	Ph (tr)	Pd(PPh₃)₄ (1.6%), LiCl, THF, reflux, 24 h	28
Me₃Sn–CH₂Ph	CH₂Ph (tr)	Pd(PPh₃)₄ (1.6%), LiCl, THF, reflux, 24 h	28
Bu₃Sn–C₆H₄CF₃	CF₃ (89)	Pd₂(dba)₃ (1%), AsPh₃ (8%), ZnCl₂, NMP, rt	30, 448
"	" (87)	Pd₂(dba)₃ (1%), AsPh₃ (8%), LiCl, NMP, 60°	30, 448
"	" (83)	Pd₂(dba)₃ (1%), AsPh₃ (8%), NMP, 60°	30, 448
"	" (54)	Pd₂(dba)₃ (1%), PPh₃ (8%), LiCl, NMP, 60°	30, 448
"	" (69)	Pd₂(dba)₃ (1%), LiCl, NMP, 60°	30, 448
Bu₃Sn–C₆H₄OMe	OMe (89)	Pd₂(dba)₃ (1%), AsPh₃ (8%), NMP, rt	30, 448
(squaric acid ester, i-PrO, Bu₃Sn)	(83)	Pd₂(dba)₃ (5%), ZnCl₂, P(2-furyl)₃ (10%), 65°, 45 min	12

TABLE I. DIRECT CROSS-COUPLING OF ALKENYL ELECTROPHILES (Continued)

Substrate	Stannane	Conditions	Product(s) and Yield(s) (%)	Refs.
[polymer]$_n$ with t-Bu	Bu_3Sn–C≡CPh (Bu, Bu)	Pd(PPh$_3$)$_4$, THF, 62°, 20 h	cyclohexenyl–C≡CPh, t-Bu (53)	376
	Bu_3Sn Ph, $E:Z = 95:5$	Pd/C (0.5%), AsPh$_3$ (20%), CuI (10%), NMP, 80°, 16 h	Ph-vinyl cyclohexenyl, t-Bu (80)	461
	MeO_2C indole Bu_3Sn	Pd(PPh$_3$)$_4$, NEt$_3$, CH$_3$CN, 100°, 7 h	MeO_2C indole, t-Bu (64)	290
	THPO indole Bu_3Sn	Pd(PPh$_3$)$_4$, NEt$_3$, LiCl, CH$_3$CN, DMF, 100°, 7 h	THPO indole, t-Bu (49)	290
n-C$_8$H$_{17}$ I, $E:Z = 83:17$	Me$_3$SnSnMe$_3$	Pd(PPh$_3$)$_4$ (1.6%), LiCl, THF, reflux, 12 h	SnMe$_3$, t-Bu (73)	28
n-C$_8$H$_{17}$ I	Bu_3Sn NHFMOC	Pd(CH$_3$CN)$_2$Cl$_2$ (5%), DMF, 20°, 24 h	n-C$_8$H$_{17}$ NHFMOC (61) $E:Z = 85:15$	466
n-C$_8$H$_{17}$ I	Me$_3$SnC≡CTMS	Pd(PPh$_3$)$_4$ (5%), THF, 50°	n-C$_8$H$_{17}$ C≡CTMS (—)	431
n-Bu vinyl, 1-Ph$^+$OTf$^-$	$Bu_3SnC≡CPh$	BnPd(PPh$_3$)$_2$Cl (5%), CuI (8%), DMF, rt, 30 min	n-Bu C≡CPh, n-Bu (77)	188
	$Bu_3SnC≡CCOBu$-t	"	n-Bu C≡CCOBu-t, n-Bu (66)	188

112

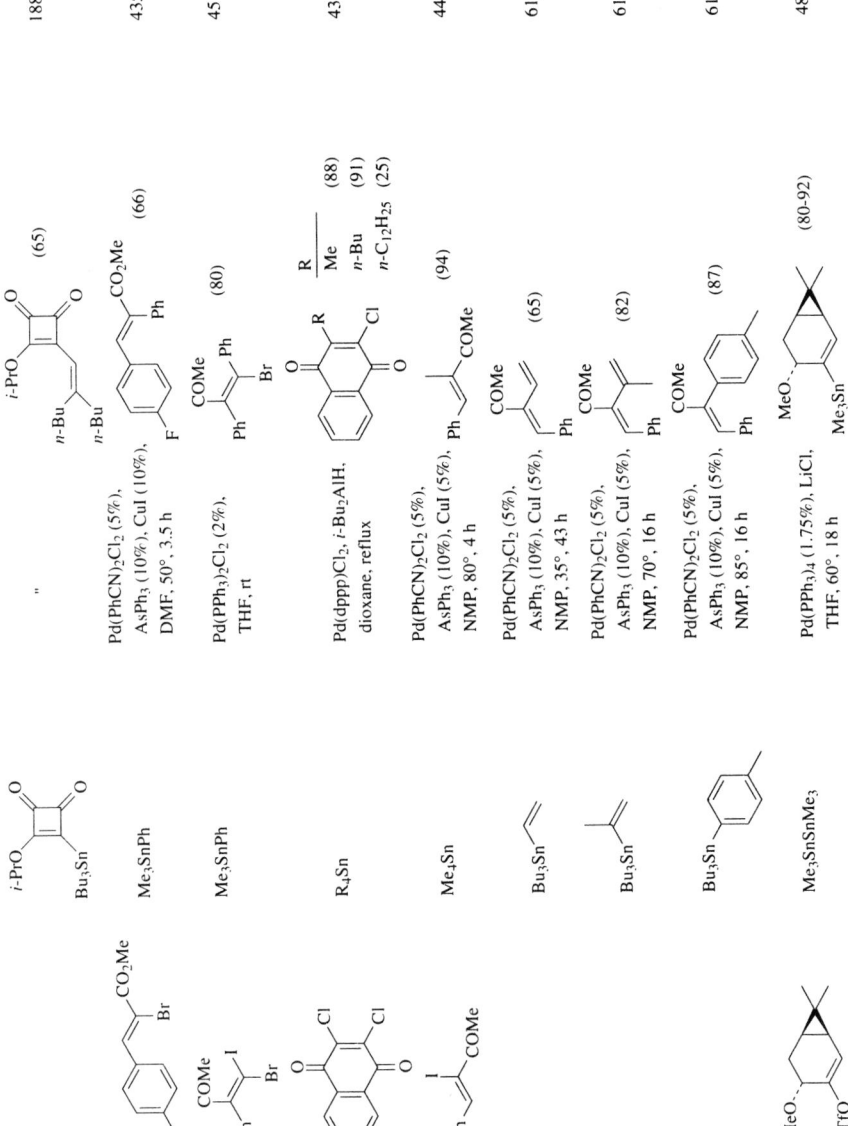

TABLE I. DIRECT CROSS-COUPLING OF ALKENYL ELECTROPHILES (*Continued*)

Substrate	Stannane	Conditions	Product(s) and Yield(s) (%)	Refs.
(cyclopentenone, C_5H_{11}-n, I)	Me_4Sn	$BnPd(PPh_3)_2Cl$ (3%), HMPA, 55°	(cyclopentenone, C_5H_{11}-n, Me) (95)	52
	Bu_3Sn (vinyl)	$BnPd(PPh_3)_2Cl$ (3%), HMPA, 55°	(cyclopentenone, C_5H_{11}-n, vinyl) (95)	52
	Bu_3Sn—C_5H_{11}-n, OH	$Pd(PPh_3)_2Cl_2$, THF, 55°, 72 h	(cyclopentenone, C_5H_{11}-n, OH, C_5H_{11}-n) (71)	53
BnO—(alkenyl)—I	$Me_3SnC{\equiv}CH$	$Pd(PPh_3)_4$, THF, 22-25°, 24 h	BnO—$C{\equiv}CH$ (59)	47
	$Me_3SnC{\equiv}CTMS$	"	BnO—$C{\equiv}CTMS$ (85)	47
(naphthoquinone, Br)	Bu_3SnPh	$Pd(PPh_3)_4$ (5%), CuBr (5%), dioxane, reflux, 12 h	(naphthoquinone, Ph) (66)	55, 57
(squarate, Ph, Br)	$R'R^2N$—(squarate)—Bu_3Sn	$BnPd(PPh_3)_2Cl$ (5%), CuI (5%), CH_3CN, 50°, 2 h	(squarate dimer, $NR'R^2$) R^1 — R^2 H — i-Pr (90) $(CH_2)_5$ (52)	64
	Ph—(squarate)—Bu_3Sn	$BnPd(PPh_3)_2Cl$ (5%), CuI (5%), CH_3CN, 50°, 3 h	(squarate dimer, Ph) (78)	64

114

Substrate	Organotin reagent	Conditions	Product (yield %)	Refs.
Ph–CH=C(Br)–CO_2Me	Bu_3Sn–CH=CH–CH$_3$ ($Z:E = 86:14$)	Pd(CH$_3$CN)$_2$Cl$_2$, DMF, 100–105°	Ph–CH=C(CO$_2$Me)–CH=CH–CH$_3$ (47)	385
cyclohexenyl OTf (dioxolane)	Bu_3Sn–CH=CH$_2$	Pd(PPh$_3$)$_4$ (3%), LiCl, THF, reflux	(65)	484
cyclohexene OTf (dioxolane)	Bu_3Sn–CH$_2$CH=CH$_2$	Pd(PPh$_3$)$_4$, LiCl, THF, reflux	(90–92)	452, 476
BnO$_2$C–CH=CH–Cl	Me$_3$SnSnMe$_3$	Pd$_2$(dba)$_3$, PPh$_3$, NMP, rt, 16 h	BnO$_2$C–CH=CH–SnMe$_3$ (61)	260
I–CH=CH–(CH$_2$)$_3$–C$_5$H$_{11}$-n, OH … CO$_2$H	Bu_3Sn–CH=CH–CH(OH)–C$_5$H$_{11}$-n	Pd(CH$_3$CN)$_2$Cl$_2$, DMF, rt, 8 h	CO$_2$H ... OH (76)	52, 53
2-bromo-5-hydroxy-1,4-naphthoquinone	Me$_4$Sn	Pd(PPh$_3$)$_4$ (5%), CuBr (5%), dioxane, reflux, 15 h	2-methyl-5-hydroxy-1,4-naphthoquinone (100)	55, 57
	Bu_4Sn (Bu-n)	Pd(PPh$_3$)$_4$ (5%), CuBr (5%), dioxane, reflux, 30 h	2-(Bu-n)-5-hydroxy-1,4-naphthoquinone (98)	55
"	Bu_4Sn	Pd(dppf)Cl$_2$ (5%), DMF, 100°, 8.5 h	" (82)	55, 57

TABLE I. DIRECT CROSS-COUPLING OF ALKENYL ELECTROPHILES (*Continued*)

Substrate	Stannane	Conditions	Product(s) and Yield(s) (%)	Refs.
	Bu₃Sn (dihydropyranyl stannane)	Pd(PPh₃)₄ (5%), dioxane, reflux, 1 h	(82)	55, 57
	Bu₃SnPh	Pd(PPh₃)₄ (5%), CuBr (5%), dioxane, reflux, 5–30 h	(100)	55, 57
	MeO / Me₃Sn / OMe aryl stannane	Pd(PPh₃)₄ (5%), dioxane, reflux, 1 h	(46) + (6)	55, 57
	BocHN / Me₃Sn aryl stannane	Pd(PPh₃)₄ (5%), dioxane, reflux, 20 h	(70)	55, 57
	Et₂NOC / Me₃Sn / OMe aryl stannane	Pd(PPh₃)₄ (2%), CuBr (8%), dioxane, reflux, 15 h	(54)	57

116

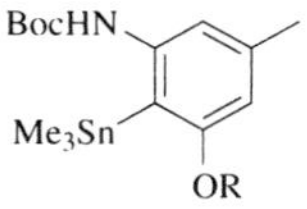

55, 57

55, 57

485, 486

356

487

R = Me (60)
R = MOM (100)

NHBoc

RO

OH

(82) +

CONHBu-*t*

MeO

OH

(5)

MeO

N–Bu-*t*

OH

(72)

···OMe

H

O

OH

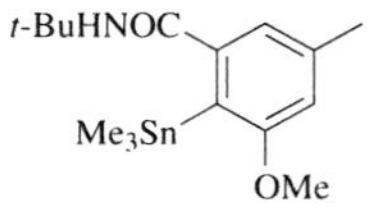

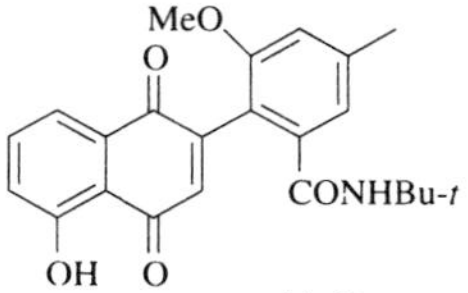

Ph

Ph

(55)

(56)

H

H

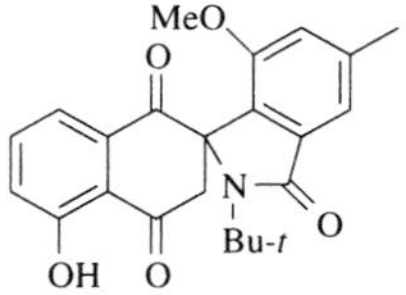

Pd(PPh₃)₄ (5%),
CuBr (5%), dioxane,
reflux, 3 h

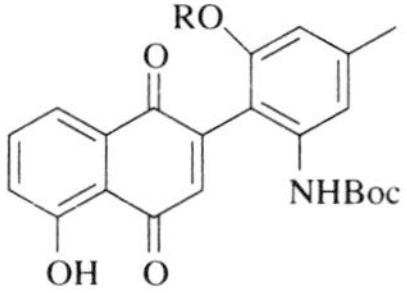

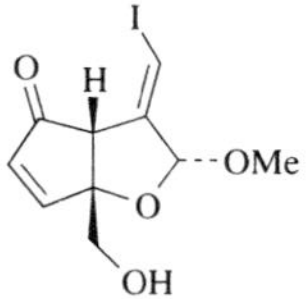

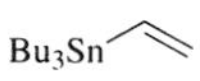

"

Pd(PPh₃)₂Cl₂, DMF,
rt, 16 h

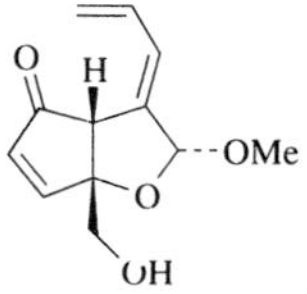

Pd(CH₃CN)₂Cl₂ (1%),
Et₄NCl, HMPA,
80°, 24 h

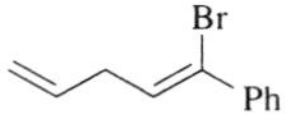

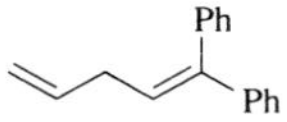

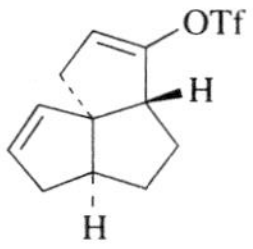

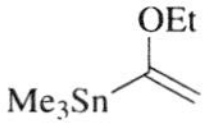

1. Pd(PPh₃)₄, LiCl,
   THF, 70°, 10 h
2. AcOH, H₂O, rt

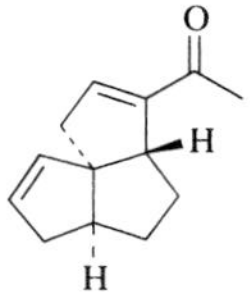

BocHN

OR

Me₃Sn

t-BuHNOC

OMe

Me₃Sn

Bu₃Sn

Bu₃SnPh

OEt

Me₃Sn

I

···OMe

H

O

OH

Br

Ph

OH

C₁₁

OTf

H

H

TABLE I. DIRECT CROSS-COUPLING OF ALKENYL ELECTROPHILES (Continued)

Substrate	Stannane	Conditions	Product(s) and Yield(s) (%)	Refs.
bicyclic OTf structure; $E:Z = 82:18$	Bu₃Sn–CH=CH₂	Pd(PPh₃)₄, LiCl, THF	vinyl bicyclic structure (69)	488
$n\text{-}C_8H_{17}$ structure with I	Bu₃Sn~~~NHFMOC	Pd(CH₃CN)₂Cl₂ (5%), DMF, 20°, 24 h	NHFMOC structure (65) $E:Z = 85:15$	466
$n\text{-}C_8H_{17}$ structure with I	Bu₃SnC≡CBu-n	Pd(PPh₃)₂Cl₂, DIBAL, THF, rt	$n\text{-}C_8H_{17}$ —C≡CBu-n (88)	442
quinone with Cl	Me₄Sn	BnPd(PPh₃)₂Cl, dioxane, reflux	dimethyl quinone (82)	43
BocHN CO₂Me structure with Br	Bu₃Sn–CH=CH₂	Pd(PPh₃)₄, PhMe	BocHN CO₂Me structure (63)	489, 490
EtO, I, TBDMS structure	Bu₃SnC≡CBu-n	Pd(PPh₃)₂Cl₂, PhMe, 50°	EtO C≡CBu-n TBDMS (97)	49
Ph CO₂Et Br structure	Bu₃Sn–CH=CH₂	Pd(PhCN)₂Cl₂ (5%), AsPh₃ (10%), CuI (10%), NMP, 20°, 29.5-46 h	Ph CO₂Et structure (85)	435
Ph CO₂Et I structure	Me₄Sn	Pd(PhCN)₂Cl₂ (5%), AsPh₃ (10%), CuI (10%), NMP, 80°, 16 h	Ph CO₂Et (91)	435

118

Bu₃Sn reagent	Tin reagent	Conditions	Product	Yield	Ref.
Bu₃Sn (allyl)		Pd(PhCN)₂Cl₂ (5%), AsPh₃ (10%), CuI (10%), NMP, 20°, 2.5 h	Ph, CO₂Et	(84)	435
Bu₃Sn		Pd(PhCN)₂Cl₂ (5%), AsPh₃ (10%), CuI (10%), NMP, 80°, 168 h	Ph, CO₂Et	(—)	435
Me₃Sn—CO₂Et		Pd(PhCN)₂Cl₂ (5%), AsPh₃ (10%), CuI (10%), NMP, 20°, 48 h	Ph, CO₂Et, EtO₂C	(30)	435
Bu₃SnPh		Pd(PhCN)₂Cl₂ (5%), AsPh₃ (10%), CuI (10%), NMP, 50°, 14 h	Ph, CO₂Et, Ph	(52)	435
Me₃Sn—C₆H₁₃-n		Pd(PhCN)₂Cl₂ (5%), AsPh₃ (10%), CuI (10%), NMP, 20°, 23 h	EtO₂C, Ph, C₆H₁₃-n	(84)	435
Bu₃Sn		Pd(OAc)₂ (7%), PPh₃ (14%), THF, 55°, 4.5 h	CO₂Bn	(78)	436
Bu₃Sn—CO₂Me		Pd(OAc)₂ (7%), PPh₃ (14%), THF, 65°, 1.5 h	CO₂Me, CO₂Bn	(68)	259

Substrate (bottom): OTf, CO₂Bn

Additional products:
EtO₂C (cyclohexene with CO₂Bn) (71) + CO₂Bn / CO₂Bn (8) — 259
EtO₂C (cyclohexene) (17) — 259

TABLE I. DIRECT CROSS-COUPLING OF ALKENYL ELECTROPHILES (Continued)

Substrate	Stannane	Conditions	Product(s) and Yield(s) (%)	Refs.
THPO⎓⎓I (structure)	Bu$_3$SnC≡CH	Pd(CH$_3$CN)$_2$Cl$_2$, DMF, 22–25°, <2 min	(68) THPO...C≡CH	47
	Bu$_3$SnC≡C–(thiophene)	Pd(CH$_3$CN)$_2$Cl$_2$, DMF, 25°	(≥93) ...OTHP	47
(OTf cyclohexene/dioxolane structure)	Bu$_3$SnC≡CH	Pd(PPh$_3$)$_4$ (10%), LiCl, THF, reflux	(69) C≡CH structure	491
	Bu$_3$Sn⎓ (vinyl)	Pd(PPh$_3$)$_4$ (5%), LiCl, THF, reflux	(91) vinyl structure	491
THPO(CH$_2$)$_4$⎓I	Me$_3$SnC≡CBu-n	Pd(CH$_3$CN)$_2$Cl$_2$, DMF, 25°	(95) n-BuC≡C⎓(CH$_2$)$_4$OTHP	47
MeO$_2$C(CH$_2$)$_7$ ... I (structure)	Bu$_3$Sn...C$_5$H$_{11}$-n, OTBDMS	Pd(PPh$_3$)$_2$Cl$_2$, DMF, 60°, 4 d	(60) C$_5$H$_{11}$-n, OTBDMS; MeO$_2$C(CH$_2$)$_7$	492
(cyclohexenone CO$_2$Me / Br structure)	Bu$_3$SnPh	Pd(PPh$_3$)$_4$	(69) CO$_2$Me / Ph structure	493
(naphthoquinone Br, OMe structure)	Me$_3$Sn aryl (MeO, NHBoc)	Pd(PPh$_3$)$_4$ (5%), CuBr (5%), dioxane, reflux, 3 h	(65) aryl structure	56

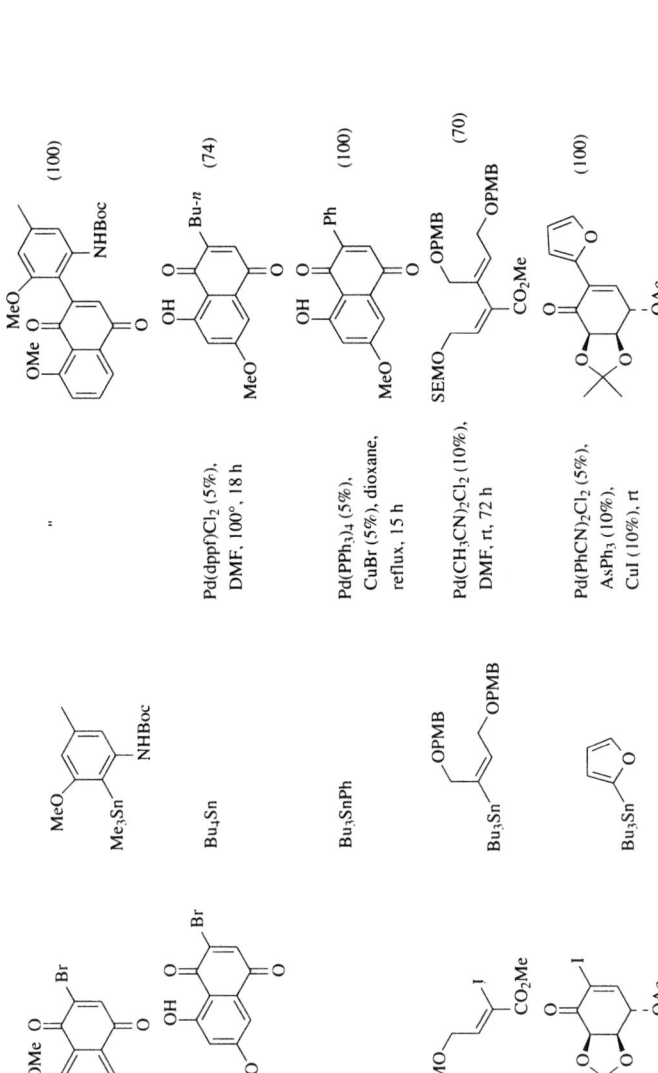

56

55

55

494

495

TABLE I. DIRECT CROSS-COUPLING OF ALKENYL ELECTROPHILES (Continued)

Substrate	Stannane	Conditions	Product(s) and Yield(s) (%)	Refs.
C$_{12}$		Pd(CH$_3$CN)$_2$Cl$_2$, DMF, rt	(26) E:Z = 1:1 + (42)	85
	Bu$_3$Sn	Pd(OAc)$_2$ (10%), CH$_2$Cl$_2$, rt	(>90)	453
	Bu$_3$Sn	Pd$_2$(dba)$_3$ (2.5%), PPh$_3$ or AsPh$_3$ or P(2-furyl)$_3$, NMP, (LiCl), 35°	" (>90)	11
	Bu$_3$Sn—⟨⟩—CO$_2$Me	Pd(PPh$_3$)$_4$, LiCl, THF	(50)	496
		Pd(PPh$_3$)$_2$Cl$_2$ (2%), THF, reflux	R = Me (90) R = SEM (61)	425

122

Substrate	Reagent	Conditions	Product (yield %)	Refs.
Bu₃Sn–(3-indolyl), N-Ts	(none)	Pd₂(dba)₃ (2.5%), AsPh₃ (10%), DMF, 60°, 1.5 h	indolyl–(4-Ph-cyclohexenyl), N-Ts (94)	291
Me₃Sn–CH=CH–S(O)–C₆H₄NO₂	(none)	Pd₂(dba)₃ (1%), P(2-furyl)₃ (4%), LiCl, ZnCl₂, NMP, rt, 2 h	(78) NO₂	260
n-C₁₀H₂₁ Br	Bu₃SnH	Pd(PPh₃)₄ (2%), C₆H₆, 75°, 3 h	n-C₁₀H₂₁ (37)	48
n-C₁₀H₂₁ I	Bu₃SnH	Pd(PPh₃)₄ (2%), C₆H₆, 25°, 3 h	" (92)	48
I—C₁₀H₂₁-n	Bu₃SnH	Pd(PPh₃)₄ (2%), C₆H₆, 25°, 5 h	" (82)	48
n-C₁₀H₂₁ D I	Bu₃SnH	Pd(PPh₃)₄ (2%), C₆H₆, 25°, 2 h	n-C₁₀H₂₁ D (—) Z:E = 95:5	48
n-C₅H₁₁ C₅H₁₁-n I	Bu₃SnH	Pd(PPh₃)₄ (2%), C₆H₆, 25°, 1 h	n-C₅H₁₁ C₅H₁₁-n (89) Z:E = 94:6	48
n-C₅H₁₁ I C₅H₁₁-n	Bu₃SnH	Pd(PPh₃)₄ (2%), C₆H₆, 75°, 5 h	n-C₅H₁₁ C₅H₁₁-n (90)	48
cyclopentenone–CH₂SO₂Ph, I	Bu₃Sn–CH=CH–CH(OH)–C₅H₁₁-n	Pd(CH₃CN)₂Cl₂ (5%), DMF, 25°, 2 h	cyclopentenone–CH₂SO₂Ph, –CH=CH–CH(OH)C₅H₁₁-n (72)	497
I–CH=CH–CH(iPr)–OTBDMS	Bu₃Sn–CH₂–CH=CH–NHFMOC	Pd(CH₃CN)₂Cl₂ (5%), DMF, 20°, 24 h	(84) NHFMOC, OTBDMS	466

TABLE I. DIRECT CROSS-COUPLING OF ALKENYL ELECTROPHILES (*Continued*)

Substrate	Stannane	Conditions	Product(s) and Yield(s) (%)	Refs.
C₁₃ t-Bu $\diagup$ C₇H₁₅-n with I	Bu₃SnH	Pd(PPh₃)₄ (2%), C₆H₆, 75°, 5 h	t-Bu $\diagup$ C₇H₁₅-n (71)	48
(ketone with OTf, Br structure)	Bu₃SnTMS	Pd(PPh₃)₂Cl₂ (3%), Bu₄NBr, Li₂CO₃, PhMe, 110°	(tricyclic ketone) (38) + (cyclohexenone with Br) (43)	417
(tricyclic OTf enone)	Me₃Sn $\diagup$ OEt	Pd(PPh₃)₄, LiCl, THF	(tricyclic OEt enone) (—)	498
(vinyl sulfone Bu-n, I, Tol-p)	Bu₃Sn $\diagup$ Ph	Pd(CH₃CN)₂Cl₂ (2%), DMF, rt	(diene sulfone Bu-n, Ph, Tol-p) (92)	59
(quinone Ph, I)	Bu₃Sn $\diagup$	Pd(PPh₃)₄, THF, 80°, 12 h	(quinone Ph, vinyl) (32) + (phenol Ph, I) (28)	58
	Bu₃SnC≡CBu-n	Pd(PPh₃)₄, THF, 90°, 12 h	(quinone Ph, C≡CBu-n) (64)	58
	Bu₃SnC≡CPh	Pd(PPh₃)₄, THF, 80°, 30 h	(quinone Ph, C≡CPh) (62)	58

124

Substrate	Reagent	Conditions	Product (%)	Refs.
(structure with I, vinyl, OBu-t, O, H)	Bu₃Sn⌬ (Bu$_3$Sn allyl)	Pd(PPh$_3$)$_4$, LiCl, THF, reflux, 5 h	(structure) (75)	486
(structure with OTf, EtO$_2$C, Ph, OBu-t)	Bu$_3$Sn–imidazole(Bu-n, SEM)	Pd(PPh$_3$)$_4$ (2%), LiCl, THF, reflux, 22 h	(two products) (28) + (33)	433
(cyclohexene OTf, Ts)	Bu$_3$SnTMS	Pd(PPh$_3$)$_2$Cl$_2$, PhMe, Bu$_4$NBr, 2,6-di-t-butyl-4-methylpyridine, 110°, 1.5 h	(95) + (5)	417
(NO$_2$, CO$_2$Me, I, cyclopentenone)	(Bu$_3$Sn, OH, C$_5$H$_{11}$-n)	Pd(CH$_3$CN)$_2$Cl$_2$ (5%), DMF, 25°, 3h	(structure) (76)	497
(I, pyran, CO$_2$H, H)	(Bu$_3$Sn, pyrrole, H, ethyl tricyclic)	Pd(PPh$_3$)$_4$ (10%), DMF, 23°, 72 h	(structure) (61)	496
(n-C$_6$H$_{13}$, I, C≡CTMS)	Me$_3$SnC≡CBu-n	Pd(CH$_3$CN)$_2$Cl$_2$ (5%)	(n-C$_6$H$_{13}$, n-BuC≡C, C≡CTMS) (74)	387
C$_{14}$ (O, Br, OTf)	Bu$_3$SnTMS	Pd(PPh$_3$)$_2$Cl$_2$ (3%), Bu$_4$NBr, Li$_2$CO$_3$, PhMe, 110°, 2 h	(fluorenone structure) (50)	417

125

TABLE I. DIRECT CROSS-COUPLING OF ALKENYL ELECTROPHILES (*Continued*)

Substrate	Stannane	Conditions	Product(s) and Yield(s) (%)	Refs.
[structure: CO₂Me ketone with Br-vinyl arene]	Bu₃Sn—CH=CH—CH(OTHP)—C₈H₁₇-*n*	Pd(PPh₃)₄, PhMe, 100°	[structure] (54)	500
[structure: OTf phenalenone]	Bu₃Sn—(butenyl)	Pd(PPh₃)₄, LiCl, THF, reflux, 20 h	[structure] (82)	501
n-C₅H₁₁—C(I)=CH—COPh	Me₄Sn	Pd(PhCN)₂Cl₂ (5%), AsPh₃ (10%), CuI (10%), NMP, 80°, 19 h	*n*-C₅H₁₁—C(CH₃)=CH—COPh (70)	446
	Bu₃Sn—CH=CH₂	Pd(PhCN)₂Cl₂ (5%), AsPh₃ (10%), CuI (10%), NMP, 45°, 20 h	*n*-C₅H₁₁—C(CH=CH₂)=CH—COPh (79)	446
n-C₅H₁₁—C(COPh)=CH—I *E:Z* = 69:31	Me₄Sn	Pd(PhCN)₂Cl₂ (5%), AsPh₃ (10%), CuI (10%), NMP, 80°, 68 h	*n*-C₅H₁₁—C(CH₃)=CH—COPh (63)	446
[structure: iodo vinylidene cyclohexane OTBDMS]	Bu₃Sn—CH=(cyclohexylidene)	Pd(CH₃CN)₂Cl₂ (5%), DMF, rt, 48 h	[structure OTBDMS] (>53)	502
[structure: TBDMSO, OTf cyclohexene]	Bu₃Sn—CH=CH₂	Pd(PPh₃)₄ (3%), LiCl, THF, reflux, 18 h	[structure OTBDMS] (88)	476

126

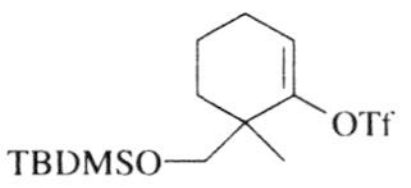

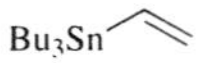

Bu₃Sn — rendered as Bu_3Sn

TBDMSO (enol triflate, OTf)	Bu_3Sn allyl	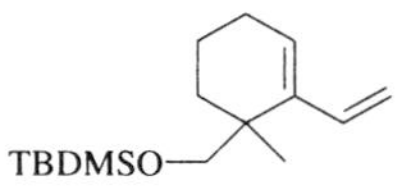 Pd(PPh₃)₄, LiCl, THF, reflux	(89) · 452
TBDMSO spiro enol triflate, Bu-*t*, OTf	Bu_3Sn allyl	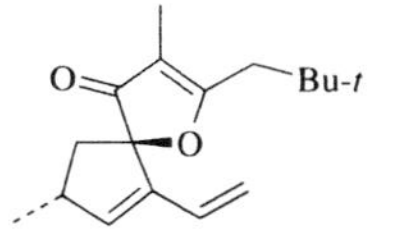 Pd(PPh₃)₄, LiCl, THF, reflux	(98) · 503
(CH₂)₆CO₂Et cyclopentenone, I	Bu_3Sn —CH=CH–CH(OH)–C_5H_{11}-*n*	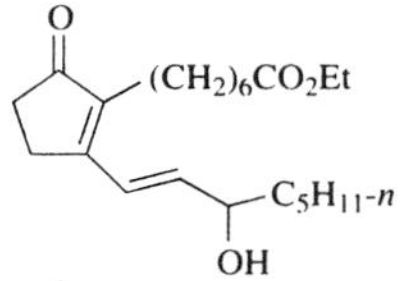 BnPd(PPh₃)₂Cl, THF, 55–60°	(70–73) · 53, 52
(CH₂)₆CO₂Et cyclopentenone, I	Bu_3Sn —CH=CH–CH(OH)–C_5H_{11}-*n*	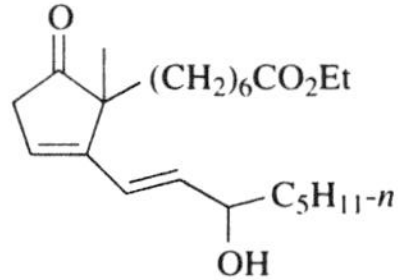 Pd(PPh₃)₄ (3%), LiCl, THF, 60°, 48 h	(60) · 53
TBDMSO bicyclic, CO_2Et, I	Bu_3Sn allyl	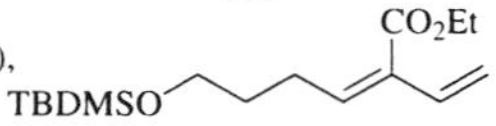 Pd(PhCN)₂Cl₂ (5%), AsPh₃ (10%), CuI (10%), DMF, 20°, 3 h	(82) · 435
H, OMe, OBu-*t*, bicyclic	Bu_3Sn allyl	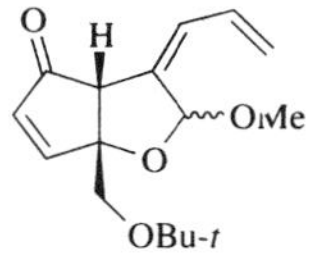 Pd(PPh₃)₂Cl₂, DMF, 23–24°, 18 h	(71–78) · 486
Ph, SO_2Ph, OTf	Bu_3Sn —CH=CH–CO_2Me	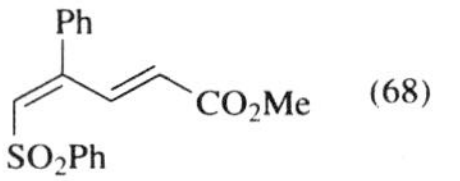 Pd(OAc)₂ (7%), PPh₃ (14%), THF, 65°	(68) · 259

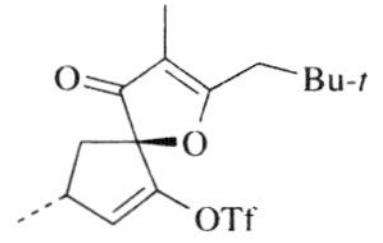

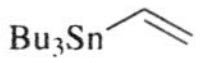

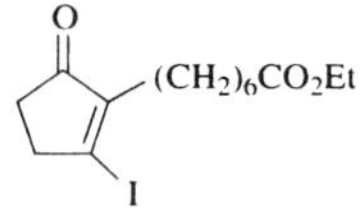

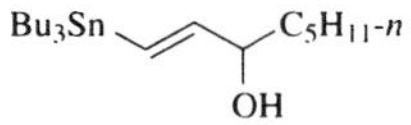

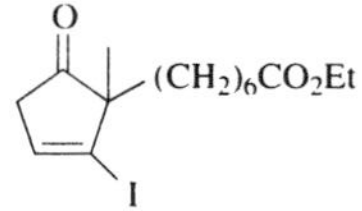

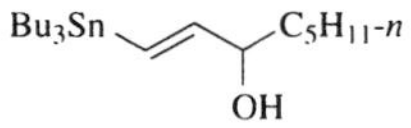

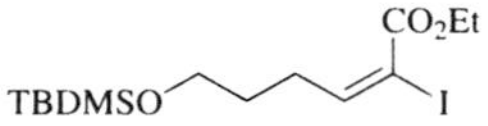

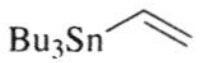

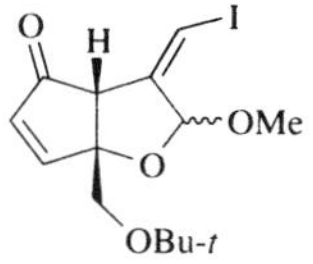

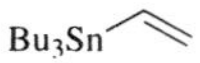

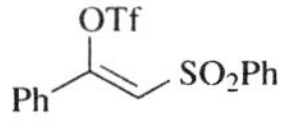

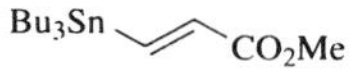

127

TABLE I. DIRECT CROSS-COUPLING OF ALKENYL ELECTROPHILES (Continued)

Substrate	Stannane	Conditions	Product(s) and Yield(s) (%)	Refs.
C$_{15}$				
(Br-decalin vinyl OTf structure)	Bu$_3$Sn (OH isopropanol vinyl)	Pd(PPh$_3$)$_4$ (3.6%), CsF, LiCl, THF, 25°, 60 h	(31)	504
(naphthylmethyl-N-Me iodo allyl structure)	Bu$_3$SnC≡CR	Pd(CH$_3$CN)$_2$Cl$_2$ (5%), DMF, 0°, 1 h	R: TMS (80); n-Bu (87); t-Bu (70)	505
(lactam lactone diol iodide structure)	Bu$_3$Sn NHFMOC	Pd(CH$_3$CN)$_2$Cl$_2$, DMF, 23°, 0.5 h	FMOCNH (84)	77
Ph COPh I	Me$_4$Sn	Pd(PhCN)$_2$Cl$_2$ (5%), AsPh$_3$ (10%), CuI (10%), NMP, 80°, 2.5 h	Ph COPh I (94)	446
	Me$_3$SnPh	Pd(PhCN)$_2$Cl$_2$ (5%), AsPh$_3$ (10%), CuI (10%), NMP, 55°, 2 h	I (43) + Ph COPh Ph (50)	446
	Me$_4$Sn	Pd(PhCN)$_2$Cl$_2$ (5%), AsPh$_3$ (10%), CuI (10%), NMP, 80°, 21 h	Ph COPh (97) E:Z = 45:55	446
(spirocyclic enone OTf alkyne structure)	Bu$_3$Sn (vinyl)	Pd(PPh$_3$)$_4$ (1%), LiCl, THF, reflux	(100)	503

Reagent (stannane)	Conditions	Product (yield)	Ref.
Me₃Sn (OTBDMS)	Pd(PPh₃)₄ (10%), LiCl, THF, reflux	(100)	503
Bu₃Sn (vinyl)	Pd(PPh₃)₄ (3%), LiCl, THF, reflux	(81)	484
Bu₃SnH	Pd(PPh₃)₄ (2%), C₆H₆, 25°, 3 h	(67) Z:E = 95:5	48
Bu₃Sn Bu-n	Pd(CH₃CN)₂Cl₂ (2-5%), DMF, 20°, 20 h	(66)	441
Bu₃Sn CO₂Et	Pd(CH₃CN)₂Cl₂ (2-5%), DMF, 20°, 20 h	(62)	441
Bu₃Sn OTHP	Pd(CH₃CN)₂Cl₂ (2-5%), DMF, 20°, 20 h	(57)	441
Bu₃Sn OPMB	Pd(CH₃CN)₂Cl₂ (2-5%), HMPA, 60°, 8 h	I + II (13) (47)	441

129

TABLE I. DIRECT CROSS-COUPLING OF ALKENYL ELECTROPHILES (*Continued*)

Substrate	Stannane	Conditions	Product(s) and Yield(s) (%)	Refs.
(Et₃SiO, O, OTf)	Bu₃Sn⁀OPMB	Pd(PPh₃)₄ (2-5%), HMPA, 60°, 8 h	**I** (34) + **II** (40)	441
	Bu₃Sn⁀OPMB	Pd(PPh₃)₄ (2-5%), HMPA, 20°, 72 h	**I** (67) + **II** (15)	441
(I, O, OTBDMS)	(Bu₃Sn…OTBDPS, O, CO₂Me)	Pd(OAc)₂ (7%), PPh₃ (14%), THF, 70°	(66) (Et₃SiO…O, OTBDPS, CO₂Me)	506
	Bu₃SnR	Pd(PhCN)₂Cl₂ (5%), AsPh₃ (10%), CuI (10%), NMP, rt, 16 h	(R, O, OTBDMS) $\dfrac{R}{CH=CH_2 \;\;(91)}$ C(Me)=CH₂ (85), C₆H₄Me-*p* (60)	61
(EtO₂C, Br, OTf)	Bu₃SnTMS	Pd(PPh₃)₂Cl₂ (3%), Bu₄NBr, Li₂CO₃, PhMe, 110°, 1.5 h	CO₂Et (70)	417
(EtO₂C, Br, OTf)	Bu₃SnTMS	Pd(PPh₃)₂Cl₂ (3%), Bu₄NBr, Li₂CO₃, PhMe, 110°, sieves, 8 h	CO₂Et (41) + EtO₂C, Br (13)	417
C₁₆ (OTf, CO₂Et)	Me₄Sn	Pd(AsPh₃)₄, CuI, NMP, 100°	CO₂Et (56)	177

130

Organotin reagent	Product (yield)	Conditions	Ref.
Bu₃SnC≡CH	(33) C≡CH, CO₂Et	Pd(AsPh₃)₄, CuI, NMP, rt	177
Et₄Sn	(77) CO₂Et	Pd(AsPh₃)₄, CuI, NMP, 100°	177
Bu₃Sn /	(51) E:Z = 1:1, CO₂Et	Pd(PPh₃)₄, THF, reflux	177
Bu₃Sn /	(69) Z:E = 16:1, CO₂Et	Pd(AsPh₃)₄, CuI, NMP, rt	177
Bu₃SnPh	(61) Ph, CO₂Et	Pd(AsPh₃)₄, CuI, NMP, 100°	177
Me₃SnC≡CPr-n	(86) n-PrC≡C—(CH₂)₉OTHP	Pd(CH₃CN)₂Cl₂, DMF, 25°	47
Me₃Sn (furan, CH₂OTBDMS, I)	(35) (furan, CH₂OTBDMS, OTBDMS, I)	Pd(PPh₃)₄, DMF, rt, 16 h	455

TABLE I. DIRECT CROSS-COUPLING OF ALKENYL ELECTROPHILES (*Continued*)

Substrate	Stannane	Conditions	Product(s) and Yield(s) (%)	Refs.
C₁₇	Bu₃SnTMS	Pd(PPh₃)₂Cl₂ (3%), Bu₄NBr, Li₂CO₃, PhMe, 100°, 3.5 h	(10) + (43)	417
	Bu₃Sn—SiMe₂Ph / TBDMSO (OMOM)	Pd(AsPh₃)₂Cl₂ (3%), DMF, rt, 1.5 h	OTBDMS ... SiMe₂Ph (94) *Z,Z:E,Z = 3:1*	51
	Bu₃Sn (OMOM, OH, CN)	Pd(AsPh₃)₂Cl₂ (2%), DMF, rt, 18 h	OMOM ... OH ... SiMe₂Ph (67)	51
	Bu₃Sn (CN)	Pd(CH₃CN)₂Cl₂ (5%), THF, DMF, rt, 40 min	CN ... OTBDMS (88)	507
	Bu₃Sn (CN)	Pd(CH₃CN)₂Cl₂ (5%), THF, DMF, rt, 40 min	CN ... OTBDMS (80)	507
	HO—C≡CSnMe₃ / I / TBDMSO	Pd(PPh₃)₄ (20%), C₆H₆, 70°, 12 h	OTBDMS ... OH ... HO ... TBDMSO (34)	508
	Bu₃Sn (N, O)	Pd(CH₃CN)₂Cl₂, DMF, 23°, 91 h	OTBDMS ... CO₂Me (79)	77

C₁₈

Substrate	Reagent	Conditions	Product (%)	Refs.
TfO-steroid (decalin with isoprenoid side chain)	Bu₃SnC≡CTMS	Pd(PPh₃)₄ (1.6%), LiCl, THF, reflux	(≥74)	509
	Bu₃SnC≡C–(cyclohexenyl-OTBDMS)	1. Pd(PPh₃)₄ (1.6%), LiCl, THF, reflux 2. TBAF	(77)	509
EtO₂C–(enol triflate) Br	Bu₃SnTMS	Pd(PPh₃)₂Cl₂ (3%), Bu₄NBr, Li₂CO₃, PhMe, 100°, 1 h	(56) CO₂Et	417
C₁₀H₂₁-n, I, n-C₆F₁₃	Bu₃SnH	Pd(PPh₃)₄	n-C₆F₁₃ ⟶ C₁₀H₂₁-n (—)	510
	Bu₃SnH	Pd(PPh₃)₄ (2%), C₆H₆, 25°, 3 h	" (97)	48
Br, Et, Ph, (4-chloroethoxyphenyl) alkene	Bu₃SnPh	[(η³-C₃H₅)PdCl]₂ (0.25%), LiBr, HMPA, rt	Ph, Et, Ph, Cl (96) Z:E = 91:9	50
TIPSO–(cyclopentenyl)–OTf	Me₃SnSnMe₃	Pd(PPh₃)₄ (5%), LiCl, THF, reflux	SnMe₃, TIPSO (90)	511
quinone (Ph, I, Ph)	Bu₃Sn–CH=CH₂	Pd(PPh₃)₄ (5%), THF, 90°, 60 h	vinyl quinone (21) + hydroquinone OH, Ph, I, Ph, OH (61)	58

TABLE I. DIRECT CROSS-COUPLING OF ALKENYL ELECTROPHILES (Continued)

Substrate	Stannane	Conditions	Product(s) and Yield(s) (%)	Refs.
	Bu₃SnPh	Pd(PPh₃)₄ (5%), THF, 80°, 24 h	(86)	58
	Bu₃SnC≡CBu-n	Pd(PPh₃)₄ (5%), THF, 70°, 16 h	(19) + (70)	58
	Bu₃SnC≡CPh	Pd(PPh₃)₄ (5%), THF, 80°, 17 h	(30) + (40)	58
		Pd(PPh₃)₂Cl₂, THF, 60°	(75)	81
		Pd(PPh₃)₂Cl₂, DMF, rt	(84)	85
		Pd(PhCN)₂Cl₂ (5%), AsPh₃ (10%), CuI (10%), NMP, rt, 4 h	(93)	61

134

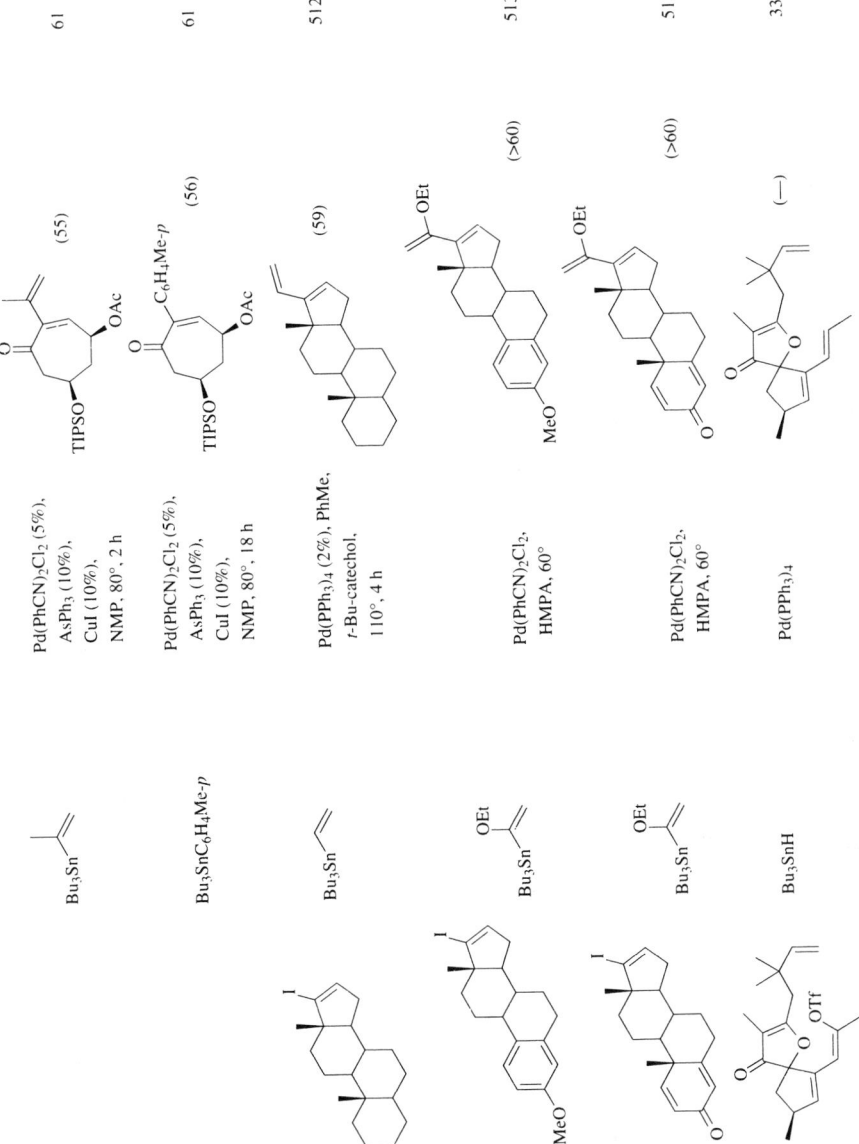

C_{19}	Bu_3Sn	Pd(PhCN)$_2$Cl$_2$ (5%), AsPh$_3$ (10%), CuI (10%), NMP, 80°, 2 h	(55)	61
	$Bu_3SnC_6H_4Me\text{-}p$	Pd(PhCN)$_2$Cl$_2$ (5%), AsPh$_3$ (10%), CuI (10%), NMP, 80°, 18 h	(56)	61
	Bu_3Sn	Pd(PPh$_3$)$_4$ (2%), PhMe, t-Bu-catechol, 110°, 4 h	(59)	512
	Bu_3Sn OEt	Pd(PhCN)$_2$Cl$_2$, HMPA, 60°	(>60)	513
	Bu_3Sn OEt	Pd(PhCN)$_2$Cl$_2$, HMPA, 60°	(>60)	513
	Bu_3SnH	Pd(PPh$_3$)$_4$	(—)	335

135

TABLE I. DIRECT CROSS-COUPLING OF ALKENYL ELECTROPHILES (*Continued*)

Substrate	Stannane	Conditions	Product(s) and Yield(s) (%)	Refs.
	Bu₃Sn⟍	Pd(PPh₃)₂Cl₂ (1%), LiCl, THF, reflux, 3.5 h	(54)	514
C₂₀	Bu₃SnTMS	Pd(PPh₃)₂Cl₂ (3%), Bu₄NBr, Li₂CO₃, PhMe, 110°, 1.5 h	(61)	417
		1. Pd(PPh₃)₄, DMF, rt; 2. I₂, CH₂Cl₂, rt	(51)	78
	Me₃SnSnMe₃	Pd(PPh₃)₄, LiCl, Li₂CO₃, THF, 60°, 12 h	(73)	515

136

Substrate	Organometallic	Conditions	Product (%)	Refs.
C21 (TIPSO chromene bromoenone, –OH, prenyl, Br)	SnBu3-chromene (TIPSO)	Pd(PPh3)4, PhMe, 90°, 2 h	(80) (TIPSO chromene enone, –OH, prenyl)	516
n-C10H21—C(SiMe2Ph)=CHBr	Bu3SnH	Pd(PPh3)4 (2%), C6H6, 75°, 5 h	(35) E:Z = 95:5 (n-C10H21, SiMe2Ph vinyl)	48
Steroid OTf (AcO)	Bu3Sn—C(OMe)=CH2	Pd(OAc)2 (5%), PPh3 (10%), LiCl, DMF, 60°, 4.5 h	(54) (steroid, C(=CH2)OMe, AcO)	517
Ph3Si—CH=CH—C(=O)—CH=CH—I	Bu3Sn—CH=CH2	Pd(CH3CN)2Cl2 (10%), DMF, rt, 1 h	(70) (Ph3Si dienone)	518
	Bu3Sn—C(OEt)=CH2	Pd(CH3CN)2Cl2 (10%), DMF, rt, 1 h	(75) (Ph3Si, OEt)	518
	Bu3Sn-(3-furyl)	Pd(CH3CN)2Cl2 (10%), DMF, rt, 1 h	(73) E:Z = 1:1 (Ph3Si, furan)	518
	Me3SnC≡CTMS	Pd(CH3CN)2Cl2 (10%), DMF, rt, 9 h	(64) (Ph3Si, C≡CTMS)	518
	Bu3Sn—CH=CH—CH2NHBoc	Pd(CH3CN)2Cl2 (10%), DMF, rt, 12 h	(86) (SiPh3, BocHN dienone)	518, 467

TABLE I. DIRECT CROSS-COUPLING OF ALKENYL ELECTROPHILES (Continued)

Substrate	Stannane	Conditions	Product(s) and Yield(s) (%)	Refs.
C_{22} [structure with OTBDPS and I]	Bu_3Sn—C(=O)—CH=CH—$SiPh_3$	$Pd(CH_3CN)_2Cl_2$ (10%), DMF, rt, 8 h	[product] $SiPh_3$ (84)	518
	Bu_3Sn [polyene with trimethylcyclohexenyl]	$Pd(PPh_3)_4$, HMPA, 65°, 3 h	[OTBDPS product] (39) + [$( )_2$ dimer] (18); + TBDPSO [dimer] $( )_2$ (19)	386
[iodo cycloheptenone, TIPSO, OTBDMS]	Bu_3Sn [vinyl]	$Pd(PhCN)_2Cl_2$ (5%), $AsPh_3$ (10%), CuI (10%), NMP, rt, 4 h	[vinyl cycloheptenone, TIPSO, OTBDMS] (95)	61
	Bu_3Sn [isopropenyl]	$Pd(PhCN)_2Cl_2$ (5%), $AsPh_3$ (10%), CuI (10%), NMP, 80°, 2 h	[isopropenyl cycloheptenone, TIPSO, OTBDMS] (80)	61
	$Bu_3SnC_6H_4Me$-p	$Pd(PhCN)_2Cl_2$ (5%), $AsPh_3$ (10%), CuI (10%), NMP, 60°, 24 h	[C_6H_4Me-p cycloheptenone, TIPSO, OTBDMS] (95)	61

138

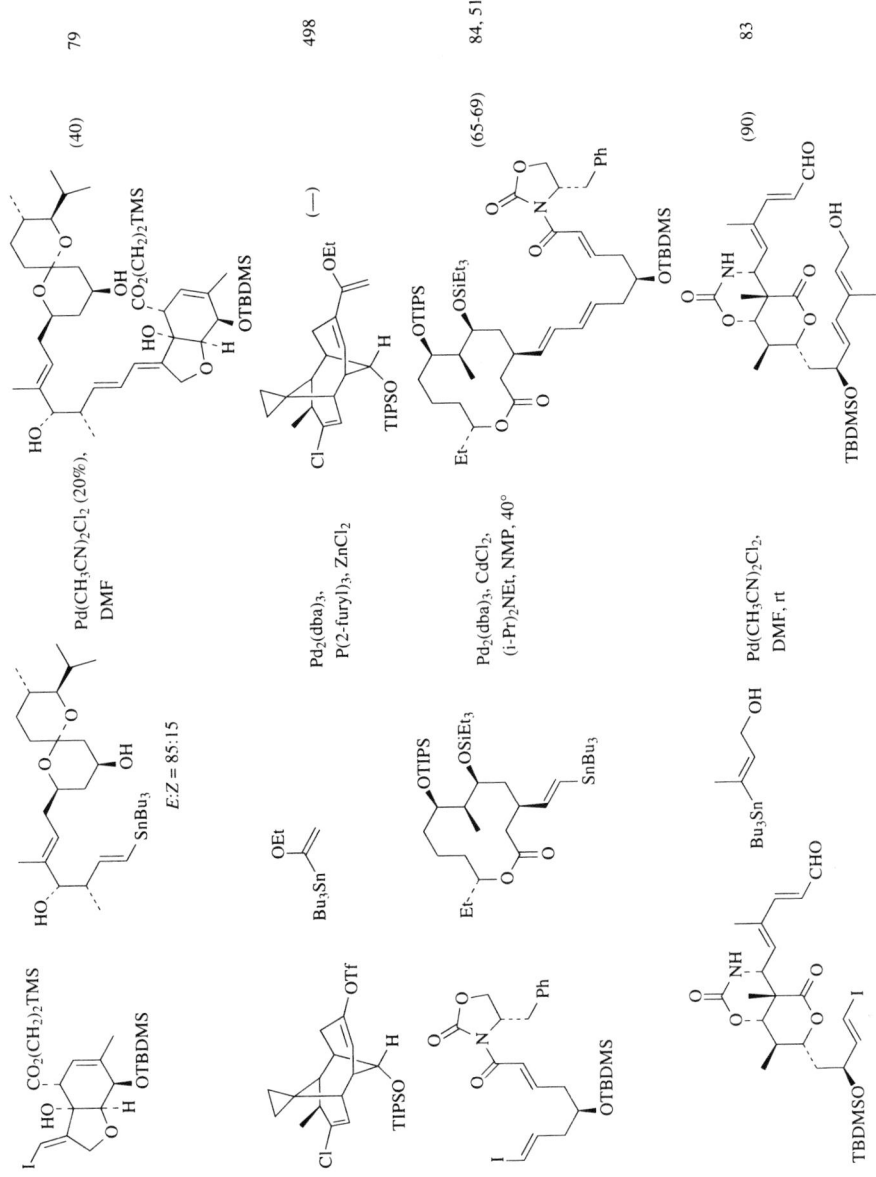

79

(40)

Pd(CH₃CN)₂Cl₂ (20%),
DMF

E:Z = 85:15

498

(—)

Pd₂(dba)₃,
P(2-furyl)₃, ZnCl₂

84, 519

(65-69)

Pd₂(dba)₃, CdCl₂,
(i-Pr)₂NEt, NMP, 40°

83

(90)

Pd(CH₃CN)₂Cl₂,
DMF, rt

C₂₃

C₂₄

C₂₅

139

TABLE I. DIRECT CROSS-COUPLING OF ALKENYL ELECTROPHILES (Continued)

Substrate	Stannane	Conditions	Product(s) and Yield(s) (%)	Refs.
C$_{26}$	Bu$_3$Sn⌒	Pd(PPh$_3$)$_4$, LiCl, DMF or dioxane, 110°	(90–100)	520
	Me$_3$SnSnMe$_3$	Pd(PPh$_3$)$_4$, LiCl, THF, 60°	(82)	521
	Bu$_3$Sn⌒	Pd(OAc)$_2$, NMP	(79)	522
C$_{27}$		Pd(dba)$_2$ (5%), P(2-furyl)$_3$ (10%), ZnCl$_2$, NMP, 65°	(95)	12
	Bu$_3$Sn⌒	Pd(PPh$_3$)$_4$	(65–80)	512, 520

140

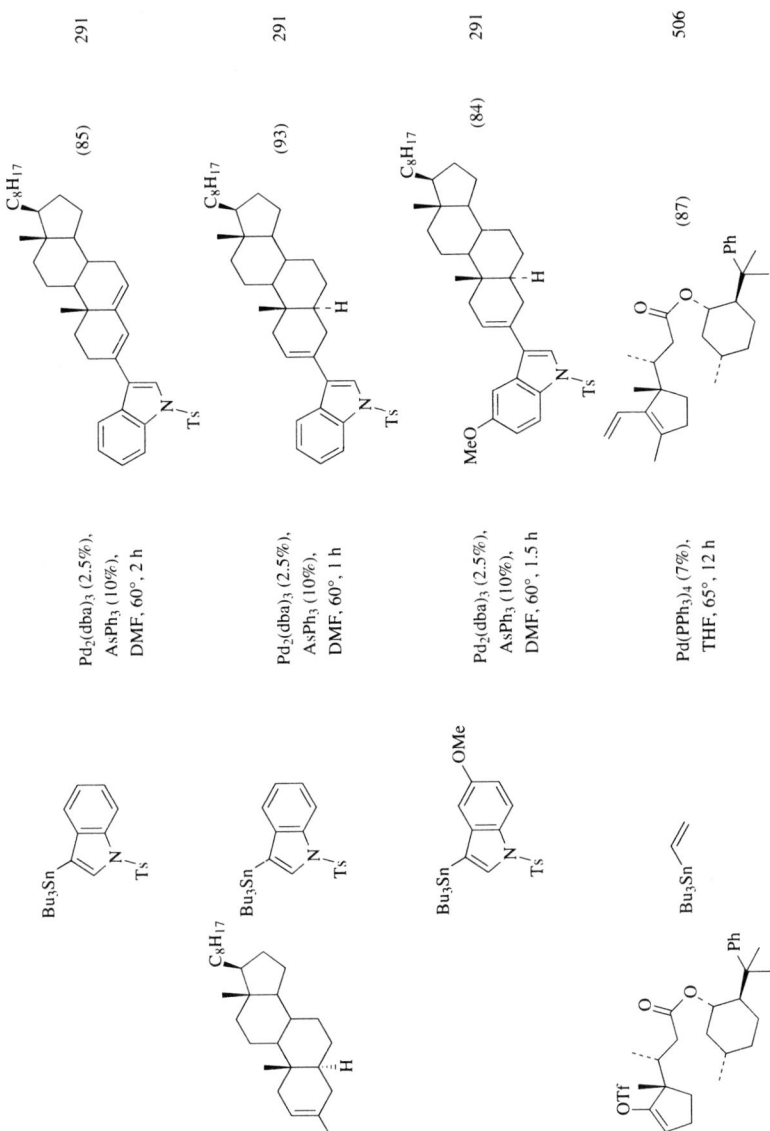

291 (85)

Pd$_2$(dba)$_3$ (2.5%),
AsPh$_3$ (10%),
DMF, 60°, 2 h

291 (93)

Pd$_2$(dba)$_3$ (2.5%),
AsPh$_3$ (10%),
DMF, 60°, 1 h

291 (84)

Pd$_2$(dba)$_3$ (2.5%),
AsPh$_3$ (10%),
DMF, 60°, 1.5 h

506 (87)

Pd(PPh$_3$)$_4$ (7%),
THF, 65°, 12 h

TABLE I. DIRECT CROSS-COUPLING OF ALKENYL ELECTROPHILES (*Continued*)

Substrate	Stannane	Conditions	Product(s) and Yield(s) (%)	Refs.
C_{28} β-lactam bearing S—SO$_2$Ph, OTf, CO$_2$PMB, Ph, N—H, O	Bu$_3$Sn—(vinyl)	Pd(OAc)$_2$, NMP, rt	(72) structure with S—SO$_2$Ph, CO$_2$PMB	522
steroidal structure with CONEt$_2$, N—Boc, Br, O	Me$_3$SnPh	Pd(PPh$_3$)$_2$Cl$_2$, LiCl, DMF, 140°, 8 h	(—) structure with CONEt$_2$, N—Boc	523
steroidal structure with CONEt$_2$, N—Boc, Br, O	Me$_3$SnPh	Pd(PPh$_3$)$_2$Cl$_2$, LiCl, DMF, 140°, 8 h	(93) structure with CONEt$_2$, N—Boc	523
	Bu$_3$Sn—(allyl)	Pd(PPh$_3$)$_2$Cl$_2$, LiCl, DMF, 140°, 4 h	(84) structure with CONEt$_2$, N—Boc	523
OTBDPS macrolactone with I *E,Z,E,E* = 5:1	Bu$_3$Sn—C$_6$H$_{13}$-*n*	Pd(CH$_3$CN)$_2$Cl$_2$, DMF, rt	(63) OTBDPS macrolactone, C$_6$H$_{13}$-*n* *E,Z,Z,E,Z,E* = 1:1:1	524

142

C_{29}

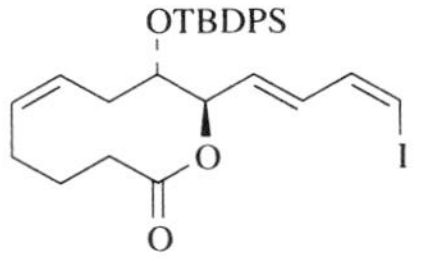

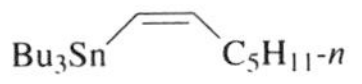

OTBDPS

$C_5H_{11}\text{-}n$ (>48)   524

Pd(CH$_3$CN)$_2$Cl$_2$, DMF, rt

Bu$_3$Sn

$C_5H_{11}\text{-}n$

OTBDPS

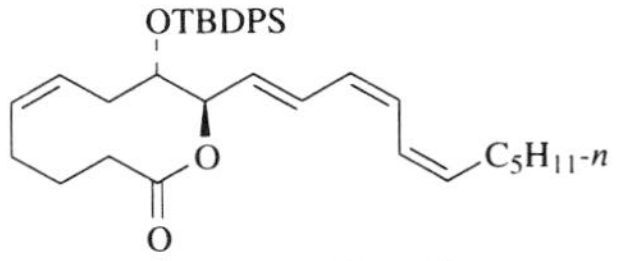

C_{49}

OMe   OH   H

OMe

OH

OMe

(28)   86

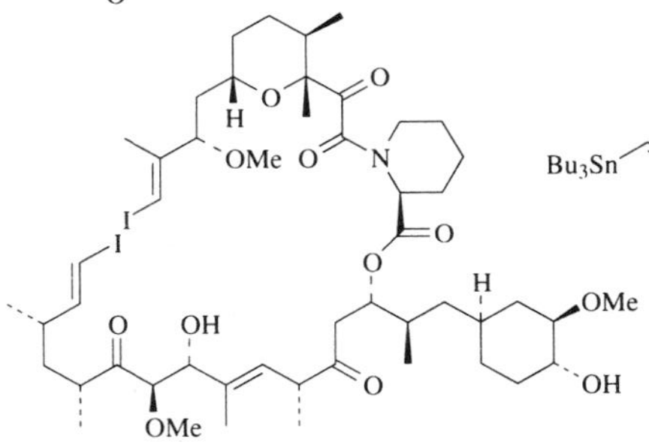

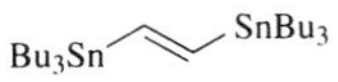

Pd(CH$_3$CN)$_2$Cl$_2$, (iPr)$_2$NEt, DMF, THF, rt

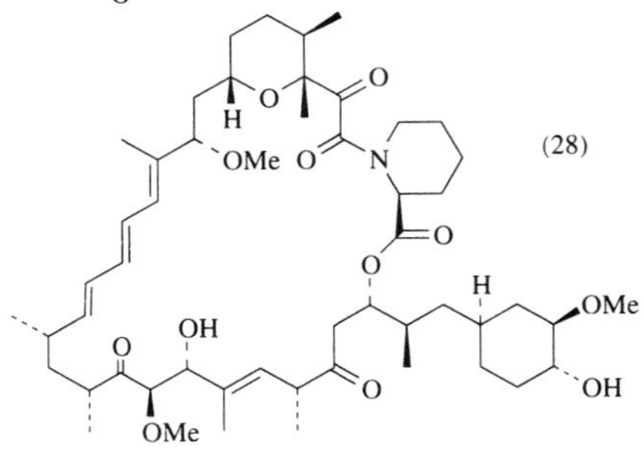

Bu$_3$Sn — SnBu$_3$

OMe   OH   H

OMe

C_{55}

OEE

OTBDPS

TBDPSO

(62)   525

Pd(PPh$_3$)$_4$ (10%), LiCl, THF, 80°, 14 h

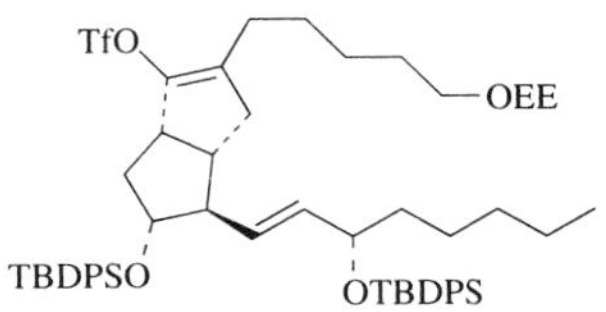

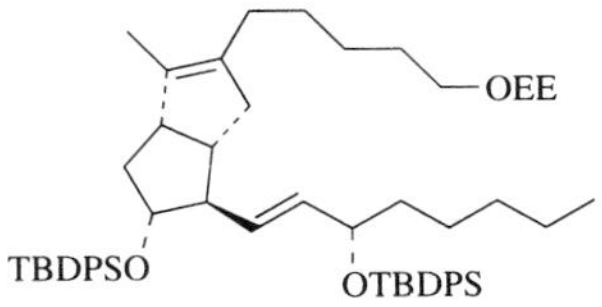

Me$_4$Sn

OEE

OTBDPS

TfO

TBDPSO

C_{75}

CN

OTBDMS

OMe

Bu$_3$Sn

Et$_3$SiO

O

N
H

OSiEt$_3$

Me$_2$N

MeO

O=P
(TMSCH$_2$CH$_2$O)$_2$

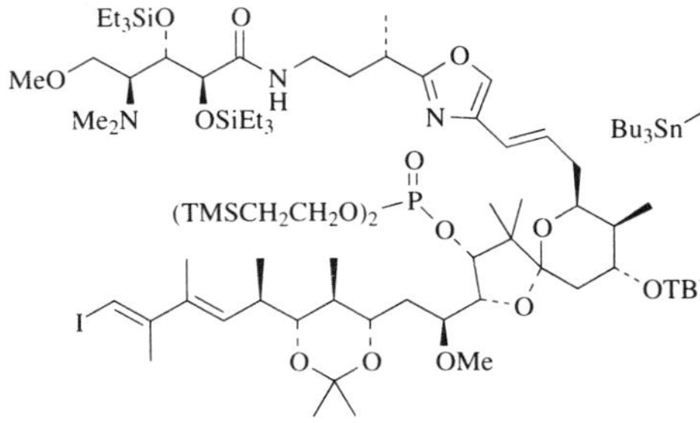

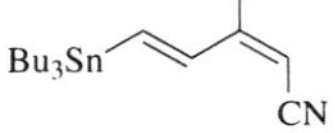

O
O

I

CN   (67)   82

Pd(PPh$_3$)$_2$Cl$_2$, DMF

OMe   OH   H

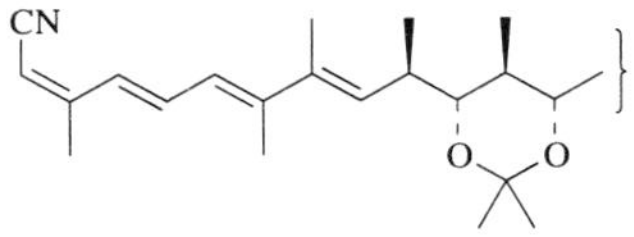

CN

O
O

143

TABLE II. INTRAMOLECULAR CROSS-COUPLING OF ALKENYL ELECTROPHILES

Substrate	Conditions	Product(s) and Yield(s) (%)	Refs.
C9 (SnMe₃, Br, EtO₂C)	Pd(PPh₃)₄ (5%), DMF, 80°, 1 h	(70–95) CO₂Et	68
C10 (SnMe₃, Br, EtO₂C)	Pd(PPh₃)₄ (5%), DMF, 80°, 1 h	(70–95) CO₂Et	68
C11 (SnMe₃, Br, CO₂Et)	Pd(PPh₃)₄ (5%), DMF, 80°, 1 h	(70–95) CO₂Et	68
C11 (SnMe₃, Br, EtO₂C, MOMO)	Pd(PPh₃)₄ (10%), DMF, Et₃N, 80°	(71) CO₂Et, MOMO	68
C11 (SnMe₃, Br, CO₂Et, MOMO)	Pd(PPh₃)₄ (10%), DMF, Et₃N, 80°	(68) CO₂Et, OMOM	68
C12 (Br, SnBu₃, MeO₂C, CO₂Me)	Pd(CH₃CN)₂Cl₂ (5%), DMF, rt, 20 h	(66) MeO₂C, CO₂Me	526
C12 (OTf, SnMe₃)	Pd(PPh₃)₄, THF	(55)	181

Substrate	Conditions	Product (Yield %)	Refs.
Bu$_3$Sn-...-O...=O, OTf (allylic stannane ester)	Pd(PPh$_3$)$_4$ (2%), LiCl, THF, reflux	lactone (57)	185
SnMe$_3$, CO$_2$Et, i-Pr, Br (diene)	Pd(PPh$_3$)$_4$, DMF	CO$_2$Et, Pr-i cyclobutene (—)	524
SnMe$_3$, CO$_2$Et, i-Pr, Br	Pd(PPh$_3$)$_4$, DMF	i-Pr CO$_2$Et (—)	524
SnMe$_3$, CO$_2$Me, OTf cyclopentene	Pd(PPh$_3$)$_4$ (5%), THF or CH$_3$CN, reflux, 11 h	CO$_2$Me bicyclic (82–85)	31, 181
O=, OTf, Br cyclohexenone + Bu$_3$SnTMS	Pd(PPh$_3$)$_2$Cl$_2$ (3%), Bu$_4$NBr, Li$_2$CO$_3$, PhMe, 110°	(38) + (43)	417
Bu$_3$Sn-...-O...=O, OTf	Pd(PPh$_3$)$_4$ (2%), LiCl, THF, reflux	lactone (56)	185
SnMe$_3$, CO$_2$Me, OTf cyclopentene (methyl)	Pd(PPh$_3$)$_4$ (5%), THF, reflux, 9 h	CO$_2$Me bicyclic (82–85)	31, 181

C$_{13}$

TABLE II. INTRAMOLECULAR CROSS-COUPLING OF ALKENYL ELECTROPHILES (*Continued*)

Substrate	Conditions	Product(s) and Yield(s) (%)	Refs.
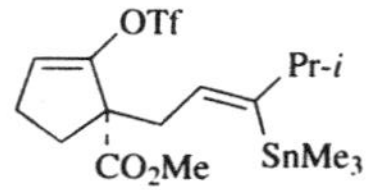	$Pd(PPh_3)_4$ (5%), CH_3CN, reflux	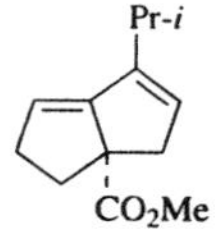 (84)	182, 184
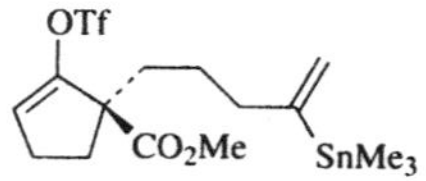	$Pd(PPh_3)_4$ (5%), THF, reflux, 23 h	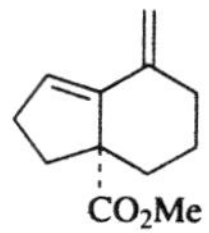 (50)	181
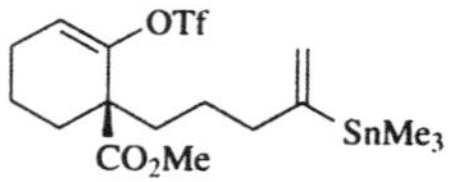	$Pd(PPh_3)_4$ (5%), THF, reflux, 3 h	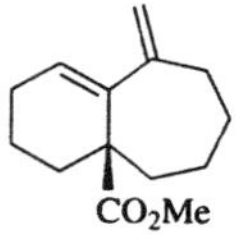 (90)	31, 181
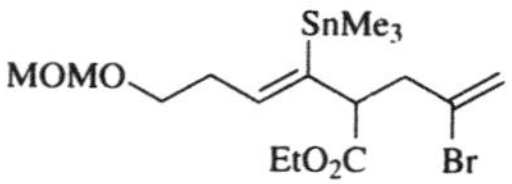	$Pd(PPh_3)_4$ (5%), DMF, 80°, 1 h	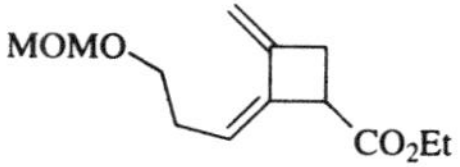 (70-95)	68
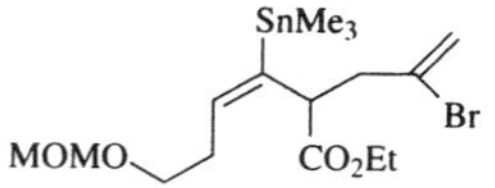	$Pd(PPh_3)_4$ (5%), DMF, 80°, 1 h	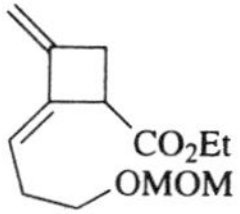 (70-95)	68
C_{14}			
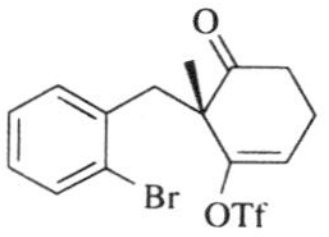 + Bu_3SnTMS	$Pd(PPh_3)_2Cl_2$ (3%), Bu_4NBr, Li_2CO_3, PhMe, 110°, 2 h	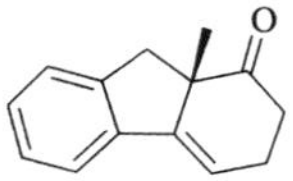 (50)	417
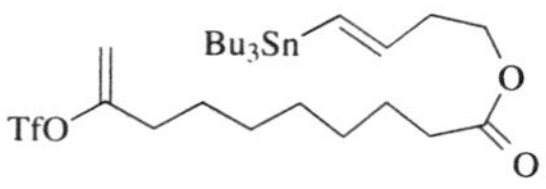	$Pd(PPh_3)_4$ (2%), LiCl, THF, reflux	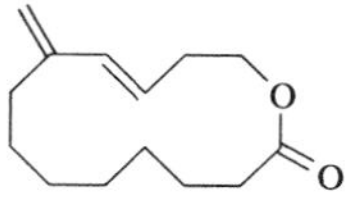 (57)	185

	Pd(PPh$_3$)$_4$ (5%), CH$_3$CN, reflux	(85)	182
	Pd(OAc)$_2$ (5%), PPh$_3$ (10%), Et$_3$N, CH$_3$CN, reflux, 2.5-4 h	(55)	69
	Pd(PPh$_3$)$_4$ (5%), CH$_3$CN, reflux	(83)	182, 184
	Pd(PPh$_3$)$_4$ (5%), THF, reflux, 19 h	(85-86)	31, 181
	Pd(PPh$_3$)$_4$ (5%), THF, reflux, 3 h	(85)	31, 181
	Pd(PPh$_3$)$_4$ (2%), LiCl, THF, reflux	(56)	185
C$_{15}$	Pd(PPh$_3$)$_4$, DMF	(—)	527

147

TABLE II. INTRAMOLECULAR CROSS-COUPLING OF ALKENYL ELECTROPHILES (*Continued*)

	Substrate	Conditions	Product(s) and Yield(s) (%)	Refs.
	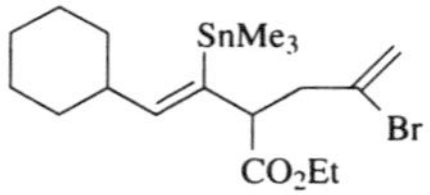	$Pd(PPh_3)_4$, DMF	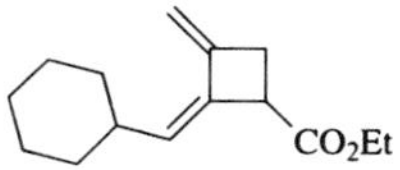 (—)	527
	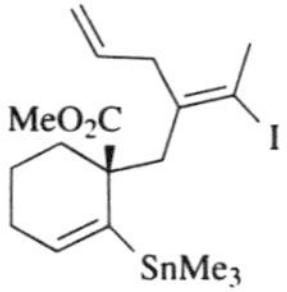	$Pd(OAc)_2$ (5%), PPh_3 (10%), Et_3N, CH_3CN, reflux, 2.5-4 h	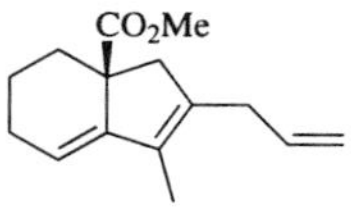 (>80)	69
	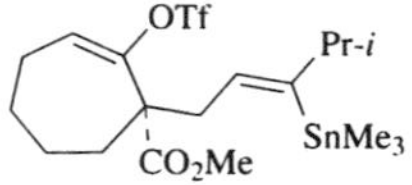	$Pd(PPh_3)_4$ (5%), CH_3CN, reflux	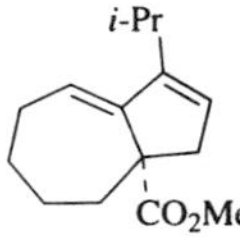 (81)	182, 184
	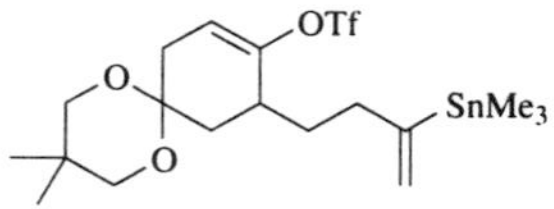	$Pd(PPh_3)_4$ (5%), THF, rt, 15 min	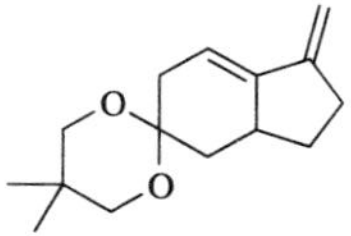 (85)	31, 181
C_{16}	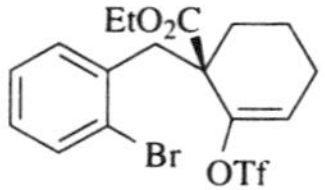 + Bu_3SnTMS	$Pd(PPh_3)_2Cl_2$ (3%), Bu_4NBr, Li_2CO_3, PhMe, 110°, 1.5 h	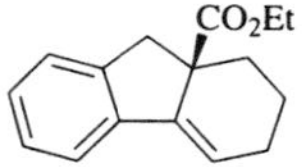 (70)	417
	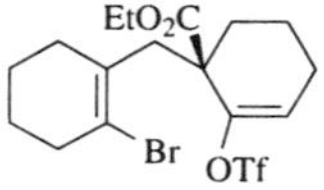 + Bu_3SnTMS	$Pd(PPh_3)_2Cl_2$ (3%), Bu_4NBr, Li_2CO_3, PhMe, 110°, sieves, 8 h	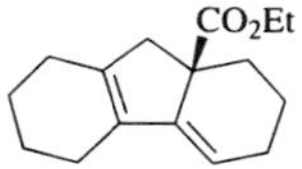 (70) + 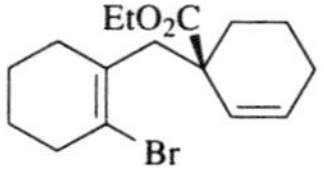 (13)	417
	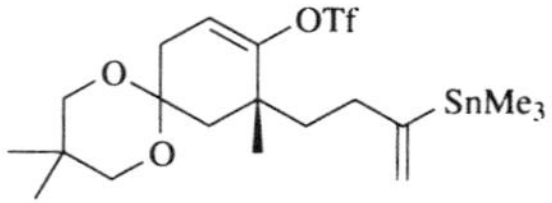	$Pd(PPh_3)_4$ (5%), THF, rt, 15 min	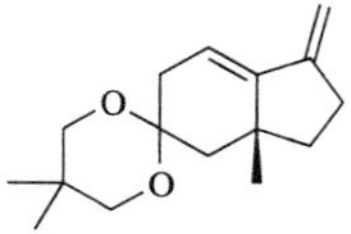 (83)	31, 181

Substrate	Conditions	Product(s) (Yield %)	Refs.
	Pd(PPh$_3$)$_4$ (5%), THF, rt, 15 min	(84)	31, 181
	Pd(OAc)$_2$ (5%), PPh$_3$ (10%), Et$_3$N, CH$_3$CN, reflux, 2.5–4 h	(>80)	69
+ Bu$_3$SnSiMe$_3$	Pd(PPh$_3$)$_2$Cl$_2$ (3%), Bu$_4$NBr, Li$_2$CO$_3$, PhMe, 100°, 3.5 h	(10) + (43)	417
	Pd(PPh$_3$)$_4$ (5%), THF, reflux, 15 min	(86)	31, 181
C$_{18}$	Pd(PPh$_3$)$_4$ (5%), DMF, 80°, 1 h	(70–95)	68
	Pd(PPh$_3$)$_4$ (5%), DMF, 80°, 1 h	(70–95)	68
C$_{19}$	Pd(PPh$_3$)$_4$, 30°, 5 min	(81)	182, 184

C$_{17}$

TABLE II. INTRAMOLECULAR CROSS-COUPLING OF ALKENYL ELECTROPHILES (*Continued*)

Substrate	Conditions	Product(s) and Yield(s) (%)	Refs.
C₂₀			
	Pd(PPh₃)₂Cl₂ (3%), Bu₄NBr, Li₂CO₃, PhMe, 110°, 1.5 h	(61)	417
	Pd(PPh₃)₄ (3%), THF, reflux	(61)	179
	Pd(PPh₃)₄ (3%), THF, reflux	(65)	179
	Pd(PPh₃)₄ (5%), THF, 50°, 86 h	(72)	71
	Pd(PPh₃)₄, THF, 50°	(32)	70
C₂₁			
	Pd(PPh₃)₄ (7%), THF, reflux	(>86)	180, 528

150

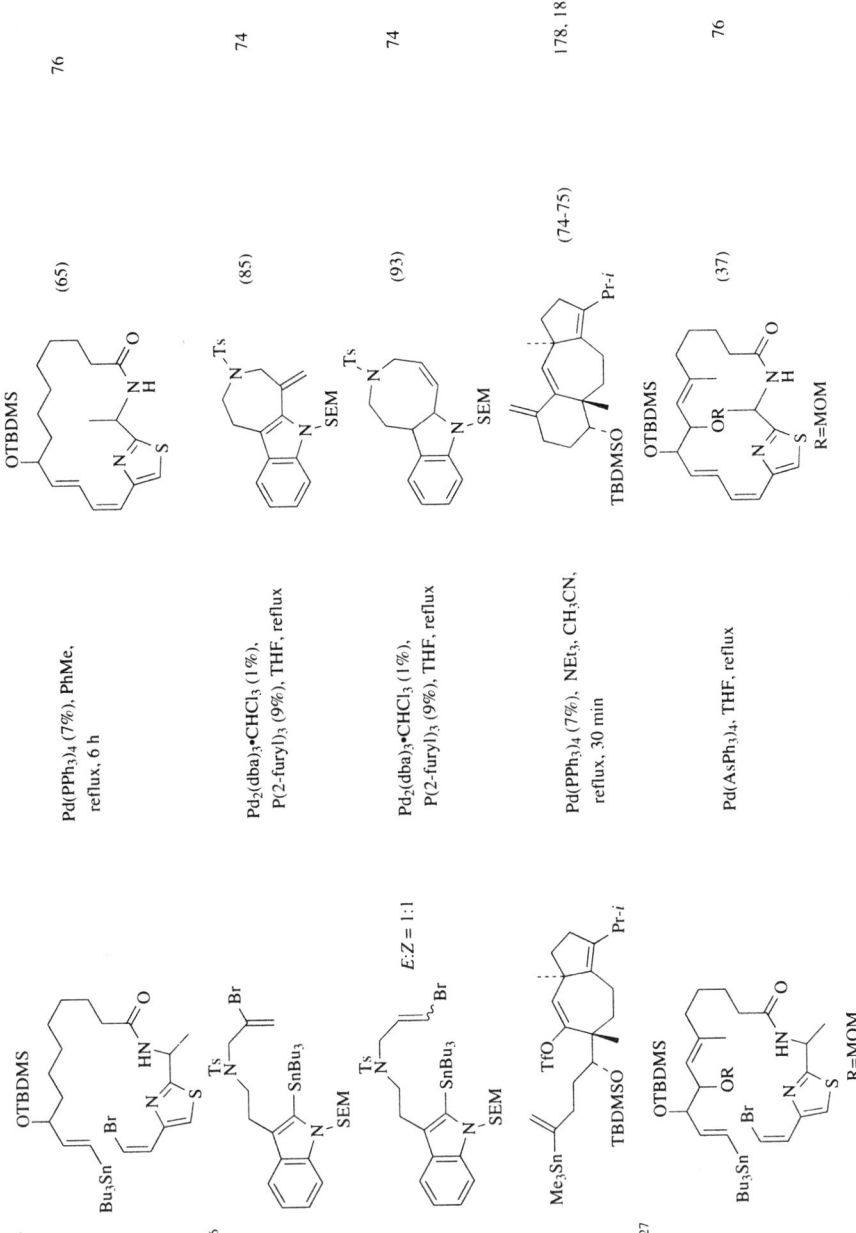

TABLE II. INTRAMOLECULAR CROSS-COUPLING OF ALKENYL ELECTROPHILES (*Continued*)

Substrate	Conditions	Product(s) and Yield(s) (%)	Refs.
	Pd(AsPh₃)₄, PhMe, 100°	(39)	75
	Pd(PPh₃)₄, THF, 60°	(60)	72
C₂₉	Pd(PPh₃)₄, 60°	(62)	73
C₄₉	Pd(CH₃CN)₂Cl₂, (*i*-Pr)₂NEt, DMF, THF, rt	(28)	86

152

TABLE III. DIRECT CROSS-COUPLING OF ARYL ELECTROPHILES

Substrate	Stannane	Conditions	Product(s) and Yield(s) (%)	Refs.
C_6 PhBr	Me₄Sn	Pd(PPh₃)₄ (0.7%), air, HMPA, 65°	PhMe (89)	19
	Bu₃Sn⌒OH	Pd(PPh₃)₄ (5%), dioxane, 80°	Ph⌒OH (60)	233
	Bu₃Sn⌒CN	Pd[P(o-Tol)₃]₂Cl₂ (1%), m-xylene, 120°, 3 h	Ph⌒CN (72)	235
	Bu₃Sn⌒OMe	Pd(PPh₃)₂Cl₂ (1%), HMPA, 80°, 20 h	Ph⌒OMe (76)	234
	Bu₃Sn⌒⫝ (allyl)	Pd(PPh₃)₄ (1%), C₆H₆, 100°, 20 h	Ph⌒⫝ (96)	3, 29
	Bu₃Sn⌒[OSnBu₃]	Pd(PPh₃)₄ (10%), C₆H₆, 100°, 20 h	Ph-C(O)-Et (19) + PhSnBu₃ (15)	529
	(isoxazole)–Bu₃Sn (Me-isoxazole-Ph)	Pd[P(o-Tol)₃]₂Cl₂ (1%), PhMe, 100°, 5 h	(Me-isoxazole)–Ph (78)	237, 238, 240
	Me-imidazole–Bu₃Sn	Pd(PPh₃)₂Cl₂ (5%), dioxane, reflux, 25 h	(imidazole)–Ph (72)	292, 530
	Bu₃Sn–(imidazole)	Pd(PPh₃)₂Cl₂ (1%), xylene, 120°, 20 h	Ph–(imidazole)–Ph (89)	531
	Bu₃Sn–(furan)	Pd(PPh₃)₂Cl₂ (4%), DMF, 70°, 1.5 h	Ph–(furan)–C(O) (67)	287
	[OSnBu₃]	Pd[P(o-Tol)₃]₂Cl₂ (1%), PhMe, 100°, 5 h	Ph-CH₂-C(O)-Et (67)	238, 240

153

TABLE III. DIRECT CROSS-COUPLING OF ARYL ELECTROPHILES (*Continued*)

Substrate	Stannane	Conditions	Product(s) and Yield(s) (%)	Refs.
	$\begin{bmatrix} \text{OSnBu}_3 \end{bmatrix}$	$Pd[P(o\text{-Tol})_3]_2Cl_2$ (1%), PhMe, 100°, 5 h	Ph (60)	238, 240
	Bu_3Sn OMe	$BnPd(PPh_3)_2Cl$, C_6H_6, 100°, 20 h	Ph OMe (81)	305, 532
	Bu_3Sn OEt	$Pd(PPh_3)_2Cl_2$ (2%), PhMe, 105°, 48 h	Ph OEt (71)	270
	Bu_3Sn OEt	1. $Pd(PPh_3)_2Cl_2$ (1%), PhMe, 100°, 20 h 2. H^+	Ph (80)	269
	Bu_3Sn OEt	$Pd(PPh_3)_2Cl_2$ (10%), Et_4NCl, DMF, 80°, 2 h	Ph OEt (78)	272, 273
	Bu_3Sn CO_2Et	$Pd[P(o\text{-Tol})_3]_2Cl_2$ (1%), $ZnBr_2$, DMF, 80°, 5 h	Ph CO_2Et (71)	236
	Bu_3Sn CO_2Me	$[(\eta^3\text{-}C_3H_5)PdOAc]_2$ (5%), DIOP (10%), TlOAc, THF, reflux, 20 h	Ph CO_2Me (72) (S) 40% ee	533
	Bu_3Sn CO_2Me	$[(\eta^3\text{-}C_3H_5)PdOAc]_2$ (5%), BPPM (10%), TlOAc, THF, 40°, 6 h	Ph CO_2Me (36) (R) 42% ee	533
	pyridine–N, Bu_3Sn	$Pd(dppb)Cl_2$, CuO, DMF, 100°, 80-90 min	pyridine Ph (82)	96
	oxazoline Bu_3Sn	$Pd(PPh_3)_2Cl_2$ (1%)	oxazoline Ph (70)	426
	Bu_3Sn OEt	$Pd(PPh_3)_4$ (1%), C_6H_6, 110°, 15 h	Ph OEt (83) E:Z = 95:5	534

Organotin reagent	Conditions	Product (%)	Refs.
[OSnBu$_3$] (enol stannane)	Pd[P(o-Tol)$_3$]$_2$Cl$_2$ (3%), C$_6$H$_6$, reflux, 7 h	(ethyl ketone, O, Ph) (0)	241
[OSnBu$_3$, Pr-i]	Pd[P(o-Tol)$_3$]$_2$Cl$_2$ (1%), PhMe, 100°, 5 h	(O, Ph, Pr-i) (87)	238, 240
[OSnBu$_3$]	Pd[P(o-Tol)$_3$]$_2$Cl$_2$ (1%), PhMe, 100°, 5 h	(O, Ph) (33)	238, 240
Me$_3$Sn—OTMS	Pd(PPh$_3$)$_4$, PhMe, 100°	Ph OTMS (74)	457
Bu$_3$SnPh	D$_{717}$-Pd(0) (polymer-supported), Me$_2$CO, reflux, 25 h	Ph-Ph (43)	535
Ph$_4$Sn	Pd(PPh$_3$)$_4$ (0.7%), air, HMPA, 65°	Ph-Ph (78)	19
Bu$_3$Sn—N (piperidide)	Pd(PPh$_3$)$_4$ (5%), THF, 65°, 4 h	Ph, N-piperidide (33)	437
[OSnBu$_3$ (cyclohexenyl)]	Pd[P(o-Tol)$_3$]$_2$Cl$_2$ (1%), PhMe, 100°, 5 h	(cyclohexanone, Ph) (54)	238, 240, 241
Bu$_3$Sn, OEt	Pd(PPh$_3$)$_4$ (1%), C$_6$H$_6$, 110°, 15 h	Ph, OEt (69)	534
[Bu$_3$SnO (dienol)]	Pd[P(o-Tol)$_3$]$_2$Cl$_2$ (3%), C$_6$H$_6$, reflux, 3 h	(O, Ph) (56)	241

TABLE III. DIRECT CROSS-COUPLING OF ARYL ELECTROPHILES (*Continued*)

Substrate	Stannane	Conditions	Product(s) and Yield(s) (%)	Refs.
	[CH₂=C(OSnBu₃)Bu-t]	Pd[P(o-Tol)₃]₂Cl₂ (1%), PhMe, 100°, 5 h	Ph–C(=O)–CH₂–Bu-t (86)	238, 240, 241
	[CH₂=C(OSnBu₃)Bu-s]	Pd[P(o-Tol)₃]₂Cl₂ (3%), C₆H₆, reflux, 10 h	Ph–C(=O)–CH₂–Bu-s (47)	241
	Me₃Sn–(dithiolylidene-dithiole)	Pd(PPh₃)₄, PhMe, reflux, 5 h	Ph–(dithiolylidene-dithiole) (59)	536
	Me₃Sn–CH₂–C(=CH₂)–CH₂–TMS	BnPd(PPh₃)₂Cl (1–2%), CHCl₃, 65°, 1 d	Ph–CH₂–C(=CH₂)–CH₂–TMS (68)	537
	Me₃Sn–C₆H₄–R	Pd(PPh₃)₄ (1%), C₆H₆, 120°, 20 h	Ph–C₆H₄–R; R = o-Me (tr); R = m-Me (75); R = p-Me (73)	538
	Bu₃Sn–(2-benzazolyl, X/N)	Pd(PPh₃)₂Cl₂ (1%), xylene, 120°, 20 h	Ph–(2-benzazolyl); X = O (75); X = S (56)	531
	Bu₃Sn–CH=CH–OEt–CMe₃	Pd(PPh₃)₄ (1%), C₆H₆, 110°, 15 h	Ph–CH=C(OEt)–CMe₃ (72) E:Z = 85:15	534
	[CH₂=C(OSnBu₃)–CH₂CH₂CH₂CH(CH₃)₂]	Pd[P(o-Tol)₃]₂Cl₂ (3%), C₆H₆, reflux, 3 h	Ph–CH₂–C(=O)–CH₂CH₂CH₂CH(CH₃)₂ (84)	241
	Bu₃Sn–CH=CH–CH(OMe)–TMS	Pd(PPh₃)₄, C₆H₆, 110°	Ph–CH=CH–CH(OMe)–TMS (76)	305, 529
	Bu₃Sn–CH₂–C=C–CO₂Et	Pd(PPh₃)₄ (5%), C₆H₆, reflux, 21 h	Ph–CH₂–C=C–CO₂Et (55) E:Z = 1:4	306, 427
	[CH₂=C(OSnBu₃)Ph]	Pd[P(o-Tol)₃]₂Cl₂ (1%), PhMe, 100°, 5 h	Ph–C(=O)–CH₂–Ph (90)	238, 240, 241

Reactant	Conditions	Product (yield)	Refs.
Bu_3Sn ... $CH(OEt)_2$, $E:Z = 85:15$	Pd(PPh₃)₄ (2%), C₆H₆, 80°, 20 h	Ph ... CH(OEt)₂ (75), $E:Z = 85:15$	539
Bu₃Sn ... TMS	Pd(PPh₃)₄ (5%), THF, 66°, 24 h	(63)	422
Bu₃Sn ... NBu-t, n-Bu	Pd(PPh₃)₂Cl₂ (1%), xylene, 120°, 20 h	Ph ... NBu-t, n-Bu (0)	531
[OSnBu₃ ... C₇H₁₅-n]	Pd[P(o-Tol)₃]₂Cl₂ (3%), C₆H₆, reflux, 4 h	O ... C₇H₁₅-n (65)	241
Bu₃Sn ... N(TMS)₂	1. Pd(PPh₃)₄ (2%), PhMe, reflux, 72 h 2. H₃O⁺	Ph ... NH₂ (72)	462, 464, 540
[OSnBu₃ ...]	Pd[P(o-Tol)₃]₂Cl₂, PhMe, 100°, 4 h	(61) + (2)	541
Bu₃Sn ... MeO—Cr(CO)₃	Pd(PPh₃)₄, THF, reflux, 4 h	(0)	542
MeO₂C ... Bu₃Sn ... N-H	Pd(PPh₃)₄, NEt₃, CH₃CN, 100°, 5 h	MeO₂C ... Ph, N-H (≥82)	290
Bu₃Sn ... Ph, O	Pd(PPh₃)₄, THF	Ph ... O (65)	543

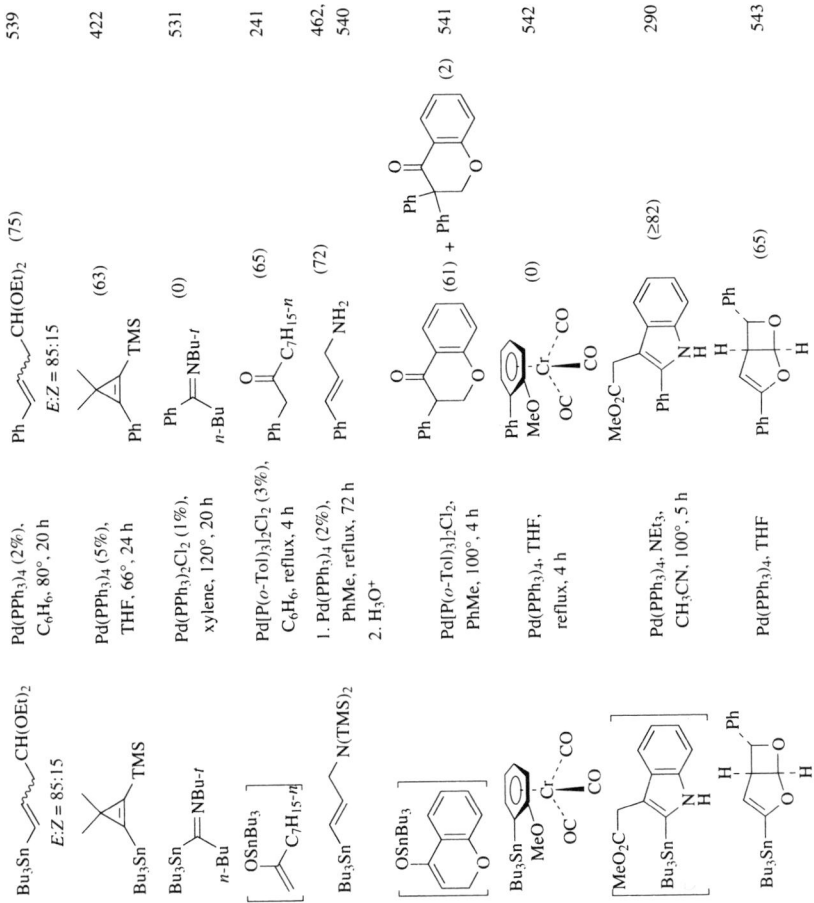

TABLE III. DIRECT CROSS-COUPLING OF ARYL ELECTROPHILES (Continued)

Substrate	Stannane	Conditions	Product(s) and Yield(s) (%)	Refs.
	Me_3Sn–[furan, OH, TBDMS]	$Pd(PPh_3)_4$ (2%), PhMe, reflux, 1 h	[furan, CH₂OH, TBDMS, Ph] (90)	371
	Bu_3Sn–[furan, OH, TBDMS]	$Pd(PPh_3)_4$ (2%), PhMe, reflux, 18 h	[furan, CH₂OH, TBDMS, Ph] (51)	371
	Me_3Sn–[TMS, Ph vinyl]	$BnPd(PPh_3)_2Cl$ (1%), 200 h	Ph—TMS, Ph (79)	256
	Bu_3Sn–[NPh imine]	$Pd(PPh_3)_2Cl_2$ (1%), xylene, 120°, 20 h	Ph, Ph, NPh (0)	531
	[indole] Bu_3Sn, $n\text{-}C_5H_{11}$, Ph	$Pd(PPh_3)_4$, NEt_3, CH_3CN, 100°, 5 h	[indole, $n\text{-}C_5H_{11}$, Ph, NH] (≥65)	290
	[indole] THPO Bu_3Sn, Ph	$Pd(PPh_3)_4$, NEt_3, CH_3CN, 100°, 7 h	[indole, THPO, Ph, NH] (≥63)	290
	Bu_3Sn—[$CO_2Bu\text{-}t$, dioxolane]	$Pd(PPh_3)_4$	[$CO_2Bu\text{-}t$, dioxolane, Ph] (84)	544
	$SnBu_3$—[pyridazine, Ph, Ph, Ph]	$Pd(PPh_3)_2Cl_2$ (5%), Et_4NCl, DMF, 80°	[pyridazine, Ph, Ph, Ph] (26)	545

Reactant	Conditions	Product(s) (%)	Refs.
(chromene OSnBu₃, OBn)	Pd[P(o-Tol)₃]₂Cl₂, PhMe, 100°, 22 h	(chromanone, Ph) (49) + (chromanone, Ph, Ph) (3) + (OBn)	541
Bu₃Sn—(sugar, OTBDMS)	Pd(PPh₃)₄, THF, reflux	(OTBDMS) (70); (OBn phenol, Ph) (6)	299, 300
Bu₃Sn—(sugar, OBn)	Pd(PPh₃)₄ (1%), PhMe, 100°, 3 h	(OBn) (88)	423
Bu₃Sn—(furan, SnBu₃)	Pd(PPh₃)₄ (5%), HMPA, 65°, 10 h	(Ph, furan, Ph) (54)	287, 546
Bu₃Sn—N=C(OMe)CCl₃	Pd(PPh₃)₂Cl₂ (1%), xylene, 120°, 20 h	Ph—N=C(OMe)CCl₃ (0)	531
Bu₃Sn—(pyrrolidine)	Pd[P(o-Tol)₃]₂Cl₂ (1%), PhMe, 100°, 3 h	Ph—N(pyrrolidine) (3)	316
Bu₃SnNHBu-n	"	PhNHBu-n (12)	90, 316
Bu₃SnNHBu-t	"	PhNHBu-t (19)	90, 316
Bu₃SnNEt₂	"	PhNEt₂ (81)	90, 316
Bu₃Sn—(morpholine)	"	Ph—N(morpholine) (61)	316

159

TABLE III. DIRECT CROSS-COUPLING OF ARYL ELECTROPHILES (*Continued*)

Substrate	Stannane	Conditions	Product(s) and Yield(s) (%)	Refs.
	Bu₃Sn–[succinimide N-Ph]	Pd[P(o-Tol₃)₂]Cl₂ (1%), PhMe, 100°, 3 h	[succinimide N-Ph] (0)	316
	Bu₃Sn–[piperidine N-Ph]	"	[piperidine N-Ph] (37)	316
	Bu₃SnNHPh	"	Ph₂NH (64)	316
	Bu₃SnN(Pr-i)₂	"	PhN(Pr-i)₂ (0)	316
	Bu₃SnN(TMS)₂	"	PhN(TMS)₂ (0)	316
	Bu₃Sn–N(Et)(Ph)–Ph	"	Ph–N(Et)(Ph)–Ph (43)	316
	Bu₃Sn–N=C(NEt₂)–Ph	Pd(PPh₃)₂Cl₂ (1%), xylene, 120°, 20 h	Ph–N=C(NEt₂)–Ph (0)	531
	Bu₃Sn–N=C=N–SnBu₃	"	Ph–N=C=N–Ph (0)	531
	Bu₃SnSBu-n	Pd(PPh₃)₄ (1%), PhMe, 120°, 20 h	PhSBu-n (86)	318
	Bu₃SnSPh	"	PhSPh (83)	318
	Bu₃SnSSnBu₃	"	PhSPh (75)	318
	Me₃SnTMS	Pd(PPh₃)₂Br₂ (1.3%), PhMe, 115°, 15 h	PhTMS (—)	547
	Bu₃SnSnBu₃	Pd(PPh₃)₄ (1.3%), PhMe, 115°, 15 h	PhSnBu₃ (79)	547, 548
PhCl	Bu₃Sn–CH₂CH=CH₂	Pd(PPh₃)₄ (1%), C₆H₆, 120°, 20 h	Ph–CH₂CH=CH₂ (0)	3

PhI	Bu₃Sn–(2-pyridyl)	Pd(dppb)Cl₂, CuO, DMF, 100°, 24 h	2-Ph-pyridine (tr)	96

Substrate	Reagent	Conditions	Product (Yield %)	Refs.
PhI	Bu₃Sn–[pyridyl]	Pd(dppb)Cl₂, CuO, DMF, 100°, 24 h	[2-Ph-pyridine] (tr)	96
	Bu₃Sn–[cyclopropenyl-TMS]	Pd(PPh₃)₄ (5%), THF, 66°, 24 h	[Ph,TMS-cyclopropene] (0)	422
	Me₃SnCN	Pd(PPh₃)₄	PhCN (—)	549
	Bu₃SnC≡CH	Pd₂(dba)₃ (4%), AsPh₃ (16%), CuI (8%), DMF, 60°, 6 h	PhC≡CH (58)	33
	Bu₃Sn–[allyl]	Pd₂(dba)₃ (2.5%), PPh₃ (10%), CuI (10%), dioxane, 50°, 72 h	Ph–[allyl] (>95)	33
	Bu₃Sn–[allyl]	Pd₂(dba)₃ (1%), P(2-furyl)₃ (8%), THF, 50°, 72 h	Ph–[allyl] (>95)	11
	Bu₃Sn–[allyl]	Pd₂(dba)₃ (1%), AsPh₃ (8%), THF, 50°, 72 h	Ph–[allyl] (>95)	11
	Bu₃Sn–[allyl]	Pd(Ph-BIAN)(dimethyl fumarate) (2%), DMF, 50°, 16 h	Ph–[allyl] (49)	415
	Me₃SnCF=CF₂	PhPd(PPh₃)₂I, HMPA or DMF, 50-70°, 3 h	PhCF=CF₂ (85-87)	265, 266
	Bu₃Sn–[allyl]	Pd(PPh₃)₄ (2%), C₆H₆, 80°	Ph–[allyl] (32)	3
	Bu₃Sn–[allyl]	Pd(PPh₃)₄ (10%), LiCl, DMF, 90°, 30 h	Ph–[allyl] (88)	29

TABLE III. DIRECT CROSS-COUPLING OF ARYL ELECTROPHILES (*Continued*)

Substrate	Stannane	Conditions	Product(s) and Yield(s) (%)	Refs.
	Bu$_3$Sn (allyl)	Pd$_2$(dba)$_3$ (4%), AsPh$_3$ (16%), CuI (8%), DMF, 60°, 6 h	Ph (allyl) (61)	33
	Bu$_3$Sn—C=C=	Pd$_2$(dba)$_3$•CHCl$_3$ (3%), PPh$_3$ (24%), LiCl, DMF, rt	Ph—C= (45)	275
	n-Bu$_4$Sn	Pd$_2$(dba)$_3$ (4%), AsPh$_3$ (16%), CuI (8%), DMF, 60°, 6 h	PhBu-*n* (34)	33
	Me$_3$Sn (2-methyloxazole)	Pd(PPh$_3$)$_2$Cl$_2$, THF, 7 h	Ph (oxazole) (80)	292
	Me$_3$Sn (methyloxazole)	Pd(PPh$_3$)$_4$ (5%), C$_6$H$_6$, reflux, 24 h	Ph (oxazole) (80)	550
	Bu$_3$Sn (isoxazole)	Pd(PPh$_3$)$_2$Cl$_2$ (5%), THF, reflux, 7 h	Ph (isoxazole) (82)	292, 530
	Bu$_3$Sn (pyrimidine)	Pd(PPh$_3$)$_2$Cl$_2$ (3%), Cl(CH$_2$)$_2$Cl, reflux, overnight	(pyrimidine) (43)	458
	Bu$_3$Sn (cyclobutenone)	BnPd(PPh$_3$)$_2$Cl (1.5%), CuI, CH$_3$CN, rt, 1 h	Ph (cyclobutenone) (59)	267
	Bu$_3$Sn (2-furan)	Pd$_2$(dba)$_3$ (4%), AsPh$_3$ (16%), CuI (8%), DMF, 60°, 6 h	Ph (furan) (58)	33
	Bu$_3$Sn (3-furan)	Pd(PPh$_3$)$_2$Cl$_2$ (4%), DMF, 70°, 2 h	Ph (furan) (75)	287

Bu$_3$SnC≡COEt	PhC≡COEt (60)	Pd(PPh$_3$)$_2$Cl$_2$ (5%), Et$_4$NCl, DMF, rt, 1 h	302
[Bu$_3$Sn, OEt alkene]	[Ph, OEt alkene] (92)	Pd(PPh$_3$)$_4$ (2%), dioxane, 95°, 96 h	270
[Bu$_3$Sn, OEt alkene]	[Ph, OEt alkene] (71)	Pd$_2$(dba)$_3$ (4%), AsPh$_3$ (16%), CuI (8%), DMF, 60°, 6 h	33
[Bu$_3$Sn furanone]	[Ph furanone] (65)	Pd(PPh$_3$)$_2$Cl$_2$ (1-2%), PhMe, reflux	261
[Bu$_3$Sn furanone]	[Ph furanone] (45)	Pd(PPh$_3$)$_2$Cl$_2$ (1-2%), PhMe, reflux	261
[Bu$_3$Sn, CO$_2$Me alkene]	[Ph, CO$_2$Me alkene] (87)	Pd(PPh$_3$)$_4$ (10%), CuI, DMF, rt, 48 h	246
[Bu$_3$Sn, CO$_2$Me alkene]	[Ph, CO$_2$Me alkene] + [Ph CO$_2$Me] (70) 1:254	Pd$_2$(dba)$_3$ (3%), AsPh$_3$ (12%), THF, 50°, 18 h	245
[Bu$_3$Sn, CO$_2$Me, D alkene]	[Ph, CO$_2$Me, D alkene] (70)	Pd$_2$(dba)$_3$ (3%), AsPh$_3$ (12%), THF, 50°, 18 h	245
Bu$_3$Sn—CO$_2$Et	Ph—CO$_2$Et (31)	Pd[P(o-Tol)$_3$]$_2$Cl$_2$ (1%), ZnBr$_2$, DMF, 80°, 5 h	236
[Bu$_3$Sn pyridine]	[Ph pyridine] (64)	Pd(dppb)Cl$_2$, CuO, DMF, 100° 70-80 min	96, 33
[Me$_3$Sn, N-Me pyrrole]	[Ph, N-Me pyrrole] (54)	Pd(PPh$_3$)$_2$Cl$_2$ (0.5%), THF, reflux, 20 h	286

TABLE III. DIRECT CROSS-COUPLING OF ARYL ELECTROPHILES (*Continued*)

Substrate	Stannane	Conditions	Product(s) and Yield(s) (%)	Refs.
	Me₃Sn (oxazoline)	Pd(PPh₃)₄ (5%), C₆H₆, reflux, 12 h	(100)	550, 551
	Bu₃SnC≡CCO₂Et	Pd(PPh₃)₂Cl₂ (1%), Et₄NCl, ZnCl₂, DMF, 50°, 2 h	PhC≡CCO₂Et (94)	552
	Bu₃Sn–(pyrrolidine amide)	Pd(PPh₃)₄ (5%), THF, reflux, 6 h	(58)	437
	Me₃SnPh	Pd(CH₃CN)₂Cl₂ (2%), DMF, 20°, 8 h	Ph-Ph (86)	553
	Me₃Sn–(piperidine amide)	Pd(PPh₃)₄ (5%), PhMe, reflux, 40-80 min	(72)	437
	Me₃Sn–(dithiole)	Pd(PPh₃)₄, PhMe, reflux, 3h	(67)	536
	Me₃Sn–(tolyl)	Pd(CH₃CN)₂Cl₂ (2%), DMF, 70°, 1 h	(90)	553
	Me₃Sn–(amide N(Pr-i)₂)	Pd(PPh₃)₄ (5%), PhMe, reflux, 40-80 min	(78)	437
	Bu₃Sn–(squarate)	BnPd(PPh₃)Cl (5%), CuI (7-10%), DMF, rt	(77)	12
	Bu₃Sn–(squarate i-PrO)	BnPd(PPh₃)₂Cl (5%), CuI (7-10%), DMF, rt	(99)	12

164

Organostannane	Conditions	Product (%)	Refs.
Bu₃Sn—(cyclopropene with TMS)	Pd(PPh₃)₄ (5%), THF, 66°, 24 h	(cyclopropene, Ph, TMS) (98)	422
Bu₃Sn—CH=C=CH—Ph	Pd₂(dba)₃•CHCl₃ (3%), PPh₃ (24%), DMF, rt	Ph—CH=C=CH—Ph (45)	275
Bu₃Sn—(CO₂Et, NHAc allyl)	Pd(PPh₃)₄ (4%), DMF, 100°, 6 h	Ph (CO₂Et, NHAc) **I** (42)	473
Bu₃Sn—(CO₂Et, NHAc allyl)	Pd₂(dba)₃•CHCl₃ (4%), AsPh₃ (30%), THF, reflux, 6 h	**I** (34) + **II** (32) E:Z = 64:36 **I** (34) + **II** (32) E:Z = 64:36 Ph (CO₂Et, NHAc) **II** (10) E:Z = 1:1	473
Bu₃Sn—(pyrazole N–Ph, Ph)	Pd(PPh₃)₂Cl₂ (5%), THF, reflux, 20 h	Ph (pyrazole N–Ph) (59)	302
Bu₃Sn—(pyrazole N–Ph)	Pd(PPh₃)₂Cl₂ (5%), THF, reflux, 24 h	(pyrazole N–Ph) (49)	302
Bu₃Sn—(pyrazole N–Ph)	Pd(PPh₃)₂Cl₂ (5%), THF, reflux, 24 h	Ph (pyrazole N–Ph) (0)	302
Bu₃Sn—(tetramethyl quinone)	Pd₂(dba)₃ (2.5%), air, CuI (50%), DMF, 60°, 0.5-1 h	(tetramethyl quinone) (84)	554
Bu₃Sn—(EtO₂C, Bu-n alkene)	Pd(PPh₃)₄ (10%), CuI, DMF, rt, 24 h	Ph (EtO₂C, Bu-n) (87)	555

TABLE III. DIRECT CROSS-COUPLING OF ARYL ELECTROPHILES (*Continued*)

Substrate	Stannane	Conditions	Product(s) and Yield(s) (%)	Refs.
		BnPd(PPh₃)₂Cl, CuI, DMF, rt	(77)	49
		Pd(PPh₃)₄, NEt₃, CH₃CN, 190°, 8 h	(≥68)	290
		BnPd(PPh₃)₂Cl, CuI, DMF, rt	(68)	49
		Pd(OAc)₂ (10%), PPh₃ (30%), DMF, rt, 72 h	(75)	555
		Pd(PPh₃)₂Cl₂ (4.8%), CH₂Cl₂, 80°, 12 h	(61)	459
		Pd(dba)₂ (5%), PhMe, reflux, 12 h	(78)	556
		Pd(PPh₃)₄ (0.7%), DMF, 110°, 5 h	(98)	289
		Pd₂(dba)₃ (2%), DMF, rt, 6 h	(50)	439

Bu₃Sn— (structure)	Pd(dba)₂ (5%), Bu₄NI, DMF, 80°, 16 h	(structure) **(60)**	382
Bu₃Sn—Ph, Ph	Pd(PPh₃)₂Cl₂, PPh₃, LiCl, DMF, heat	Ph—Ph **(35)**	251
MOMO ... SnBu₃ (uracil/acetonide)	Pd(PPh₃)₄ (10%), CuI (20%), DMF, 80°, 15 min	MOMO ... Ph **(88)**	170
Bu₃Sn—O—Bu-n, BnO	Pd₂(dba)₃ (2%), DMF, rt, 6 h	Ph—O—Bu-n, BnO **(89)**	439
H₂N Bu₃Sn (indole, SEM)	Pd(PPh₃)₄ (10%), DMF, 90°, 3 h	H₂N Ph (indole, SEM) **(93)**	74
SnBu₃, Ph, Ph (pyridazine)	Pd(PPh₃)₂Cl₂ (5%), Et₄NCl, DMF, 80°, 30 h	Ph, Ph, Ph (pyridazine) **(39)**	545
O=C Bu-t, N-H, F, R₃Sn, Ph	Pd(PPh₃)₄, PhMe, reflux	O=C Bu-t, N-H, F, Ph **R = Me or Bu (25-30)**	556

167

TABLE III. DIRECT CROSS-COUPLING OF ARYL ELECTROPHILES (*Continued*)

Substrate	Stannane	Conditions	Product(s) and Yield(s) (%)	Refs.
	Bu₃Sn〔furan〕SnBu₃	[(η³-C₃H₅)PdCl₂]₂ (13%), DMF, 65°, 2 h	Ph〔furan〕Ph (65)	287, 546
	〔cyclopropene〕 Bu₃Sn, SnBu₃	Pd(PPh₃)₄ (5%), THF, 60°, 16 h	Ph〔cyclopropene〕Ph (28)	422
	Et₂NOC〔benzene〕SnBu₃, CONEt₂, Bu₃Sn	Pd(0)	Et₂NOC〔benzene〕Ph, CONEt₂, Ph (low)	390
	Et₃SnSPh	PhPd(PPh₃)₂I (1%), DMSO, 100°, 1 h	Ph₂S (94)	320
	Et₃SnSSnEt₃	PhPd(PPh₃)₂I (5%), DMSO, 100°, 4 h	Ph₂S (96)	320
	Me₃SnTMS	Pd(PPh₃)₂Br₂ (1.3%), PhMe, 115°, 15 h	Ph-Ph (—)	547
	Me₃SnSnMe₃	Pd(PPh₃)₂Br₂ (1.3%), PhMe, 115°, 15 h	PhSnMe₃ (96)	547, 557
	Et₃SnSnEt₃	Pd(CH₃CN)₂Cl₂ (1%), HMPA, 20°, 2 h	PhSnEt₃ (99)	313
	Bu₃SnSnBu₃	Pd(PPh₃)₄ (1.3%), PhMe, 115°, 15 h	PhSnBu₃ (96)	547, 313
PhN₂⁺ BF₄⁻	Me₄Sn	Pd(OAc)₂ (10%), CH₃CN, rt, 2 h	PhMe (55)	204
	Bu₃Sn〔vinyl〕	Pd(dba)₂ (10%), CH₃CN, rt, 5 min	Ph〔vinyl〕 (80)	204
	Ph, Me₃Sn	Pd(dba)₂ (5%), CH₃CN, Et₂O, rt	Ph〔〕Ph (97) *E:Z* = 40:60	249

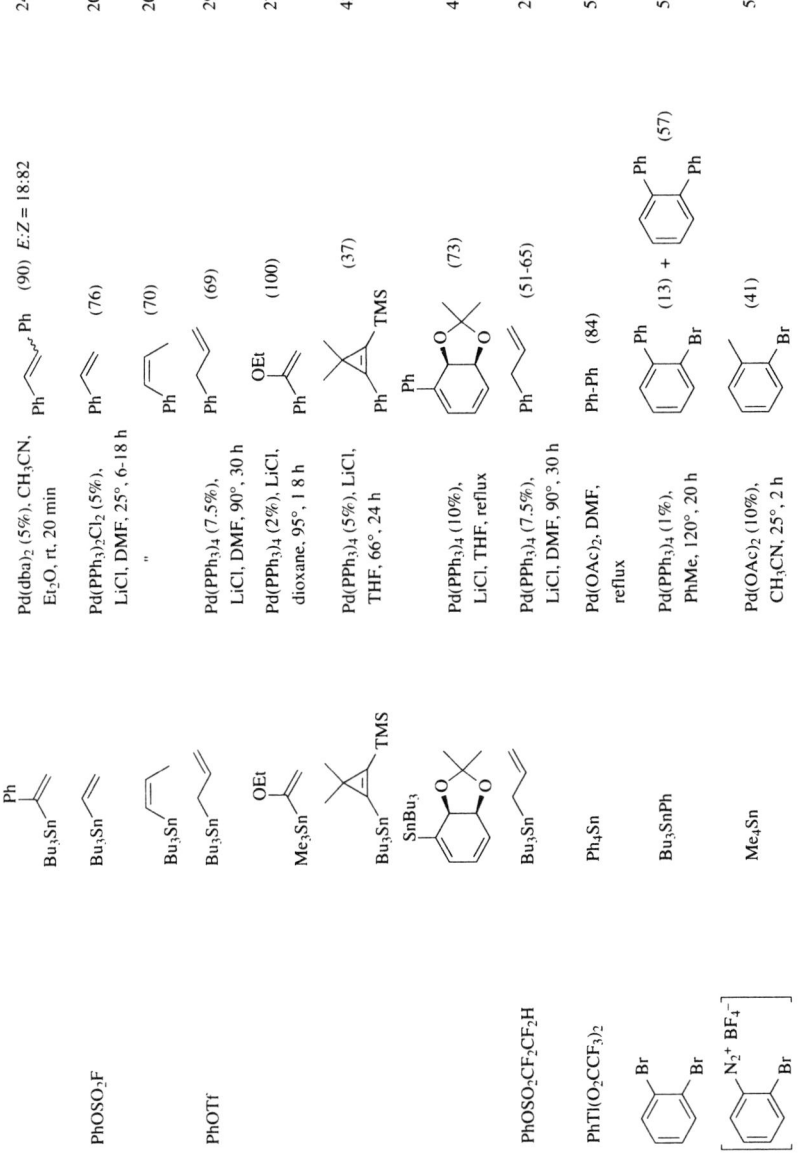

				Ref.
PhOSO₂F	Ph–C(=CH₂)SnBu₃	Pd(dba)₂ (5%), CH₃CN, Et₂O, rt, 20 min	Ph～～Ph (90) E:Z = 18:82	249
	Bu₃Sn–allyl	Pd(PPh₃)₂Cl₂ (5%), LiCl, DMF, 25°, 6-18 h	Ph (76)	203
	Bu₃Sn–(cis)	"	(70)	203
PhOTf	Bu₃Sn–(butenyl)	Pd(PPh₃)₄ (7.5%), LiCl, DMF, 90°, 30 h	Ph (69)	29, 201
	Me₃Sn–C(OEt)=CH₂	Pd(PPh₃)₄ (2%), LiCl, dioxane. 95°, 18 h	OEt Ph (100)	270
	Bu₃Sn (cyclopropene TMS)	Pd(PPh₃)₄ (5%), LiCl, THF, 66°, 24 h	TMS Ph (37)	422
	SnBu₃ (dioxole)	Pd(PPh₃)₄ (10%), LiCl, THF, reflux	(73)	475
PhOSO₂CF₂CF₂H	Bu₃Sn–allyl	Pd(PPh₃)₄ (7.5%), LiCl, DMF, 90°, 30 h	Ph (51-65)	29, 201
PhTl(O₂CCF₃)₂	Ph₄Sn	Pd(OAc)₂, DMF, reflux	Ph-Ph (84)	558
o-Br-C₆H₄-Br	Bu₃SnPh	Pd(PPh₃)₄ (1%), PhMe, 120°, 20 h	Ph Br (13) + Ph Ph (57)	538
o-Br-C₆H₄-N₂⁺ BF₄⁻	Me₄Sn	Pd(OAc)₂ (10%), CH₃CN, 25°, 2 h	Br (41)	559

TABLE III. DIRECT CROSS-COUPLING OF ARYL ELECTROPHILES (*Continued*)

Substrate	Stannane	Conditions	Product(s) and Yield(s) (%)	Refs.
(1,3-dibromobenzene)	Bu₃Sn—(glycal, Ph, OTBDMS)	Pd(PPh₃)₄ (10%), PhMe, reflux	(77)	423
(4-bromobenzene)	Bu₃Sn—CH=CH₂	Pd(PPh₃)₄ (2%), BHT, PhMe, reflux, 1 h	(63)	88
	Bu₃Sn—CH₂CH=CH₂	Pd(PPh₃)₄ (4%), BHT, PhMe, reflux, 1 h	(73)	88
	Bu₃Sn—CH=CH—OEt	Pd(PPh₃)₂Cl₂ (10%), Et₄NCl, DMF, 80°, 1.5 h	OEt (56)	272, 273
	SnMe₃—(4-pyridyl)	Pd(PPh₃)₄ (10%), PhMe, reflux	(50)	560, 561
	Bu₃SnNEt₂	Pd[P(o-Tol)₃]₂Cl₂ (1%), PhMe, 100°, 3 h	NEt₂ (30)	90, 316
(1-iodo-4-bromobenzene)	Bu₃Sn—C(=CH₂)CO₂Me	Pd(PPh₃)₄ (10%), CuI (75%), DMF, rt, 24 h	CO₂Me (92)	246
(N₂⁺ PF₆⁻ / Br)	Me₄Sn	Pd(OAc)₂ (10%), CH₃CN, rt, 2 h	(76)	204

170

Aryl halide / substrate:

4-bromophenyl fluorosulfonate (OSO₂F)

4-bromophenyl triflate (OTf)

3,5-dibromobenzene (Br, Br)

Stannane	Conditions	Product (yield)	Ref.
Bu₃Sn–CH=CH₂	Pd(PPh₃)₂Cl₂ (5%), LiCl, DMF, 25°, 6–18 h	(68)	203
Bu₃Sn–CH=CH₂	Pd(PPh₃)₄ (2%), BHT, dioxane, 98°, 2.5 h	(77)	189
Bu₃Sn–CH=CH₂	Pd(PPh₃)₄ (2%), LiCl, BHT, dioxane, 98°, 7 h	I + II, (75) I:II = 6:1	189
Bu₃Sn–CH=CH₂	Pd(PPh₃)₂Cl₂ (2%), LiCl, BHT, DMF, 70°, 3 h	I + II, (45) I:II = 1:5	189
Me₃Sn– (2,6-difluorophenyl)	Pd(PPh₃)₂Cl₂ (10–15%), PPh₃ (10%), LiCl, CuBr, BHT, dioxane, reflux	(66)	191
Bu₃Sn– (4-methylphenyl)	Pd(PPh₃)₂Cl₂ (3%), PPh₃ (10%), CuBr, BHT, dioxane, reflux	(89)	191
Bu₃Sn– (4-methylphenyl)	Pd(PPh₃)₄ (20%), LiCl, CuBr (20%), BHT, dioxane, reflux	(47)	191
Me₃Sn– (2,6-dimethoxy-4-methylphenyl)	"	(11)	191
Me₃Sn–(2-pyridyl)	Pd(PPh₃)₂Cl₂ (5%), THF, reflux, 23 h	(35)	562

I + II, (75) I:II = 6:1

I + II, (45) I:II = 1:5

TABLE III. DIRECT CROSS-COUPLING OF ARYL ELECTROPHILES (*Continued*)

Substrate	Stannane	Conditions	Product(s) and Yield(s) (%)	Refs.
Br, NH$_2$, Br (2,6-dibromoaniline)	Bu$_3$Sn–CH=CH–OEt	1. Pd(PPh$_3$)$_2$Cl$_2$ (5%), Et$_4$NCl, CH$_3$CN, reflux, 4 h 2. H$^+$, reflux, 3 h	indole with Br (N–H) (96)	553
Br, Cl	Bu$_3$Sn–CH$_2$OH	Pd(PPh$_3$)$_4$ (5%), dioxane, 80°	CH$_2$OH with Cl (71)	233
	Bu$_3$Sn–CH$_2$CN	Pd[P(o-Tol)$_3$]$_2$Cl$_2$ (1%), m-xylene, 120°, 24 h	CH$_2$CN with Cl (66)	235
	Bu$_3$Sn–CH$_2$OMe	Pd(PPh$_3$)$_2$Cl$_2$ (5%), HMPA, 80°, 20 h	CH$_2$OMe with Cl (61)	234
	[OSnBu$_3$ isopropenyl]	Pd[P(o-Tol)$_3$]$_2$Cl$_2$ (1%), PhMe, 100°, 5 h	CH$_2$COCH$_3$ with Cl (80)	240
	Bu$_3$Sn–CH$_2$CO$_2$Et	Pd[P(o-Tol)$_3$]$_2$Cl$_2$ (1%), ZnBr$_2$, DMF, 80°, 5 h	CH$_2$CO$_2$Et with Cl (66)	236
	(2-BocHN-phenyl)–Bu$_3$Sn	Pd(PPh$_3$)$_2$Cl$_2$ (5%), DMF, 90°, 25 h	biphenyl, NHBoc, Cl (76)	564
	Bu$_3$SnSPh	Pd(PPh$_3$)$_4$ (1%), PhMe, 120°, 20 h	SPh with Cl (73)	318
N$_2$$^+$ BF$_4$$^-$, Cl	Me$_4$Sn	Pd(OAc)$_2$ (10%), CH$_3$CN, rt, 2 h	CH$_3$ with Cl (64)	204

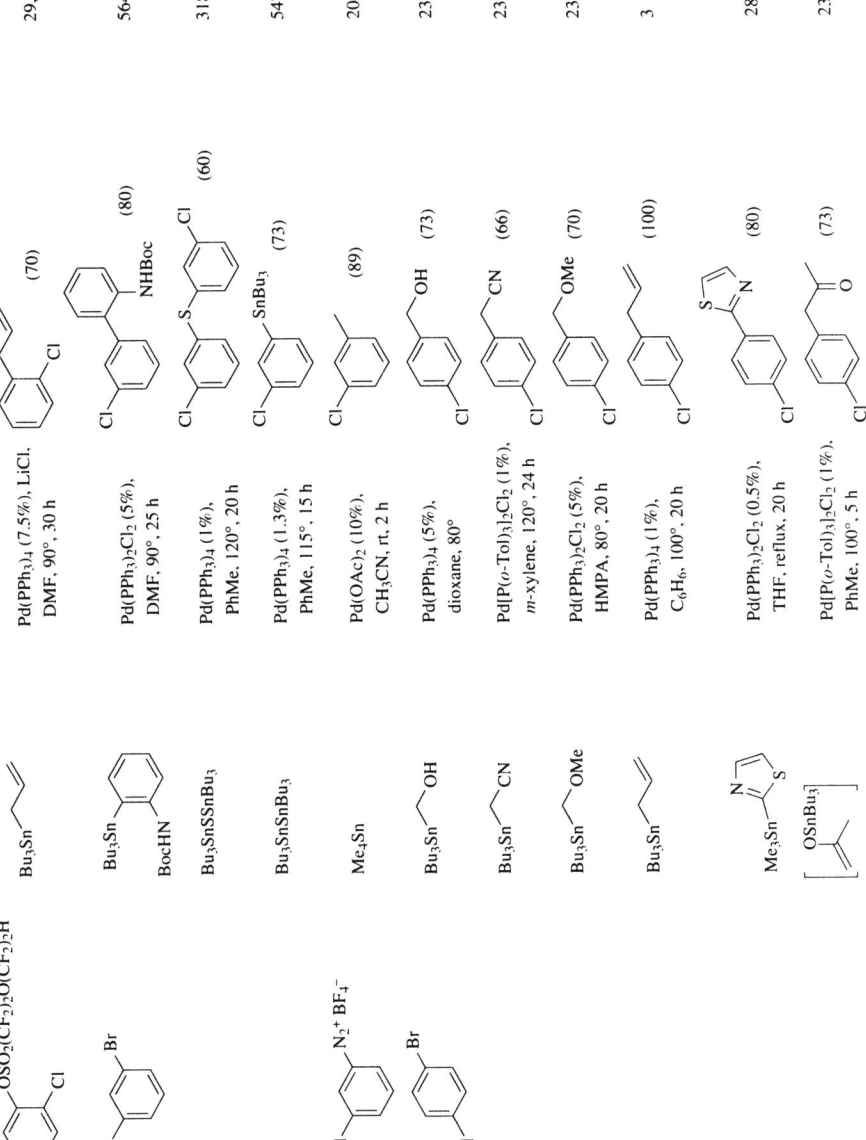

TABLE III. DIRECT CROSS-COUPLING OF ARYL ELECTROPHILES (*Continued*)

Substrate	Stannane	Conditions	Product(s) and Yield(s) (%)	Refs.
	Bu₃Sn⌇OMe	BnPd(PPh₃)₂Cl, C₆H₆, 100°, 20 h	(aryl)⌇OMe, Cl (79)	305
	OEt Me₃Sn	1. Pd(PPh₃)₂Cl₂ (1%), PhMe, 100°, 20 h 2. H⁺	(73)	269
	Bu₃Sn⌇CO₂Et	Pd[P(o-Tol)₃]₂Cl₂ (1%), ZnBr₂, DMF, 80°, 5 h	(aryl)⌇CO₂Et, Cl (89)	236
	Bu₃Sn / OEt	Pd(PPh₃)₄ (1%), C₆H₆, 110°, 15 h	OEt, Cl (76) E:Z = 65:35	534
	Bu₃Sn—C₆H₄—OMe	Pd(PPh₃)₄ (1%), PhMe, 120°, 20 h	MeO—C₆H₄—C₆H₄—Cl (66)	538
	TMS Bu₃Sn⌇OMe	Pd(PPh₃)₄, C₆H₆, 110°	TMS, OMe, Cl (72)	305, 532
	Bu₃Sn—furan—OBu-t	BnPd(PPh₃)₂Cl, DMF, 70°, 16 h	OBu-t furan, Cl (94)	296
	Me₃Sn—furan(CH₂OH)—TBDMS	Pd(PPh₃)₄ (2%), PhMe, reflux, 1 h	OH furan, TBDMS, Cl (88)	371

174

Stannane	Product (yield %)	Conditions	Refs.
Bu₃Sn–(N-TIPS-pyrrole)	4-Cl-C₆H₄–(N-TIPS-pyrrole) (89)	Pd(PPh₃)₄ (17%), dioxane, reflux. 40 h	565
Bu₃Sn–glycal(OTBDMS)₃	4-Cl-C₆H₄–glycal(OTBDMS)₃ (49)	Pd(PPh₃)₂Cl₂, PhMe	299, 300
Bu₃SnNEt₂	4-Cl-C₆H₄–NEt₂ (55)	Pd[P(o-Tol)₃]₂Cl₂ (1%), PhMe, 100°, 3 h	90, 316
Bu₃SnSPh	4-Cl-C₆H₄–SPh (74)	Pd(PPh₃)₄ (1%), PhMe, 120°, 20 h	318
Bu₃SnSnBu₃	4-Cl-C₆H₄–S–C₆H₄–Cl (86)	Pd(PPh₃)₄ (1%), PhMe, 120°, 20 h	318
Me₃SnTMS	4-Cl-C₆H₄–TMS (34)	Pd(PPh₃)₄ (1.3%), PhMe, 115°, 15 h	547
Bu₃SnSnBu₃	4-Cl-C₆H₄–SnBu₃ (59)	Pd(PPh₃)₄ (1.3%), PhMe, 115°, 15 h	547
Bu₃Sn–CH₂CH=CH₂	4-Cl-C₆H₄–CH₂CH=CH₂ (4)	Pd(PPh₃)₄ (1%), C₆H₆, 120°, 20 h	3
Me₃Sn–CH=CH₂	4-Cl-C₆H₄–CH=CH₂ (94)	Pd(CH₃CN)₂Cl₂ (2%), DMF, 20°, 10 min	553

TABLE III. DIRECT CROSS-COUPLING OF ARYL ELECTROPHILES (Continued)

Substrate	Stannane	Conditions	Product(s) and Yield(s) (%)	Refs.
4-Cl-C₆H₄-OSO₂(CF₂)₂O(CF₂)₂H	Bu₃Sn⌒	Pd(PPh₃)₄ (1%), C₆H₆, 120°, 20 h	4-Cl-C₆H₄-allyl (29)	3
	Me₃SnC≡CPh	Pd(CH₃CN)₂Cl₂ (2%), DMF, 20°, 25 min	4-Cl-C₆H₄-C≡CPh (90)	553
	Bu₃Sn-(furanyl)-OBu-t	BnPd(PPh₃)₂Cl, DMF, 70°, 16 h	4-Cl-C₆H₄-(furanyl)-OBu-t (94)	296
	Me₃SnSnMe₃	Pd(PPh₃)₄ (1.3%), PhMe, 115°, 15 h	4-Cl-C₆H₄-SnMe₃ (74)	547
	Bu₃SnSnBu₃	NiBr₂ (10%), HMPA, 135°, 17 h	4-Cl-C₆H₄-SnBu₃ (92)	566, 313
	Bu₃Sn⌒	Pd(PPh₃)₄ (7.5%), LiCl DMF, 90°, 30 h	4-Cl-C₆H₄-allyl (77)	29, 201
	Bu₃Sn⌒⌒	Pd(PPh₃)₄ (10%), LiCl THF, 90°, 30 h	4-Cl-C₆H₄-allyl (39) + 4-Cl-C₆H₄-CH=CHCH₃ (33)	29
4-Cl-C₆H₄-N₂⁺ BF₄⁻	Me₄Sn	Pd(OAc)₂ (10%), CH₃CN, rt, 2 h	4-Cl-C₆H₄-CH₃ (88)	204
4-F-C₆H₄-Br	Me₄Sn	Pd(PPh₃)₄ (0.7%), air, HMPA, 65°	4-F-C₆H₄-CH₃ (89)	19

Organostannane		Halide/Electrophile	Conditions	Product (%)	Refs.

Me₃Sn—(oxazoline with 4-F-phenyl) ; (4,4-dimethyloxazoline) ; Pd(PPh₃)₄ (5%), C₆H₆, reflux, 24 h ; (100) ; 550, 551

Bu₃Sn—CH(OEt)—CH=CH₂ ; Pd(PPh₃)₄ (1%), C₆H₆, 110°, 15 h ; (78) *E:Z* = 60:40 ; 534

Bu₃Sn—CH=CH—CH(OEt)(OEt), *E:Z* = 85:15 ; Pd(PPh₃)₄ (2%), C₆H₆, 80°, 20 h ; (60) *E:Z* = 85:15 ; 539

SnMe₃—(2,6-dimethylpyridine) ; Pd(PPh₃)₂Cl₂ (7%), HMPA, dioxane, reflux, 24 h ; (47) ; 567

Bu₃Sn—CH₂—C(=CH₂)—CH(NHAc)(CO₂Et) ; Pd₂(dba)₃·CHCl₃ (4%), P(2-furyl)₃ (30%), THF, reflux, 6 h ; (56) ; 473

Bu₃Sn—(4-F-phenyl) ; Pd(PPh₃)₂Cl₂, PhMe, 100°, 16 h ; (—) ; 568

Bu₃Sn—(tetrathiafulvalene unit) ; Pd(PPh₃)₄, PhMe, reflux ; (61) ; 536

Bu₃Sn—C(=CH—Ph)—Ph (1-phenyl vinyl) ; Pd(dba)₂ (5%), CH₃CN, Et₂O, rt, 2 h ; (86) *E:Z* = 8:92 ; 249

Halides (electrophiles):
- 2-bromo-1,3-difluorobenzene
- 1-iodo-3,5-difluorobenzene
- 1-bromo-3-iodo-2,4,5,6-tetrafluorobenzene
- 1,4-diiodobenzene
- 4-iodobenzenediazonium tetrafluoroborate (N₂⁺ BF₄⁻)

TABLE III. DIRECT CROSS-COUPLING OF ARYL ELECTROPHILES (*Continued*)

Substrate	Stannane	Conditions	Product(s) and Yield(s) (%)	Refs.
OTf / I (4-iodophenyl triflate)	Bu₃Sn–CH=CH₂	Pd(PPh₃)₄ (2%), LiCl, BHT, dioxane, 98°, 16 h	TfO–CH=CH₂ styrene (73)	189
HO, OH diiodo hydroquinone	isoxazolyl-(3,4,5-trimethoxyphenyl), Bu₃Sn	Pd₂(dba)₃ (5%), AsPh₃ (40%), dioxane, 50°, 48 h	bis-isoxazole product, OH (61)	569
OTf / TfO	Bu₃Sn–CH₂CH=CH₂	Pd(PPh₃)₂Cl₂ (20%), dppf (80%), LiCl, DMF, reflux	bis-allyl product (100)	191
Br, NH₂ (2-bromoaniline)	Bu₃Sn–CH=CH–OEt	Pd(PPh₃)₂Cl₂ (1%), Et₄NCl, DMF, 100°, 1.5 h	aniline–CH=CH–OEt (—)	273
I, NH₂ (2-iodoaniline)	Cl₃Sn–CH₂CH₂CH₂–CO₂H	PdCl₂ (0.8%), PPh₂(C₆H₄SO₃Na-*m*) (3.2%), KOH, H₂O, 100°, 4 h	dihydroquinolinone (57)	282
	Cl₃SnPh	PdCl₂ (0.8%), PPh₂(C₆H₄SO₃Na-*m*) (1.6%), KOH, H₂O, 100°, 3 h	2-aminobiphenyl (88)	282
Br, H₂N (4-bromoaniline)	Bu₃Sn–CH=CH–OEt	Pd(PPh₃)₂Cl₂ (1%), Et₄NCl, DMF, 80°, 18 h	H₂N–C₆H₄–CH=CH–OEt (0)	272, 273

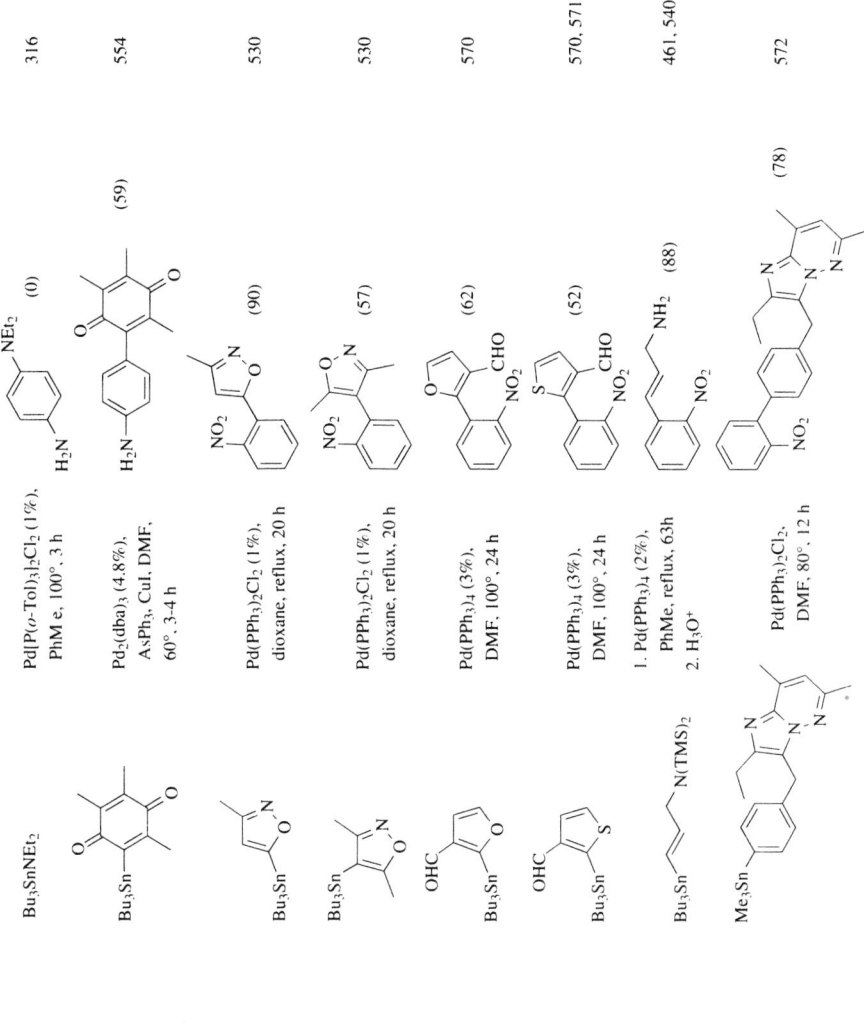

				316
Bu₃SnNEt₂		Pd[P(o-Tol)₃]₂Cl₂ (1%), PhMe, 100°, 3 h	(0)	
		Pd₂(dba)₃ (4.8%), AsPh₃, CuI, DMF, 60°, 3-4 h	(59)	554
		Pd(PPh₃)₂Cl₂ (1%), dioxane, reflux, 20 h	(90)	530
		Pd(PPh₃)₂Cl₂ (1%), dioxane, reflux, 20 h	(57)	530
		Pd(PPh₃)₄ (3%), DMF, 100°, 24 h	(62)	570
		Pd(PPh₃)₄ (3%), DMF, 100°, 24 h	(52)	570, 571
		1. Pd(PPh₃)₄ (2%), PhMe, reflux, 63h 2. H₃O⁺	(88)	461, 540
		Pd(PPh₃)₂Cl₂, DMF, 80°, 12 h	(78)	572

179

TABLE III. DIRECT CROSS-COUPLING OF ARYL ELECTROPHILES (*Continued*)

Substrate	Stannane	Conditions	Product(s) and Yield(s) (%)	Refs.
	$Bu_3SnSnBu_3$	$Pd(PPh_3)_4$ (3%), PhMe, 80°	SnBu₃ / NO₂ (59)	310
	Bu₃Sn (dimethyl quinone)	$Pd_2(dba)_3$ (2.5%), $AsPh_3$ (20%), CuI (50%), DMF, 60°, 3–4 h	(82) quinone with NO₂	554
	Bu₃Sn-indole SEM	$Pd(PPh_3)_4$ (0.7%), DMF, 110°, 6 h	(97) indole N–SEM, NO₂	289
	$Me_3SnSnMe_3$	$Pd(CH_3CN)_2Cl_2$ (2%), DMF, 20°, 5 min	SnMe₃ / NO₂ (100)	312, 573
	$Me_3SnSnMe_3$	$[(\eta^3\text{-}C_3H_5)PdCl]_2$, CH_2Cl_2, 20°	" (75)	557
	$Bu_3SnSnBu_3$	$Pd(PPh_3)_4$ (1%), PhMe, 80°, 72 h	SnBu₃ / NO₂ (98)	311
	Me_4Sn	$Pd(OAc)_2$ (10%), CH_3CN, rt, 2 h	NO₂ (94)	204
O_2N — $N_2^+\ PF_6^-$ / NO_2	Me—Sn (azabicycle)	$Pd(dppf)Cl_2$, PhMe, 105°, 2 h	O_2N — CH₃ (93)	41
	MOMO—CH₂—Sn (azabicycle)	$Pd(PPh_3)_4$, PhMe, 105°, 3 h	O_2N — CH₂OMOM (80)	41
— Br	n-Bu—Sn (azabicycle)	$Pd(dppf)Cl_2$, PhMe, 105°, 12 h	O_2N — Bu-n (86)	41

180

Substrate	Reagent	Conditions	Product (%)	Refs.
3-iodonitrobenzene (O_2N–C$_6$H$_4$–I)	TMS-CH$_2$-Sn (azabicyclic)	Pd(PPh$_3$)$_4$, PhMe, 105°, 20 h	O_2N–C$_6$H$_4$–CH$_2$TMS (85)	41
	Me$_3$SnSnMe$_3$	Pd(PPh$_3$)$_4$ (1.3%), PhMe, 80°, 1 h	O_2N–C$_6$H$_4$–SnMe$_3$ (41)	547
	Bu$_3$Sn-quinone	Pd$_2$(dba)$_3$ (2.5%), AsPh$_3$ (20%), CuI (50%), DMF, 60°, 3-4 h	(80)	554
	Bu$_3$Sn-OMe-quinone (n-Bu)	Pd$_2$(dba)$_3$ (2.5%), air, CuI (50%), DMF, 60°, 0.5-1 h	(88)	554
	Me$_3$SnSnMe$_3$	Pd(CH$_3$CN)$_2$Cl$_2$ (2%), DMF, 20°, 10 min	O_2N–C$_6$H$_4$–SnMe$_3$ (98)	312, 573
	Me$_3$SnSnMe$_3$	Pd(PPh$_3$)$_4$ (1%), PhMe, 60°, 72 h	" (65)	310, 311
	Me$_3$SnSnMe$_3$	[(η^3-C$_3$H$_5$)PdCl]$_2$, CH$_2$Cl$_2$, 20°	" (55)	557
	Me$_4$Sn	Pd(OAc)$_2$ (2%), DMF, 60-70°, 2.5 h	(77)	206
Bis(iodonium) salt [O_2N–C$_6$H$_4$–I$^+$BF$_4^-$]$_2$	Me$_3$Sn-tolyl	Pd(OAc)$_2$ (2%), DMF, 60-70°, 2.5 h	(83)	206
Diazonium salt (O_2N–C$_6$H$_4$–N$_2^+$PF$_6^-$)	Me$_4$Sn	Pd(OAc)$_2$ (10%), CH$_3$CN, rt, 2 h	(62)	204

TABLE III. DIRECT CROSS-COUPLING OF ARYL ELECTROPHILES (Continued)

Substrate	Stannane	Conditions	Product(s) and Yield(s) (%)	Refs.
4-O₂N-C₆H₄-Br	Bu₃Sn–C(=CH₂)Ph	Pd(dba)₂ (5%), CH₃CN, Et₂O, rt, 2 h	(54) E:Z = 8:92	249
	Bu₃Sn–CH₂OH	Pd(PPh₃)₄ (5%), dioxane, 80°	OH (0)	233
	Bu₃Sn–CH=CH₂	Pd(PPh₃)₄ (2%), BHT, PhMe, reflux, 4 h	(80)	88
	Bu₃Sn–CH₂CN	Pd[P(o-Tol)₃]Cl₂ (1%), m-xylene, 120°, 3 h	CN (tr)	235
	Bu₃Sn–CH₂OMe	Pd(PPh₃)₂Cl₂ (1%), HMPA, 80°, 20 h	OMe (65)	234
	Bu₃Sn–CH₂CH=CH₂	Pd(PPh₃)₄ (1%), C₆H₆, 100°, 20 h	(72)	3
	[isopropenyl OSnBu₃]	Pd[P(o-Tol)₃]₂Cl₂ (10%), PhMe, 100°, 5 h	(0)	240
	Bu₃Sn–(2-furyl)	Pd(PPh₃)₂Cl₂, 70°	(99)	574
	Bu₃Sn–CH₂CH=CH–CH₂OMe	BnPd(PPh₃)₂Cl, C₆H₆, 100°, 20 h	OMe (75)	305
	Bu₃Sn–C(OEt)=CH₂	Pd(PPh₃)₂Cl₂ (2%), PhMe, 105°, 48 h	(82)	270

182

Organostannane	Conditions	Product (%)	Ref.
Bu₃Sn–C(OEt)=CH₂	1. Pd(PPh₃)₂Cl₂ (1%), PhMe, 100°, 20 h 2. H⁺	4-O₂N–C₆H₄–C(=O)CH₃ (91)	269
Bu₃Sn–CH=CH–OEt	Pd(PPh₃)₂Cl₂ (10%), Et₄NCl, DMF, 80°, 2 h	4-O₂N–C₆H₄–CH=CH–OEt (86)	272, 273
Bu₃Sn–CH₂–CO₂Et	Pd[P(o-Tol)₃]₂Cl₂ (1%), ZnBr₂, DMF, 80°, 5 h	4-O₂N–C₆H₄–CH₂–CO₂Et (34)	236
Bu₃Sn–(2-thienyl)	Pd(PPh₃)₂Cl₂, 70°	4-O₂N–C₆H₄–(2-thienyl) (93)	574
Bu₃Sn–(1-Me-2-pyrrolyl)	Pd(PPh₃)₂Cl₂, 70°	4-O₂N–C₆H₄–(1-Me-2-pyrrolyl) (77)	574
Bu₃Sn–C(OTMS)=CH₂	Pd(PPh₃)₄, PhMe, 100°	4-O₂N–C₆H₄–C(OTMS)=CH₂ (63)	457
Bu₃SnPh	D₇₁₇–Pd(0)-polymer, Me₂CO, reflux, 25 h	4-O₂N–C₆H₄–Ph (57)	535
Bu₃Sn–(4-MeC₆H₄)	Pd(PPh₃)₄ (1%), PhMe, 120°, 20 h	4-O₂N–C₆H₄–(4-MeC₆H₄) (68)	538
Bu₃Sn–CH=CH–CH(OMe)(TMS)	Pd(PPh₃)₄, C₆H₆, 110°	4-O₂N–C₆H₄–CH=CH–CH(TMS)(OMe) (58)	305

TABLE III. DIRECT CROSS-COUPLING OF ARYL ELECTROPHILES (Continued)

Substrate	Stannane	Conditions	Product(s) and Yield(s) (%)	Refs.
	Bu$_3$Sn—furan—OBu-t	BnPd(PPh$_3$)$_2$Cl, DMF, 70°, 16 h	OBu-t furan with O$_2$N-phenyl (98)	296
	Bu$_3$Sn—N(TMS)$_2$	1. Pd(PPh$_3$)$_4$, PhMe, reflux, 63 h 2. H$_3$O$^+$	NH$_2$ with O$_2$N-phenyl (59)	461
	Bu$_3$Sn (OMe, OMe, OMe)	Pd(PPh$_3$)$_4$ (5%), CuI (8%), NMP, 70°, 48 h	OMe OMe OMe with O$_2$N-phenyl (45)	575
	Ph—C(O)—furan—Bu$_3$Sn	Pd(PPh$_3$)$_4$ (5%), HMPA, 80°, 23 h	(81) Ph—C(O)—furan with O$_2$N-phenyl	287, 546
	Bu$_3$Sn—pyrrole—N—TIPS	Pd(PPh$_3$)$_4$ (10%), dioxane, reflux, 24 h	N—TIPS pyrrole with O$_2$N-phenyl (77)	565
	Bu$_3$Sn sugar OTBDMS, OTBDMS, OTBDMS	Pd(PPh$_3$)$_2$Cl$_2$, PhMe	OTBDMS, OTBDMS, OTBDMS sugar with O$_2$N-phenyl (78)	299, 300
	Bu$_3$Sn—furan—SnBu$_3$	Pd(PPh$_3$)$_4$ (5%), HMPA, 80°, 24 h	(85) furan with two O$_2$N-phenyl (NO$_2$)	287, 546

184

Organotin reagent	Product (%)	Conditions	Refs.
Bu₃SnNEt₂	O₂N–C₆H₄–NEt₂ (24)	Pd[P(o-Tol)₃]₂Cl₂ (1%), PhMe, 100°, 3 h	90, 316
Bu₃SnSPh	O₂N–C₆H₄–SPh (52)	Pd(PPh₃)₄ (1%), PhMe, 120°, 20 h	318
Bu₃SnSSnBu₃	(O₂N–C₆H₄)₂S, p-NO₂ (44)	Pd(PPh₃)₄ (1%), PhMe, 120°, 20 h	318
Me₃SnSnMe₃	O₂N–C₆H₄–C₆H₄–NO₂ (58)	Pd(PPh₃)₄ (5%), PhMe, 120°, 40 h	548
Me₃SnSnMe₃	O₂N–C₆H₄–SnMe₃ (37)	Pd(PPh₃)₄ (5%), PhMe, 80–120°, 1–15 h	547
Bu₃SnSnBu₃	O₂N–C₆H₄–SnBu₃ (48)	Pd(CH₃CN)₂Cl₂ (1%), HMPA, 20°, 1 h	313
Bu₃SnSnBu₃	" (38)	Pd(PPh₃)₄ (1%), PhMe, 120°, 20 h	310
Bu₃SnSnBu₃	O₂N–C₆H₄–C₆H₄–NO₂ (26)	Pd(PPh₃)₄ (1.3%), PhMe, 115°, 15 h	547, 548
Bu₃Sn⌣ (allyl)	O₂N–C₆H₄–CH₂CH=CH₂ (59)	Pd(PPh₃)₄ (1%), C₆H₆, 120°, 20 h	3
Bu₃SnSnBu₃	O₂N–C₆H₄–SnBu₃ (0)	Pd(PPh₃)₂Br₂ (1.3%), PhMe, 115°, 15 h	547
Me₄Sn	O₂N–C₆H₄–CH₃ (87)	PhPd(PPh₃)₂I (2%), HMPA, 70°, 30 min	463
Me₃Sn⌣ (vinyl)	O₂N–C₆H₄–CH=CH₂ (98)	Pd(CH₃CN)₂Cl₂ (2%), DMF, 20°, <1 min	553, 463

Substrates (bottom): O₂N–C₆H₄–Cl, O₂N–C₆H₄–F, O₂N–C₆H₄–I

185

TABLE III. DIRECT CROSS-COUPLING OF ARYL ELECTROPHILES (Continued)

Substrate	Stannane	Conditions	Product(s) and Yield(s) (%)	Refs.
	Bu₃Sn (cyclobutenone)	BnPd(PPh₃)₂Cl (1.5%), THF, 78°, 2 h	(50) cyclobutenone-aryl-NO₂	267
	Bu₃SnC≡COEt	Pd(PPh₃)₂Cl₂ (5%), Et₄NCl, DMF, rt, 1 h	(52) O₂N-C₆H₄-C≡COEt	302
	Bu₃Sn (butenolide)	Pd(PPh₃)₂Cl₂ (1-2%), PhMe, reflux	(0)	261
	Bu₃Sn (CO₂Me vinyl)	Pd(PPh₃)₄ (10%), CuI, DMF, rt, 12 h	(76)	246
	Me₃Sn (thiophene)	Pd(CH₃CN)₂Cl₂ (2%), DMF, 20°, 5 min	(96)	553
	Bu₃SnC≡CCO₂Et	Pd(PPh₃)₂Cl₂ (1%), Et₄NCl, ZnCl₂, C₆H₆, 50°, 1 h	(47) O₂N-C₆H₄-C≡CCO₂Et	552
	Bu₃Sn (methylthiophene)	Pd/C (5%), CuI (10%), AsPh₃ (20%), NMP, 80°, 24 h	(85)	458
	Me₃SnPh	Pd(CH₃CN)₂Cl₂ (2%), DMF, 20°, 3 h	(100) O₂N-C₆H₄-Ph	553

186

Reagent	Conditions	Product(s) (%)	Refs.
Me₃SnPh	PhPd(PPh₃)₂I (2%), HMPA, 20°, 20 min	(4-O₂N-C₆H₄–Ph) **I** (92) + Ph-Ph **II** (8)	463, 576
Me₃SnPh	(p-O₂NC₆H₄)Pd(PPh₃)₂I (2%), Cl(CH₂)₂Cl, 120-130°, 2 h	**I** (83) + **II** (17)	87
Me₃Sn–C₆H₄–Cl	PhPd(PPh₃)₂I (2%), HMPA, 20°, 20 min	(O₂N–C₆H₄–C₆H₄–Cl) **I** (86) + (Cl-isomer) (8)	463
"	Pd(CH₃CN)₂Cl₂ (2%), DMF, 20°, 5 h	**I** (94)	553
Me₃Sn–C₆H₄–NO₂	PhPd(PPh₃)₂I (2%), HMPA, 20°, 5 h	(O₂N–C₆H₄–C₆H₄–NO₂) (97)	463
Me₃Sn–(m-tolyl)	PhPd(PPh₃)₂I (2%), Cl(CH₂)₂Cl, 120°, 2 h	(76)	87
Me₃Sn–(p-tolyl)	PhPd(PPh₃)₂I (2%), HMPA, 20°, 20 min	**I** (87) + (12)	463
"	Pd(CH₃CN)₂Cl₂ (2%), DMF, 20°, 3 h	**I** (96)	553
Me₃Sn–C₆H₄–OMe	Pd(CH₃CN)₂Cl₂ (2%), DMF, 20°, 3 h	(O₂N–C₆H₄–C₆H₄–OMe) (93)	463, 553
Me₃SnC≡CPh	PhPd(PPh₃)₂I (2%), HMPA, 20°, 10 h	(O₂N–C₆H₄–C≡CPh) **I** (57)	463

TABLE III. DIRECT CROSS-COUPLING OF ARYL ELECTROPHILES (*Continued*)

Substrate	Stannane	Conditions	Product(s) and Yield(s) (%)	Refs.
	Me$_3$SnC≡CPh	Pd(CH$_3$CN)$_2$Cl$_2$ (2%), DMF, 20°, 5 min	**I** (93)	553
	Me$_3$SnC≡CPh	PhPd(PPh$_3$)$_2$I (2%), Cl(CH$_2$)$_2$Cl, 120°, 2 h	**I** (94)	87
	Bu$_3$Sn	Pd$_2$(dba)$_3$ (2.5%), air, CuI (50%), DMF, 60°, 0.5–1 h	(85)	554
	Bu$_3$Sn	"	(76)	554
	Bu$_3$Sn	Pd(PPh$_3$)$_4$ (10%), DMF, 90°, 3 h	(86)	74
	Me$_3$Sn	Pd(CH$_3$CN)$_2$Cl$_2$ (2%), DMF, 90°	(66) + (—)	577
			(—)	

	$Et_3SnSSnEt_3$	$PhPd(PPh_3)_2I$ (5%), DMSO, 100°, 4 h	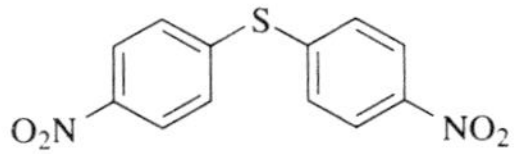 (100)	320
	$Me_3SnSnMe_3$	$Pd(CH_3CN)_2Cl_2$ (2%), DMF, 20°, 5 min	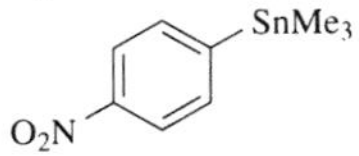 (100)	312
	$Me_3SnSnMe_3$	$[(\eta^3\text{-}C_3H_5)PdCl]_2$ (1%), CH_2Cl_2, 20°	" (75)	557
	$Et_3SnSnEt_3$	$Pd(CH_3CN)_2Cl_2$ (1%), HMPA, 20°, 5 min	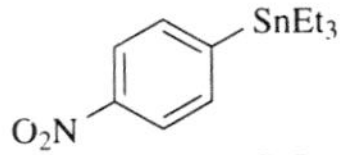 (81)	313
	$Bu_3SnSnBu_3$	$Pd(CH_3CN)_2Cl_2$ (1%), HMPA, 20°, 5 min	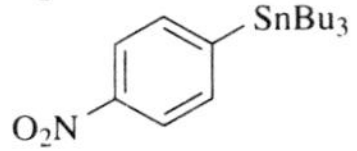 (94)	313
	$Bu_3SnSnBu_3$	$NiBr_2$ (10%), HMPA, 135°, 3 h	" (72)	315
	$Bu_3SnSnBu_3$	$Pd(PPh_3)_2Br_2$ (1.3%), PhMe, 115°, 15 h	" (0)	547
	$Bu_3SnSnBu_3$	$Pd(PPh_3)_4$ (1%), PhMe, 60°, 72 h	" (63)	311
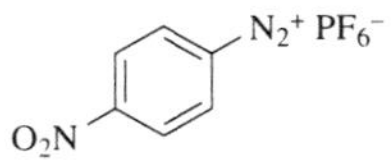	Me_4Sn	$Pd(OAc)_2$ (10%), CH_3CN, rt, 2 h	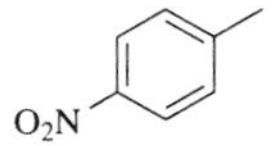 (95)	204
	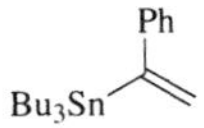	$Pd(dba)_2$ (5%), Et_2O CH_3CN, rt, 1 h	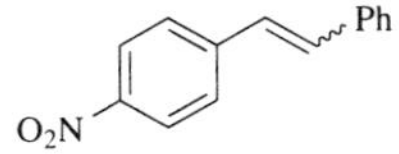 (60) E:Z = 16:84	249
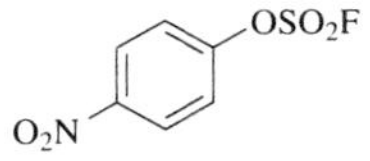	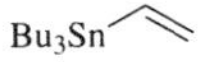	$Pd(PPh_3)_2Cl_2$ (5%), LiCl, DMF, 25°, 6-18 h	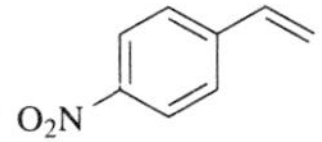 (60)	203
	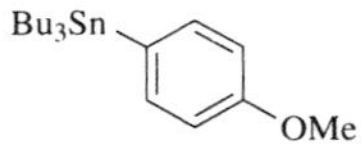	"	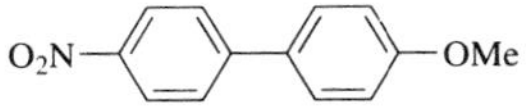 (50)	203

TABLE III. DIRECT CROSS-COUPLING OF ARYL ELECTROPHILES (*Continued*)

Substrate	Stannane	Conditions	Product(s) and Yield(s) (%)	Refs.
4-O₂N-C₆H₄-OTf	(TMSCH₂)₄Sn	Pd(PPh₃)₂Cl₂ (2%), LiCl, BHT, dioxane, 98°, 9 h	4-O₂N-C₆H₄-Cl (25)	189
	Bu₃Sn-C₆H₄-OMe	Pd(PPh₃)₄ (2%), LiCl, BHT, dioxane, 98°, 36 h	MeO-C₆H₄-C₆H₄-NO₂ (74)	189, 420
	Bu₃Sn-CH=CH-CO₂Bn (cis)	Pd(PPh₃)₂Cl₂ (2%), LiCl, BHT, DMF, 100°, 5 h	O₂N-C₆H₄-CH=CH-CO₂Bn (47) $E:Z = 1:2$	189
	Bu₃Sn-CH=CH-CO₂Bn (trans)	Pd(PPh₃)₄ (2%), LiCl, BHT, dioxane, 98°, 9 h	O₂N-C₆H₄-CH=CH-CO₂Bn (82)	189
OTf / Br / NO₂ benzene	Bu₃Sn-CH=CH₂	Pd(PPh₃)₄ (2%), PhMe, reflux, 48 h	OTf, vinyl, NO₂ benzene (91)	89
Br / O₂N / NO₂ benzene	Bu₃Sn-CH₂CH=CH₂	Pd(PPh₃)₂Cl₂, PPh₃, DMF	O₂N, allyl, NO₂ benzene (96)	578
Br / O₂N / NO₂ (H₂N) benzene	Me₃SnPh	PhPd(PPh₃)₂I (2%), Cl(CH₂)₂Cl, 120–130°, 2 h	Ph, NO₂, H₂N benzene (93)	87
I / O₂N / NO₂ benzene	Me₃SnPh	[(η³-C₃H₅)PdCl₂]₂ (1%), Me₂CO, 20°, 24 h	" (99)	553, 557
	Me₃SnPh	(p-O₂NC₆H₄)Pd(PPh₃)₂I (2%), Cl(CH₂)₂Cl, 120–130°, 2 h	" (94)	87
	Me₃SnSnMe₃	Pd(CH₃CN)₂Cl₂ (2%), DMF, 20°, 5 min	O₂N, SnMe₃, NO₂ benzene (93)	312, 573

190

Me₃SnSnMe₃ $[(\eta^3\text{-}C_3H_5)PdCl]_2$ (1%), CH₂Cl₂, 20°   " (70)   573

Bu₃SnSnBu₃   Pd(CH₃CN)₂Cl₂ (1%), HMPA, 20°, 1 h   (79)   313

Me₃SnPh   PhPd(PPh₃)₂I, Cl(CH₂)₂Cl, 120–130°, 1.5 h   R = Cl (95), R = I (96)   87, 576 / 87

Me₃SnC≡CPh   "   R = I (85)   87

(structure: Bu₃Sn–isoxazole–OH)   Pd₂(dba)₃ (2.5%), AsPh₃ (20%), dioxane, 50°   (95)   471, 579

(structure: MeO, OMe, OMe Bu₃Sn–isoxazole)   "   (81)   471

Cl₃SnPh   PdCl₂ (0.5–3%), PPh(C₆H₄SO₃Na-m)₂, KOH, H₂O, 90°   (89)   281, 282

Bu₃Sn OEt   Pd(PPh₃)₂Cl₂, Et₄NCl, DMF, 80°, 12 h   (0)   272, 273

191

TABLE III. DIRECT CROSS-COUPLING OF ARYL ELECTROPHILES (*Continued*)

Substrate	Stannane	Conditions	Product(s) and Yield(s) (%)	Refs.
HO–C₆H₄–I	Cl_3SnPh	$PdCl_2$ (0.5-3%), PPh(C₆H₄SO₃Na-*m*)₂, KOH, H₂O, 90°	Ph–C₆H₄–OH (<5)	281
	Bu_3SnNEt_2	Pd[P(*o*-Tol)₃]₂Cl₂ (1%), PhMe, 100°, 3 h	NEt₂–C₆H₄–OH (0)	316
	Cl_3Sn–CO₂H	PPh₂(C₆H₄SO₃Na-*m*) (6.4%), PdCl₂ (1.6%), KOH, H₂O, 100°, 3 h	CO₂H derivative (<10)	282
	Cl_3SnPh	PdCl₂ (0.8%), KOH, H₂O, 100°, 3 h	Ph–C₆H₄–OH (87)	282
HO / OH (iodo resorcinol)	isoxazole-Bu_3Sn (3-Ph)	Pd₂(dba)₃ (2.5%), AsPh₃ (20%), dioxane, 50°	isoxazole product (62)	569
	3-(4-hydroxyphenyl)isoxazol-5-yl Bu_3Sn	''	(78)	471
	3-(3,4,5-trimethoxyphenyl)isoxazol-5-yl Bu_3Sn	''	(67)	471

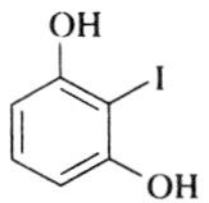

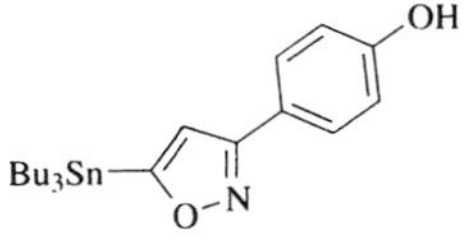

471

(42)

471, 579

(35)

471

(43)

233   (80)

235   (74)

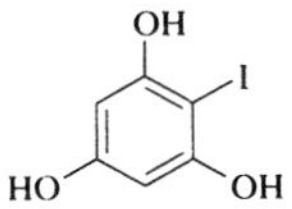

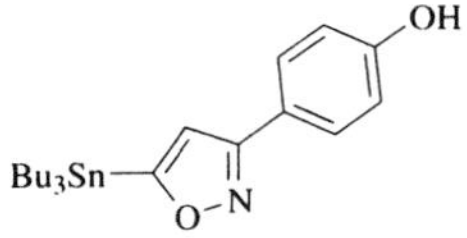

234   (80)

237, 240   (91)

269   (78)

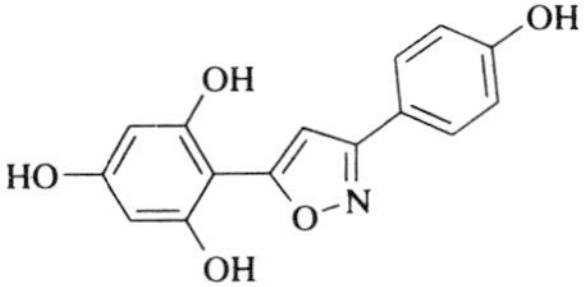

PdCl$_2$ (5%), dioxane, 101°

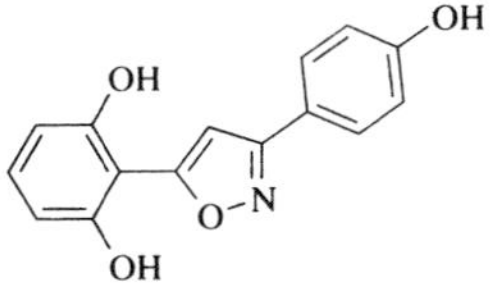

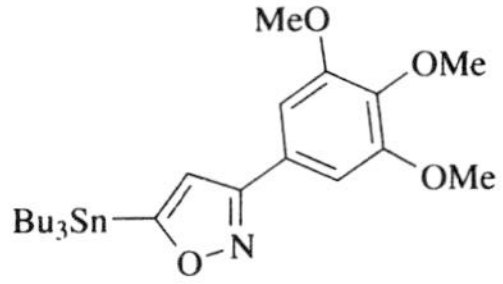

PdCl$_2$ (5%), dioxane, 105°

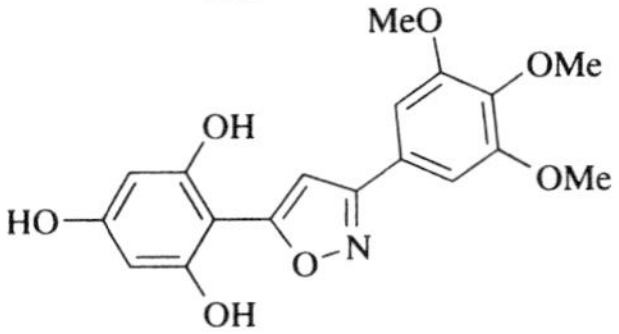

PdCl$_2$ (5%), dioxane, 105°

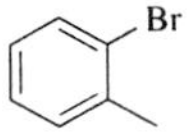

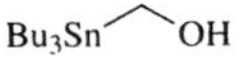

Pd(PPh$_3$)$_4$ (5%), dioxane, 80°

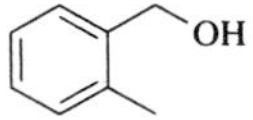

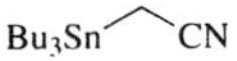

Pd[P(o-Tol)$_3$]$_2$Cl$_2$ (1%), m-xylene, 120°, 3 h

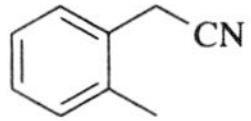

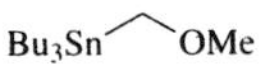

Pd(PPh$_3$)$_2$Cl$_2$ (1%), HMPA, 80°, 70 h

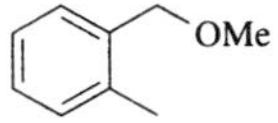

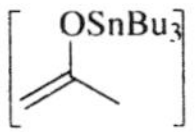

Pd[P(o-Tol)$_3$]$_2$Cl$_2$ (1%), PhMe, 100°, 5 h

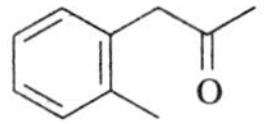

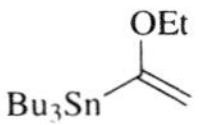

1. Pd(PPh$_3$)$_2$Cl$_2$ (1%), PhMe, 100°, 20 h
2. H$^+$

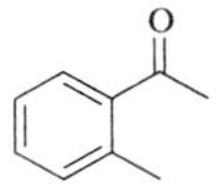

Bu$_3$Sn

Bu$_3$Sn—OH

Bu$_3$Sn—CN

Bu$_3$Sn—OMe

[OSnBu$_3$]

OEt
Bu$_3$Sn

C$_7$

193

TABLE III. DIRECT CROSS-COUPLING OF ARYL ELECTROPHILES (Continued)

Substrate	Stannane	Conditions	Product(s) and Yield(s) (%)	Refs.
	Bu_3Sn—CO_2Et	Pd[P(o-Tol)$_3$]$_2$Cl$_2$ (1%), ZnBr$_2$, DMF, 80°, 5 h	CO_2Et (71)	236
	Me_3Sn OTMS	Pd(PPh$_3$)$_4$, PhMe, 100°	OTMS (63)	457
	OSnBu$_3$	Pd[P(o-Tol)$_3$]$_2$Cl$_2$ (3%), C$_6$H$_6$, reflux, 3 h	(59)	241
	Bu_3SnPh	Pd(PPh$_3$)$_4$ (1%), C$_6$H$_6$, 120°, 20 h	Ph (93)	538
	Bu_3Sn— (m-tolyl)	Pd(PPh$_3$)$_4$ (1%), C$_6$H$_6$, 120°, 20 h	(97)	538
	Bu_3Sn— (p-tolyl)	Pd(PPh$_3$)$_4$ (1%), C$_6$H$_6$, 120°, 20 h	(87)	538
	Me_3Sn (furan) OBu-t	BnPd(PPh$_3$)$_2$Cl, DMF, 70°, 16 h	OBu-t (64)	296
	Me_3Sn (furan) OH, TBDMS	Pd(PPh$_3$)$_4$, PhMe, reflux, 4 h	OH, TBDMS (80)	371

194

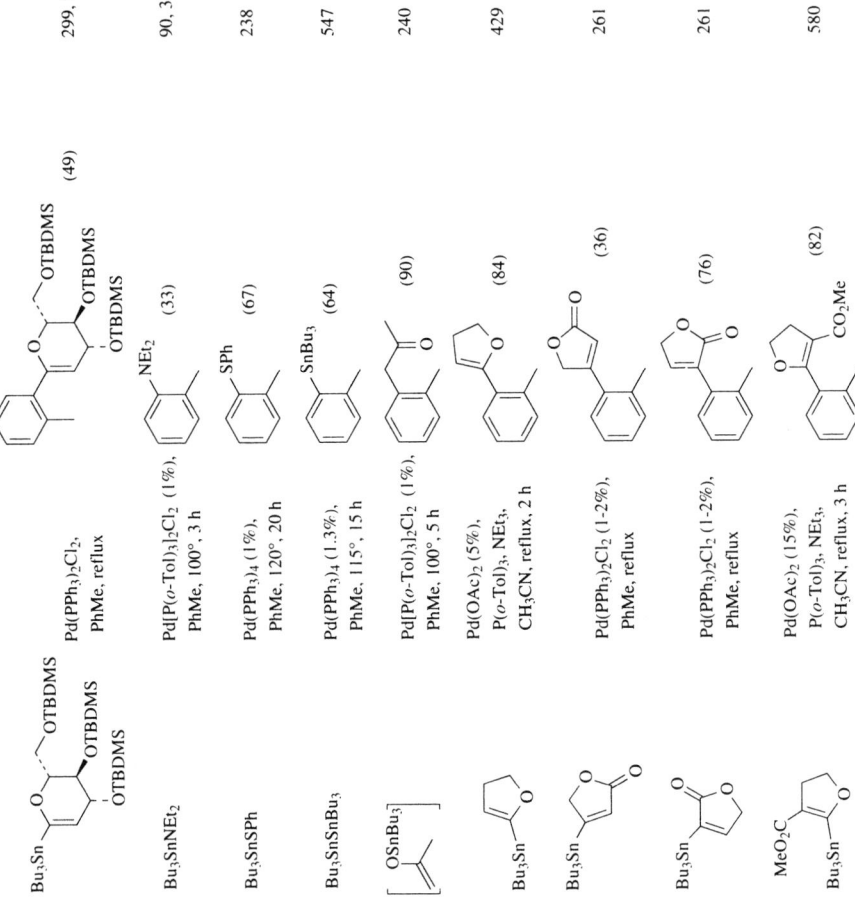

Bu₃Sn [OTBDMS structure]	Pd(PPh₃)₂Cl₂, PhMe, reflux	[o-tolyl OTBDMS structure] (49)	299, 300
Bu₃SnNEt₂	Pd[P(o-Tol)₃]₂Cl₂ (1%), PhMe, 100°, 3 h	[o-tolyl NEt₂] (33)	90, 316
Bu₃SnSPh	Pd(PPh₃)₄ (1%), PhMe, 120°, 20 h	[o-tolyl SPh] (67)	238
Bu₃SnSnBu₃	Pd(PPh₃)₄ (1.3%), PhMe, 115°, 15 h	[o-tolyl SnBu₃] (64)	547
[OSnBu₃ structure]	Pd[P(o-Tol)₃]₂Cl₂ (1%), PhMe, 100°, 5 h	[o-tolyl ketone] (90)	240
Bu₃Sn [furan]	Pd(OAc)₂ (5%), P(o-Tol)₃, NEt₃, CH₃CN, reflux, 2 h	[furan o-tolyl] (84)	429
Bu₃Sn [butenolide]	Pd(PPh₃)₂Cl₂ (1-2%), PhMe, reflux	[butenolide o-tolyl] (36)	261
Bu₃Sn [butenolide]	Pd(PPh₃)₂Cl₂ (1-2%), PhMe, reflux	[butenolide o-tolyl] (76)	261
MeO₂C [furan] Bu₃Sn	Pd(OAc)₂ (15%), P(o-Tol)₃, NEt₃, CH₃CN, reflux, 3 h	[furan o-tolyl CO₂Me] (82)	580

TABLE III. DIRECT CROSS-COUPLING OF ARYL ELECTROPHILES (*Continued*)

Substrate	Stannane	Conditions	Product(s) and Yield(s) (%)	Refs.
(3-bromotoluene)	Bu$_3$Sn—CH=C(PhS)—CO$_2$Me	Pd(OAc)$_2$ (15%), P(o-Tol)$_3$, NEt$_3$, CH$_3$CN, reflux, 3 h	(62)	580
	(2-SEM-indolyl)SnBu$_3$	Pd(PPh$_3$)$_4$ (0.7%), DMF, 110°, 2 h	(93)	289
	Et$_3$SnSnEt$_3$	PhPd(PPh$_3$)$_2$I (5%), DMSO, 100°, 4 h	(82)	320
	Bu$_3$Sn—CH$_2$OH	Pd(PPh$_3$)$_4$ (5%), dioxane, 80°	(62)	233
	Bu$_3$Sn—CH$_2$CN	Pd[P(o-Tol)$_3$]$_2$Cl$_2$ (1%), m-xylene, 120°, 3 h	(74)	235
	Bu$_3$Sn—CH$_2$OMe	Pd(PPh$_3$)$_2$Cl$_2$ (1%), HMPA, 80°, 20 h	(72)	234
	[CH$_2$=C(CH$_3$)OSnBu$_3$]	Pd[P(o-Tol)$_3$]$_2$Cl$_2$ (1%), PhMe, 100°, 5 h	(88)	237, 240
	Bu$_3$Sn—CH$_2$CO$_2$Et	Pd[P(o-Tol)$_3$]$_2$Cl$_2$ (1%), Z nBr$_2$, DMF, 80°, 5 h	(60)	236
	CH$_2$=C(OEt)SnBu$_3$	1. Pd(PPh$_3$)$_2$Cl$_2$ (1%), PhMe, 100°, 20 h 2. H$^+$	(70)	269

196

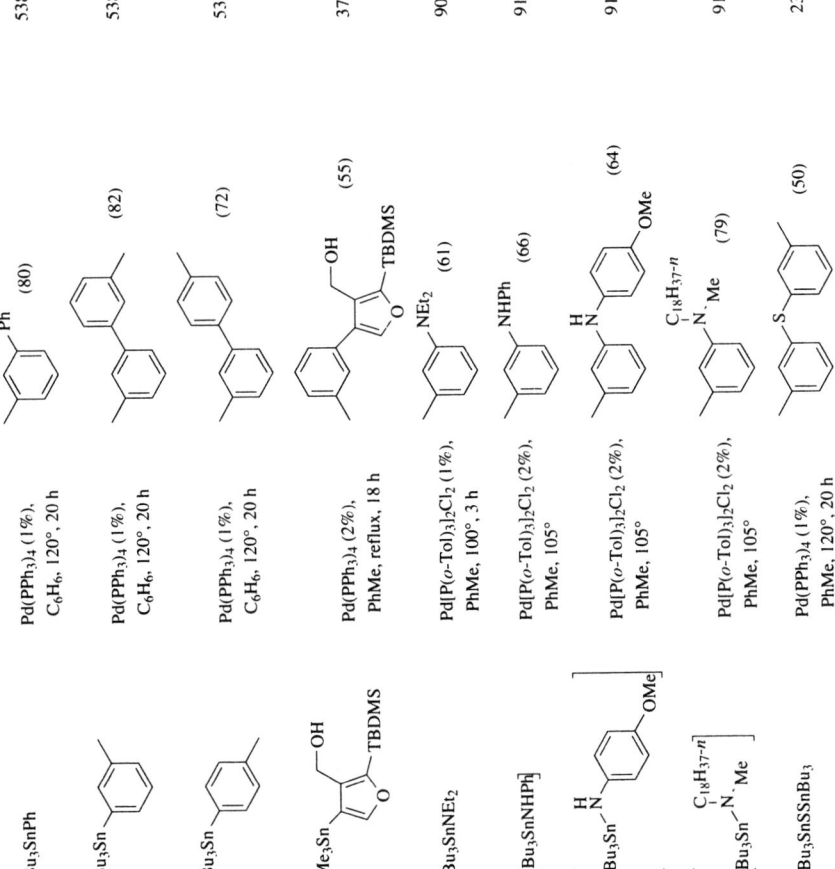

Reagent	Conditions	Product (Yield)	Refs.
Bu$_3$SnPh	Pd(PPh$_3$)$_4$ (1%), C$_6$H$_6$, 120°, 20 h	(80)	538
Bu$_3$Sn	Pd(PPh$_3$)$_4$ (1%), C$_6$H$_6$, 120°, 20 h	(82)	538
Bu$_3$Sn	Pd(PPh$_3$)$_4$ (1%), C$_6$H$_6$, 120°, 20 h	(72)	538
Me$_3$Sn (furan-OH, TBDMS)	Pd(PPh$_3$)$_4$ (2%), PhMe, reflux, 18 h	(55)	371
Bu$_3$SnNEt$_2$	Pd[P(o-Tol)$_3$]$_2$Cl$_2$ (1%), PhMe, 100°, 3 h	(61)	90, 316
[Bu$_3$SnNHPh]	Pd[P(o-Tol)$_3$]$_2$Cl$_2$ (2%), PhMe, 105°	(66)	91
[Bu$_3$Sn–NH–C$_6$H$_4$OMe]	Pd[P(o-Tol)$_3$]$_2$Cl$_2$ (2%), PhMe, 105°	(64)	91
[Bu$_3$Sn–N(C$_{18}$H$_{37}$-n)Me]	Pd[P(o-Tol)$_3$]$_2$Cl$_2$ (2%), PhMe, 105°	(79)	91
Bu$_3$SnSSnBu$_3$	Pd(PPh$_3$)$_4$ (1%), PhMe, 120°, 20 h	(50)	238

TABLE III. DIRECT CROSS-COUPLING OF ARYL ELECTROPHILES (Continued)

Substrate	Stannane	Conditions	Product(s) and Yield(s) (%)	Refs.
3-iodotoluene	Et$_3$SnSSnEt$_3$	PhPd(PPh$_3$)$_2$I (5%), DMSO, 100°, 4 h	di(m-tolyl) sulfide (70)	320
4-fluorophenyl (m-tolyl sulfonate)	Bu$_3$SnPh	Pd(OAc)$_2$ (5%), Ph$_2$PMe (11%), LiCl, DMF, 110°, 76 h	Ph-biaryl (41)	202
	Bu$_3$SnPh	Pd(OAc)$_2$ (5%), dppp (5.5%), LiCl, DMF, 110°, 72 h	Me$_2$N—C$_6$H$_4$—SO$_2$O—m-tolyl (70)	202
4-bromotoluene	Me$_4$Sn	Pd(PPh$_3$)$_4$ (0.7%), air, HMPA, 65°	p-xylene (84)	19
	Bu$_3$Sn$\frown$OH	Pd(PPh$_3$)$_4$ (5%), dioxane, 80°	CH$_2$OH (52)	233
	Bu$_3$Sn$\frown$CN	Pd[P(o-Tol)$_3$]$_2$Cl$_2$ (1%), m-xylene, 120°, 3 h	CH$_2$CN (78)	235
	Bu$_3$Sn$\frown$OMe	Pd(PPh$_3$)$_2$Cl$_2$ (1%), HMPA, 80°, 20 h	CH$_2$OMe (67)	234
	isopropenyl OSnBu$_3$	Pd[P(o-Tol)$_3$]$_2$Cl$_2$ (1%), PhMe, 100°, 5 h	CH$_2$C(O)CH$_3$ (80)	237, 240
	Bu$_3$Sn, OEt (vinyl ether)	1. Pd(PPh$_3$)$_2$Cl$_2$ (1%), PhMe, 100°, 20 h 2. H$^+$	C(O)CH$_3$ (67)	269
	Bu$_3$Sn$\frown$CO$_2$Et	Pd[P(o-Tol)$_3$]$_2$Cl$_2$ (1%), ZnBr$_2$, DMF, 80°, 5 h	CH$_2$CO$_2$Et (93)	236

Organotin reagent	Conditions	Product	Ref.
Me₃Sn (4,4-dimethyloxazoline, m-tolyl)	Pd(PPh₃)₄ (5%), C₆H₆, reflux, 12 h	(70)	550
Bu₃Sn—CH=CH₂, OEt	Pd(PPh₃)₄ (1%), C₆H₆, 110°, 15 h	(71) E:Z = 75:25, OEt	534
Bu₃Sn, OEt	Pd(PPh₃)₄ (1%), C₆H₆, 110°, 15 h	(74) E:Z = 80:20, OEt	534
Bu₃Sn, OEt	1. Pd(PPh₃)₄ (5%), C₆H₆, 120°, 20 h 2. H⁺	(71) CHO	581
Me₃Sn, OTMS	1. Pd(PPh₃)₄, PhMe, 100° 2. H⁺	(66)	457
Bu₃SnPh	Pd(PPh₃)₄ (1%), C₆H₆, 120°, 20 h	(61) Ph	538
Me₃Sn, OTMS	1. Pd(PPh₃)₄, PhMe, 100° 2. H⁺	(35)	457
Bu₃Sn (m-tolyl)	Pd(PPh₃)₄ (1%), C₆H₆, 120°, 20 h	(77)	538
Bu₃Sn (p-tolyl)	Pd(PPh₃)₄ (1%), C₆H₆, 120°, 20 h	(64)	538

TABLE III. DIRECT CROSS-COUPLING OF ARYL ELECTROPHILES (*Continued*)

Substrate	Stannane	Conditions	Product(s) and Yield(s) (%)	Refs.
	OSnBu₃ structure	Pd[P(*o*-Tol)₃]₂Cl₂ (3%), C₆H₆, reflux, 3 h	(62)	241
	Me₃Sn / TMSO	1. Pd(PPh₃)₄, PhMe, 100° 2. H⁺	(31)	457
	Bu₃Sn—OMe / TMS	1. Pd(PPh₃)₄, C₆H₆, 100° 2. Bu₄NF, THF	OMe (87)	532
	indole Bu₃Sn, N–R	Pd(PPh₃)₂Cl₂, THF, reflux, 2 h	R = Me (82) R = SEM (66)	425
	OSnBu₃ chromene	Pd[P(*o*-Tol)₃]₂Cl₂, PhMe, 100°, 20 h	(38)	541
	Bu₃Sn—naphthalene—OMe	Pd(PPh₃)₄ (5%), DMF, 105°	(36) + (22)	378
	Bu₃Sn—pyrrole, N–TIPS	Pd(PPh₃)₄ (16%), dioxane, reflux, 40 h	N–TIPS (69)	565

The aryl halide substrate (4-iodotoluene):

Organotin reagent	Conditions	Product (%)	Refs.
Bu₃SnNEt₂	Pd[P(o-Tol)₃]₂Cl₂ (1%), PhMe, 100°, 3 h	4-MeC₆H₄–NEt₂ (79)	90, 316
Bu₃Sn–N(Ph)Me	Pd[P(o-Tol)₃]₂Cl₂ (2%), PhMe, 105-100°	(73)	91
Bu₃Sn–N(tetrahydroisoquinoline)	Pd[P(o-Tol)₃]₂Cl₂ (2%), PhMe, 105°	(55)	91
Bu₃SnSPh	Pd(PPh₃)₄ (1%), PhMe, 120°, 20 h	4-MeC₆H₄–SPh (76)	318
Bu₃SnSSnBu₃	Pd(PPh₃)₄ (1%), PhMe, 120°, 20 h	4-MeC₆H₄–S–C₆H₄Me (60)	312
Me₃SnTMS	Pd(PPh₃)₄ (1.3%), PhMe, 115°, 15 h	4-MeC₆H₄–TMS (60)	547
Bu₃SnSnBu₃	Pd(PPh₃)₄ (1.3%), PhMe, 115°, 15 h	4-MeC₆H₄–SnBu₃ (75)	547
Me₄Sn	Pd(Ph-BIAN)(dimethyl fumarate) (1%), DMF, 50°, 18 h	(41)	415
Bu₃Sn-furan	Pd(PPh₃)₂Cl₂ (4%), CuI, DMF, rt, 2 h	(72)	287, 546
Bu₃SnC≡COEt	Pd(PPh₃)₂Cl₂ (5%), Et₄NCl, DMF, rt, 15 h	4-MeC₆H₄–C≡COEt (45)	302

TABLE III. DIRECT CROSS-COUPLING OF ARYL ELECTROPHILES (Continued)

Substrate	Stannane	Conditions	Product(s) and Yield(s) (%)	Refs.
	Bu₃Sn— (CO₂Me)	Pd(PPh₃)₂Cl₂ (10%), CuI, DMF, rt, 48 h	CO₂Me structure (71)	246
	Bu₃SnC≡CCO₂Et	Pd(PPh₃)₂Cl₂ (1%), Et₄NCl, ZnCl₂, DMF, rt, 40 h	C≡CCO₂Et (23)	552
	Cl₃SnPh	PdCl₂ (0.8%), PPh₂(C₆H₄SO₃Na-m) (1.6%), KOH, H₂O, 100°, 3 h	Ph (86)	282
	Bu₃Sn—	Pd(Ph-BIAN) (dimethyl fumarate) (1%), DMF, 50°, 18 h	(35)	415
	furan (Bu₃Sn, SnBu₃)	Pd(PPh₃)₂Cl₂ (7%), DMF, 65°, 10 h	(45)	287, 546
	Me₃SnPPh₂	Pd(CH₃CN)₂Cl₂ (2.5%), C₆H₆, 60°, 36 h	PPh₂ (74)	321
	Et₃SnSSnEt₃	PhPd(PPh₃)₂I (5%), DMSO, 100°, 4 h	S (88)	320
	Me₃SnSnMe₃	Pd(PPh₃)₂Br₂ (1.3%), PhMe, 115°, 15 h	SnMe₃ (86)	547

202

Aryl halide	Organotin reagent	Conditions	Product (yield)	Ref.
tolyl–N$_2^+$ BF$_4^-$	Bu$_3$SnSnBu$_3$	Pd(PPh$_3$)$_2$Br$_2$ (1.3%), PhMe, 115°, 15 h	tolyl–SnBu$_3$ (81)	547
2-CF$_3$-bromobenzene	Ph-C(=CH$_2$)SnBu$_3$	Pd(dba)$_2$ (5%), CH$_3$CN, Et$_2$O, rt, 15 min	Ph-CH=CH-tolyl (97) $E{:}Z$ = 7:93	249
3-CF$_3$-iodobenzene	furan-OBu-t SnBu$_3$	BnPd(PPh$_3$)$_2$Cl, DMF, 70°, 16 h	furan-OBu-t / 2-CF$_3$ (90)	296
	butenolide SnBu$_3$	Pd(PPh$_3$)$_2$Cl$_2$ (1-2%), PhMe, reflux	butenolide / 3-CF$_3$ (61)	261
	furanone SnBu$_3$	Pd(PPh$_3$)$_2$Cl$_2$ (1-2%), PhMe, reflux	furanone / 3-CF$_3$ (23)	261
	spiro dioxolane-cyclobutenone SnBu$_3$	Pd/C (0.5%), AsPh$_3$ (20%), CuI (10%), NMP, 80°, 24 h	spiro product / 3-CF$_3$ (67)	461
	CO$_2$Me-C(=CH$_2$)SnBu$_3$	Pd(PPh$_3$)$_4$ (10%), CuI (75%), DMF, rt, 24 h	CO$_2$Et-C(=CH$_2$) / 4-CF$_3$ (72)	246
4-CF$_3$-iodobenzene	naphthalene-OMe SnBu$_3$	Pd(PPh$_3$)$_4$ (5%), DMF, 105°	naphthalene-OMe / 4-CF$_3$C$_6$H$_4$ (49) + naphthalene-OMe-Ph (2)	378

203

TABLE III. DIRECT CROSS-COUPLING OF ARYL ELECTROPHILES (Continued)

Substrate	Stannane	Conditions	Product(s) and Yield(s) (%)	Refs.
(aryl bromide, F_3C, NO_2)	Bu_3Sn (pyridine, Cl, Cl)	$Pd(PPh_3)_4$, CuI	(>61)	582
(2-bromobenzonitrile)	Bu_3Sn (4,4-dimethyloxazoline)	$Pd(PPh_3)_4$ (2%), C_6H_6, 80°, 20 h	(98)	550
	Bu_3Sn (indanone)	$Pd(PPh_3)_2Cl_2$, dioxane, 105°	(—)	583
	Bu_3Sn (tetralone)	$Pd(PPh_3)_2Cl_2$, dioxane, 105°	(—)	583
	Me_3Sn (thienyl-OTBDMS)	$Pd(PPh_3)_4$, PhMe, reflux	OTBDMS (65)	584
	Bu_3Sn (benzosuberone)	$Pd(PPh_3)_2Cl_2$, dioxane, 105°	(—)	584

Organotin reagent	Conditions	Product (yield %)	Refs.
Bu₃Sn–(pyridyl)–CH₂OTBDMS	Pd(PPh₃)₂Cl₂, CuI, THF, reflux, 48 h	(2-cyanophenyl)pyridyl-CH₂OTBDMS (67)	585
Me₃Sn–(tricyclic imidazopyridazine)	Pd(PPh₃)₄Cl₂, DMF, 100°, 12 h	product (83)	572
Bu₃SnSnBu₃	Pd(PPh₃)₄ (1%), PhMe, 80°	o-NC–C₆H₄–SnBu₃ (42)	310
Bu₃Sn–C₆H₄–CH₂OTBDMS	Pd(PPh₃)₂Cl₂, CuI, THF, reflux, 48 h	(2-cyanophenyl)–C₆H₄–CH₂OTBDMS (67)	585
Bu₃Sn–CH=CH–CH₂–N(TMS)₂	1. Pd(PPh₃)₄ (2%), PhMe, reflux, 72 h 2. H⁺	NC–C₆H₄–CH=CH–CH₂NH₂ (72)	464
Bu₃SnSnBu₃	Pd(PPh₃)₄ (1%), PhMe, 80°	NC–C₆H₄–SnBu₃ (31)	310
Bu₃Sn–CH₂–OH	Pd(PPh₃)₄ (5%), dioxane, 80°	NC–C₆H₄–CH₂OH (0)	233
Bu₃Sn–CH₂–CN	Pd[P(o-Tol)₃]₂Cl₂ (1%), m-xylene, 120°, 20 h	NC–C₆H₄–CH₂CN (tr)	235
Bu₃Sn–CH₂–OMe	Pd(PPh₃)₂Cl₂ (1%), HMPA, 80°, 20 h	NC–C₆H₄–CH₂OMe (57)	234

Substrates: 2-iodobenzonitrile; 3-bromobenzonitrile; 4-bromobenzonitrile

TABLE III. DIRECT CROSS-COUPLING OF ARYL ELECTROPHILES (*Continued*)

Substrate	Stannane	Conditions	Product(s) and Yield(s) (%)	Refs.
	[OSnBu₃ isopropenyl]	Pd[P(o-Tol)₃]₂Cl₂ (1%). PhMe, 100°, 5 h	NC–C₆H₄–CH₂C(O)CH₃ (0)	237
	Bu₃Sn⌁OMe	BnPd(PPh₃)₂Cl, C₆H₆, 100°, 20 h	NC–C₆H₄–CH=CH–CH₂OMe (73)	305
	Bu₃Sn⌁CO₂Et	Pd[P(o-Tol)₃]₂Cl₂ (1%), ZnBr₂, DMF, 80°, 5 h	NC–C₆H₄–CH₂CO₂Et (67)	236
	CH₂=C(OEt)SnBu₃	1. Pd(PPh₃)₂Cl₂ (1%), PhMe, 100°, 20 h 2. H⁺	NC–C₆H₄–C(O)CH₃ (81)	269
	Bu₃SnSPh	Pd(PPh₃)₄ (1%), PhMe, 120°, 20 h	NC–C₆H₄–SPh (72)	318
	CH₂=C(OTMS)SnMe₃	Pd(PPh₃)₄ (1%), PhMe, 100°	NC–C₆H₄–C(OTMS)=CH₂ (67)	457
	TTF–SnR₃	Pd(PPh₃)₄, PhMe, reflux, 3 h	NC–C₆H₄–TTF R = Me (98) R = Bu (95)	536
	Me₃Sn–C(OTMS)=CHCH₃	1. Pd(PPh₃)₄, PhMe, 100° 2. H⁺	NC–C₆H₄–C(O)CH₂CH₃ (47)	457
	Bu₃Sn–C₆H₄–CH₃	Pd(PPh₃)₄ (1%), PhMe, 120°, 20 h	NC–C₆H₄–C₆H₄–CH₃ (92)	538

Organostannane	Conditions	Product (% yield)	Refs.
Me₃Sn–C(=CH₂ type)OTMS (Me_3Sn silyl enol ether)	1. $Pd(PPh_3)_4$, PhMe, 100° 2. H⁺	4-cyanophenyl isobutyryl ketone (43)	457
Bu_3Sn–CH=CH–C(OMe)(TMS)	$Pd(PPh_3)_4$, C₆H₆, 110°	NC–C₆H₄–CH=CH–C(TMS)=OMe (65)	305
Bu_3Sn–CH=CH–C(OMe)(TMS)	1. $Pd(PPh_3)_4$, C₆H₆, 110° 2. Bu₄NF, THF	NC–C₆H₄–CH=CH–CH=OMe (77)	532
Bu_3Sn–(indol-2-yl), N–R	$Pd(PPh_3)_2Cl_2$, THF, reflux, 2 h	NC–C₆H₄–(indol-2-yl), N–R: R = Me (91), R = Boc (66)	425
Bu_3Sn–CH₂–CH=CH–N(TMS)₂	1. $Pd(PPh_3)_4$ (2%), PhMe, reflux, 24 h 2. H⁺	NC–C₆H₄–CH=CH–CH₂–NH₂ (79)	464, 540
Bu_3Sn–(pyrrol-3-yl), N–TIPS	$Pd(PPh_3)_4$ (15%), dioxane, reflux, 38 h	NC–C₆H₄–(pyrrol-3-yl), N–TIPS (81)	565
Bu_3Sn–(glycal OTBDMS, OTBDMS, OTBDMS)	$Pd(PPh_3)_2Cl_2$, PhMe, 20 min	NC–C₆H₄–(glycal OTBDMS, OTBDMS, OTBDMS) (81)	299, 300
Bu_3SnNEt_2	$Pd[P(o\text{-}Tol)_3]_2Cl_2$ (1%), PhMe, 100°, 3 h	NC–C₆H₄–NEt₂ (25)	90, 316
$Bu_3SnSnBu_3$	$Pd(PPh_3)_4$ (1%), PhMe, 120°, 20 h	NC–C₆H₄–S–C₆H₄–CN (57)	318

TABLE III. DIRECT CROSS-COUPLING OF ARYL ELECTROPHILES (Continued)

Substrate	Stannane	Conditions	Product(s) and Yield(s) (%)	Refs.
NC–C$_6$H$_4$–I	Me$_3$SnSnMe$_3$	Pd(PPh$_3$)$_4$ (1.3%), PhMe, 115°, 15 h	NC–C$_6$H$_4$–SnMe$_3$ (64)	547
	Bu$_3$SnSnBu$_3$	Pd(PPh$_3$)$_2$Br$_2$ (1.3%), PhMe, 110°, 15 h	NC–C$_6$H$_4$–SnBu$_3$ (57)	547, 310
	Me$_3$Sn–CH=CH$_2$	Pd(CH$_3$CN)$_2$Cl$_2$ (2%), DMF, 20°, 5 min	NC–C$_6$H$_4$–CH=CH$_2$ (99)	553
	Me$_3$Sn-thienyl	Pd(CH$_3$CN)$_2$Cl$_2$ (2%), DMF, 20°, 10 min	NC–C$_6$H$_4$-thienyl (91)	553
	Bu$_3$SnC≡CCO$_2$Et	Pd(PPh$_3$)$_2$Cl$_2$ (1%), Et$_4$NCl, ZnCl$_2$, DMF, rt, 72h	NC–C$_6$H$_4$–C≡CCO$_2$Et (8)	552
	Me$_3$SnPh	Pd(CH$_3$CN)$_2$Cl$_2$ (2%), DMF, 20°, 4 h	NC–C$_6$H$_4$–Ph (92)	553
	Me$_3$Sn–C$_6$H$_4$–OMe	Pd(CH$_3$CN)$_2$Cl$_2$ (2%), DMF, 20°, 4 h	NC–C$_6$H$_4$–C$_6$H$_4$–OMe (94)	553
	Me$_3$SnC≡CPh	Pd(CH$_3$CN)$_2$Cl$_2$ (2%), DMF, 20°, 10 min	NC–C$_6$H$_4$–C≡CPh (95)	553
	Me$_3$SnSnMe$_3$	Pd(CH$_3$CN)$_2$Cl$_2$ (2%), DMF, 20°, 10 min	NC–C$_6$H$_4$–SnMe$_3$ (98)	312
	Bu$_3$SnSnBu$_3$	Pd(PPh$_3$)$_4$ (1%), PhMe, 60°, 72 h	NC–C$_6$H$_4$–SnBu$_3$ (85)	311

Aryl halide	Organostannane	Conditions	Product (%)	Refs.
(bromotoluene, NO₂)	(Bu₃Sn–aryl OMe,OMe,OMe)	Pd(PPh₃)₄ (5%), CuI (8%), NMP, 70°, 48 h	(51)	575
(OTf, OMe, NO₂)	(Bu₃Sn–CH₂CH=CH₂)	Pd(PPh₃)₂Cl₂ (10-15%), PPh₃ (40%), LiCl, DMF, reflux	(0)	190
(Br, NO₂, MeO)	(Bu₃Sn–pyridine Cl, Cl)	Pd(PPh₃)₄, CuI	(≥55)	582
(I, HO₂C, N₃)	Me₃SnSnMe₃	Pd(PPh₃)₄, dioxane, 70°, 3 h	(70)	586
(CHO, Br)	(Me₃Sn–furan, OH, TBDMS)	Pd(PPh₃)₄ (2%), PhMe, reflux, 1 h	(85)	371
(CHO, OTf)	(Bu₃Sn–N-Me, CH(TMS)₂ amide)	Pd(II), LiCl, DMF, 100°	(40)	587
(Br, OHC)	(Bu₃Sn–CH=CH₂)	Pd(PPh₃)₄ (2%), BHT, PhMe, reflux, 3 h	(78)	88

TABLE III. DIRECT CROSS-COUPLING OF ARYL ELECTROPHILES (*Continued*)

Substrate	Stannane	Conditions	Product(s) and Yield(s) (%)	Refs.
	Bu₃Sn—C₆H₄—Cl	Pd(PPh₃)₄ (1%), PhMe, 120°, 20 h	NC—C₆H₄—C₆H₄—Cl (75)	538
	Bu₃Sn—CH=CH—CH(OEt)— , *E:Z* = 85:15	Pd(PPh₃)₄ (1%), C₆H₆, 110°, 15 h	—OEt (80) *E:Z* = 75:25	534, 588
	Me₃Sn— (OTMS)	Pd(PPh₃)₄, PhMe, 100°	OTMS / OHC (71)	457
	Me₃Sn— TMS	BnPd(PPh₃)₂Cl (1–2%), CHCl₃, 65°, 1 d	TMS / OHC (40)	537
	Me₃Sn—(furan)—OBu-*t*	BnPd(PPh₃)₂Cl, DMF, 70°, 16 h	OBu-*t* (96)	296
	Bu₃Sn—CH=CH—CH(OEt)₂ , *E:Z* = 85:15	Pd(PPh₃)₄ (2%), C₆H₆, 80°, 20 h	CH(OEt)₂ (85) *E:Z* = 85:15	539
	Bu₃Sn—(indole N-Boc)	Pd(PPh₃)₂Cl₂ (2%), THF, reflux	(indole N-Boc / OHC) (62)	425
	Bu₃Sn—(pyrrole N-TIPS)	Pd(PPh₃)₄ (16%), dioxane, reflux, 38 h	N-TIPS pyrrole / OHC (85)	565

Substrate	Reagent	Conditions	Product(s) (%)	Refs.
	Me₃SnSnMe₃	Pd(PPh₃)₄, LiCl, dioxane, reflux, 24 h	(92)	445
	Bu₃Sn⌇	Pd(PPh₃)₄ (2%), LiCl, BHT, dioxane, 98°, 3 h	(90)	189
	Bu₃Sn⌇OH	Pd(PPh₃)₄ (5%), dioxane, 80°	(83)	233
	Bu₃Sn⌇CN	Pd[P(o-Tol)₃]₂Cl₂ (1%), m-xylene, 120°, 3 h	(70)	235
	Bu₃Sn⌇OMe	Pd(PPh₃)₂Cl₂ (5%), HMPA, 80°, 70 h	(tr)	234
	Bu₃Sn⌇CO₂Et	Pd[P(o-Tol)₃]₂Cl₂ (1%), ZnBr₂, DMF, 80°, 5 h	(82)	236
	Bu₃SnSPh	Pd(PPh₃)₄ (1%), PhMe, 120°, 20 h	(66)	318
	OSnBu₃	Pd[P(o-Tol)₃]₂Cl₂, PhMe, 100°, 25 h	(33) + (8)	541
	Me₃Sn (furan, TBDMS)	Pd(PPh₃)₄ (2%), PhMe, 110°, 18 h	(0)	371

TABLE III. DIRECT CROSS-COUPLING OF ARYL ELECTROPHILES (Continued)

Substrate	Stannane	Conditions	Product(s) and Yield(s) (%)	Refs.
(2-iodoanisole)	Bu₃Sn–(tetramethyl-p-benzoquinonyl)	Pd₂(dba)₃ (2.5%), AsPh₃ (20%), air, CuI (50%), DMF, 60°, 3–4 h	(67)	554
(3-bromoanisole) MeO–Br	Bu₃SnC≡CSnBu₃	Pd(PPh₃)₄ (10%), LiCl, BHT, dioxane, reflux, 5 h	(85)	589
	Bu₃Sn–N(Me)(CH₂CH₂Ph)	Pd[P(o-Tol)₃]₂Cl₂ (2%), PhMe, 105°	(79)	91
	Bu₃Sn–(CH₂CH₂)N(Me)–(3,4-dimethoxyphenyl)	Pd[P(o-Tol)₃]₂Cl₂ (2%), PhMe, 105°	(84)	91
(3-iodoanisole) MeO–I	Bu₃Sn–(2,6-dimethyl-p-benzoquinonyl)	Pd₂(dba)₃ (2.5%), AsPh₃ (20%), air, CuI (50%), DMF, 60°, 3–4 h	(67)	554
	Bu₃SnC≡CSnBu₃	Pd(PPh₃)₄ (10%), LiCl, BHT, dioxane, reflux, 5 h	(70)	589
(3-methoxyphenyl triflate) MeO–OTf	Bu₃SnCH=CH₂	Pd₂(dba)₃ (2%), P(2-furyl)₃ (8%), LiCl, CuI (10%), DMF, 80°, 2 h	(70)	276

212

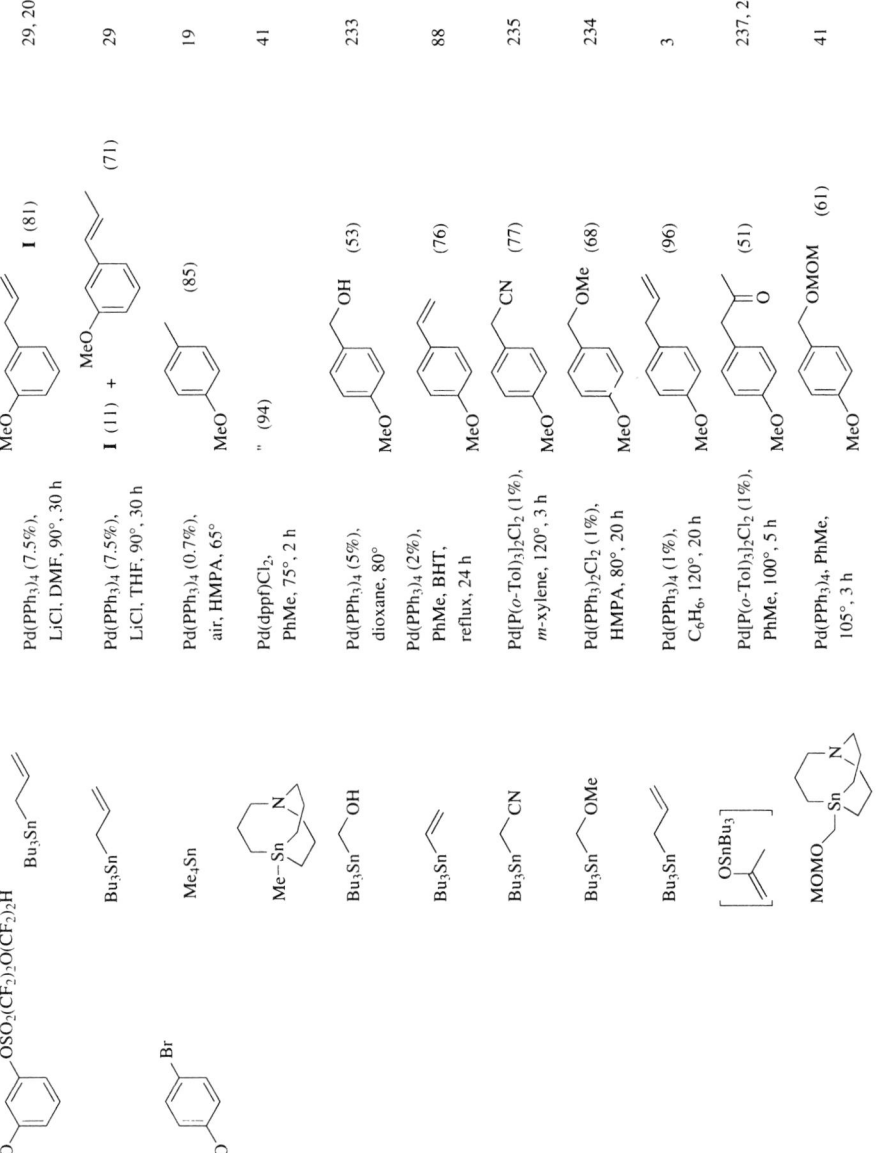

TABLE III. DIRECT CROSS-COUPLING OF ARYL ELECTROPHILES (*Continued*)

Substrate	Stannane	Conditions	Product(s) and Yield(s) (%)	Refs.
	n-Bu–Sn (bicyclic amine)	Pd(dppf)Cl$_2$, PhMe, 105°, 48 h	Bu-*n* aryl (64)	41
	Bu$_3$Sn—OMe	BnPd(PPh$_3$)$_2$Cl, C$_6$H$_6$, 100°, 20 h	MeO—aryl—OMe (82)	305
	Bu$_3$Sn—CO$_2$Et	Pd[P(*o*-Tol)$_3$]$_2$Cl$_2$ (1%), ZnBr$_2$, DMF, 80°, 5 h	MeO—aryl—CO$_2$Et (47)	236
	OEt (vinyl)	1. Pd(PPh$_3$)$_2$Cl$_2$ (1%), PhMe, 100°, 20 h 2. H$^+$	MeO—aryl—C(O)CH$_3$ (54)	269
	Bu$_3$Sn—OEt	Pd(PPh$_3$)$_2$Cl$_2$, Et$_4$NCl, DMF, 80°, 1 h	MeO—aryl—OEt (67)	272, 273
	Me$_3$Sn (oxazoline)	Pd(PPh$_3$)$_4$ (5%), C$_6$H$_6$, reflux, 12 h	MeO—aryl—oxazoline (85)	550, 551
	Bu$_3$Sn—OEt (allyl)	Pd(PPh$_3$)$_4$ (1%), C$_6$H$_6$, 110°, 15 h	MeO—aryl—OEt (63) *E:Z* = 50:50	534
	Me$_3$Sn—OTMS	1. Pd(PPh$_3$)$_4$, PhMe, 100° 2. H$^+$	MeO—aryl—C(O)CH$_3$ (75)	457
	Bu$_3$Sn—C$_6$H$_4$CF$_3$	Pd(PPh$_3$)$_4$ (1%), PhMe, 120°, 20 h	MeO—aryl—aryl—CF$_3$ (61)	538

214

	Conditions	Product (Yield)	Refs.
OSnBu3 (branched enolate)	Pd[P(o-Tol)3]2Cl2 (3%), C6H6, reflux, 3 h	(86) [p-MeOC6H4 ketone]	241
Me3Sn / OTMS	1. Pd(PPh3)4, PhMe, 100°; 2. H+	(18) [p-MeO aryl propiophenone]	457
Bu3Sn—OMe / TMS	Pd(PPh3)4, C6H6, 110°	TMS OMe (75)	305
Bu3Sn—OMe / TMS	1. Pd(PPh3)4, C6H6, 110°; 2. Bu4NF, THF	OMe (56)	532
Bu3Sn—CO2Et	Pd(OAc)2 (5%), PPh3 (20%), C6H6, reflux, 1 d	CO2Et (60) E:Z = 1:3	306, 427
Bu3Sn—CH(OEt)2, E:Z = 85:15	Pd(PPh3)4 (2%), C6H6, 80°, 20 h	CH(OEt)2 (55) E:Z = 85:15	539
OSnBu3 / C7H15-n	Pd[P(o-Tol)3]2Cl2 (3%), C6H6, reflux, 3 h	C7H15-n O (62)	241
Bu3Sn—N(TMS)2	1. Pd(PPh3)4 (2%), PhMe, reflux, 48 h; 2. H+	NH2 (78)	462, 464, 540
OSnBu3 (chromene)	Pd[P(o-Tol)3]2Cl2, PhMe, 100°, 20 h	(10) + (26) [flavanones]	541

TABLE III. DIRECT CROSS-COUPLING OF ARYL ELECTROPHILES (*Continued*)

Substrate	Stannane	Conditions	Product(s) and Yield(s) (%)	Refs.
(naphthalene–OMe)	Bu$_3$Sn–(naphthalene)–OMe	Pd(PPh$_3$)$_4$ (5%), DMF, 105°	p-MeOC$_6$H$_4$–(naphthalene)–OMe **I** (22) + Ph–(naphthalene)–OMe **II** (55)	378
	"	Pd(CH$_3$CN)$_2$Cl$_2$ (5%), AsPh$_3$, DMF, 105°	**I** (25) + **II** (49)	378
	"	Pd(CH$_3$CN)$_2$Cl$_2$ (5%), P(2-furyl)$_3$, DMF, 105°	**I** (6) + (furan)–OMe (60)	378
	Bu$_3$Sn–(furan)–C(Ph)(Ph)OH	Pd(PPh$_3$)$_4$ (6%), DMF/HMPA (10:1), 70°, 24 h	p-MeOC$_6$H$_4$–(furan)–C(Ph)(Ph)OH (61)	287
	Bu$_3$Sn–(glycal)–OBn, OBn, OBn	Pd(PPh$_3$)$_4$ (10%), PhMe, reflux 3 h	p-MeOC$_6$H$_4$–(glycal)–OBn, OBn, OBn (70)	423, 424
	Bu$_3$Sn–(glycal)–OTBDMS, OTBDMS, OTBDMS	Pd(PPh$_3$)$_2$Cl$_2$, PhMe, 2 h	p-MeOC$_6$H$_4$–(glycal)–OTBDMS, OTBDMS, OTBDMS (30)	299, 300
	Bu$_3$SnNEt$_2$	Pd[P(o-Tol)$_3$]$_2$Cl$_2$ (1%), PhMe, 100°, 3 h	Et$_2$N–C$_6$H$_4$–OMe (39)	90, 316
	Bu$_3$SnSPh	Pd(PPh$_3$)$_4$ (1%), PhMe, 120°, 20 h	PhS–C$_6$H$_4$–OMe (100)	238

216

Substrate: 4-iodoanisole (MeO–C₆H₄–I)

Organotin reagent	Conditions	Product (yield)	Refs.
Bu₃SnSnBu₃	Pd(PPh₃)₄ (1%), PhMe, 120°, 20 h	MeO–C₆H₄–S–C₆H₄–OMe (62)	316
Me₃SnTMS	Pd(PPh₃)₄ (1.3%), PhMe, 115°, 15 h	MeO–C₆H₄–TMS (50)	547
Me₃SnSnMe₃	Pd(PPh₃)₄ (5%), PhMe, 120°, 40 h	MeO–C₆H₄–SnMe₃ (52)	548
Bu₃SnSnBu₃	Pd(PPh₃)₄ (1.3-5%), PhMe, 115°, 15 h	MeO–C₆H₄–SnBu₃ (81)	547, 548
Bu₃Sn–(cyclobutenone)	BnPd(PPh₃)₂Cl (1.5%), CH₃CN, rt, 5.75 h	cyclobutenone–C₆H₄–OMe (52)	267
Bu₃SnC≡COEt	Pd(PPh₃)₂Cl₂ (5%), Et₄NCl, DMF, rt, 1.5 h	MeO–C₆H₄–C≡COEt (60)	552
Me₃Sn–C(=CH₂)CO₂Me	Pd(PPh₃)₄ (10%), CuI (75%), DMF, rt, 48 h	MeO–C₆H₄–C(=CH₂)CO₂Me (42)	246
Ph₃Sn–C(=O)–(pyrrolidine)	Pd(PPh₃)₄ (5%), PhMe, reflux	MeO–C₆H₄–C(=O)N(pyrrolidine) (62)	437
Bu₃SnC≡CO₂Et	Pd(PPh₃)₂Cl₂ (5%), Et₄NCl, ZnCl₂, DMF, rt, 1.5 h	MeO–C₆H₄–C≡CO₂Et (0)	552
Me₃SnPh	Pd/C (0.5%), AsPh₃ (20%), CuI (10%), NMP, 100°, 21 min	MeO–C₆H₄–Ph (88)	461, 33, 590

217

TABLE III. DIRECT CROSS-COUPLING OF ARYL ELECTROPHILES (*Continued*)

Substrate	Stannane	Conditions	Product(s) and Yield(s) (%)	Refs.
	Me_3Sn (piperidine carbonyl)	Pd(PPh$_3$)$_4$ (5%), PhMe, reflux	(72)	437
	Me_3Sn N(Pr-i)$_2$ carbonyl	Pd(PPh$_3$)$_4$ (5%), PhMe, reflux, 40-80 min	N(Pr-i)$_2$ (83)	437
	Me_3Sn (dithiole)	Pd(PPh$_3$)$_4$ (5%), PhMe, reflux, 5 h	(33) p-MeOC$_6$H$_4$	536
	Bu_3Sn (cyclobutenone dioxolane)	BnPd(PPh$_3$)$_2$Cl (5%), CuI (7-10%), DMF, rt	(72) p-MeOC$_6$H$_4$	12
	Bu_3Sn i-PrO (cyclobutenedione)	BnPd(PPh$_3$)$_2$Cl (5%), CuI (7-10%), DMF, rt	(80) p-MeOC$_6$H$_4$, i-PrO	12
	Bu_3Sn Ph $E:Z = 95:5$	Pd/C (0.5%), AsPh$_3$ (20%), CuI (10%), NMP, 80°, 12 h	Ph (82) MeO	461
	Bu_3Sn (benzofuran)	Pd$_2$(dba)$_3$ (2.5%), AsPh$_3$ (20%), CuI (50%), DMF, 60°, 3-4 h	(30) MeO	392
	Bu_3Sn (tetramethyl quinone)	Pd(PPh$_3$)$_4$ (0.7%), DMF, 110°, 3 h	(91) p-MeOC$_6$H$_4$	554

218

Organometallic reagent	Conditions	Product (yield)	Refs.
Bu$_3$Sn-(indole, N-SEM, 2-position)	Pd(PPh$_3$)$_4$, DMF, 110°, 3 h	MeO-C$_6$H$_4$-(indole, N-SEM) (56)	289
Bu$_3$Sn-(indole, 3-CH$_2$CH$_2$NH$_2$, N-SEM) H$_2$N	Pd(PPh$_3$)$_4$ (10%), DMF, 90°, 1.5 h	p-MeOC$_6$H$_4$-(indole, N-SEM) H$_2$N (75)	74
Bu$_3$Sn-(benzofuran, 6-OTBDMS) OTBDMS	Pd(PPh$_3$)$_4$ (4%)	MeO-C$_6$H$_4$-(benzofuran) OTBDMS (40)	392
Bu$_3$Sn-(indole, 3-position, N-Ts)	Pd$_2$(dba)$_3$ (5%), AsPh$_3$ (10%), CuI (10%), DMF, 60°, 4 h	p-MeOC$_6$H$_4$-(indole, N-Ts) (63)	291
Bu$_3$Sn-(dihydrofuran, OH, CH$_2$OCPh$_3$) OH OCPh$_3$	Pd(OAc)$_2$ (10%), AsPh$_3$ (20%), CH$_3$CN/THF (2:1), 40°, 8 h	MeO-C$_6$H$_4$-(dihydrofuran) OH OCPh$_3$ (59)	301
Bu$_3$SnC≡CSnBu$_3$	Pd(PPh$_3$)$_4$ (10%), LiCl, BHT, dioxane, reflux, 5 h	MeO-C$_6$H$_4$-C≡C-C$_6$H$_4$-OMe (71)	589
R$_3$SnSnR$_3$	Pd(PPh$_3$)$_4$ (1.3%), PhMe, 115°, 15 h	MeO-C$_6$H$_4$-SnR$_3$ R = Me (96) R = Bu (53)	547, 312
Bu$_3$Sn-CH=CH$_2$	Pd(PPh$_3$)$_2$Cl$_2$ (5%), LiCl, DMF, 25°, 6-18 h	MeO-C$_6$H$_4$-CH=CH$_2$ (70)	203
Me$_4$Sn	Pd(PPh$_3$)$_2$Cl$_2$ (5%), LiCl, DMF, 25°, 6-18 h	MeO-C$_6$H$_4$-CH$_3$ (84)	189

MeO-C$_6$H$_4$-OSO$_2$F

MeO-C$_6$H$_4$-OTf

TABLE III. DIRECT CROSS-COUPLING OF ARYL ELECTROPHILES (*Continued*)

Substrate	Stannane	Conditions	Product(s) and Yield(s) (%)	Refs.
MeO–C₆H₄–OSO₂(CF₂)₂O(CF₂)₄H	Bu₃Sn–CH=CH₂	Pd(PPh₃)₂Cl₂ (2%), LiCl, BHT, dioxane 98°, 6.5 h	MeO–C₆H₄–CH=CH₂ (74)	189
MeO–C₆H₄–Tl(O₂CCF₃)₂	Bu₃Sn–CH₂CH=CH₂	Pd(PPh₃)₄ (7.5%), LiCl, DMF, 90°, 30 h	MeO–C₆H₄–CH₂CH=CH₂ (58)	29, 201
	Bu₃Sn–(3-pyridyl)	Pd(OAc)₂, DMF, reflux	MeO–C₆H₄–(3-pyridyl) (40)	558
	Ph₄Sn	Pd(OAc)₂, DMF, reflux	MeO–C₆H₄–Ph (49)	558
2-bromobenzyl alcohol	Bu₃Sn–(2,3-dihydrofuran)	Pd(OAc)₂ (10%), P(o-Tol)₃ (20%), NEt₃, CH₃CN, reflux	(35) + (12)	54
	glycal (OBn, Ph-dioxane) stannane	Pd(PPh₃)₄ (10%), PhMe, reflux	(75)	423
2-iodobenzyl alcohol	cyclobutenone dioxolane methyl stannane	BnPd(PPh₃)₂Cl (5%), CuI (7-10%), DMF, rt	(57)	12
	cyclobutenedione i-PrO stannane	BnPd(PPh₃)₂Cl (5%), CuI (7-10%), DMF, rt	(57)	12

Pd(dba)$_2$, Bu$_4$NI, DMF, 80°, 16 h	(55)	382
Pd(PPh$_3$)$_4$ (2%), LiCl, BHT, dioxane, 98°, 4 h	(73)	189
Pd$_2$(dba)$_3$ (2%), P(2-furyl)$_3$ (8%), LiCl, CuI (10%), DMF, 80°, 2 h	(24)	276
Pd(PPh$_3$)$_4$ (2%), LiCl, BHT, dioxane, 98°, 4 h	(23) + (61)	189
PdCl$_2$ (5%), dioxane, 105°	(44)	471
PdCl$_2$ (0.8%), KOH, H$_2$O, 100°, 3 h	(88)	282
PdCl$_2$ (0.8%), KOH, H$_2$O, 100°, 6 h; PdCl$_2$ (1.6%), PPh$_2$(C$_6$H$_4$SO$_3$Na-m) (6.4%), KOH, H$_2$O, 100°, 3 h	(98) (71)	282
PdCl$_2$ (0.8%), KOH, H$_2$O, 25°, 2 h	(83)	282

TABLE III. DIRECT CROSS-COUPLING OF ARYL ELECTROPHILES (*Continued*)

Substrate	Stannane	Conditions	Product(s) and Yield(s) (%)	Refs.
HO$_2$C—⟨I⟩	Cl$_3$SnMe	PdCl$_2$ (0.5-3%), KOH, H$_2$O, 90°	HO$_2$C—⟨ ⟩—CH$_3$ (82)	281
	Cl$_3$Sn—⟍	PdCl$_2$ (0.5-3%), PPh(C$_6$H$_4$SO$_3$Na-*m*)$_2$, KOH, H$_2$O, 90°	HO$_2$C—⟨ ⟩—CH=CH$_2$ (97)	281
	Cl$_3$SnPr-*i*	PdCl$_2$ (0.5-3%), KOH, H$_2$O, 90°	HO$_2$C—⟨ ⟩—Pr-*i* (<5)	281
	Cl$_3$Sn—CO$_2$H	PdCl$_2$ (0.5-3%), PPh(C$_6$H$_4$SO$_3$Na-*m*)$_2$, KOH, H$_2$O, 90°	HO$_2$C—⟨ ⟩—CH=CH—CO$_2$H (76)	281
	Cl$_3$SnBu-*n*	"	HO$_2$C—⟨ ⟩—Bu-*n* (78)	281
	Cl$_3$SnPh	"	HO$_2$C—⟨ ⟩—Ph (95)	281
C$_8$ MsO—⟨ ⟩—Br	Bu$_3$Sn—⟨ ⟩(OMe)$_3$	Pd(PPh$_3$)$_4$ (5%), CuI (8%), NMP, 70°, 48 h	MsO—⟨ ⟩—⟨ ⟩(OMe)$_3$ (17)	575
MeS—⟨ ⟩—Br	Me$_3$Sn—oxazoline	Pd(PPh$_3$)$_4$ (5%), C$_6$H$_6$, reflux, 15 h	MeS—⟨ ⟩—oxazoline (100)	550, 551
2,6-dimethyl-⟨ ⟩—Br	Me$_3$Sn—furan—OBu-*t*	BnPd(PPh$_3$)$_2$Cl, DMF, 70°, 16 h	⟨ ⟩—furan—OBu-*t* (23)	296

222

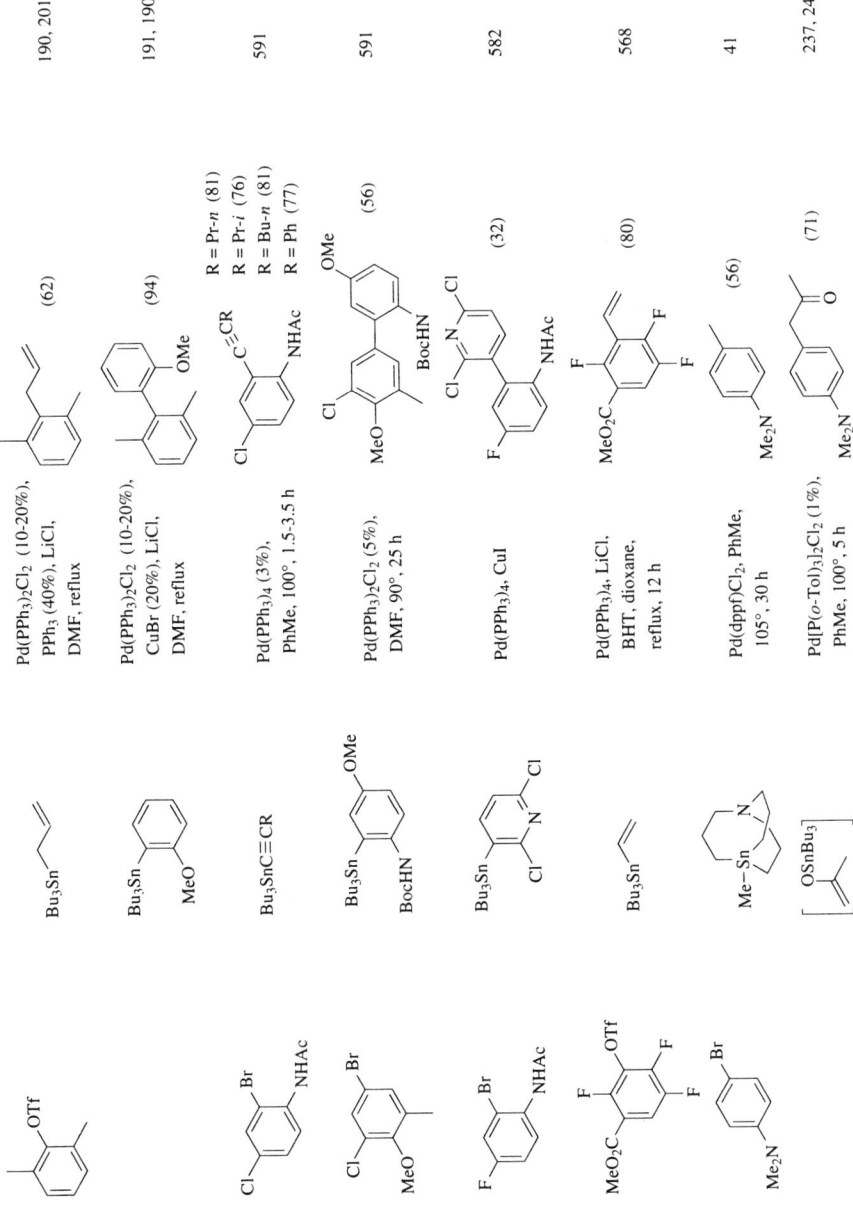

	Pd(PPh₃)₂Cl₂ (10-20%), PPh₃ (40%), LiCl, DMF, reflux	(62)	190, 201
	Pd(PPh₃)₂Cl₂ (10-20%), CuBr (20%), LiCl, DMF, reflux	(94)	191, 190
	Pd(PPh₃)₄ (3%), PhMe, 100°, 1.5-3.5 h	R = Pr-n (81) R = Pr-i (76) R = Bu-n (81) R = Ph (77)	591
	Pd(PPh₃)₂Cl₂ (5%), DMF, 90°, 25 h	(56)	591
	Pd(PPh₃)₄, CuI	(32)	582
	Pd(PPh₃)₄, LiCl, BHT, dioxane, reflux, 12 h	(80)	568
	Pd(dppf)Cl₂, PhMe, 105°, 30 h	(56)	41
	Pd[P(o-Tol)₃]₂Cl₂ (1%), PhMe, 100°, 5 h	(71)	237, 240

TABLE III. DIRECT CROSS-COUPLING OF ARYL ELECTROPHILES (*Continued*)

Substrate	Stannane	Conditions	Product(s) and Yield(s) (%)	Refs.
(4-Me$_2$N-C$_6$H$_4$-I)	MOMO–CH$_2$–Sn(N-bicyclic)	Pd(PPh$_3$)$_4$, PhMe, 105°, 3 h	(CH$_2$OMOM aryl, Me$_2$N) (80)	41
	Bu$_3$Sn–CH=CH–OEt	Pd(PPh$_3$)$_2$Cl$_2$, Et$_4$NCl, DMF, 80°, 18 h	(CH=CH–OEt aryl, Me$_2$N) (22)	272, 273
	Bu$_3$SnNEt$_2$	Pd[P(o-Tol)$_3$]$_2$Cl$_2$ (1%), PhMe, 100°, 3 h	(NEt$_2$ aryl, Me$_2$N) (36)	90, 316
	[Bu$_3$Sn–N(Bn)Me]	Pd[P(o-Tol)$_3$]$_2$Cl$_2$ (2%), PhMe, 105°	(N(Bn)Me aryl, Me$_2$N) (81)	91
	Me$_3$SnSnMe$_3$	Pd(PPh$_3$)$_4$ (1.3%), PhMe, 115°, 15 h	(SnMe$_3$ aryl, Me$_2$N) (0)	547
	Bu$_3$SnC≡CCOEt	Pd(PPh$_3$)$_2$Cl$_2$ (5%), Et$_4$NCl, DMF, 50°, 12 h	(C≡CCOEt aryl, Me$_2$N) (0)	302
(4-NH$_2$-CH$_2$CH$_2$-C$_6$H$_4$-Br)	Bu$_3$SnSnBu$_3$	Pd(PPh$_3$)$_4$ (1%), PhMe, 95°, 48 h	(SnBu$_3$ aryl, CH$_2$CH$_2$NH$_2$) (78)	592
	Bu$_3$SnC≡C–CH$_2$OMe	Pd(PPh$_3$)$_4$ (3%), PhMe, 100°, 3.5 h	(C≡C–CH$_2$OMe aryl, NHAc) (60)	591
(2-Br-C$_6$H$_4$-NHAc)	Bu$_3$Sn–CH=CH–OEt	Pd(PPh$_3$)$_2$Cl$_2$ (1%), dioxane, 100°, 4 h	(CH=CH–OEt aryl, NHAc) (72)	273
	Bu$_3$SnC≡CR	Pd(PPh$_3$)$_4$ (3%), PhMe, 100°, 2.5 h	(C≡CR aryl, NHAc) R = Pr-n (77) R = Pr-i (68)	591

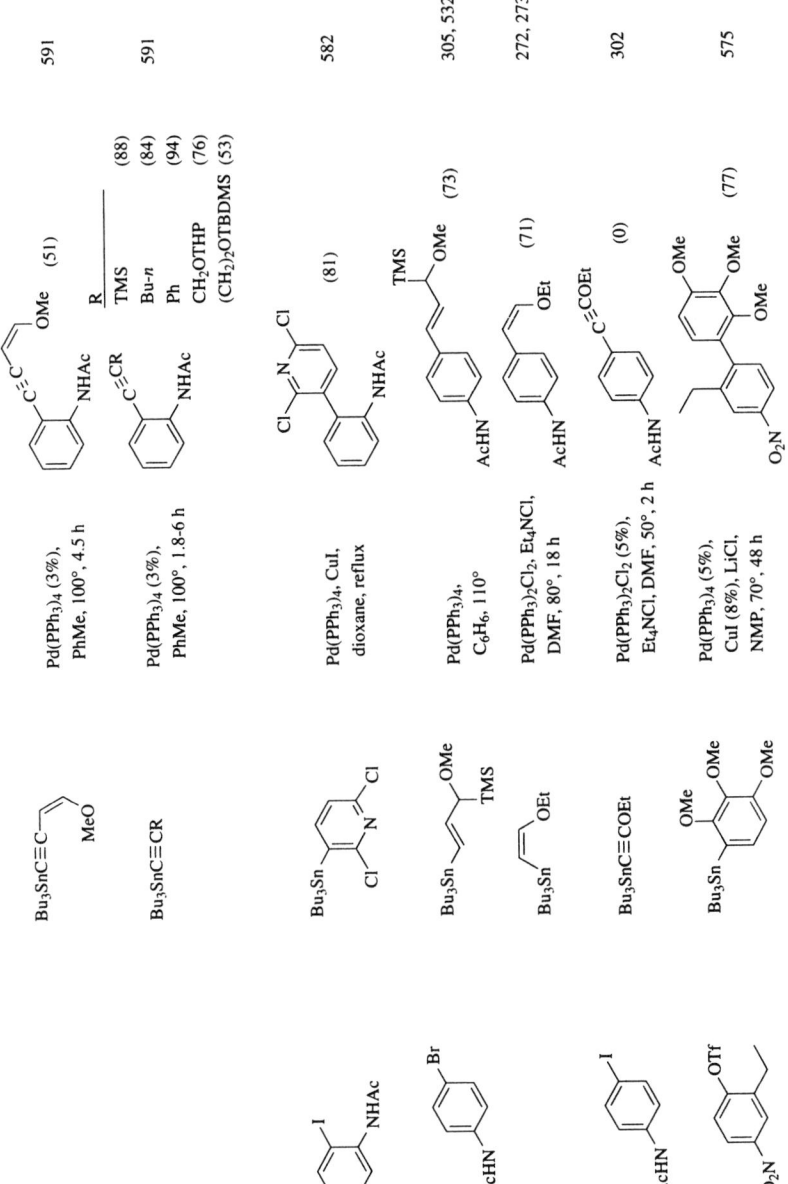

		R	
		TMS	(88)
		Bu-*n*	(84)
		Ph	(94)
		CH$_2$OTHP	(76)
		(CH$_2$)$_2$OTBDMS	(53)

591

591

582

305, 532

272, 273

302

575

TABLE III. DIRECT CROSS-COUPLING OF ARYL ELECTROPHILES (*Continued*)

Substrate	Stannane	Conditions	Product(s) and Yield(s) (%)	Refs.
(aryl iodide, O_2N, NHAc)	Bu_3Sn (dichloropyridine)	Pd(PPh$_3$)$_4$, CuI, dioxane, reflux	(81)	582
(2'-bromoacetophenone)	Bu_3Sn (vinyl)	Pd$_2$(dba)$_3$ (2%), P(2-furyl)$_3$ (4%), LiCl, NMP, rt, 2 h	(90)	40
	Bu$_3$SnSnBu$_3$	Pd(PPh$_3$)$_4$ (1%), PhMe, 80°	(25)	310
(4-bromoacetophenone)	Me$_4$Sn	Pd(PPh$_3$)$_4$ (0.7%), air, HMPA, 65°	(95)	19
	Bu$_3$Sn⌣OH	Pd(PPh$_3$)$_4$ (5%), dioxane, 80°	(0)	233
	Bu$_3$Sn (vinyl)	Pd(PPh$_3$)$_4$ (2%), BHT, PhMe, reflux, 4 h	(82)	88
	Bu$_3$Sn⌣CN	Pd[P(o-Tol)$_3$]$_2$Cl$_2$ (1%), m-xylene, 120°, 20 h	(tr)	235
	Bu$_3$Sn⌣OMe	Pd(PPh$_3$)$_2$Cl$_2$ (1%), HMPA, 80°, 20 h	(64)	234

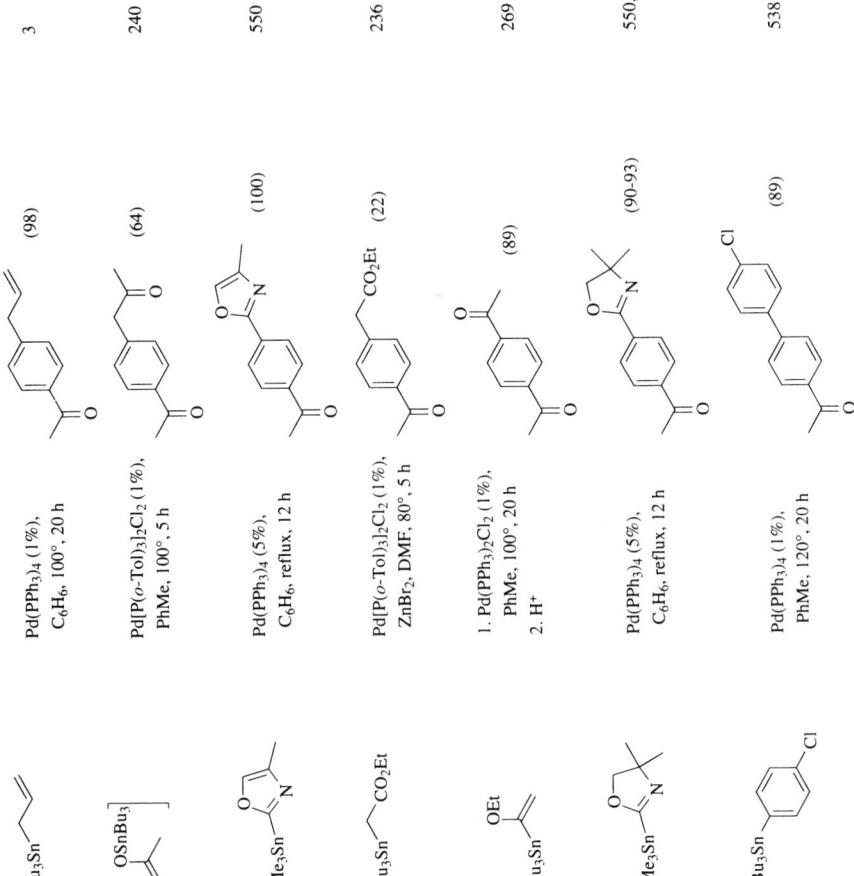

Bu₃Sn〜	Pd(PPh₃)₄ (1%), C₆H₆, 100°, 20 h	(98)	3
[OSnBu₃]	Pd[P(o-Tol)₃]₂Cl₂ (1%), PhMe, 100°, 5 h	(64)	240
Me₃Sn-oxazole	Pd(PPh₃)₄ (5%), C₆H₆, reflux, 12 h	(100)	550
Bu₃Sn〜CO₂Et	Pd[P(o-Tol)₃]₂Cl₂ (1%), ZnBr₂, DMF, 80°, 5 h	(22)	236
OEt	1. Pd(PPh₃)₂Cl₂ (1%), PhMe, 100°, 20 h 2. H⁺	(89)	269
Me₃Sn-oxazoline	Pd(PPh₃)₄ (5%), C₆H₆, reflux, 12 h	(90-93)	550, 551
Bu₃Sn-C₆H₄-Cl	Pd(PPh₃)₄ (1%), PhMe, 120°, 20 h	(89)	538

TABLE III. DIRECT CROSS-COUPLING OF ARYL ELECTROPHILES (*Continued*)

Substrate	Stannane	Conditions	Product(s) and Yield(s) (%)	Refs.
	Bu$_3$Sn— (m-tolyl)	Pd(PPh$_3$)$_4$ (1%), PhMe, 120°, 20 h	(98)	538
	Bu$_3$Sn— (p-tolyl)	Pd(PPh$_3$)$_4$ (1%), PhMe, 120°, 20 h	(90)	538
	Bu$_3$Sn— OEt	Pd(PPh$_3$)$_4$ (1%), C$_6$H$_6$, 115°, 15 h	(55) E:Z = 63:37	534
	[OSnBu$_3$]	Pd[P(o-Tol)$_3$]$_2$Cl$_2$ (3%), C$_6$H$_6$, reflux, 3 h	(70)	241
	Bu$_3$Sn— CO$_2$Et	Pd(PPh$_3$)$_4$ (5%), C$_6$H$_6$, reflux, 21 h	CO$_2$Et (60) E:Z = 1:3	306
	Bu$_3$Sn— CH(OEt)$_2$ E:Z = 85:15	Pd(PPh$_3$)$_4$ (2%), C$_6$H$_6$, 80°, 20 h	CH(OEt)$_2$ (78) E:Z = 85:15	539
	Bu$_3$Sn— N(TMS)$_2$	1. Pd(PPh$_3$)$_4$ (2%) PhMe, reflux, 48 h 2. H$^+$	NH$_2$ (30)	540

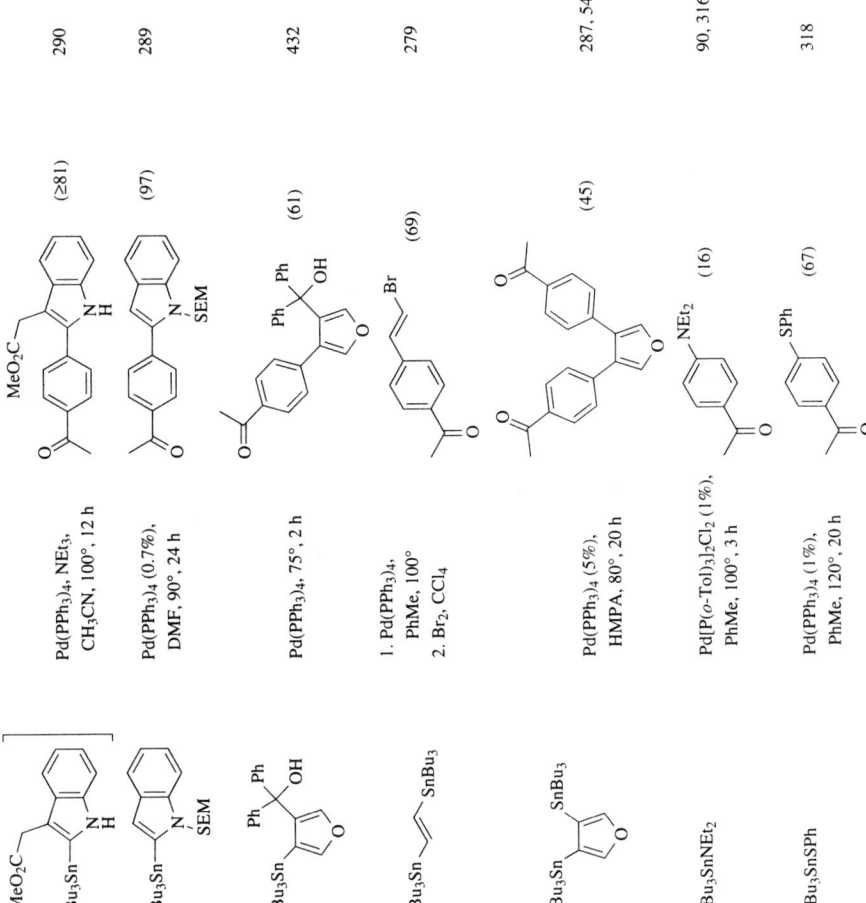

	Pd(PPh₃)₄, NEt₃, CH₃CN, 100°, 12 h	(≥81)	290
Bu₃Sn indole (N-SEM)	Pd(PPh₃)₄ (0.7%), DMF, 90°, 24 h	(97)	289
furan, Ph₂C(OH), Bu₃Sn	Pd(PPh₃)₄, 75°, 2 h	(61)	432
Bu₃Sn–CH=CH–SnBu₃	1. Pd(PPh₃)₄, PhMe, 100° 2. Br₂, CCl₄	(69)	279
Bu₃Sn furan SnBu₃	Pd(PPh₃)₄ (5%), HMPA, 80°, 20 h	(45)	287, 546
Bu₃SnNEt₂	Pd[P(o-Tol)₃]₂Cl₂ (1%), PhMe, 100°, 3 h	(16)	90, 316
Bu₃SnSPh	Pd(PPh₃)₄ (1%), PhMe, 120°, 20 h	(67)	318

TABLE III. DIRECT CROSS-COUPLING OF ARYL ELECTROPHILES (*Continued*)

Substrate	Stannane	Conditions	Product(s) and Yield(s) (%)	Refs.
4-iodoacetophenone	$Bu_3SnSnBu_3$	Pd(PPh₃)₄ (1%), PhMe, 120°, 20 h	(52)	318
	R_3SnSnR_3	Pd(PPh₃)₄ (1.3%), PhMe, 115°, 15 h	R = Me (56) R = Bu (57)	547, 310
	Me_3Sn–CH=CH₂	Pd(CH₃CN)₂Cl₂ (2%), DMF, 70°, 10 min	(96)	553, 461
	Bu_3Sn (F-vinyl)	Pd(PPh₃)₄, THF, 65°, 1 h	(66)	264
	Bu_3Sn (CO₂Me-vinyl)	Pd(PPh₃)₄ (10%), CuI (75%), DMF, rt, 12 h	(78)	246
	Cl_3SnPh	PdCl₂ (0.8%), KOH, PPh₂(C₆H₄SO₃Na-m) (1.6%), H₂O, 100°, 3 h	**I** (80)	282
	Me_3SnPh	PhPd(PPh₃)₂I (2%), HMPA, 70°, 30 min	**I** (76) + Ph-Ph (24)	463
	Bu_3Sn (benzothiophene)	Pd/C (5%), CuI (10%), AsPh₃ (20%), NMP, 80°, 24 h	(60)	461

Substrate	Organotin reagent	Conditions	Product	Yield	Refs.
4-acetylphenyl-N$_2$$^+$ BF$_4$$^-$	Me$_3$SnSnMe$_3$	Pd(CH$_3$CN)$_2$Cl$_2$ (2%), DMF, 20°, 15 min	4-acetylphenyl–SnMe$_3$	(95)	312
	Bu$_3$SnSnBu$_3$	Pd(PPh$_3$)$_4$ (1%), PhMe, 60°, 72 h	4-acetylphenyl–SnBu$_3$	(83)	311
	Ph–C(SnBu$_3$)=CH$_2$	Pd(dba)$_2$ (5%), CH$_3$CN, Et$_2$O, rt, 1h	4-acetylphenyl–CH=CH–Ph	(80) E:Z = 17:83	249
	Bu$_3$Sn–CH=CH$_2$	Pd(PPh$_3$)$_2$Cl$_2$ (5%), LiCl, DMF, 25°, 6-18 h	4-acetylphenyl–CH=CH$_2$	(70)	203
4-acetylphenyl–OSO$_2$F	Me$_4$Sn	Pd(PPh$_3$)$_4$ (2%), LiCl, BHT, dioxane, 100°, 16 h	4-acetyltoluene	(75)	189
	Me$_4$Sn	"	"	(>95)	11
	Bu$_3$Sn–CH=CH$_2$	Pd(PPh$_3$)$_4$ (2%), BHT, LiCl, dioxane, 98°, 4 h	4-acetylphenyl–CH=CH$_2$	I (95)	189
4-acetylphenyl–OTf	"	Pd$_2$(dba)$_3$ (1%), LiCl, L (8%), NMP, 35°			
		L = PPh$_3$		I (>95)	11
		L = P(2-furyl)$_3$		I (95)	11, 40
		L = AsPh$_3$		I (>95)	11

Additional condition noted: Pd$_2$(dba)$_3$ (1%), LiCl, AsPh$_3$ (8%), NMP, 60°

231

TABLE III. DIRECT CROSS-COUPLING OF ARYL ELECTROPHILES (Continued)

Substrate	Stannane	Conditions	Product(s) and Yield(s) (%)	Refs.
	Bu$_3$Sn (F-substituted vinyl)	Pd(PPh$_3$)$_4$, LiCl, THF, 65°, 1 h	(65)	264
	Bu$_3$SnCH=C=CH$_2$	Pd$_2$(dba)$_3$ (2%), P(2-furyl)$_3$ (8%), LiCl, CuI (8%), DMF, 80°, 1 h	CH=C=CH$_2$ (60)	276
	Bu$_3$Sn	Pd(PPh$_3$)$_4$ (2%), LiCl, BHT, dioxane, 98°, 43 h	I + (72) 1:3	189
	Bu$_3$Sn	Pd$_2$(dba)$_3$ (1%), P(2-furyl)$_3$ (4%), LiCl, NMP, rt, 2 h	I (78)	11
	Bu$_3$Sn, E:Z = 2:1	Pd(PPh$_3$)$_4$ (2%), LiCl, BHT, dioxane, 98°, 31 h	I + II E:Z = 4:1 + III I + II + III (11) I:II:III = 65:20:15	189
	Bu$_4$Sn	Pd$_2$(dba)$_3$ (1%), AsPh$_3$ (12%), LiCl, NMP, 80°	I (>98) IV (29)	30, 189
	Me$_3$Sn—TMS	Pd(PPh$_3$)$_4$ (2%), LiCl, BHT, dioxane, 98°, 65 h	(83)	189

Reagent	Conditions	Product(s)	Ref.
Me₃SnPh	Pd(PPh₃)₄ (2%), LiCl, BHT, dioxane, 98°, 23 h	**I** (85)	189
Me₃SnPh	Pd(PPh₃)₄ (2%), LiCl, BHT, DMF, 90°, 1 h	**I** (54) + **II** (16)	189
Me₃SnPh	Pd₂(dba)₃ (1%), LiCl, AsPh₃ (8%), NMP, 80°	**I** (54) + **II** (21)	30
Bu₃SnPh	Pd₂(dba)₃ (1%), LiCl, PPh₃ (8%), dioxane, 65°, 40 h	**I** (81) + **III** (2)	189
Bu₃SnPh	Pd₂(dba)₃ (1%), AsPh₃ (8%), NMP, 65°, 40 h	**I** (81) + **III** (1)	30
Bu₃SnPh	Pd₂(dba)₃ (1%), LiCl, AsPh₃ (8%), NMP, 65°, 40 h	**I** (92) + **III** (6)	30
Bu₃SnR	Pd₂(dba)₃ (1%), LiCl, AsPh₃ (8%), NMP, 80°	**I** + **II**	30

R	
$p\text{-}CF_3C_6H_4$	**I** (89) + **II** (9)
$o\text{-}OHCC_6H_4$	**I** (72) + **II** (4)
$o\text{-}MeOC_6H_4$	**I** (88) + **II** (10)
$p\text{-}MeOC_6H_4$	**I** (92) + **II** (7)
$o\text{-}CH_2OH$	**I** (50) + **II** (24)
$p\text{-}CH_2OH$	**I** (76) + **II** (6)
$o\text{-}EtC_6H_4$	**I** (46) + **II** (27)

TABLE III. DIRECT CROSS-COUPLING OF ARYL ELECTROPHILES (Continued)

Substrate	Stannane	Conditions	Product(s) and Yield(s) (%)	Refs.
	Bu$_3$Sn, Me$_2$N	Pd$_2$(dba)$_3$ (1%), LiCl, AsPh$_3$ (8%), NMP, 65°	OH (45) + Bu-n (6)	30
	Bu$_3$Sn (cyclohexene), Bu-t	Pd(PhCN)$_2$Cl$_2$ (2%), AsPh$_3$ (8%), CuI (5%), LiCl, NMP, 80°, 6 h	Bu-t (84)	33
	MeO$_2$C, Bu$_3$Sn (indole) N H	Pd(PPh$_3$)$_4$, NEt$_3$, CH$_3$CN/DMF (9:1), 100°, 3 h	MeO$_2$C (indole) N H (≥75)	290
	Bu$_3$Sn (naphthalene) NMe$_2$	Pd$_2$(dba)$_3$ (1%), LiCl, AsPh$_3$ (8%), NMP, 65°	OH (39) + Me$_2$N (0)	30
	Bu$_3$Sn, CO$_2$Et, NHAc	Pd$_2$(dba)$_3$•CHCl$_3$ (4%), AsPh$_3$ (30%), LiCl, THF, reflux, 6 h	CO$_2$Et NHAc (66) + CO$_2$Et AcHN (3)	473
	Me$_3$SnSnMe$_3$	Pd(PPh$_3$)$_4$ (2%), LiCl, BHT, dioxane, 98°, 24 h	(94)	189

Substrate	Reagent	Conditions	Product(s) (%)	Refs.
4-(OSO₂Ph)C₆H₄COCH₃	Bu₃Sn–CH=CH₂	Pd(OAc)₂ (5%), dppp (5.5%), LiCl, DMF, 90°, 24 h	4-vinylacetophenone (90)	202
4-(OSO₂C₆H₄F-p)C₆H₄COCH₃	Bu₃Sn–CH=CH₂	"	4-vinylacetophenone (86)	202
	Bu₃Sn–CH₂CH=CH₂	Pd(OAc)₂ (5%), dppp (5.5%), LiCl, DMF, 100°, 26 h	4-(1-propenyl)acetophenone (52) + 4-allylacetophenone (28)	202
	Bu₄Sn	Pd(OAc)₂ (5%), Ph₂PMe (11%), LiCl, DMF, 120°, 48 h	4-(n-Bu)acetophenone (67)	202
	Bu₃SnPh	Pd(OAc)₂ (5%), dppp (5.5%), LiCl, DMF, 110°, 24 h	4-Ph-acetophenone (85)	202
	Bu₃Sn–CH=CH–Ph	Pd(OAc)₂ (5%), dppp (5.5%), LiCl, DMF, 100°, 20 h	styryl acetophenone (69)	202
4-(OSO₂R)C₆H₄COCH₃	Bu₃Sn–CH=CH₂	Pd(OAc)₂ (5%), dppp (5.5%), LiCl, DMF, 90°, 24 h	R = C₆H₄NO₂-p (78); C₆H₂Me₃-2,4,6 (53)	202
4-(OSO₂C₆H₄F-p)C₆H₄OEt	Bu₃Sn–CH=CH₂	Pd(OAc)₂ (5%), Ph₂PMe (5.5%), LiCl, DMF, 110°, 88 h	EtO–C₆H₄–OSO₂C₆H₄(NMe₂)-p I (34)	202
	Bu₃Sn–CH=CH₂	Pd(OAc)₂ (5%), dppp (5.5%), LiCl, DMF, 110°, 96 h	EtO-vinyl (12) + I (13)	202

235

TABLE III. DIRECT CROSS-COUPLING OF ARYL ELECTROPHILES (*Continued*)

Substrate	Stannane	Conditions	Product(s) and Yield(s) (%)	Refs.
[2-bromophenethyl alcohol; Br, CH_2CH_2OH]	[Bu_3Sn–dihydropyranyl]	$Pd(OAc)_2$, NEt_3, CH_3CN	[dihydropyran-phenethyl alcohol product] (71)	54
[aryl bromide; Br, CO_2Me]	[Bu_3Sn–glycal with three OTBDMS]	$Pd(PPh_3)_2Cl_2$, PhMe, 15 h	[MeO_2C-aryl glycal, OTBDMS, OTBDMS, OTBDMS] (48)	300
[aryl iodide; I, CO_2Me]	[Bu_3Sn–cyclobutenone]	$BnPd(PPh_3)_2Cl$ (1.5%), THF, 65°, 2 h	[cyclobutenone–aryl, CO_2Me] (50)	267
	[Bu_3Sn–furan-2(5H)-one]	$Pd(PPh_3)_2Cl_2$ (1-2%), PhMe, reflux	[furanone–aryl, CO_2Me] (72)	261
	[Bu_3Sn–furan-2(5H)-one isomer]	$Pd(PPh_3)_2Cl_2$ (1-2%), PhMe, reflux	[butenolide–aryl, CO_2Me] (65)	261
	[Bu_3Sn–thiophene]	$Pd(PPh_3)_2Cl_2$ (0.5%), THF, reflux, 20 h	[thienyl–aryl, CO_2Me] (81)	286
	[Bu_3Sn–methyl cyclobutenone dioxolane]	$BnPd(PPh_3)_2Cl$ (5%), CuI (7-10%), DMF, rt	[dioxolane cyclobutenone–aryl, MeO_2C] (70)	12

236

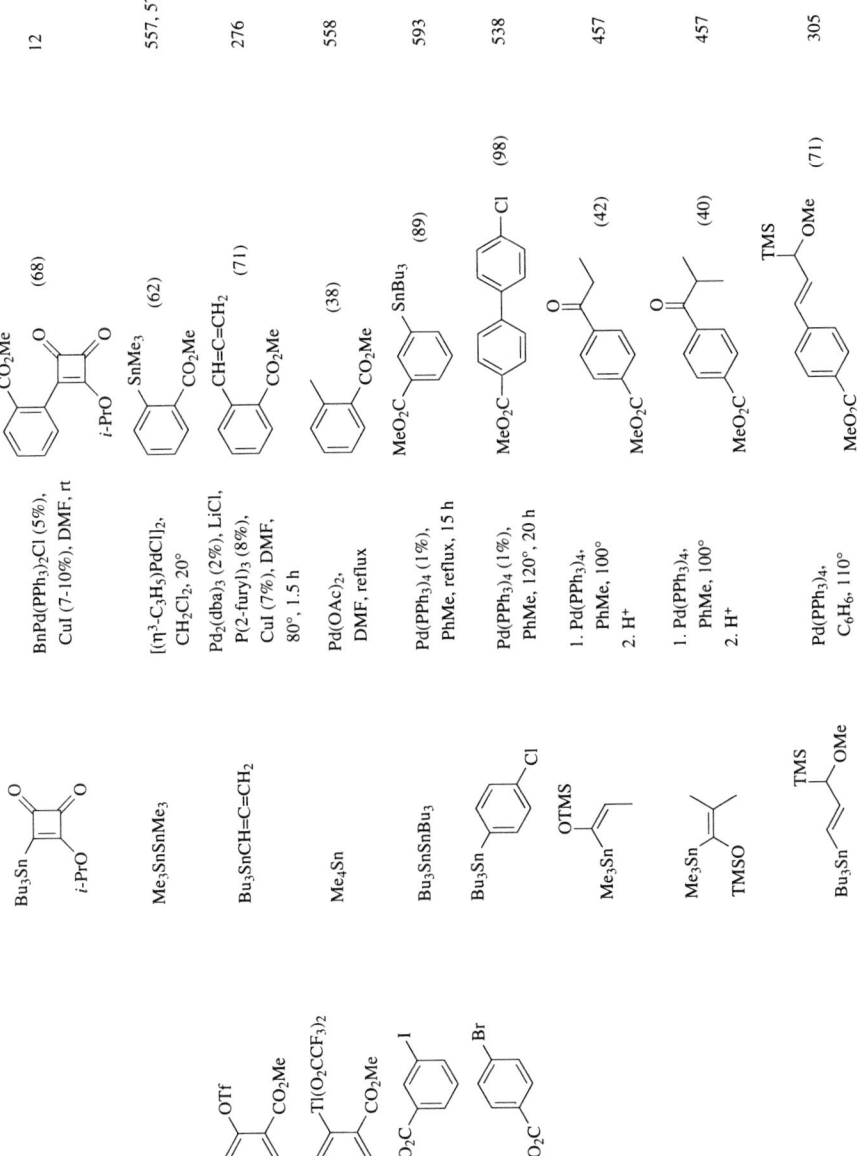

557, 573

276

558

593

538

457

457

305

12

TABLE III. DIRECT CROSS-COUPLING OF ARYL ELECTROPHILES (*Continued*)

Substrate	Stannane	Conditions	Product(s) and Yield(s) (%)	Refs.
MeO2C–C6H4–I (4-iodo)	Bu3Sn / TMS / OMe (alkenyl stannane)	1. Pd(PPh3)4, C6H6, 110° 2. Bu4NF, THF	OMe styryl product (82)	532
	Bu3Sn sugar, OTBDMS ×3	Pd(PPh3)2Cl2, PhMe, 15 h	OTBDMS glycal product (56)	299, 300
	Me3Sn–furan (O)	Pd(PPh3)2Cl2 (0.5%), THF, reflux, 20 h	(73)	286
	Bu3Sn–C(=CH2)CO2Me	Pd(PPh3)4 (10%), CuI (75%), DMF, rt, 12 h	CO2Me (78)	246
	Me3Sn–thiophene (S)	Pd(CH3CN)2Cl2 (2%), DMF, 70°, 20 min	(98)	553
	Me3Sn–pyridine (N)	Pd(PPh3)2Cl2 (0.5%), THF, reflux, 20 h	(95)	286
	Bu3SnC≡CCO2Et	Pd(PPh3)2Cl2 (1%), Et4NCl, ZnCl2, C6H6, rt, 48 h	C≡CCO2Et (6)	552

Organostannane	Conditions	Product (Yield)	Refs.
Me₃SnPh	Pd(CH₃CN)₂Cl₂ (2%), DMF, 20°, 5 h	p-MeO₂C-C₆H₄-Ph **I** (97)	553
Me₃SnPh	PhPd(PPh₃)₂I (2%), HMPA, 70°, 30 min	**I** (74) + Ph-Ph (21)	463
Me₃Sn–(thiophene, TMS, Ph₃C–N–N=N)	Pd(PPh₃)₂Cl₂, DMF, heat	p-MeO₂CC₆H₄... X = O (56), X = S (61)	584
Me₃SnSnMe₃	Pd(CH₃CN)₂Cl₂ (2%), DMF, 20°, 15 min	MeO₂C-C₆H₄-SnMe₃ (97)	312
Bu₃SnSnBu₃	Pd(PPh₃)₂Cl₂ (2%), PhMe, reflux, 55 min	MeO₂C-C₆H₄-SnBu₃ **I** (75)	594
Bu₃SnSnBu₃	Pd(CH₃CN)₂Cl₂ (2%), HMPA, 20°, 10 min	**I** (92)	313
Bu₃SnSnBu₃	NiBr₂ (10%), HMPA, 135°, 7 h	**I** (72)	566
Bu₃Sn–(indole, Ts)	Pd₂(dba)₃ (5%), AsPh₃ (10%), DMF, 60°, 4h	p-MeO₂CC₆H₄-(indole, Ts) (80)	291
Bu₃Sn–(imidazole, Bu-n, SEM)	Pd(PPh₃)₄ (2%), LiCl, BHT, dioxane, 100°, 3.5 h	MeO₂C-C₆H₄-(imidazole, Bu-n, SEM) (60)	433
Bu₃Sn–(glycal, OTBDMS, OTBDMS, OTBDMS)	Pd(PPh₃)₂Cl₂, PhMe, 2 h	(OTBDMS, OTBDMS, OTBDMS / OAc) (40)	299, 300

MeO₂C–C₆H₄–OTf

Br–C₆H₄–OAc

TABLE III. DIRECT CROSS-COUPLING OF ARYL ELECTROPHILES (*Continued*)

Substrate	Stannane	Conditions	Product(s) and Yield(s) (%)	Refs.
(AcO–C6H4–Br)	Bu3Sn⟶ (vinyl)	Pd(PPh3)4 (3%), BHT, PhMe, reflux, 8 h	(62)	88
(AcO–C6H4–I)	Bu3Sn–(2-thienyl)	Pd(PPh3)2Cl2 (0.5%), THF, reflux	(65)	286
(CHO, OTf, OMe arene)	Bu3Sn–(allyl)	Pd(PPh3)2Cl2 (10–15%), PPh3 (40%), LiCl, DMF, reflux	(0) + (34)	190
(Br, OMOM arene)	(3-methylisoxazol-5-yl)–SnBu3	Pd(PPh3)2Cl2 (1%), dioxane, reflux, 4 h	(68)	530
	Bu3Sn–CH=CH–OEt	Pd(PPh3)2Cl2 (1%), dioxane, 100°, 5 h	(79)	530
	(glycal)–OTBDMS stannane	Pd(PPh3)2Cl2, PhMe, 100°, 1 h	(65)	299, 300
(MeO, Br, OMe arene)				
(OMe, OTf, OMe arene)	Me4Sn	Pd(PPh3)2Cl2 (10–15%), PPh3 (40%), LiCl, DMF, reflux	(92)	190, 376

Organometallic	Conditions	Product (%)	Refs.
Bu₃Sn (vinyl)	Pd(PPh₃)₂Cl₂ (10-15%), PPh₃ (40%), LiCl, DMF, reflux	2-vinyl-1,3-dimethoxybenzene (OMe, OMe) (85)	190, 376
Bu₃Sn (allyl)	Pd(PPh₃)₂Cl₂ (10-15%), PPh₃ (40%), LiCl, DMF, reflux	(OMe, OMe) (84)	190, 379
Bu₄Sn	Pd(PPh₃)₂Cl₂ (10-15%), PPh₃ (40%), LiCl, DMF, reflux	Bu-n (OMe, OMe) (0) + (OMe, OMe) (86)	190, 379
Bu₃SnPh	Pd(PPh₃)₂Cl₂ (10-15%), PPh₃ (40%), LiCl, DMF, reflux	Ph (OMe, OMe) (74)	190, 379
MeO (aryl) Bu₃Sn	Pd(PPh₃)₂Cl₂ (10-20%), PPh₃ (15-20%), LiCl, CuBr (20%), DMF, reflux	MeO, MeO biphenyl (OMe) (49)	191, 190
Bu₃SnC≡CPh	Pd(PPh₃)₂Cl₂ (10-15%), PPh₃ (40%), LiCl, DMF, reflux	C≡CPh (OMe) (50)	190, 379
bicyclic stannane H, CO₂Bu-t, O, Bu₃Sn	Pd(PPh₃)₄, THF	bicyclic CO₂Bu-t, H, O, MeO, MeO (55)	595
Br₃SnMe	PdCl₂ (0.8%), KOH. H₂O, 100°, 3 h	CO₂H (98)	282

Aryl halide substrates (left column):

MeO, MeO — Br (3,4-dimethoxybromobenzene)

CO₂H — I (4-iodo, CO₂H)

241

TABLE III. DIRECT CROSS-COUPLING OF ARYL ELECTROPHILES (Continued)

Substrate	Stannane	Conditions	Product(s) and Yield(s) (%)	Refs.
C9				
[methylenedioxybenzene with Br and CHO]	[indoline, N–Boc, SnBu3]	Pd(OAc)2 (10%), P(o-Tol)3 (20%), Et3N, DMF, 70°, 50 h	[indoline, Boc–N, CHO] (63)	596
[benzene with Br, Me, MsO]	Bu3Sn [OMe OMe OMe]	Pd(PPh3)4 (5%), CuI (8%), NMP, 70°, 48 h	[biaryl OMe OMe OMe, CHO, Me, MsO] (33)	575
[benzene with OH, OMe, I, OH, MeO]	Me3Sn [isoxazole O–N, OMe OMe OMe]	Pd(PPh3)4 (10%), dioxane, 101°	[OMe OMe OMe, OH, MeO, OH, O–N] (80) and (33)	471
[2-allylphenyl OTf]	Bu3SnC≡CPr-n	Pd(PPh3)2Cl2 (2%), LiCl, BHT, DMF, 60°, 7 h	[2-allylphenyl C≡CPr-n] (65)	189
[bromomesitylene, Br]	[OSnBu3 isopropenyl]	Pd[P(o-Tol)3]2Cl2 (1%), PhMe, 100°, 5 h	[mesityl CH2C(O)] (94)	240
	Bu3Sn [3-chlorophenyl, Cl]	Pd(PPh3)4 (1%), PhMe, 120°, 20 h	[mesityl, 3-Cl-phenyl, Cl] (58)	538
	Bu3SnSPh	Pd(PPh3)4 (1%), PhMe, 120°, 20 h	[mesityl SPh] (34)	318

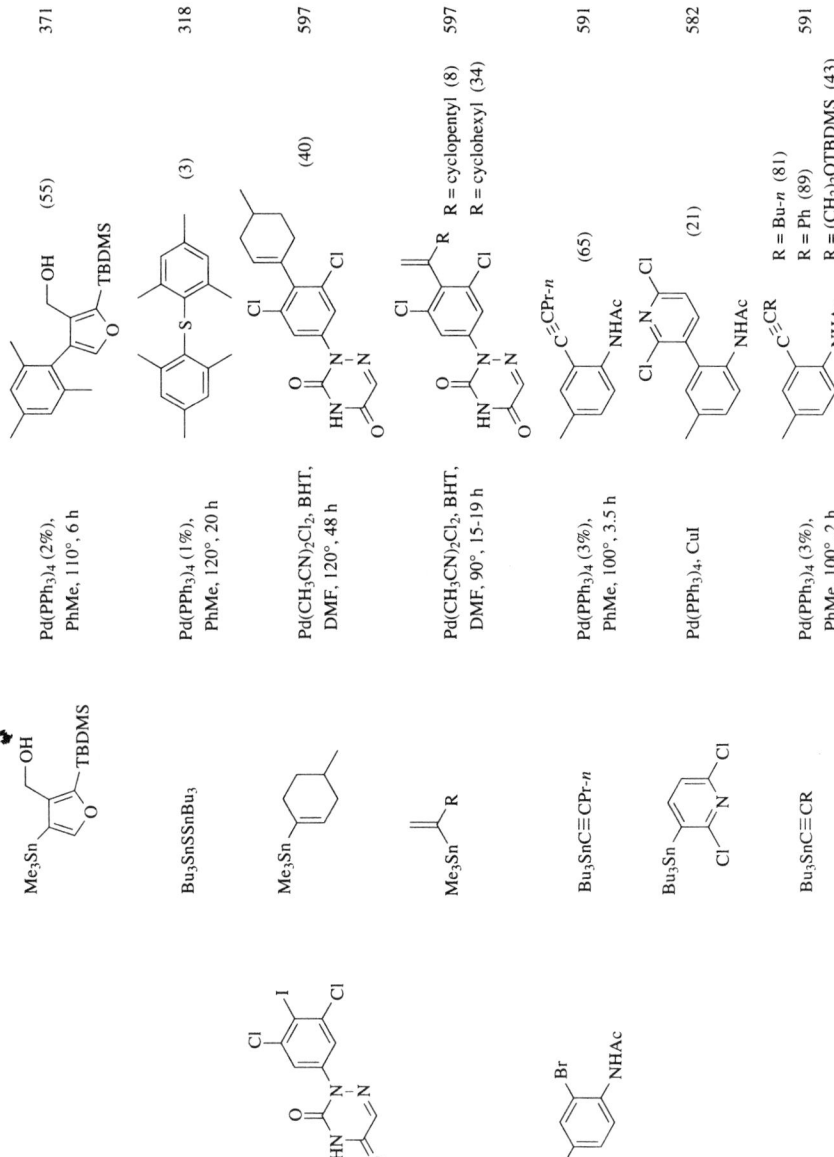

371

(55)

Pd(PPh₃)₄ (2%),
PhMe, 110°, 6 h

318

(3)

Pd(PPh₃)₄ (1%),
PhMe, 120°, 20 h

597

(40)

Pd(CH₃CN)₂Cl₂, BHT,
DMF, 120°, 48 h

597

R = cyclopentyl (8)
R = cyclohexyl (34)

Pd(CH₃CN)₂Cl₂, BHT,
DMF, 90°, 15-19 h

591

(65)

Pd(PPh₃)₄ (3%),
PhMe, 100°, 3.5 h

582

(21)

Pd(PPh₃)₄, CuI

591

R = Bu-n (81)
R = Ph (89)
R = (CH₂)₂OTBDMS (43)

Pd(PPh₃)₄ (3%),
PhMe, 100°, 2 h

243

TABLE III. DIRECT CROSS-COUPLING OF ARYL ELECTROPHILES (*Continued*)

Substrate	Stannane	Conditions	Product(s) and Yield(s) (%)	Refs.
(structure: OTf, Br, NHAc aryl)	Bu₃SnC≡C—(chain)—OTBDMS	Pd(PPh₃)₄ (3%), PhMe, 100°, 1.5 h	(structure with OTBDMS, NHAc) (71)	591
	Bu₃SnC≡C—(chain)—C≡C—TMS	Pd(PPh₃)₄ (3%), PhMe, 100°, 1.5 h	(structure, TMS, NHAc) (52)	591
(TfO, Br, NHAc aryl)	Bu₃SnC≡CR	Pd(PPh₃)₄ (3%), PhMe, 100°, 3 h	(OTf, C≡CR, NHAc) R = Bu-n (36) R = Ph (67)	591
	Bu₃SnC≡CR	Pd(PPh₃)₄ (3%), PhMe, 100°, 2 h	(TfO, C≡CR, NHAc) R = Pr-i (80) R = Bu-n (84)	591
	Bu₃SnC≡CPh	Pd(PPh₃)₄ (3%), PhMe, 100°, 3 h	(TfO, C≡CPh, NHAc) (96)	591
(EtO, O, N-H, Br aryl)	Bu₃Sn—(vinyl)—OEt	Pd(PPh₃)₂Cl₂, DMF, Et₄NCl, 80°, 18 h	(structure, OEt, EtO-NHC(O)) (65)	272, 273
(MeO, Br, NHAc aryl)	Bu₃SnC≡CR	Pd(PPh₃)₄ (3%), PhMe, 100°	(MeO, C≡CR, NHAc)	591

R	Time (h)	
Pr-n	6	(60)
Pr-i	3.5	(57)
Bu-n	6	(62)
Ph	5	(76)

Aryl substrate	Organostannane	Conditions	Product (yield %)	Refs.
MOM(Ms)N–C₆H₄–I (p)	Bu₃SnC≡COEt	Pd(PPh₃)₂Cl₂ (5%), Et₄NCl, DMF, 50°, 3 h	MOM(Ms)N–C₆H₄–C≡COEt (62)	302
Indanone–OTf	Bu₃SnSnBu₃	Pd(PPh₃)₂Cl₂, DMF, 105°	Indanone–SnBu₃ (—)	583
2-Br–C₆H₄–CO₂Et	Bu₃Sn–CH=CH–OEt	Pd(PPh₃)₂Cl₂ (1%), Et₄NCl, DMF, 80°	CO₂Et, –CH=CH–OEt (80)	273
3-Br–C₆H₄–CO₂Et	Bu₃Sn–CH=CH–CH₂–N(TMS)₂	1. Pd(PPh₃)₄, PhMe, reflux, 72 h; 2. H⁺	EtO₂C–C₆H₄–CH=CH–CH₂–NH₂ (66)	464, 540
4-Br–C₆H₄–CO₂Et	Bu₃Sn–CH=CH–OEt	Pd(PPh₃)₂Cl₂, Et₄NCl, DMF, 80°, 1 h	EtO₂C–C₆H₄–CH=CH–OEt (73)	272, 273
N₂⁺ BF₄⁻–C₆H₄–CO₂Et	[Bu₃Sn–N(R)–Me]	Pd[P(o-Tol)₃]Cl₂ (2%), PhMe, 105°	R(Me)N–C₆H₄–CO₂Et, R = Ph (83), R = Bn (88)	91
I–C₆H₄–CO₂Et	Ph–C(=CH₂)–SnBu₃	Pd(dba)₂ (5%), CH₃CN, Et₂O, rt, 1 h	EtO₂C–C₆H₄–CH=CH–Ph (97) $E:Z$ = 7:93	249
	Bu₃SnC≡COEt	Pd(PPh₃)₂Cl₂ (5%), Et₄NCl, DMF, rt, 5 h	EtO₂C–C₆H₄–C≡COEt (59)	302
	2-(Bu₃Sn)-1-methylindole	Pd(PPh₃)₂Cl₂ (4%), P(2-furyl)₃ (8%), THF, rt, 2 d	EtO₂C–C₆H₄–(1-methylindol-2-yl) (89)	425

TABLE III. DIRECT CROSS-COUPLING OF ARYL ELECTROPHILES (*Continued*)

Substrate	Stannane	Conditions	Product(s) and Yield(s) (%)	Refs.
	Bu_3Sn—CH=CH—CH(OH)—$C_8H_{17}-n$	Pd(dba)$_2$ (4%), P(2-furyl)$_3$ (8%), THF, 25°, 2 d	(82)	447
	(indole, N-Boc 2-stannane)	Pd(PPh$_3$)$_2$Cl$_2$ (2%), THF, reflux	(68)	425
	Bu_3Sn—, TBDMSO, OTBDMS	Pd(PPh$_3$)$_2$Cl$_2$ (8%), ZnCl$_2$, LiCl, dioxane, reflux, 48 h	TBDMSO—, OTBDMS, CO$_2$Me (43)	252, 598
	Bu_3Sn—O—, OTBDMS, OTBDMS, OTBDMS	Pd(PPh$_3$)$_2$Cl$_2$, THF, reflux, 2 h	OAc, OTBDMS, OTBDMS, OTBDMS (46)	299, 300
	Bu_3Sn—(2-pyridyl)	Pd(PPh$_3$)$_4$ (3%), Ag$_2$O, DMF, reflux, 24 h	(18)	32
	Me$_4$Sn	Pd(PPh$_3$)$_2$Cl$_2$ (10–20%), PPh$_3$ (40%), LiCl, DMF, reflux	OMe, OMe (96)	190, 379
	Bu_3Sn—	Pd(PPh$_3$)$_2$Cl$_2$ (10–20%), PPh$_3$ (40%), LiCl, DMF, reflux	OMe, OMe (90)	190, 379

Reactant	Conditions	Product	Yield	Ref.
Bu$_3$Sn–allyl	Pd(PPh$_3$)$_2$Cl$_2$ (10-20%), PPh$_3$ (40%), LiCl, DMF, reflux	(aryl: OMe, CH$_2$OMe, allyl)	(98)	190, 376
Bu$_3$SnPh	Pd(PPh$_3$)$_2$Cl$_2$ (10-20%), PPh$_3$ (40%), LiCl, DMF, reflux	(aryl: OMe, CH$_2$OMe, Ph)	(79)	190, 376
Bu$_3$Sn–(2-MeO-aryl)	Pd(PPh$_3$)$_2$Cl$_2$ (10-20%), CuBr (20%), LiCl, DMF, reflux	(biaryl: MeO, MeO, CH$_2$OMe)	(93)	191, 190
Bu$_3$SnC≡CPh	Pd(PPh$_3$)$_2$Cl$_2$ (10-20%), PPh$_3$ (40%), LiCl, DMF, reflux	(aryl: MeO, CH$_2$OMe, C≡CPh)	(56)	190
R$_3$Sn–(aryl: OMe, MeO, Me)	Pd(PPh$_3$)$_2$Cl$_2$ (10-20%), PPh$_3$ (40%), LiCl, DMF, reflux	(biaryl: MeO, MeO, CH$_2$OMe)	R = Me (26) R = Bu (13)	190
Bu$_3$Sn–allyl	Pd(PPh$_3$)$_2$Cl$_2$ (10-20%), PPh$_3$ (40%), LiCl, DMF, reflux	(aryl: CO$_2$Me, OMe, allyl)	(97)	190
Bu$_3$SnPh	Pd(PPh$_3$)$_2$Cl$_2$ (10-20%), PPh$_3$ (40%), LiCl, DMF, reflux	(aryl: CO$_2$Me, OMe, Ph)	(87)	190

Substrate: (aryl: CO$_2$Me, OTf, OMe)

TABLE III. DIRECT CROSS-COUPLING OF ARYL ELECTROPHILES (Continued)

Substrate	Stannane	Conditions	Product(s) and Yield(s) (%)	Refs.
(aryl bromide, CHO, MeO, MeO)	Bu₃Sn—(aryl, MeO)	Pd(PPh₃)₂Cl₂ (10-20%), PPh₃ (10-15%), LiCl, CuBr (20%), DMF, reflux	(30)	191, 190
(aryl OMe, OTf, OMe, OHC)	(indoline, N–Boc, SnBu₃)	Pd(OAc)₂ (10%), P(o-Tol)₃ (20%), Et₃N, DMF, 70°, 50 h	(65)	596
	Bu₃Sn—CH₂CH=CH₂	Pd(PPh₃)₂Cl₂ (10-15%), PPh₃ (40%), LiCl, DMF, reflux	(78)	190
	Me₃Sn (OMe, MeO)	Pd(PPh₃)₂Cl₂ (10-20%), P(o-Tol)₃ (10-15%), LiCl, CuBr (20%), BHT, DMF, reflux	(26)	191
(aryl Br, Et, MsO)	Bu₃Sn (OMe, OMe, OMe)	Pd(PPh₃)₄ (5%), CuI (8%), NMP, 70°, 48 h	(21-31)	575
(Br, OTf, OMe, HO)	Bu₃Sn—CH₂CH=CH₂	Pd(PPh₃)₂Cl₂ (10-15%), PPh₃ (40%), LiCl, DMF, reflux	(67)	190

248

C$_{10}$

Bu$_3$Sn$\diagup\!\!\diagdown$	Pd(PPh$_3$)$_2$Cl$_2$ (10-15%), PPh$_3$ (40%), LiCl, DMF, reflux	(41)	190
Me$_3$Sn— (oxazoline)	Pd(PPh$_3$)$_4$ (5%), C$_6$H$_6$, reflux, 48 h	(25)	550
Bu$_3$Sn$\diagup\!\!\diagdown$N(TMS)$_2$	1. Pd(PPh$_3$)$_4$ (2%), PhMe, reflux, 120 h 2. H$^+$	NH$_2$ (74)	464, 540
Me$_3$Sn$\diagup\!\!\diagdown$TMS	BnPd(PPh$_3$)$_2$Cl (1%), HMPA, 65°, 2 d	TMS (45)	537
Bu$_3$Sn$\diagup\!\!\diagdown$CO$_2$Et	Pd(OAc)$_2$ (5%), PPh$_3$ (20%), C$_6$H$_6$, reflux, 1 d	CO$_2$Et (50) E:Z = 1:10	306, 427
Bu$_3$Sn ... OTBDMS OTBDMS OTBDMS	Pd(PPh$_3$)$_2$Cl$_2$, PhMe, 100°, 2 h	OTBDMS OTBDMS OTBDMS (59)	299, 300

TABLE III. DIRECT CROSS-COUPLING OF ARYL ELECTROPHILES (Continued)

Substrate	Stannane	Conditions	Product(s) and Yield(s) (%)	Refs.
1-iodonaphthalene	Bu$_3$SnNEt$_2$	Pd[P(o-Tol)$_3$]$_2$Cl$_2$ (1%), PhMe, 100°, 3 h	NEt$_2$-naphthalene (38)	90
	Bu$_3$SnH	Pd(PPh$_3$)$_4$ (2%), C$_6$H$_6$, 25°, 5 h	naphthalene (94)	48
	3-Bu$_3$Sn-(N-Ts-indole)	Pd$_2$(dba)$_3$ (5%), AsPh$_3$ (10%), CuI (10%), DMF, 60°, 5 h	(92)	291
	BnOCH$_2$O−C(SnBu$_3$)=CH−CH(CH$_3$)CH$_2$CH$_3$	Pd$_2$(dba)$_3$ (2%), DMF, rt, 6 h	(45)	439
	Et$_3$SnSnEt$_3$	PhPd(PPh$_3$)$_2$I (5%), DMSO, 100°, 4 h	(98)	320
	Bu$_3$Sn−CH=CH$_2$	Pd(PPh$_3$)$_2$Cl$_2$ (5%), LiCl, DMF, 25°, 6-18 h	(91)	203
1-naphthyl OSO$_2$F	Bu$_3$Sn−C$_6$H$_4$−OMe	Pd(PPh$_3$)$_2$Cl$_2$ (5%), LiCl, DMF, 25°, 6-18 h	(69)	203

	Reagent	Conditions	Product	Yield (%)	Refs.
![OTf naphthalene]	Bu₃SnH	Pd(dba)₂ (5%), PPh₃ (10%), LiCl, THF, 60°, 16 h	naphthalene	(78)	192
	Bu₃Sn⌁	Pd(dba)₂ (5%), PPh₃ (10%), LiCl, THF, 60°, 16 h	1-vinylnaphthalene	(65)	192
	Bu₃Sn~~~OH	Pd(PPh₃)₂Cl₂ (2%), LiCl, BHT, DMF, 60°, 3 h	naphthyl-CH=CH-CH₂OH	(82)	189
	Bu₃Sn~~~OH	Pd(PPh₃)₂Cl₂ (2%), LiCl, BHT, DMF, 60°, 6 h	naphthyl-CH=CH-CH₂OH	(62)	189
	Me₃SnPh	Pd(dba)₂ (5%), PPh₃ (10%), LiCl, THF, 60°, 16 h	1-phenylnaphthalene	(68)	192
	Bu₃SnC≡CPh	Pd(dba)₂ (5%), PPh₃ (10%), LiCl, THF, 60°, 16 h	naphthyl-C≡CPh	(75)	192
	Bu₃Sn—CH=CH—Ph	Pd(dba)₂ (5%), PPh₃ (10%), LiCl, THF, 60°, 16 h	naphthyl-CH=CH-Ph	(61)	192

TABLE III. DIRECT CROSS-COUPLING OF ARYL ELECTROPHILES (*Continued*)

Substrate	Stannane	Conditions	Product(s) and Yield(s) (%)	Refs.
$OSO_2C_6H_4F$-p	Bu$_3$Sn—CH=CH$_2$	Pd(OAc)$_2$ (5%), dppp (5.5%), LiCl, DMF, 100°, 30 h	(50)	202
	Me$_4$Sn	Ni(acac)$_2$, PPh$_3$, DIBAL, THF, 3.3 h	(80)	205
Br (naphthalene)	Me$_3$Sn (methyloxazole)	Pd(PPh$_3$)$_4$ (5%), C$_6$H$_6$, reflux, 24 h	(92)	550
	Me$_3$Sn (dimethyloxazoline)	Pd(PPh$_3$)$_4$ (5%), C$_6$H$_6$, reflux, 12 h	(90–95)	550, 551
	Bu$_3$Sn (furan-OH, TBDMS)	Pd(PPh$_3$)$_4$ (2%), PhMe, reflux, 18 h	(46)	371
OTf (naphthalene)	Bu$_3$SnR	Pd(dba)$_2$ (5%), PPh$_3$ (10%), LiCl, THF, 60°, 16 h		192

R
H	(75)
CH=CH$_2$	(75)
2-thienyl	(70)
Ph	(72)
(*E*)-CH=CHPh	(67)
C≡CPh	(68)

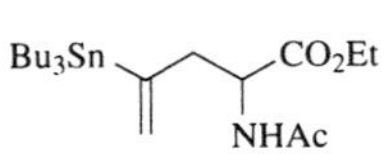

Reagent	Substrate	Conditions	Product	Ref.

Row 1:

Bu₃Sn—(=CH₂)—CH₂—CH(CO₂Et)(NHAc)

Bu_3Sn ... CO_2Et, $NHAc$

Pd(PPh₃)₄, LiCl, DMF, 100°, 6 h

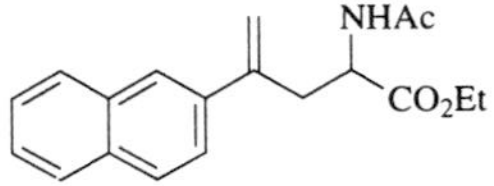

(60) + (10)

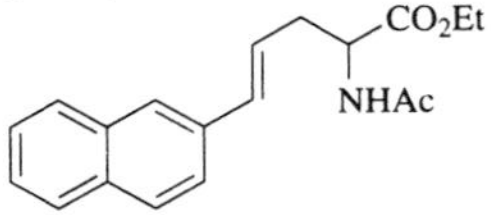

473

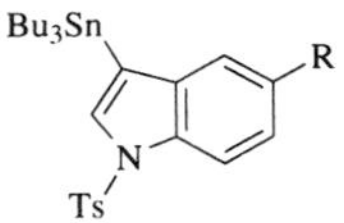

Row 2:

Bu_3Sn (indole, R, Ts)

Pd₂(dba)₃ (5%), AsPh₃ (10%), DMF, 60°, 1.5 h

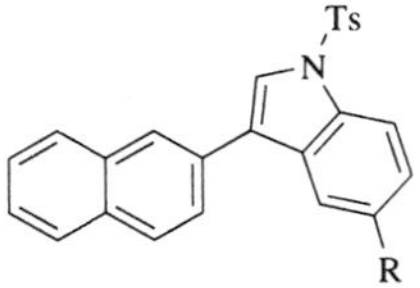

R = H (87)
R = OMe (91)

291

Row 3:

Bu₃SnSnBu₃

Pd(dba)₂ (5%), PPh₃ (10%), LiCl, THF, 60°, 16 h

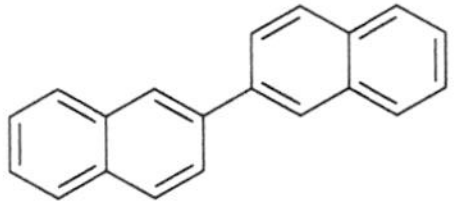

(69)

192

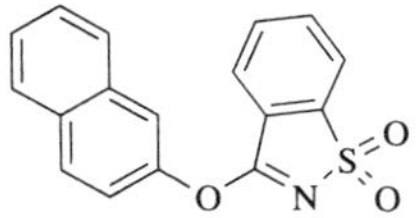

Row 4:

Me₄Sn

Pd(PPh₃)₄ LiCl, dioxane, 23 h

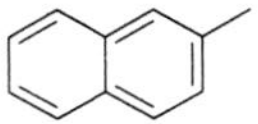

(48)

205

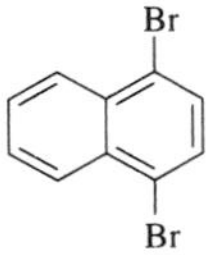

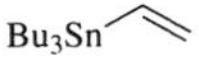

Row 5:

Bu_3Sn ⌁

Pd(PPh₃)₄ PhMe

(—)

599

Row 6:

Bu_3Sn ⌁

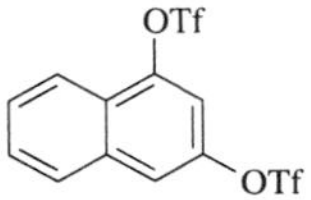

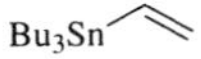

Pd(PPh₃)₄ (2%), LiCl, BHT, dioxane, reflux

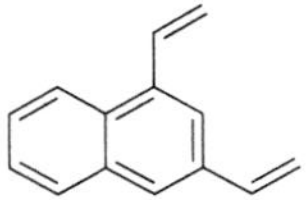

(74-86)

599

Substrates (bottom):

- naphthalene–O–benzisothiazole S,S-dioxide (with O, N, S, O, O)
- 1,4-dibromonaphthalene (Br, Br)
- naphthalene bis-triflate (OTf, OTf)

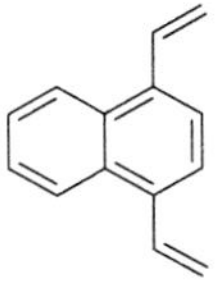

TABLE III. DIRECT CROSS-COUPLING OF ARYL ELECTROPHILES (*Continued*)

Substrate	Stannane	Conditions	Product(s) and Yield(s) (%)	Refs.
naphthalene-1,4-diyl bis(triflate)	Bu$_3$Sn—vinyl	Pd(PPh$_3$)$_4$ (2%), LiCl, BHT, dioxane, reflux	1,4-divinylnaphthalene (74-86)	599
naphthalene-1,5-diyl bis(triflate)	Bu$_3$Sn—vinyl	Pd(PPh$_3$)$_4$ (2%), LiCl, BHT, dioxane, reflux	1,5-divinylnaphthalene (75)	599
naphthalene-1,6-diyl bis(triflate)	Bu$_3$Sn—vinyl	Pd(PPh$_3$)$_4$ (2%), LiCl, BHT, dioxane, reflux	divinylnaphthalene (74-86)	599
naphthalene-1,5-diyl bis(triflate)	Bu$_3$Sn—vinyl	Pd(PPh$_3$)$_4$ (2%), LiCl, BHT, dioxane, reflux	divinylnaphthalene (74-86)	599
naphthalene-1,2-diyl bis(triflate)	Bu$_3$Sn—vinyl	Pd(PPh$_3$)$_4$ (2%), LiCl, BHT, dioxane, reflux	divinylnaphthalene (74-86)	599
naphthalene-2,6-diyl bis(triflate)	Bu$_3$Sn—vinyl	Pd(PPh$_3$)$_4$ (2%), LiCl, BHT, dioxane, reflux	divinylnaphthalene (74-86)	599
naphthalene-2,7-diyl bis(triflate)	Bu$_3$Sn—vinyl	Pd(PPh$_3$)$_4$ (2%), LiCl, BHT, dioxane, reflux	divinylnaphthalene (74-86)	599

Substrate	Reagent	Conditions	Product(s) (Yield %)	Refs.
4-*t*-Bu-C₆H₄Br	Bu₃Sn–C(OEt)=CH₂	Pd(PPh₃)₂Cl₂ (2%), PhMe, 105°, 48 h	aryl enol ether (82)	270
MeO₂C-aryl(OTf)(NHAc)	Bu₃SnC≡CR	Pd(PPh₃)₄ (3%), LiCl, dioxane, 100°	alkynyl product (97)	591
(OMe/OTf/OMe, Br, CH₂OMe arene)	Bu₃Sn–CH₂CH=CH₂ (butenylstannane)	Pd(PPh₃)₂Cl₂ (10-15%), PPh₃ (40%), LiCl, DMF, reflux	diallyl arene (92)	190, 379
(OMe/OTf/OMe, CH₂OMe arene)	Bu₃Sn–CH₂CH=CH₂	Pd(PPh₃)₂Cl₂ (10-15%), PPh₃ (40%), LiCl, DMF, reflux	diallyl arene (47-67)	190, 379
(OMe/OTf/OMe, Cl phthalide)	Bu₃Sn–CH₂CH=CH₂	Pd(PPh₃)₂Cl₂ (10-15%), PPh₃ (40%), LiCl, DMF, reflux	allyl-Cl phthalide (0) + OH/OMe/Cl isobenzofuranone (34)	190
(OMe/OTf/OMe, Cl, CH₂OMe arene)	Bu₃Sn–CH₂CH=CH₂	Pd(PPh₃)₂Cl₂ (10-15%), PPh₃ (40%), LiCl, DMF, reflux	allyl-Cl arene (63)	190, 379

Time (h) table for the MeO₂C alkynylation:

R	Time (h)	(Yield)
TMS	1	(97)
Pr-*n*	3	(73)
Pr-*i*	1.75	(77)
Bu-*n*	3	(88)
Ph	2	(84)

TABLE III. DIRECT CROSS-COUPLING OF ARYL ELECTROPHILES (*Continued*)

Substrate	Stannane	Conditions	Product(s) and Yield(s) (%)	Refs.
	Bu₃Sn⌒	Pd(PPh₃)₂Cl₂ (10-15%), PPh₃ (40%), LiCl, DMF, reflux	(28)	190
	Bu₃SnCH=C=CH₂	Pd₂(dba)₃ (2%), P(2-furyl)₃ (8%), LiCl, CuI (10%), DMF, 80°, 1.5 h	CH=C=CH₂ (60)	276
	Bu₃SnSnBu₃	Pd(PPh₃)₂Cl₂, DMF, 105°	SnBu₃ (—)	583
	Bu₃Sn⌒	Pd(PPh₃)₄ (2%), BHT, PhMe, reflux, 2 h	(72)	88
	Me₃Sn	Pd(PPh₃)₂Cl₂, DMF, 100°, 12 h	(73)	572
	Bu₃Sn⌒	Pd(PPh₃)₂Cl₂ (10-15%), PPh₃ (40%), LiCl, DMF, reflux	(34)	190

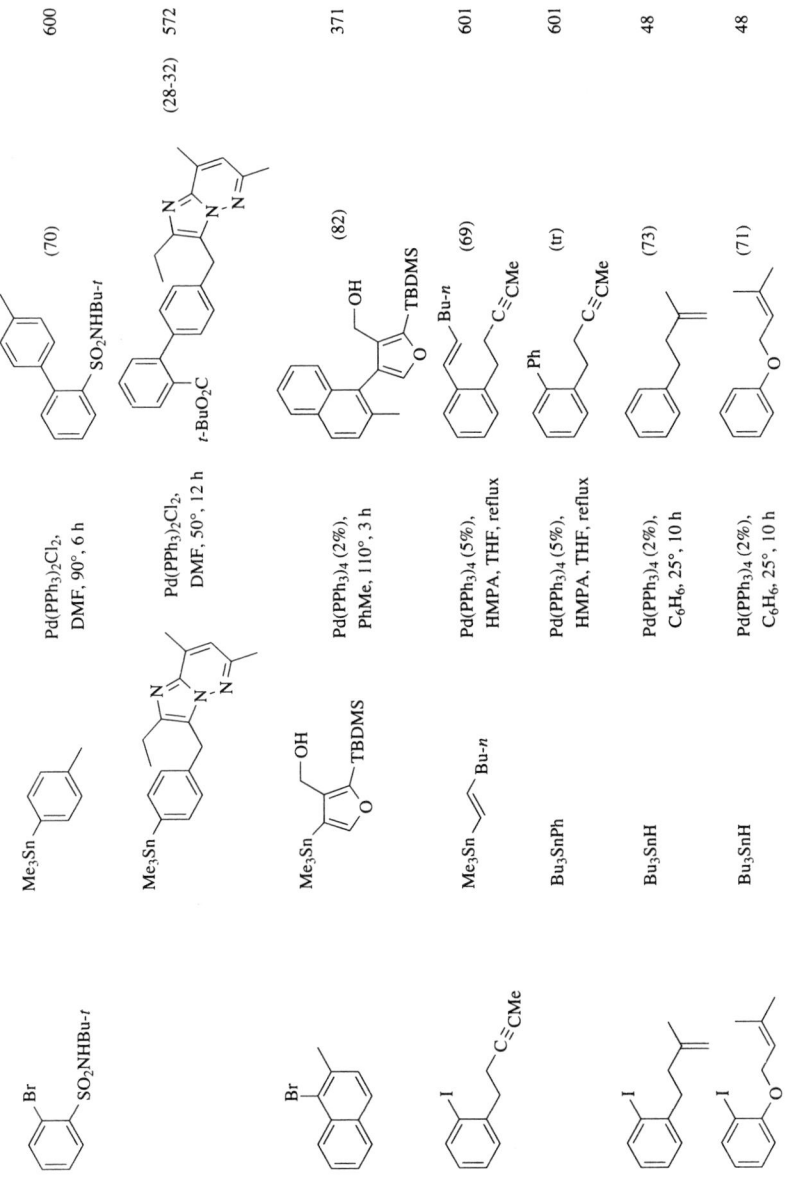

Pd(PPh₃)₂Cl₂, DMF, 90°, 6 h	(70)		600
Pd(PPh₃)₂Cl₂, DMF, 50°, 12 h	(28-32)		572
Pd(PPh₃)₄ (2%), PhMe, 110°, 3 h	(82)		371
Pd(PPh₃)₄ (5%), HMPA, THF, reflux	(69)		601
Pd(PPh₃)₄ (5%), HMPA, THF, reflux	(tr)		601
Pd(PPh₃)₄ (2%), C₆H₆, 25°, 10 h	(73)		48
Pd(PPh₃)₄ (2%), C₆H₆, 25°, 10 h	(71)		48

C₁₁

257

TABLE III. DIRECT CROSS-COUPLING OF ARYL ELECTROPHILES (Continued)

Substrate	Stannane	Conditions	Product(s) and Yield(s) (%)	Refs.
(EtO$_2$C-C(O)CH$_2$- trifluorobromophenyl)	2,6-dimethylpyridin-4-yl-SnMe$_3$	Pd(PPh$_3$)$_2$Cl$_2$ (7%), dioxane, reflux, 16 h	(84)	567, 602
8-cyano-1-iodonaphthalene	Bu$_3$Sn-C$_6$H$_4$-CO$_2$Et	Pd(PPh$_3$)$_2$Cl$_2$ (5%), DMF, 100°, 18 h	(80)	594
	Bu$_3$Sn-C$_6$H$_4$-CH$_2$CH$_2$OTHP	Pd(PPh$_3$)$_2$Cl$_2$ (5%), DMF, 100°, 22 h	(≥76)	594
I, NHBoc, Cl aryl	Me$_3$SnSnMe$_3$	Pd(CH$_3$CN)$_2$Cl$_2$, NaH, NMP, reflux	SnMe$_3$, NHBoc, Cl (57)	603
Br, MeO naphthalene	Bu$_3$Sn-CH=CH$_2$	Pd(PPh$_3$)$_4$ (2%), BHT, PhMe, reflux, 1 h	(83)	88
OTf, MeO naphthalene	Bu$_3$Sn-C(=CH$_2$)CO$_2$Me	Pd(PPh$_3$)$_4$ (10%), CuI (75%), LiCl, DMF, rt, 48 h	CO$_2$Me (71)	246
OTf, MeO naphthalene	Bu$_3$SnCH=C=CH$_2$	Pd$_2$(dba)$_3$ (2%), P(2-furyl)$_3$ (8%), LiCl, CuI (10%), DMF, 80°, 1.5 h	CH=C=CH$_2$ (67)	276

Substrate	Stannane	Conditions	Product (yield)	Refs.
(OTf naphthalenone bicyclic)	Bu₃SnSnBu₃	Pd(PPh₃)₂Cl₂, DMF, 105°	(—)	583
(4-bromophenyl thienyl ketone)	Bu₃Sn⌇ (vinyl)	Pd(PPh₃)₄ (2%), BHT, PhMe, reflux, 1 h	(85)	88
(bromo dimethoxy indandione)	Bu₃Sn–benzothiazole	Pd(PPh₃)₄ (2%), PhMe, heat	(81)	604
(bromo dimethoxy indandione)	Bu₃Sn–benzothiophene	Pd(PPh₃)₄ (2%), PhMe, heat	(73)	604
(bromo dimethoxy indandione)	Bu₃Sn–bithiophene	Pd(PPh₃)₄ (2%), PhMe, heat	(91)	604
(MeS, OMe, OTf, CH₂OMe arene)	Me₄Sn	Pd(PPh₃)₂Cl₂ (10–15%), PPh₃ (40%), LiCl, DMF, reflux	(58)	190, 379
(Br, OMOM, OMOM, OHC arene)	Bu₃Sn⌇ (vinyl)	Pd(PPh₃)₄ (2%), PhMe, 100°, 8 h	(87)	605

259

TABLE III. DIRECT CROSS-COUPLING OF ARYL ELECTROPHILES (*Continued*)

Substrate	Stannane	Conditions	Product(s) and Yield(s) (%)	Refs.
C₁₂ (dibenzothiophenium BF₄⁻)	Bu₃Sn⌁OTBDMS	Pd(PPh₃)₄ (2%), PhMe, 100°, 2 h	OHC...OTBDMS, OMOM, OMOM (88)	605
(biphenyl-OTf, Ph)	Me₄Sn	Pd(OAc)₂ (2%), DMF, 60-20°, 1 h	(2,2'-dimethylbiphenyl) (65)	206
(4-bromobiphenyl)	Bu₃SnCH=C=CH₂	Pd₂(dba)₃ (2%), P(2-furyl)₃ (8%), LiCl, CuI (8%), DMF, 80°, 5 h	Ph...CH=C=CH₂ (20)	276
	Bu₃Sn—furan—OBu-*t*	BnPd(PPh₃)₂Cl, DMF, 70°, 16 h	Ph...furan...OBu-*t* (93)	296
	Me₃Sn—(OH, TBDMS furan)	Pd(PPh₃)₄, PhMe, 110°, 3 h	Ph...OH...TBDMS furan (70)	371
(biphenyl-OTf, Ph)	Bu₃SnCH=C=CH₂	Pd₂(dba)₃ (2%), P(2-furyl)₃ (8%), LiCl, CuI (8%), DMF, 80°, 5 h	Ph...CH=C=CH₂ (31)	276
(3,5-dibromobiphenyl)	Me₃SnSnMe₃	Pd(PPh₃)₄ (4%), PhMe, 110-120°, 4 h	Me₃Sn...SnMe₃...Ph (90)	606

260

Substrate	Reagent	Conditions	Product (yield)	Refs.
(acenaphthene, Br)	Bu₃Sn–(indol-2-yl, N-SEM)	Pd(PPh₃)₄ (0.7%), DMF, 110°, 5 h	(acenaphthene–indole, N-SEM) (97)	289
TBDMSO–(aryl, Br, Cl)	Bu₃Sn–(phenyl, BocHN)	Pd(PPh₃)₂Cl₂ (5%), DMF, 90°, 25 h	TBDMSO–(biphenyl, Cl, NHBoc) (65)	564
MeO, MeO–(naphthalene, OTf)	Me₃Sn–(aryl, OMe, CO₂Me)	Pd(PPh₃)₄ (5%), LiCl, dioxane, 105°, overnight	MeO, MeO–(naphthalene–aryl, OMe, CO₂Me) (46)	607
(4-bromophenyl O–SO₂–C₆H₄F)	Bu₃Sn–(allyl)	Pd(OAc)₂ (5%), dppp (5.5%), LiCl, DMF, 90°, 19 h	(4-vinylphenyl O–SO₂–C₆H₄F) (50)	202
I–(C₆H₄)–(C₆H₄)–NO₂	Bu₃Sn–(vinyl, F, TMS)	Pd(PPh₃)₂Cl₂, THF, reflux, 20 h	(vinyl F, TMS)–(C₆H₄)–(C₆H₄)–NO₂ (74)	263
MeO–(aryl, CONEt₂, OTf)	Me₄Sn	Pd(PPh₃)₂Cl₂ (20%), DMF, 120°, 8 h	MeO–(aryl, CONEt₂, CH₃) (49)	608
TfO, TfO–(anthraquinone)	Bu₃Sn–(allyl)	Pd(PPh₃)₄, LiCl, dioxane, 90-95°, 2.5-4 h	(divinyl anthraquinone) (74)	370

261

TABLE III. DIRECT CROSS-COUPLING OF ARYL ELECTROPHILES (*Continued*)

Substrate	Stannane	Conditions	Product(s) and Yield(s) (%)	Refs.
(3-BzO-phenyl iodide)	Bu$_4$Sn	Pd(dppf)Cl$_2$, LiCl, DMF, 90-95°, 2.5-4 h	(2,6-di-*n*-Bu anthraquinone) (63)	370
	Me$_3$SnPh	Pd(dppf)Cl$_2$, LiCl, DMF, 90-95°, 2.5-4 h	(2,6-di-Ph anthraquinone) (67)	370
	Bu$_3$SnC≡CSnBu$_3$	Pd(PPh$_3$)$_4$ (10%), LiCl, BHT, dioxane, reflux, 5 h	(3-OBz, 3-BzO diphenylacetylene) (28)	589
(anthraquinone OH/OTf)	Bu$_3$Sn–CH=CH$_2$	Pd(dppf)Cl$_2$, LiCl, DMF, 90-95°, 2 h	(1-OH-2-vinyl anthraquinone) (70)	370
	Me$_3$SnPh	Pd(dppf)Cl$_2$, LiCl, DMF, 90-95°, 2.5 h	I (56) + (1-OH-2-Ph anthraquinone) ; (1-OH-2-methyl anthraquinone) (28)	370

Reagent	Conditions	Product (%)	Refs.
Bu₃SnPh	Pd(dppf)Cl₂, LiCl, DMF, 90-95°, 3 h	**I** (56) + product with OH, Bu-*n* (14)	370
Bu₃SnSnBu₃	Pd(dppf)Cl₂, LiCl, DMF, 90-95°, 17 h	OH-substituted anthraquinone (93)	370
Bu₄Sn	Pd(dppf)Cl₂, LiCl, DMF, 90-95°, 20 h	OH, Bu-*n* anthraquinone (97)	370
Bu₃Sn–CH=CH–SnBu₃	Pd(PPh₃)₄, PhMe, 100°	SnBu₃ / CO₂Me vinyl ketone product (≥57)	500
Bu₃Sn–CH₂CH=CH₂	Pd(PPh₃)₄, LiCl, dioxane, 90-95°, 1 h	OH, OH vinyl anthraquinone (70)	370
Bu₄Sn	Pd(PPh₃)₂Cl₂, LiCl, DMF, 90-95°, 2.5 h	OH, Bu-*n*, OH anthraquinone (74)	370

Substrates:

2-bromophenyl ketone with CO₂Me chain (Br, CO₂Me)

anthraquinone with OH, OTf, OH substituents

TABLE III. DIRECT CROSS-COUPLING OF ARYL ELECTROPHILES (Continued)

Substrate	Stannane	Conditions	Product(s) and Yield(s) (%)	Refs.
	Bu₃Sn (dihydropyran)	Pd(PPh₃)₄, LiCl, dioxane, 90-95°, 8.5 h	(100)	370
	Bu₃SnPh	Pd(PPh₃)₄, LiCl, dioxane, 90-95°, 17 h	(80)	370
	Bu₃Sn—Ph	Pd(PPh₃)₄, LiCl, dioxane, 90-95°, 2 h	(100)	370
C₁₃ (4'-methylbiphenyl-4-yl OTf)	Me₃Sn (2,6-difluorophenyl)	Pd(PPh₃)₂Cl₂ (10-15%), PPh₃ (30-40%), LiCl, CuBr, DMF, reflux	(93)	191
	Bu₃Sn (4-methylphenyl)	Pd(PPh₃)₂Cl₂ (10-15%), PPh₃ (30-40%), LiCl, CuBr, DMF, reflux	(62)	191

264

Aryl halide/triflate	Organostannane	Conditions	Product (yield)	Ref.
2,4-dibromo-*N*-Ts-aniline (Br, Br, NHTs)	Bu₃Sn–aryl (OMe, MeO, Me)	Pd(PPh₃)₂Cl₂ (10–15%), AsPh₃ (30–40%), LiCl, CuBr, DMF, reflux	MeO, OMe, Me biaryl (21)	191
Br, NHTs arene	Bu₃Sn–CH=CH₂	Pd(PPh₃)₄ (2%), PhMe, reflux	divinyl NHTs (52)	89
Br, Br, NHTs arene	Bu₃Sn–CH=CH₂	Pd(PPh₃)₄ (2%), PhMe, reflux	divinyl NHTs (50)	89
Cl, Br, NHTs arene	Bu₃Sn–CH₂CH=CH₂	Pd(PPh₃)₄ (2%), PhMe, reflux	Cl, vinyl NHTs (82)	89
EtO₂C, MeOCH₂, OMe, OTf, OMe, Cl arene	Bu₃Sn–CH₂CH=CH₂	Pd(PPh₃)₂Cl₂ (10–15%), PPh₃ (40%), LiCl, DMF, reflux	allyl product (0) + phenol product (40)	190
OTf, NHTs arene	Bu₃Sn–CH=CH₂	Pd(PPh₃)₄ (2%), LiCl, BHT, DMF, rt, 4 h	vinyl NHTs (78)	189
OTf, NHTs arene	Bu₃Sn–CH=CH–TMS	Pd(PPh₃)₄ (2%), LiCl, BHT, DMF, dioxane, 98°, 11 h	TMS-allyl NHTs (81)	189
TsHN, OTf arene	Bu₃Sn–CH=CH–CH₂OTHP (*E:Z* = 7:1)	Pd(PPh₃)₄ (2%), LiCl, BHT, DMF, dioxane, 98°, 12 h	OTHP styryl TsHN (72) *E:Z* = 3:1	189

TABLE III. DIRECT CROSS-COUPLING OF ARYL ELECTROPHILES (Continued)

Substrate	Stannane	Conditions	Product(s) and Yield(s) (%)	Refs.
4-bromophenyl ketone, HN–CH$_2$CH$_2$–NEt$_2$	Bu$_3$SnSnBu$_3$	Pd(PPh$_3$)$_4$, NEt$_3$, reflux, 12 h	aryl–SnBu$_3$ (44)	609
2-bromophenyl OBn glycal	Bu$_3$Sn– (OTBDMS sugar)	Pd(PPh$_3$)$_2$Cl$_2$, PhMe, reflux, 1 h	(44) OTBDMS/OTBDMS/OTBDMS, OBn	299, 300
2-bromobenzyl OTBDMS	Bu$_3$Sn– (dihydropyran, O)	Pd(OAc)$_2$ (5%), P(o-Tol)$_3$, NEt$_3$, CH$_3$CN, reflux	(33) OTBDMS	429
EtO$_2$C, i-Pr, OTf aryl	F, methyl aryl, Bu$_3$Sn	Pd(PPh$_3$)$_2$Cl$_2$, LiCl, DMF, 100°	F / EtO$_2$C / i-Pr (>88)	610
OMe, OTf, OMe, EtO$_2$C, MeO aryl	allyl Bu$_3$Sn	Pd(PPh$_3$)$_2$Cl$_2$ (10–15%), PPh$_3$ (40%), LiCl, DMF, reflux	OMe / OMe / EtO$_2$C / MeO (68)	190
C$_{14}$ 9-bromoanthracene	indole (N-SEM) Bu$_3$Sn	Pd(PPh$_3$)$_4$ (0.7%), DMF, 110°, 6 h	anthracenyl-indole (N-SEM) (95)	289

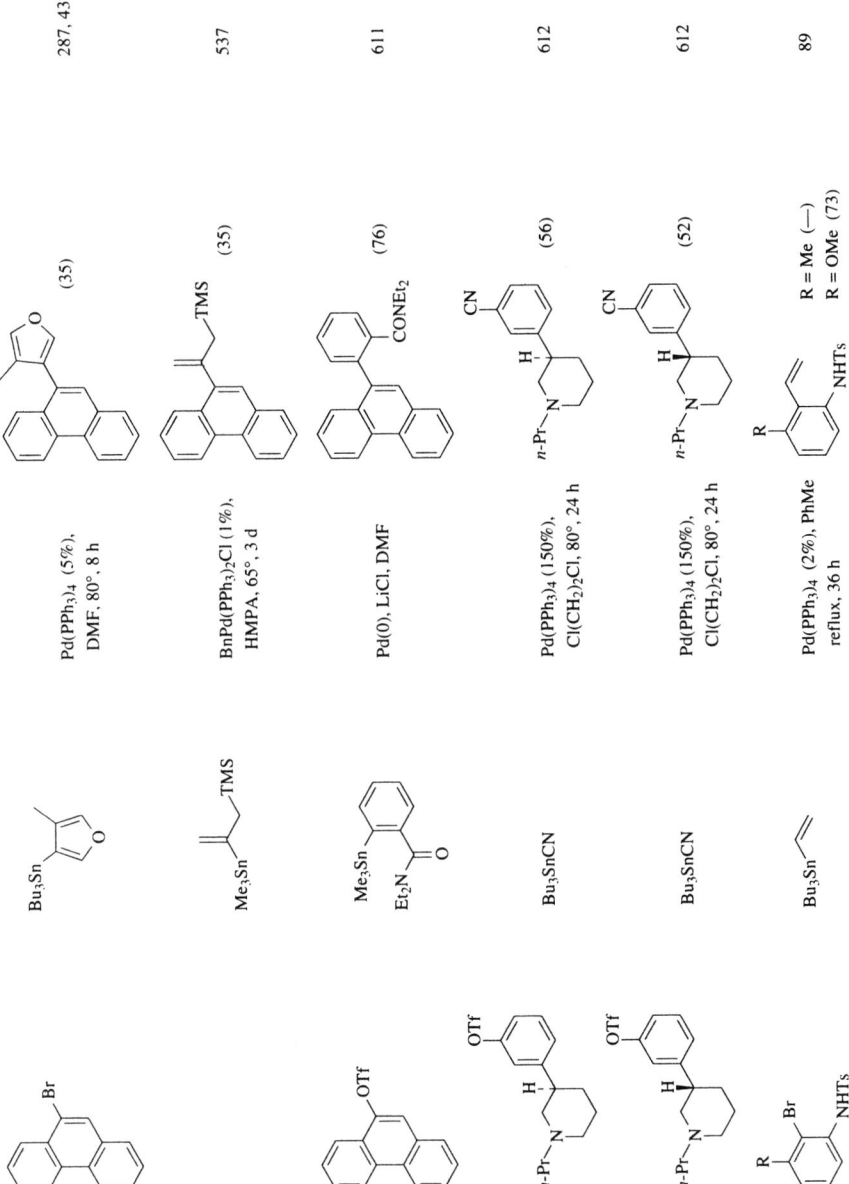

TABLE III. DIRECT CROSS-COUPLING OF ARYL ELECTROPHILES (*Continued*)

Substrate	Stannane	Conditions	Product(s) and Yield(s) (%)	Refs.
anthraquinone–I	$Me_3SnSnMe_3$	$PhPd(PPh_3)_2I$ (1%), DMF, 50°, 2 h	anthraquinone–$SnMe_3$ (52)	613
4-Br-benzoate (BnO)	Bu_3SnTMS	$Pd(PPh_3)_2Cl_2$ (3.5%), HMPA, 80°, 5 h	$SnBu_3$-benzoate (BnO) (36) + TMS-benzoate (BnO) (20) + $(BnO)_2$ biaryl (—) + PhC(O)OBn (—)	614
C_{15} 2,6-dichloro-4-iodophenyl isoxazole	thiophene–$SnMe_3$ (R)	$Pd(PPh_3)_2Cl_2$ (5%), THF, reflux, 4 h	thiophene-coupled isoxazole (R)	615

R	Time (h)	
H	4	(73)
Me	20	(82)

Substrate	Stannane	Conditions	Product(s) and Yield(s) (%)	Refs.
4-Br-2-(PhC(O))-N-(COCF$_3$)aniline	Bu_3Sn–CH=CH–$SnBu_3$	$Pd(PPh_3)_4$ (2%), PhMe, reflux, 4 h	bis-aryl vinylene product (78)	616

Me$_3$SnSnMe$_3$ — Pd(PPh$_3$)$_4$ (4.5%), dioxane, reflux, 6.5 h — (61) — 617

Bu$_3$Sn — Pd$_2$(dba)$_3$ (2.5%), AsPh$_3$ (20%), dioxane, 50°, 24 h — (97) — 569

Bu$_3$Sn — Pd(PPh$_3$)$_4$ (2%), PhMe, reflux, 36 h — R = CO$_2$Me (60), R = OAc (57) — 89

Bu$_3$Sn — Pd(dppf)Cl$_2$ — (—) — 370

Bu$_4$Sn — Pd(dppf)Cl$_2$ — (—) — 370

Bu$_3$Sn — Pd(PPh$_3$)$_2$Cl$_2$, LiCl, DMF, 70° — (83) — 618

NHTs OTf OH OMe OAc HO Br R Ph N O MeO OMe Bu-n CF$_3$ OEt SnBu$_3$

TABLE III. DIRECT CROSS-COUPLING OF ARYL ELECTROPHILES (*Continued*)

Substrate	Stannane	Conditions	Product(s) and Yield(s) (%)	Refs.
C₁₆ [structure: n-Pr₂N-tetralin-OTf]	Bu₃Sn—C₆H₄—OMe	Pd(PPh₃)₄, LiCl, dioxane, 100°	[structure with OMe groups] (91)	618
[structure: n-Pr₂N-tetralin-OTf]	Me₄Sn	Pd(PPh₃)₂Cl₂ (5%), LiCl, BHT, dioxane, DMF, 110°, 24 h	[structure: n-Pr₂N-tetralin-Me] (57)	619, 620
[structure: n-Pr₂N-tetralin-OTf]	Me₄Sn	Pd(PPh₃)₂Cl₂ (5%), LiCl, BHT, dioxane, DMF, 110°, 24 h	[structure: n-Pr₂N-tetralin-Me] (56)	619, 620
[structure: n-Pr₂N-tetralin-OTf]	Bu₃SnPh	Pd(PPh₃)₂Cl₂ (5%), LiCl, BHT, dioxane, DMF, 120°, 24 h	[structure: n-Pr₂N-tetralin-Ph] (83)	619, 620
[structure: n-Pr₂N-tetralin-OTf]	Bu₃SnPh	Pd(PPh₃)₂Cl₂ (5%), LiCl, BHT, dioxane, DMF, 120°, 24 h	[structure: n-Pr₂N-tetralin-Ph] (88)	619, 620
[structure: EtO₂C, i-Pr, Bu-t, OTf]	Bu₃Sn—C₆H₄—F	Pd(PPh₃)₂Cl₂, LiCl, DMF, 100°	[structure: EtO₂C, i-Pr, Bu-t, C₆H₄F] (>88)	610

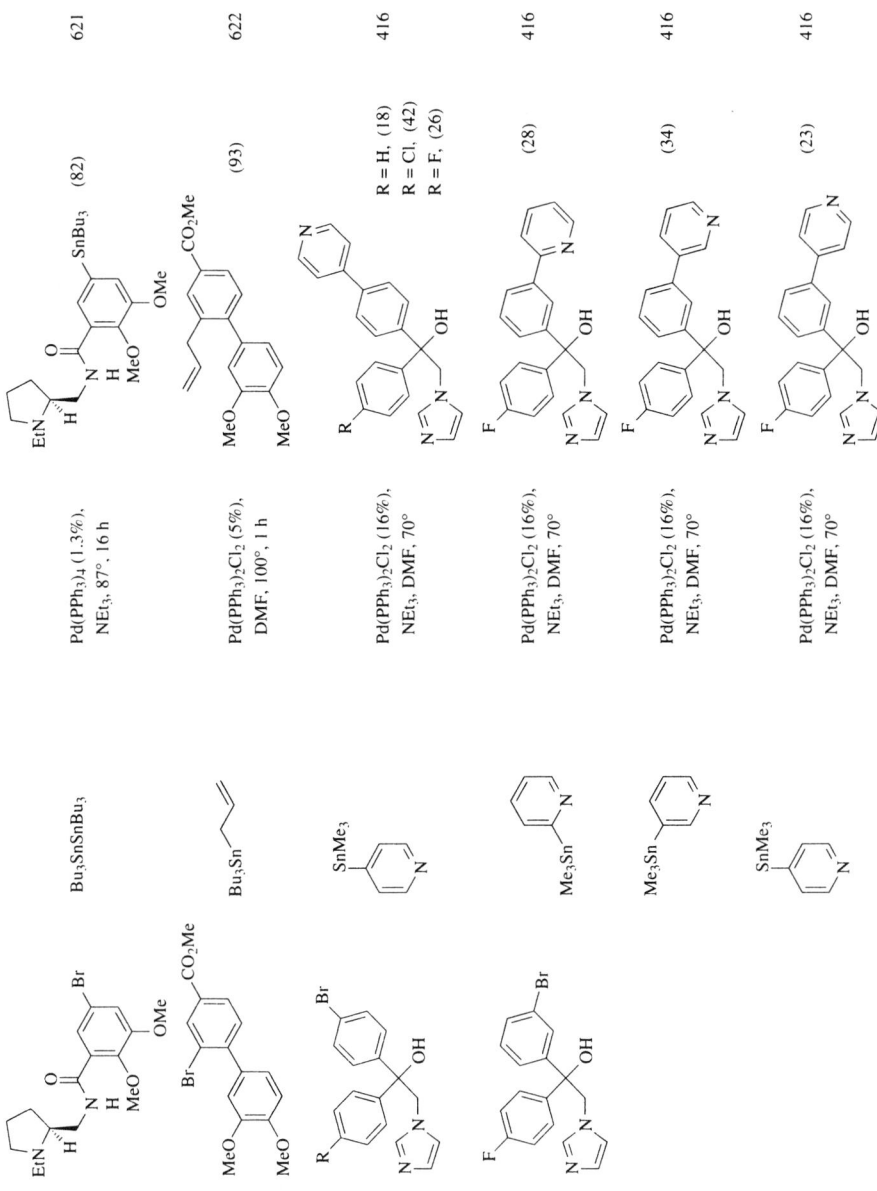

621

622

416

416

416

416

TABLE III. DIRECT CROSS-COUPLING OF ARYL ELECTROPHILES (Continued)

Substrate	Stannane	Conditions	Product(s) and Yield(s) (%)	Refs.
	Me$_3$Sn	Pd(PPh$_3$)$_2$Cl$_2$ (16%), NEt$_3$, DMF, 70°	(20)	416
	Me$_3$Sn	Pd(PPh$_3$)$_2$Cl$_2$ (16%), NEt$_3$, DMF, 70°	(13)	416
	Me$_3$Sn	Pd(PPh$_3$)$_2$Cl$_2$ (16%), NEt$_3$, DMF, 70°	(17)	416
	Me$_3$SnSnMe$_3$	Pd(PPh$_3$)$_4$, LiCl, dioxane, 60°, 24 h	(47)	572
C$_{18}$	Bu$_3$Sn	Pd(PPh$_3$)$_4$ (10%), LiCl, dioxane, reflux, 4 h	(61)	623

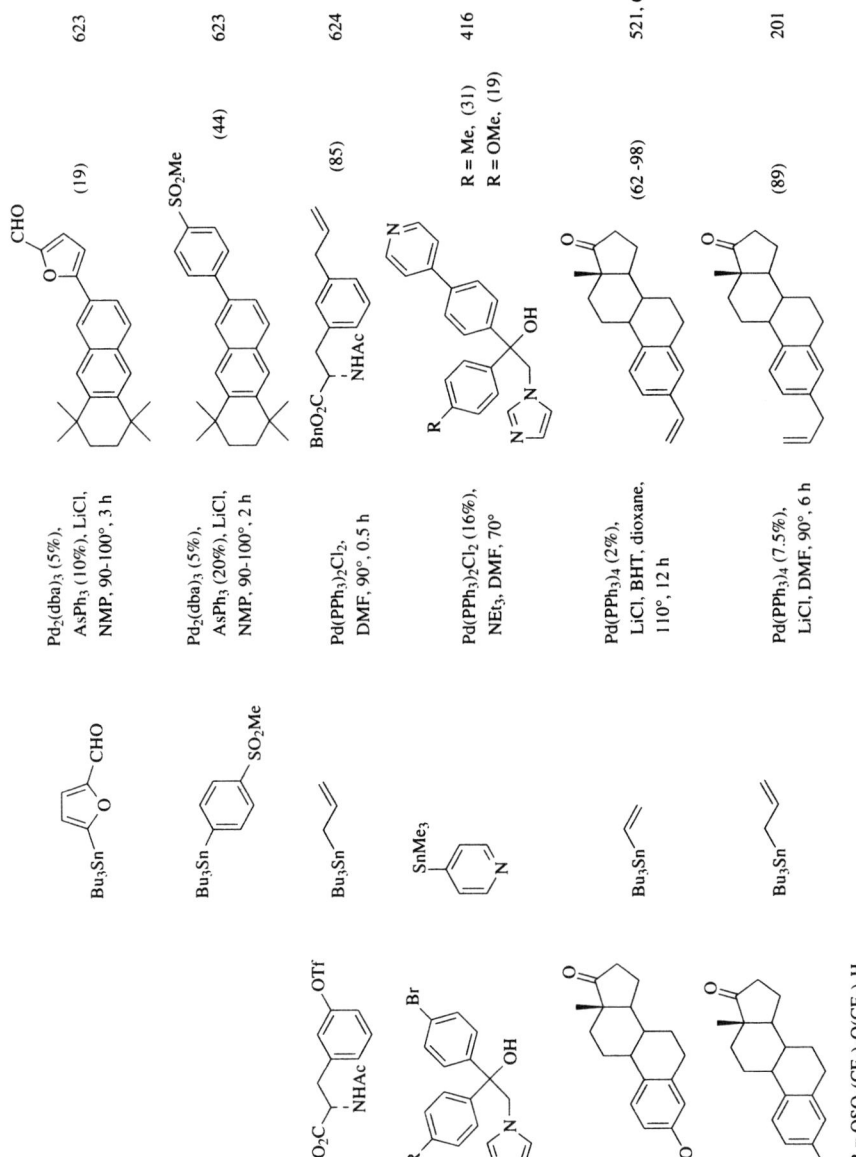

				623
			(19)	
				623
			(44)	
				624
			(85)	
			R = Me, (31)	416
			R = OMe, (19)	
			(62 - 98)	521, 625
			(89)	201

R = OSO₂(CF₂)₂O(CF₂)₂H

$R = OSO_2(CF_2)_2O(CF_2)_2H$

TABLE III. DIRECT CROSS-COUPLING OF ARYL ELECTROPHILES (*Continued*)

Substrate	Stannane	Conditions	Product(s) and Yield(s) (%)	Refs.
		1. Pd(PPh$_3$)$_4$ (4%), LiCl, dioxane, reflux, overnight 2. Bu$_4$NF, THF	(81)	392
		Pd(PPh$_3$)$_2$Cl$_2$, LiCl, DMF, 100°	(>88)	610
C$_{19}$		Pd(CH$_3$CN)$_2$Cl$_2$, DMF, 80°, 21 h	(31)	389
		Pd(PPh$_3$)$_2$Cl$_2$, LiCl, DMF, 100°	(60)	610
		Pd(PPh$_3$)$_2$Cl$_2$, LiCl	(80)	626

274

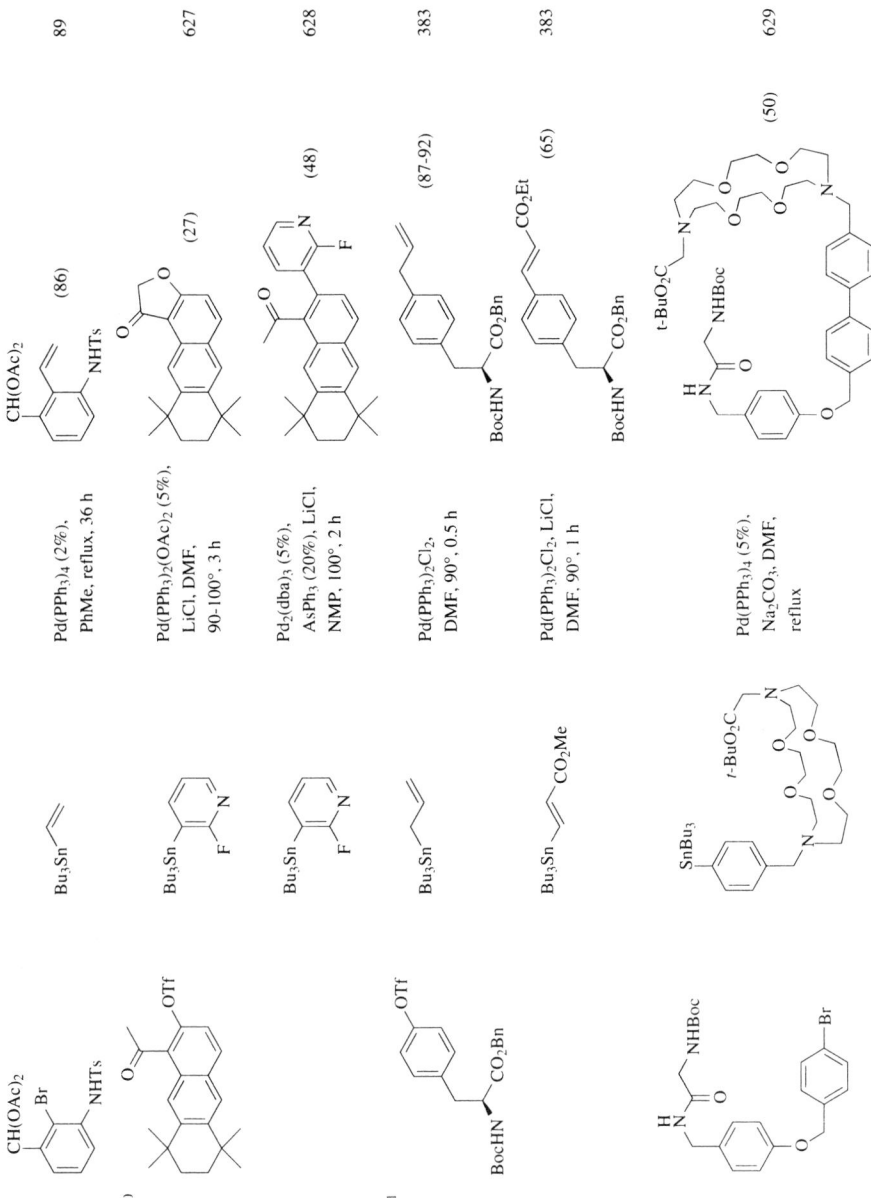

89

627

628

383

383

629

(86)

(27)

(48)

(87-92)

(65)

(50)

Pd(PPh₃)₄ (2%),
PhMe. reflux, 36 h

Pd(PPh₃)₂(OAc)₂ (5%),
LiCl, DMF,
90-100°, 3 h

Pd₂(dba)₃ (5%),
AsPh₃ (20%), LiCl,
NMP, 100°, 2 h

Pd(PPh₃)₂Cl₂,
DMF, 90°, 0.5 h

Pd(PPh₃)₂Cl₂, LiCl,
DMF, 90°, 1 h

Pd(PPh₃)₄ (5%),
Na₂CO₃, DMF,
reflux

C₂₀

C₂₁

275

TABLE III. DIRECT CROSS-COUPLING OF ARYL ELECTROPHILES (*Continued*)

Substrate	Stannane	Conditions	Product(s) and Yield(s) (%)	Refs.
(aryl triflate: F–C6H4, CO2Et, OTf, Ph)	Et4Sn	Pd(PPh3)2Cl2, LiCl, DMF, 100°	(>88) (Et, CO2Et, Ph, F-C6H4)	610
(aryl triflate: F–C6H4, CO2Et, OTf, Ph)	Bu3Sn— (vinyl)	Pd(PPh3)2Cl2, LiCl, DMF, 85°, 1.5 h	(90) (vinyl, CO2Et, Ph, F-C6H4)	610
(terpenoid: OMe, OTf, MeO)	Me4Sn	Pd(PPh3)2Cl2, DMF, 140°	(75) (terpenoid: OMe, MeO)	630
(4-iodophenoxy, CH–Ph, O, O)	Me3SnSnMe3	Pd(PPh3)2Cl2 (3%), dioxane, reflux, 2 h	SnMe3 (100)	631
(BnO, CH2OH, Br, OBn)	Bu3Sn (glycal: OBn, OBn, OBn)	Pd(PPh3)4 (10%), Na2CO3, PhMe, reflux, 4 h	(70) (OH, OBn, OBn, OBn, BnO, OBn)	423, 424
(BnO, CH2OH, Br, OBn)	Bu3Sn (glycal: Ph, O, O, OTBDMS)	Pd(PPh3)4 (10%), Na2CO3, PhMe, reflux, 3 h	(78) (OH, BnO, OBn, Ph, O, O, OTBDMS)	297

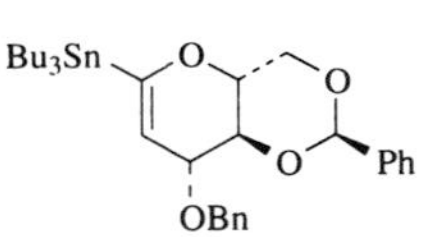

423, 424 (73)

632 (71)

633 (79)

299 (85)

300 (85)

634 (—)

Pd(PPh₃)₄ (10%), PhMe, reflux

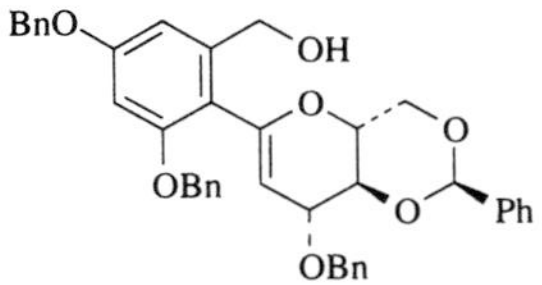

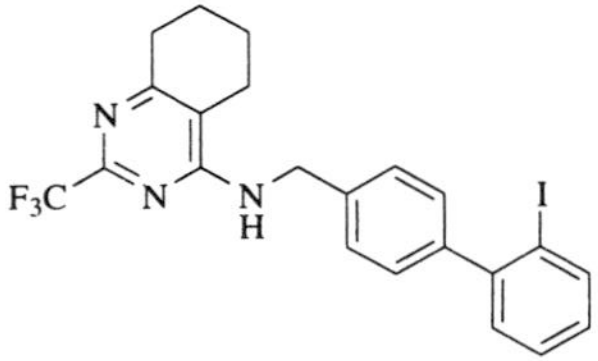

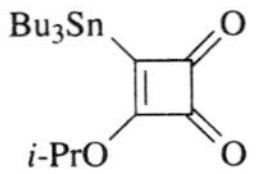

BnPd(PPh₃)₂Cl, CuI, CH₃CN, 70°

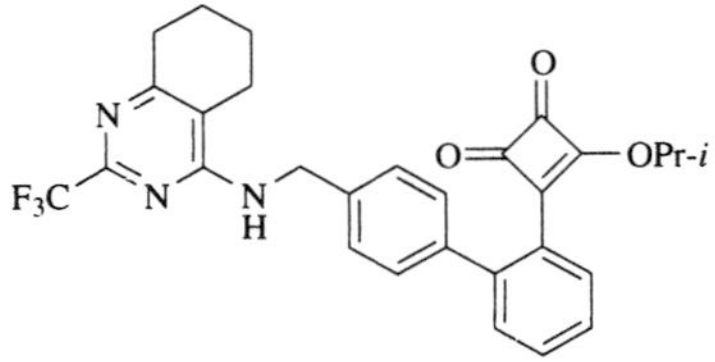

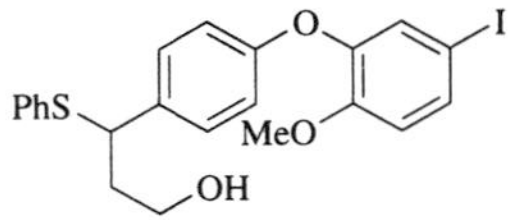

Pd(OAc)₂ (10%), NMP, rt, 40 h

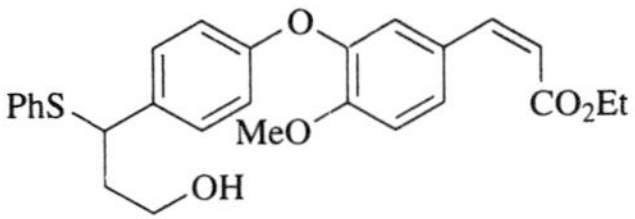

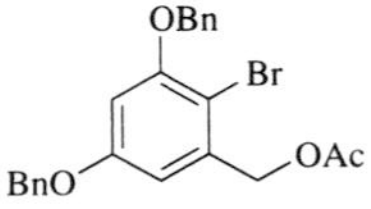

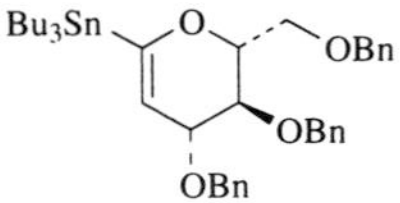

Pd(PPh₃)₂Cl₂ (5%), PhMe, reflux

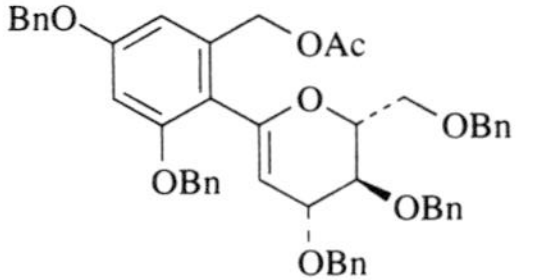

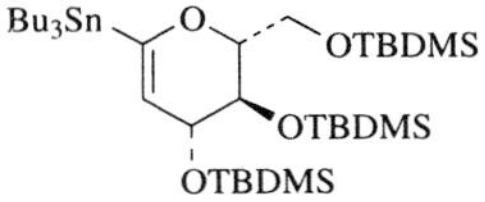

Pd(PPh₃)₂Cl₂ (2.5%), C₆H₆, reflux, 15 h

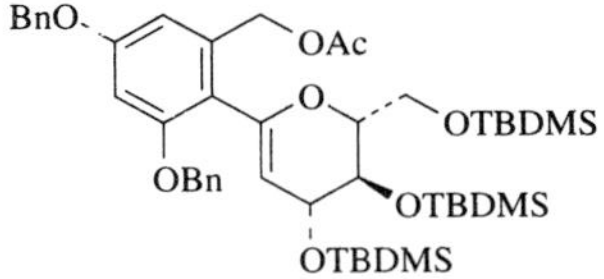

Pd(PPh₃)₂Cl₂, dioxane

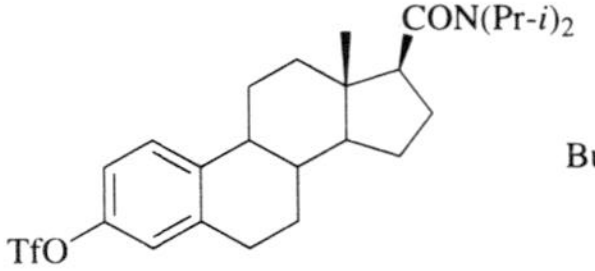

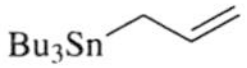

C₂₂

C₂₃

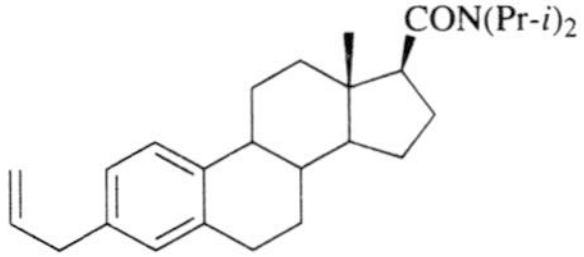

C₂₄

TABLE III. DIRECT CROSS-COUPLING OF ARYL ELECTROPHILES (Continued)

Substrate	Stannane	Conditions	Product(s) and Yield(s) (%)	Refs.
C$_{25}$	Bu$_3$Sn	Pd(PPh$_3$)$_4$ (10%), LiCl, BHT, dioxane, 110°, 12 h	(—)	625
	Me$_3$Sn (benzofuran, R)	Pd(PPh$_3$)$_4$ (4%), THF, reflux, 16 h	R = H (73), R = OTBDPS (90)	392, 391
	Bu$_3$SnSnBu$_3$	Pd(PPh$_3$)$_4$	(—)	635
C$_{26}$	Me$_4$Sn	Pd(OAc)$_2$, P(o-Tol)$_3$, NEt$_3$, DMF, 100°, 8 h	(45-80)	636
C$_{27}$	Bu$_3$Sn	Pd(PPh$_3$)$_4$ (2%), LiCl, dioxane, reflux, 12 h	(80)	637
	Bu$_3$Sn	Pd(PPh$_3$)$_2$Cl$_2$ (2.5%), mesitylene, reflux, 30 min	(66)	300

C$_{29}$

419 (63)

638 (68)

639

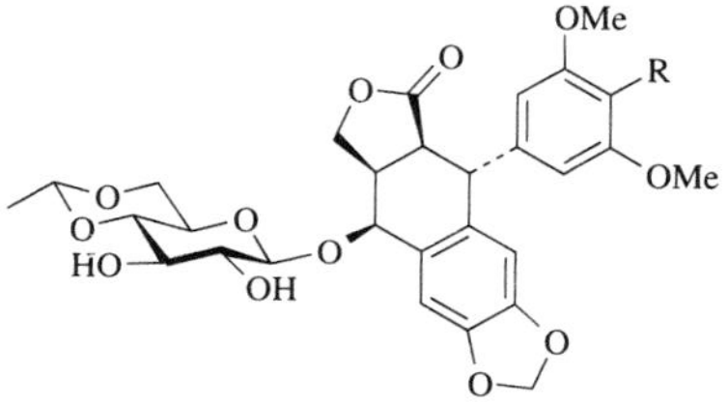

R
Me (83)
CH=CH$_2$ (48)
CH$_2$CH=CH$_2$ (63)

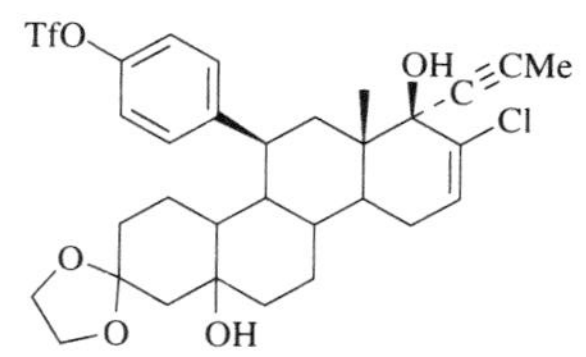

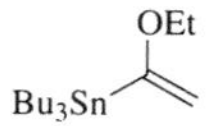

1. Pd(PPh$_3$)$_4$ (5%), LiCl, pyridine, dioxane, reflux, 2 h
2. H$^+$, acetone

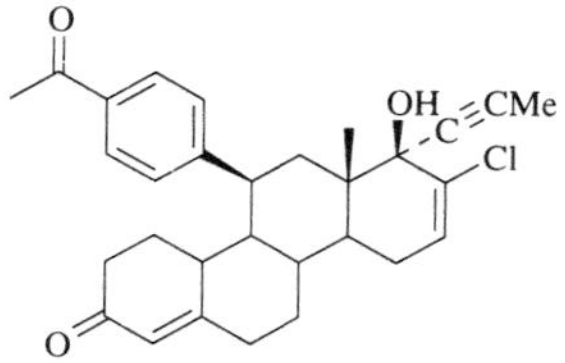

Pd(PPh$_3$)$_4$

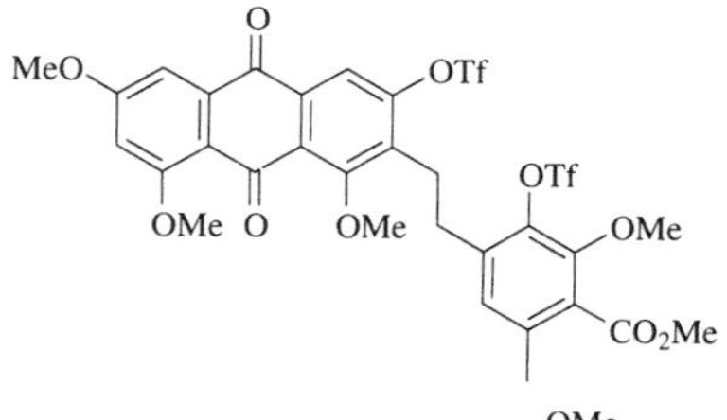

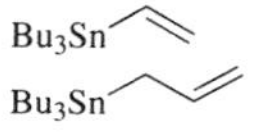

Pd(PPh$_3$)$_2$Cl$_2$ (15%), PPh$_3$ (40%), LiCl, DMF, reflux

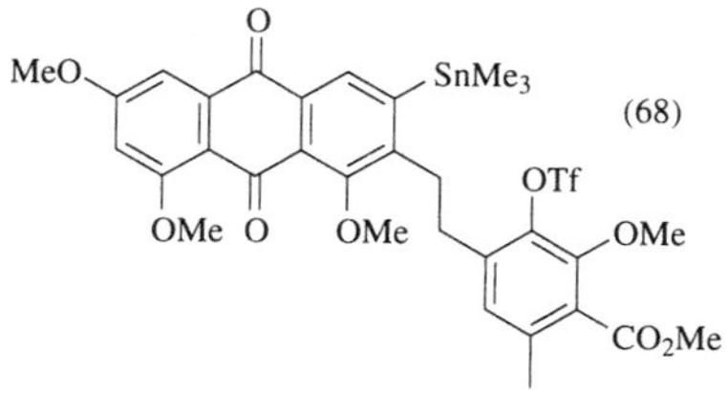

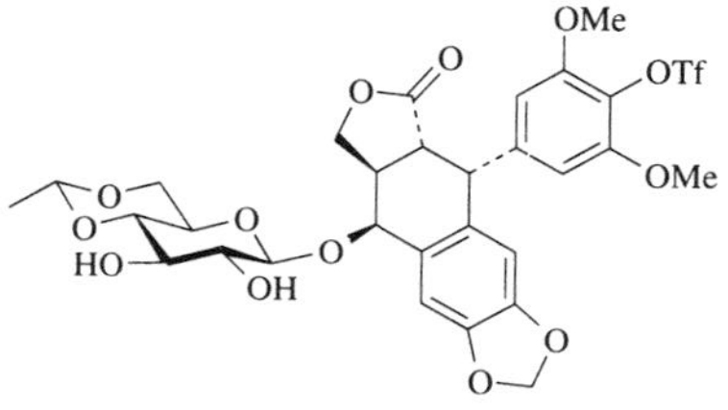

OEt
Bu$_3$Sn

Me$_3$SnSnMe$_3$

Me$_4$Sn
Bu$_3$Sn
Bu$_3$Sn

TABLE III. DIRECT CROSS-COUPLING OF ARYL ELECTROPHILES (*Continued*)

Substrate	Stannane	Conditions	Product(s) and Yield(s) (%)	Refs.
C$_{31-32}$	Bu$_3$Sn~	Pd(PPh$_3$)$_4$, C$_6$H$_6$	 n = 3 (20) n = 4 (19)	640
C$_{32}$	Bu$_3$Sn~ Bu$_3$Sn~	Pd(PPh$_3$)$_2$Cl$_2$ (15%), PPh$_3$ (40%), LiCl, DMF, reflux	 R = CH=CH$_2$ (49) R = CH$_2$CH=CH$_2$ (70)	639

C_{37}

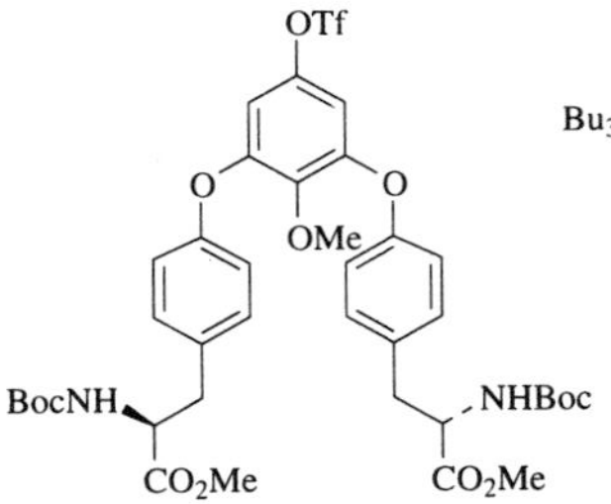

Bu_3Sn—CH=CH$_2$

$Pd(PPh_3)_4$ (5%), LiCl, BHT, dioxane, 90°, 1 h

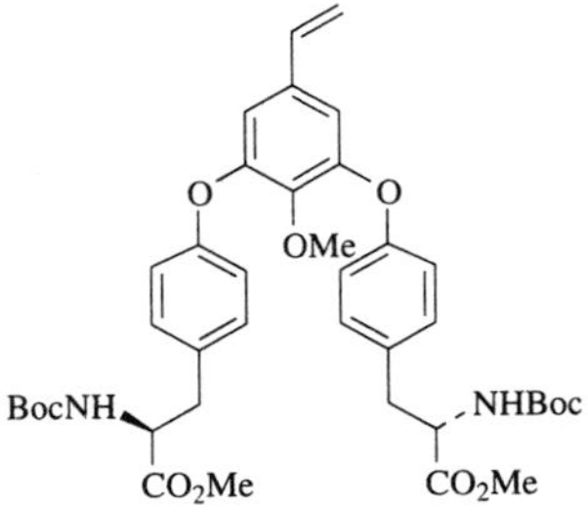

(86)

641

C_{47}

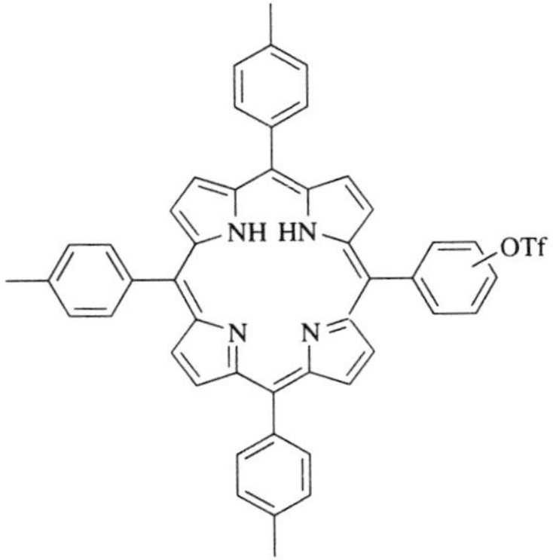

$Bu_3SnC{\equiv}CR$

$Pd(PPh_3)_4$ (14%), LiCl, THF, 90°, 2 d

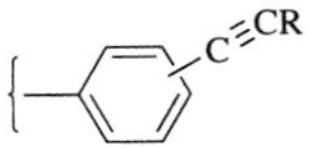

642

OTf	R	
m	TMS	(74)
m	Bu-*n*	(76)
m	Ph	(72)
p	TMS	(76)
p	Bu-*n*	(76)
p	Ph	(73)

TABLE III. DIRECT CROSS-COUPLING OF ARYL ELECTROPHILES (*Continued*)

Substrate	Stannane	Conditions	Product(s) and Yield(s) (%)	Refs.
(Zn tetraaryl porphyrin bearing OTf)	$Bu_3SnC{\equiv}CTMS$	$Pd(PPh_3)_4$ (14%), LiCl, THF, 90°, 2 d	—C≡CTMS (75)	642
C$_{53}$ (bis-TBDMS/Cl/OMe aryl ditriflate substrate)	Bu_3Sn	$Pd(PPh_3)_2Cl_2$ (5%), dioxane, 90°, 6 h	(85)	643

282

TABLE IV. INTRAMOLECULAR CROSS-COUPLING OF ARYL ELECTROPHILES

Substrate	Conditions	Product(s) and Yield(s) (%)	Refs.
C₁₁	Pd(PPh₃)₄ (4.9%), THF, reflux, 20 h	(61)	644
C₁₂ + Me₃SnSnMe₃	Pd(OAc)₂ (10%), PPh₃ (20%), PhMe, 110°, 18 h	(90)	645
	Pd(PPh₃)₄ (3%), syringe pump, PhMe, 105°, 5.5 h	(65)	646
C₁₄ + Me₃SnSnMe₃ X = Br X = I E:Z = 12:88	Pd(PPh₃)₄ (5-20%), dioxane, 100-105°, 24 h	(97) (87)	647
	Pd(PPh₃)₄	I (78)	647
+ Me₃SnSnMe₃	Pd(PPh₃)₄ (5-20%), dioxane, 100-105°, 24 h	I (82)	647

283

TABLE IV. INTRAMOLECULAR CROSS-COUPLING OF ARYL ELECTROPHILES (*Continued*)

Substrate	Conditions	Product(s) and Yield(s) (%)	Refs.
+ Me₃SnSnMe₃	Pd(PPh₃)₄ (5-20%), dioxane, LiCl, 100-105°, 24 h	**I** (95)	647
X = Br, X = I + Me₃SnSnMe₃	Pd(PPh₃)₄ (5-20%), dioxane, 100-105°, 24 h	**I** (80) **I** (87)	647
+ Me₃SnSnMe₃	Pd(PPh₃)₄ (5-20%), dioxane, LiCl, 100-105°, 24 h	**I** (88)	647
C₁₅	Pd(PPh₃)₂Cl₂ (2.2%), PPh₃ (8.8%), PhMe, reflux, 2 h	(70) E:Z = 1.5:1	648
	Pd(PPh₃)₄ (3%), syringe pump, PhMe, 105°, 5.5 h	**I** (37)	646
	Pd(PPh₃)₂Cl₂ (2-3%), LiCl, DMF, 60°, 69 h	**I** (22)	646

		Reagent	Conditions	Product (Yield)	Ref.
C$_{16}$	(methylenedioxy-bromobenzoyl indole, Br)	+ Bu$_3$SnSnBu$_3$	Pd(PPh$_3$)$_2$Cl$_2$ (5%), Et$_4$NCl, Li$_2$CO$_3$, PhMe, reflux, 12 h	**I** (68)	563
	(methylenedioxy-iodobenzoyl indole, I)	+ Me$_3$SnSnMe$_3$	Pd(0), xylene, 140°, 24 h	**I** (60)	645
	[methylenedioxy vinylstannane bromide]		Pd(PPh$_3$)$_2$Cl$_2$ (2.2%), PPh$_3$ (8.8%), PhMe, reflux, 2 h	(58) E:Z=1.1:1	648
	(MeO, OH, Ph vinylstannane bromide)		Pd(PPh$_3$)$_4$ (5%), 2,6-di-tert-butylphenol, PhMe, reflux, 4.5 h	(45)	648
C$_{17}$	(benzoate ester, SnBu$_3$, I)		Pd(PPh$_3$)$_4$ (3%), syringe pump, PhMe, 105°, 5.5 h	**I** (67)	646
	(benzoate ester, SnBu$_3$, OTf)		Pd(PPh$_3$)$_2$Cl$_2$ (2-3%), LiCl, DMF, 70°, 72 h	**I** (<5)	646

TABLE IV. INTRAMOLECULAR CROSS-COUPLING OF ARYL ELECTROPHILES (*Continued*)

Substrate	Conditions	Product(s) and Yield(s) (%)	Refs.
	Pd(PPh₃)₄, PhMe, 110°	(87)	646
	Pd(PPh₃)₂Cl₂ (2.2%), PPh₃ (8.8%), PhMe, reflux, 2 h	(49) *E*:*Z*=1.5:1	648
	Pd(PPh₃)₄ (3%), syringe pump, PhMe, 105°, 5.5 h	(66)	646
C₁₈	Pd(PPh₃)₄ (5%), BHT, PhMe, reflux, 4.5 h	(50)	648
C₁₉	BnPd(PPh₃)₂Cl, air, HMPA, 115°	(12)	649

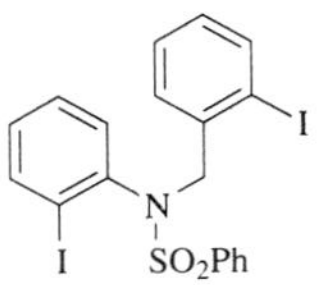

645

650
650

645

646

646

638

I (39)

CO₂Me

I (56)

(77)

SO₂Ph

CO₂Me

(43)
(53)

(45)

N–SO₂Ph

O

MEMO

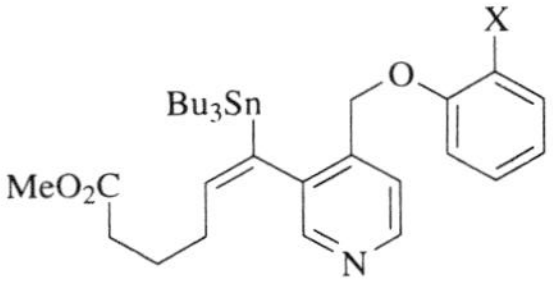

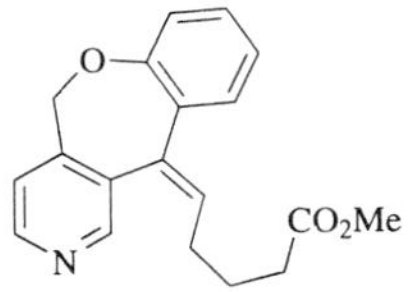

O

MEMO

I (54)

MeO   OMe

MeO

O   O

OMe

MeO   OMe

Pd(OAc)₂ (10%), PPh₃ (20%),
PhMe, 110°, 18 h

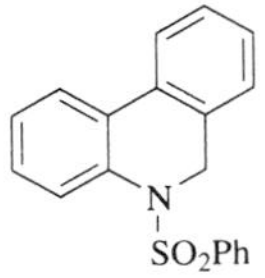

Pd(PPh₃)₄, PhMe, reflux, 24 h
Pd(PPh₃)₂Cl₂, PPh₃, LiCl, DMF

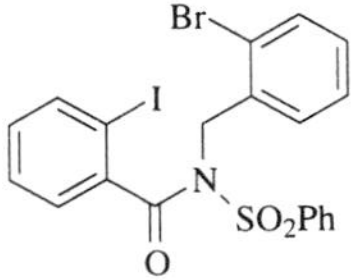

Pd(OAc)₂ (10%), PPh₃ (20%),
PhOMe, 120°, 18 h

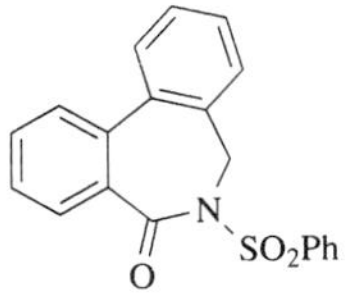

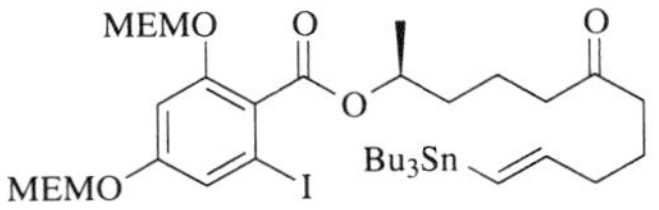

Pd(PPh₃)₄, syringe pump,
PhMe, 105°, 14 h

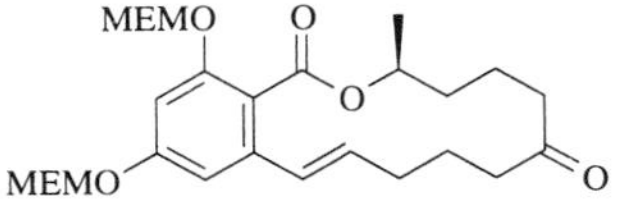

Polymer complexed Pd(PPh₃)₄,
PhMe, 105°, 61 h

Pd(PPh₃)₄

+ Me₃SnSnMe₃

I

SO₂Ph

X

O

Bu₃Sn

MeO₂C

X = Br
X = I

C₂₀

+ Me₃SnSnMe₃

Br

I

N–SO₂Ph

O

O

Bu₃Sn

MEMO

I

MEMO   O

C₂₉

SnMe₃   OTf   OMe   CO₂Me

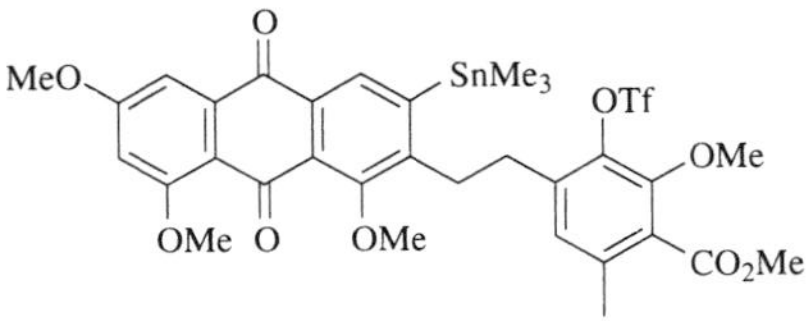

O   OMe

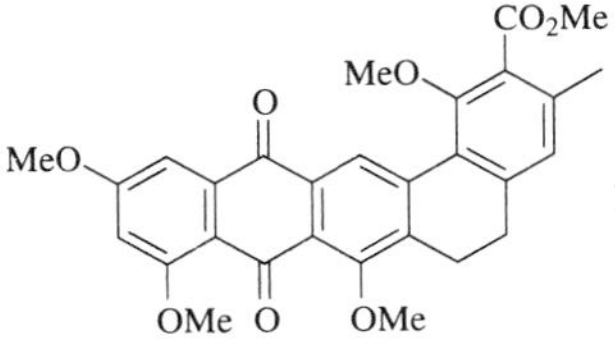

MeO   OMe   O

TABLE IV. INTRAMOLECULAR CROSS-COUPLING OF ARYL ELECTROPHILES (*Continued*)

Substrate	Conditions	Product(s) and Yield(s) (%)	Refs.
	Pd(PPh₃)₄ (15%), 100°, 7 d	**I** (44)	638
	Pd(PPh₃)₄ (5–10%), K₂CO₃, DMF, reflux	(15)	629

288

TABLE V. DIRECT CROSS-COUPLING OF FURAN AND BENZOFURAN ELECTROPHILES

Substrate	Stannane	Conditions	Product(s) and Yield(s) (%)	Refs.
C₄				
2-bromofuran	5-(Bu₃Sn)-2-(SMe)pyrimidine	Pd(PPh₃)₂Cl₂ (3%), Cl(CH₂)₂Cl, reflux, 4.5 h	furan–pyrimidine–SMe (89)	651
3-bromofuran	Me₃Sn–C(=O)–piperidine	Pd(PPh₃)₄ (5%), PhMe, reflux	furanoyl piperidine (84)	437
	4,4-dimethyl-2-(Me₃Sn)oxazoline	Pd(PPh₃)₄ (5%), C₆H₆, 80°, 18 h	furan–oxazoline (80)	550, 551
	5-(Bu₃Sn)-2-(SMe)pyrimidine	Pd(PPh₃)₂Cl₂ (3%), Cl(CH₂)₂Cl, reflux, 10 h	furan–pyrimidine–SMe (72)	651
	2-(Bu₃Sn)furan	Pd(PPh₃)₄ (5%), PhMe, reflux	bifuran (48)	100
	4-(SnMe₃)pyridine	Pd(PPh₃)₄ (10%), PhMe, reflux, 5 h	pyridyl–furan–pyridyl (50)	560, 561
C₅				
OHC–furan–Br	4-(Bu₃Sn)-2-(SMe)pyrimidine	Pd(PPh₃)₂Cl₂ (5%), DMF, 80°, 3 h	OHC–furan–pyrimidine–SMe (82)	458

289

Substrate	Stannane	Conditions	Product(s) and Yield(s) (%)	Refs.
C7 HO2C–(furan)–Br	Bu3Sn–(pyrimidine)–SMe	Pd(PPh3)2Cl2 (3%), Cl(CH2)2Cl, reflux, 4.5 h	OHC–(furan)–(pyrimidine)–SMe (100)	651
AcO–(furan)–Br	Cl3SnPh	PdCl2, PPh(C6H4SO3Na-*m*)2, aq. KOH, 90°	HO2C–(furan)–Ph (96)	281
	Bu3Sn–(pyrimidine)–SMe	Pd(PPh3)2Cl2 (3%), Cl(CH2)2Cl, reflux, 4 h	AcOCH2–(furan)–(pyrimidine)–SMe (82)	651
C11 (OTf bicyclic)	Bu3Sn–CH=CH–C8H17-*n*	Pd(PPh3)4, LiCl, THF, 50–60°, 2.5 h	*n*-C8H17 CH=CH (bicyclic)–CO2Me (50)	652
(OTf bicyclic)–CO2Me	Bu3Sn–CH=CH–C10H21-*n*	Pd(PPh3)4, LiCl, THF, 50–60°, 2.5 h	*n*-C10H21 (bicyclic)–CO2Me (50)	652
C13 Br–(benzofuran)–CO2Et, *n*-C5H11	Bu3Sn–CH=CH2	Pd(PPh3)4 (3%), PhMe, reflux, 3 h	CH=CH2 (benzofuran)–CO2Et (97)	653
I–(benzofuran)–C5H11-*n*	Bu3Sn–CH=CH2	Pd(0)	*n*-C5H11 (benzofuran)–CH=CH2 (80)	654
	Bu3Sn–(furan)	Pd(0)	*n*-C5H11 (benzofuran)–(furan) (75)	654

C_{22}	$Bu_3SnSnBu_3$	$Pd(OAc)_2$ (10%) NEt$_3$, 100°, 2.5 h	(43)	653
C_{23}	Bu_3Sn— (prenyl)	$Pd(PPh_3)_2Cl_2$ (5%), LiCl, DMF, 105°, 24 h	(65)	655
C_{24}	Bu_3Sn— (allyl)	$Pd(PPh_3)_4$ (2.8%), LiCl, DMF 100°, 2 h	(98)	656
C_{25}	Me_4Sn	$Pd(PPh_3)_2Cl_2$ (10%), dioxane, 100°, 48 h	R = Me (14)	657, 658
	Bu_3Sn—CH=CH$_2$	$Pd(PPh_3)_2Cl_2$ (10%), DMF, 60°, 1 h	CH$_2$=CH (37)	
	Ph_4Sn	$Pd(PPh_3)_2Cl_2$ (10%), dioxane, 100°, 24 h	Ph (39)	

TABLE V. DIRECT CROSS-COUPLING OF FURAN AND BENZOFURAN ELECTROPHILES (Continued)

Substrate	Stannane	Conditions	Product(s) and Yield(s) (%)	Refs.
C_{26}				
	Bu$_3$Sn⟍⟍ Bu$_3$Sn⟍⟍⟍	Pd(PPh$_3$)$_2$Cl$_2$ (10%), DMF, 70°, 2 h Pd(PPh$_3$)$_4$ (10%), C$_6$H$_6$, 100°, 18 h	 $\dfrac{R}{\text{CH}_2=\text{CH}}$ (72) CH$_2$=CHCH$_2$ (97)	658
	Bu$_3$Sn⟍⟍ Bu$_3$Sn⟍⟍⟍ Ph$_4$Sn	Pd(PPh$_3$)$_2$Cl$_2$ (10%), DMF, 70°, 2 h Pd(PPh$_3$)$_4$ (10%), C$_6$H$_6$, 100°, 18 h Pd(PPh$_3$)$_2$Cl$_2$ (10%), dioxane, 100°, 24 h	 $\dfrac{R}{\text{CH}_2=\text{CH}}$ (41) CH$_2$=CHCH$_2$ (21) Ph (0)	658

TABLE VI. DIRECT CROSS-COUPLING OF PYRROLE AND INDOLE ELECTROPHILES

Substrate	Stannane	Conditions	Product(s) and Yield(s) (%)	Refs.
C₈				
5-Bromoindole (Br, N–H)	Me₃Sn-furan-OBu-t	BnPd(PPh₃)₂Cl (1.7%), DMF, 70°, 16 h	indole–furan–t-BuO (60)	296
2-Iodoindole (N–H)	2-(Bu₃Sn)indole (SEM)	Pd(PPh₃)₄ (10%), DMF, 110°	2,2′-biindole (SEM / N–H) (65)	289
C₉				
2,5-dibromopyrrole (N–Boc)	2-(Bu₃Sn)pyrrole (N–Boc)	Pd(PPh₃)₄ (2%), C₆H₆, aq. Na₂CO₃, reflux, 2 d	terpyrrole (Boc / Boc / Boc) (50)	659
2-Bromo-3-formylindole (CHO, N–H)	Me₄Sn	Pd(OAc)₂, Bu₄NCl	2-methyl-3-formylindole (CHO) (39)	109
	Bu₃Sn–CH=CH–CO₂Me	Pd(OAc)₂, Bu₄NCl	3-formyl-2-(CH=CH–CO₂Me)indole (CHO) (67)	109
	2-(Me₃Sn)pyridine	Pd(OAc)₂, Bu₄NCl	3-formyl-2-(2-pyridyl)indole (CHO) (5)	109
	3-(Me₃Sn)pyridine	Pd(OAc)₂, Bu₄NCl	3-formyl-2-(3-pyridyl)indole (CHO) (38)	109
	Bu₃Sn–CH=CH–C(CH₃)₂OH	Pd(OAc)₂, Bu₄NCl	3-formyl-2-(CH=CH–C(CH₃)₂OH)indole (CHO) (87)	109

293

TABLE VI. DIRECT CROSS-COUPLING OF PYRROLE AND INDOLE ELECTROPHILES (*Continued*)

Substrate	Stannane	Conditions	Product(s) and Yield(s) (%)	Refs.
[EtO$_2$C, CO$_2$Me, OTf pyrroline structure]	Ph$_4$Sn	Pd(OAc)$_2$, Bu$_4$NCl	[2-Ph indole-3-CHO] (68)	109
[3-Br indole, N-Ms]	Bu$_3$SnH	Pd(PPh$_3$)$_4$ (2%), THF, reflux, 3 h	[EtO$_2$C, CO$_2$Me pyrroline] (70)	660
C$_{10}$ [3-I indole, N-Ms]	Bu$_3$Sn—OEt	Pd(PPh$_3$)$_2$Cl$_2$ (3%), Et$_4$NCl, DMF, 80°, 9 h	[indole-3-CH=CH-OEt, N-Ms] (83)	272, 273
	Bu$_3$SnC≡COEt	Pd(PPh$_3$)$_2$Cl$_2$ (5%), Et$_4$NCl, DMF, 50°, 2 h	[indole-3-C≡C-COEt, N-Ms] (60)	302
C$_{11}$ [5-Br indole-3-CH$_2$CO$_2$H, N-H]	Cl$_3$SnPh	PdCl$_2$, KOH, PPh(*m*-C$_6$H$_4$SO$_3$Na)$_2$, 90°	[5-Ph indole-3-CH$_2$CO$_2$H, N-H] (79)	281
[3-Br indole-2-CO$_2$Et, N-H]	Bu$_3$Sn—OEt	Pd(PPh$_3$)$_2$Cl$_2$ (3%), Et$_4$NCl, DMF, 80°, 18 h	[indole-3-CH=CH-OEt-2-CO$_2$Et, N-H] (65)	273
C$_{14}$ [2-I indole, N-SEM]	Bu$_3$Sn—[indol-2-yl, N-SEM]	Pd(PPh$_3$)$_4$ (10%), DMF, 110°	[2,2'-biindole, N-SEM, N-SEM] (96)	289

294

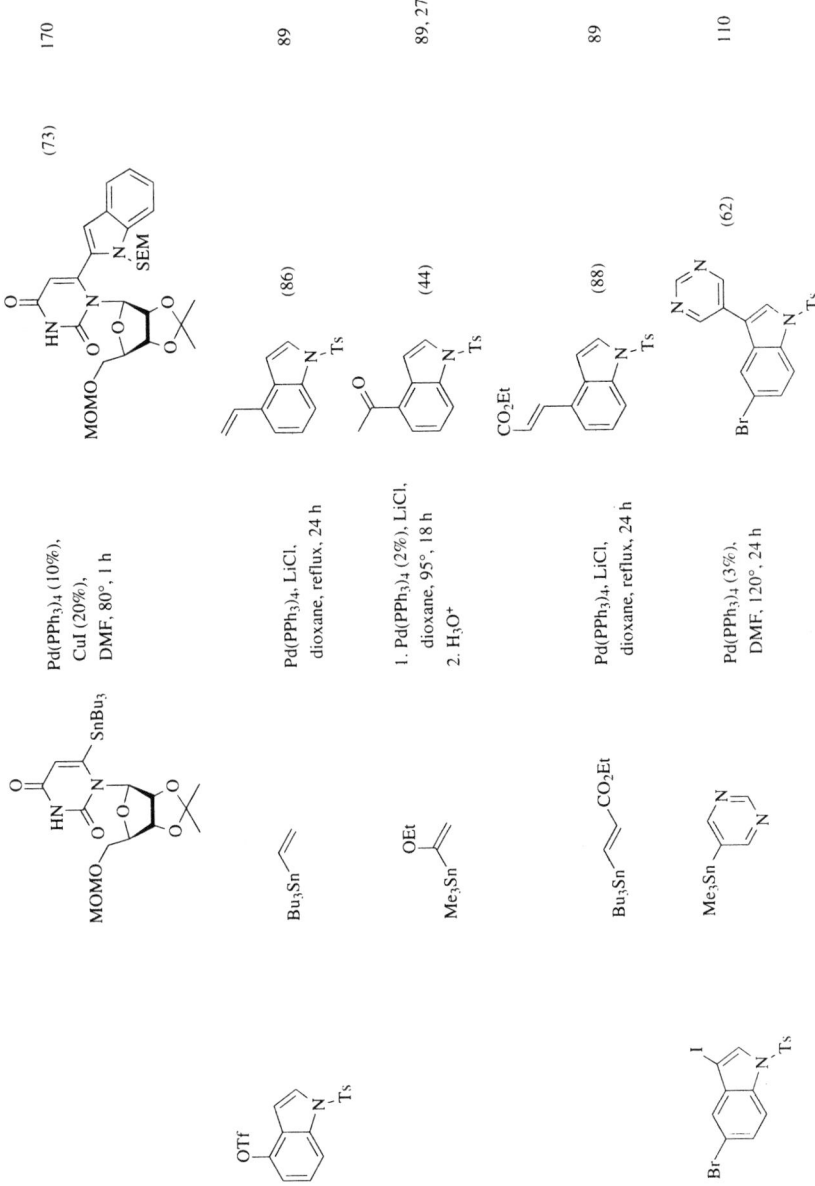

170

89

89, 270

89

110

295

TABLE VI. DIRECT CROSS-COUPLING OF PYRROLE AND INDOLE ELECTROPHILES (Continued)

Substrate	Stannane	Conditions	Product(s) and Yield(s) (%)	Refs.
	Me$_3$Sn-(pyrazine)	Pd(PPh$_3$)$_4$ (3%), DMF, 120°, 24 h	Br-substituted N-Ts indole with pyrazinyl group (56)	110
	Me$_3$Sn-(pyridine)	Pd(PPh$_3$)$_4$ (3%), DMF, 120°, 24 h	Br-substituted N-Ts indole with pyridin-3-yl group (55)	110
(bicyclic N–Ts iodo compound, with I)	Me$_4$Sn; Bu$_3$Sn-(allyl); Bu$_3$Sn-(cinnamyl, Ph)	Pd(OAc)$_2$, P(o-tolyl)$_3$, n-Bu$_3$N, 100°, 24 h	R-substituted bicyclic N–Ts product; R: Me (17), CH=CH$_2$ (36), (E)-CH=CHPh (26)	661
C$_{17}$ (indole, N–CO$_2$Et, with CO$_2$Me, CO$_2$Me side chain, I)	Bu$_3$Sn-(vinyl)	Pd(OAc)$_2$ (0.9%), THF, 85°, 15 h	4-vinyl indole product with CO$_2$Me, CO$_2$Me, CO$_2$Et (87)	662

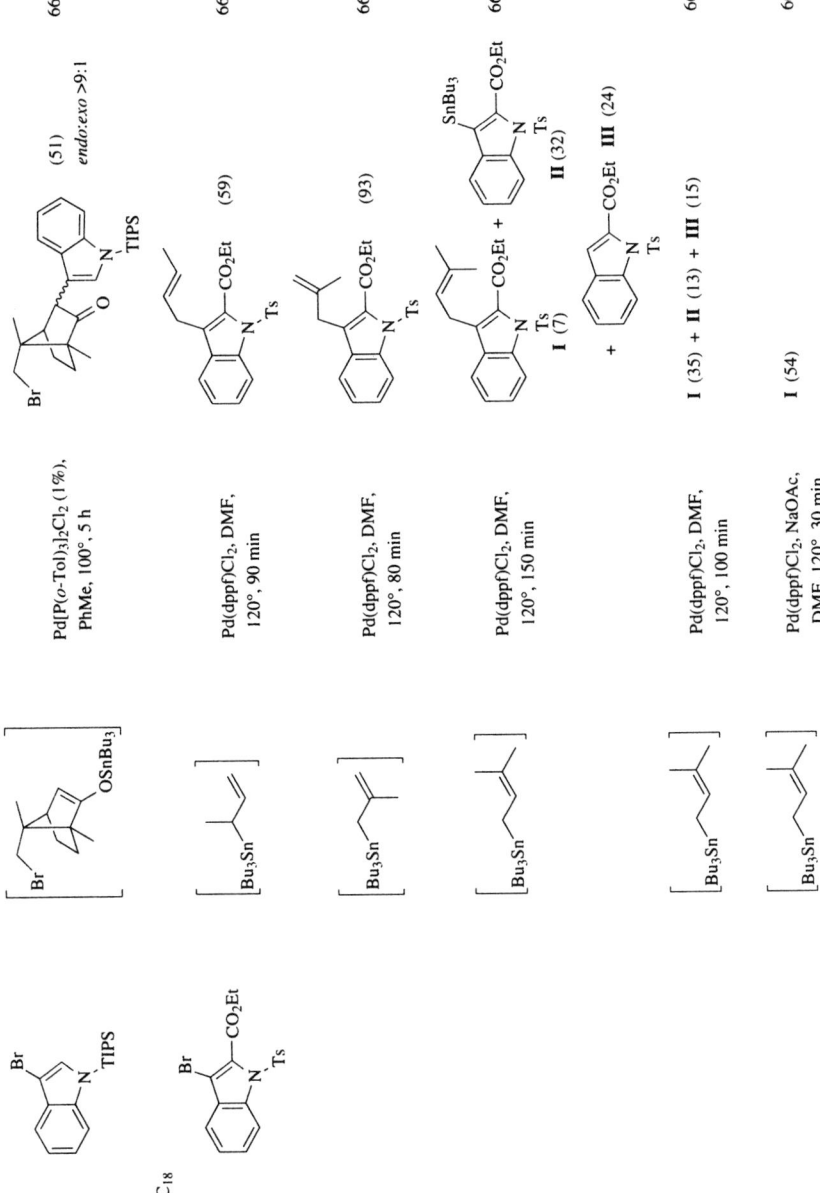

663

(51)
endo:exo >9:1

664

(59)

664

(93)

664

I (7)

II (32)

CO₂Et **III** (24)

I (35) + **II** (13) + **III** (15)

664

I (54)

664

C₁₈

TABLE VI. DIRECT CROSS-COUPLING OF PYRROLE AND INDOLE ELECTROPHILES (Continued)

Substrate	Stannane	Conditions	Product(s) and Yield(s) (%)	Refs.
	[Bu₃Sn— structure]	Pd(dppf)Cl₂, NaOAc, DMF, 120°, 100 min	CO₂Et (60)	664
	[Bu₃Sn—Pr-n structure]	Pd(dppf)Cl₂, DMF, 120°, 100 min	**I** (72)	664
	[Bu₃Sn—cyclohexene structure]	Pd(dppf)Cl₂, DMF, 120°, 110 min	(43)	664
	[Bu₃Sn—Ph structure]	Pd(dppf)Cl₂, DMF, 120°, 35 min	CO₂Et (52) + (38)	664
	Bu₃Sn (Ph structure)	Pd(dppf)Cl₂, DMF, 120°, 150 min	CO₂Et (73)	664
C₁₉ [MeO, MeO, OTf, N–Boc carbazole structure]	Bu₃SnPh	1. Pd(PPh₃)₄ (5%), LiCl, dioxane, 94°, 38 h 2. 150°, 6 h	(63)	665

(Product structures are Ts-protected indole-2-carboxylate and carbazole derivatives; CO_2Et, MeO, Ph substituents as drawn.)

C23 [structure: MeO2C, N–Me, H ergoline with Br and SEM-indole] Bu3Sn–[indole]–SEM Pd(PPh3)4 (10%), DMF, 110° [product structure] (94) 289

C25 [structure: NMe2, Boc-indole with Br, thiazolidine-pyridine] Me3SnSnMe3 Pd(PPh3)4 (7%), PhMe, reflux, 7 h [product] (85) 666

[pyrrole with Br, Boc]n [pyrrole–SnBu3, Boc] Pd(PPh3)4 (2%), C6H6, aq. Na2CO3, reflux, 2 d [product] 659

n
C27 3
C45 5
C63 7

299

TABLE VII. DIRECT CROSS-COUPLING OF THIOPHENE AND BENZOTHIOPHENE ELECTROPHILES

Substrate	Stannane	Conditions	Product(s) and Yield(s) (%)	Refs.
C₄ 2-bromothiophene (S-Br)	Bu₃Sn–CH=CH–CH₂NH₂	Pd(PPh₃)₄ (2%), PhMe, 110°, 24 h	(thienyl)–CH=CH–CH₂NH₂ (70)	462
	Bu₃Sn–CH=CH–OEt	Pd(PPh₃)₂Cl₂ (3%), Et₄NCl, DMF, 80°, 2.5 h	(thienyl)–CH=CH–OEt (68)	272, 273
	Me₃Sn–(4,4-dimethyloxazolin-2-yl)	Pd(PPh₃)₄ (5%), C₆H₆, 80°, 12 h	thienyl-oxazoline (70)	550
	Bu₃Sn–(2-SMe-pyrimidin-5-yl)	Pd(PPh₃)₂Cl₂ (3%), THF, reflux, 10 h	thienyl-(2-SMe-pyrimidine) (71)	651
	Me₃Sn–C(=O)–N(piperidine)	Pd(PPh₃)₄ (5%), PhMe, reflux	thienyl–C(=O)–N(piperidine) (57)	437
	Me₃Sn–(tetrathiafulvalenyl)	Pd(PPh₃)₄ (10%), PhMe, reflux, 3 h	thienyl-tetrathiafulvalene (62)	536
	Bu₃Sn–CH=CH–CH₂–N(TMS)₂	1. Pd(PPh₃)₄ (2%), PhMe, reflux, 48 h 2. HCl (1 N)	(thienyl)–CH=CH–CH₂NH₂ (82)	461, 540
	Bu₃Sn–(1-SEM-indol-2-yl)	Pd(PPh₃)₄ (10%), DMF, 110°	thienyl-(1-SEM-indole) (88)	289

Reagent / Stannane	Conditions	Product (Yield)	Ref.
(indole-SEM, H₂N, Bu₃Sn)	Pd(PPh₃)₄ (10%), DMF, 90°, 4 h	(89)	74
MeO-bithiophene-SnBu₃	Pd(PPh₃)₂Cl₂ (10%), THF, reflux, 20 h	(6)	667
Bu₃SnC≡CH	Pd(PPh₃)₄ (10%), DMF, 25°	(84)	47
H–C≡C=CH₂, Bu₃Sn	Pd₂(dba)₃ (3%) PPh₃ (24%), DMF, rt, 20 h	(42)	275
Bu₃Sn-cyclobutenone	BnPd(PPh₃)₂Cl (1.5%), CH₃CN, rt, 2 h	(52)	267
Bu₃SnC≡COEt	Pd(PPh₃)₂Cl₂ (5%), Et₄NCl, DMF, rt, 5 h	(47)	302
Bu₃Sn-furanone	Pd(PPh₃)₂Cl₂ (2%), PhMe, reflux	(58)	261
thiophene-SnBu₃	Pd(PPh₃)₂Cl₂ (5%), THF, 60°, 16 h	(80)	103
Bu₃SnPh	PhPd(PPh₃)₂I (2%), HMPA, 70°, 30 min	I (94)	463
Bu₃SnPh	Pd/C (0.5%), CuI (10%), AsPh₃ (20%), NMP, 80°, 16 h	I (77)	461

(2-iodothiophene)

Substrate	Stannane	Conditions	Product(s) and Yield(s) (%)	Refs.
	Me_3Sn–C(O)–N(piperidine)	$Pd(PPh_3)_4$ (5%), PhMe, reflux	(24) + (32)	437
	Bu_3Sn (i-PrO squarate)	$BnPd(PPh_3)_2Cl$ (5%), CuI, DMF, rt	(75)	12
	Bu_3Sn (dioxolane squarate)	$BnPd(PPh_3)_2Cl$ (5%), CuI, DMF, rt	(67)	12
	Bu_3Sn CO_2Et NHAc	$Pd_2(dba)_3$ (5%), P(2-furyl)$_3$ (40%), THF, 65°	**I** (30) + **II** (74)	375
	Bu_3Sn CO_2Et NHAc	$Pd_2(dba)_3$ (5%), AsPh$_3$ (40%), THF, 65°	**II** (23)	375
	Bu_3Sn (quinone)	$Pd_2(dba)_3$ (2.5%), CuI, AsPh$_3$, DMF, 60°	(61)	554

Substrate	Organostannane	Conditions	Product (%)	Refs.
3-bromothiophene	Bu_3Sn–CH=C(Ph)(CO_2Et)	Pd(OAc)$_2$, CuI, PPh$_3$, DMF, rt, 23 h	thienyl–CH=C(Ph)(CO_2Et) (79)	435
	Bu_3Sn–C(OBn)=CH–Bu-n	Pd$_2$(dba)$_3$ (2%), DMF, rt	(79)	439
	Bu_3Sn–CH=CH–SnBu$_3$	Pd(OAc)$_2$, P(2-furyl)$_3$, DME, reflux, 5 h	thienyl–CH=CH–thienyl (—)	668
	Me$_3$SnSnMe$_3$	Pd(CH$_3$CN)$_2$Cl$_2$ (1%), HMPA, rt, 80 min	thienyl–SnMe$_3$ (95)	313
	2-(Bu$_3$Sn)pyridine	Pd(dppb)Cl$_2$ (5%), DMF, 100°, 24 h	3-(2-pyridyl)thiophene (74)	96, 669
	2-(Me$_3$Sn)-4,4-dimethyloxazoline	Pd(PPh$_3$)$_4$ (5%), C$_6$H$_6$, 80°, 12 h	(100)	550, 551
	5-(Bu$_3$Sn)-2-(SMe)pyrimidine	Pd(PPh$_3$)$_2$Cl$_2$ (3%), THF, reflux, 24 h	(90)	651
3-iodothiophene	2-(Bu$_3$Sn)pyridine	Pd(dppb)Cl$_2$ (5%), CuO, DMF, 100°, 1.5 h	3-(2-pyridyl)thiophene (63)	96, 669

Substrate	Stannane	Conditions	Product(s) and Yield(s) (%)	Refs.
		Pd$_2$(dba)$_3$ (2.5%) AsPh$_3$ (10%), CuI (10%), DMF, 60°, 4 h	(73)	291
		Pd(PPh$_3$)$_2$Cl$_2$ (10%), THF, reflux, 20 h	(52)	667
		Pd(PPh$_3$)$_4$ (10%), PhMe, reflux, 5 h	(64)	560, 561
		Pd(PPh$_3$)$_4$ (10%), PhMe, reflux, 5 h	(64)	560
		Pd(PPh$_3$)$_4$ (10%), PhMe, reflux, 5 h	(55)	560
		Pd(PPh$_3$)$_4$ (20%), PhMe, reflux, 3 h	(52)	536
		Pd(PPh$_3$)$_4$ (10%), PhMe, reflux, 5 h	(30)	560
		Pd(PPh$_3$)$_2$Cl$_2$ (5%), THF, 60°, 16 h	(61)	103

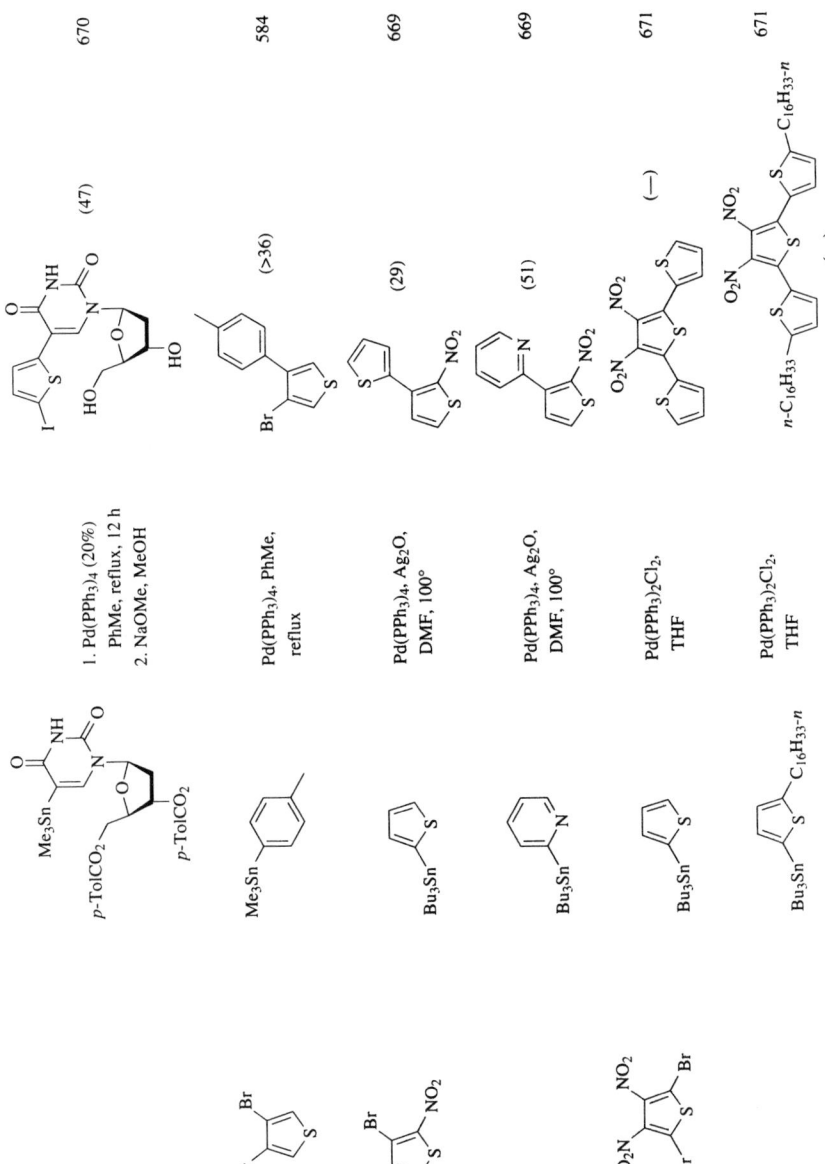

670

584

669

669

671

671

TABLE VII. DIRECT CROSS-COUPLING OF THIOPHENE AND BENZOTHIOPHENE ELECTROPHILES (*Continued*)

Substrate	Stannane	Conditions	Product(s) and Yield(s) (%)	Refs.
C₅	Bu₃Sn⎓⎓SnBu₃	Pd(OAc)₂, P(2-furyl)₃, DME, reflux, 5 h	(46)	668
		Pd(PPh₃)₂Cl₂ (3%), DMF, 80°, 4.5 h	(87)	651
		Pd₂(dba)₃ (1%) P(2-furyl)₃ (4%), dioxane	(—)	672
	Bu₃Sn	Pd(PPh₃)₂Cl₂ (5%), THF, 60°, 16 h	(71)	103
	Bu₃Sn	Pd(PPh₃)₂Cl₂ (5%), THF, 60°, 16 h	(79)	103
		Pd(PPh₃)₂Cl₂ (2%), THF, reflux	(89)	425
	Me₃Sn	Pd(PPh₃)₂Cl₂ (3%), THF, reflux, 24 h	(9)	104
C₆	Me₃Sn	Pd(PPh₃)₄, Ag₂O, DMF, 100°	(25)	669

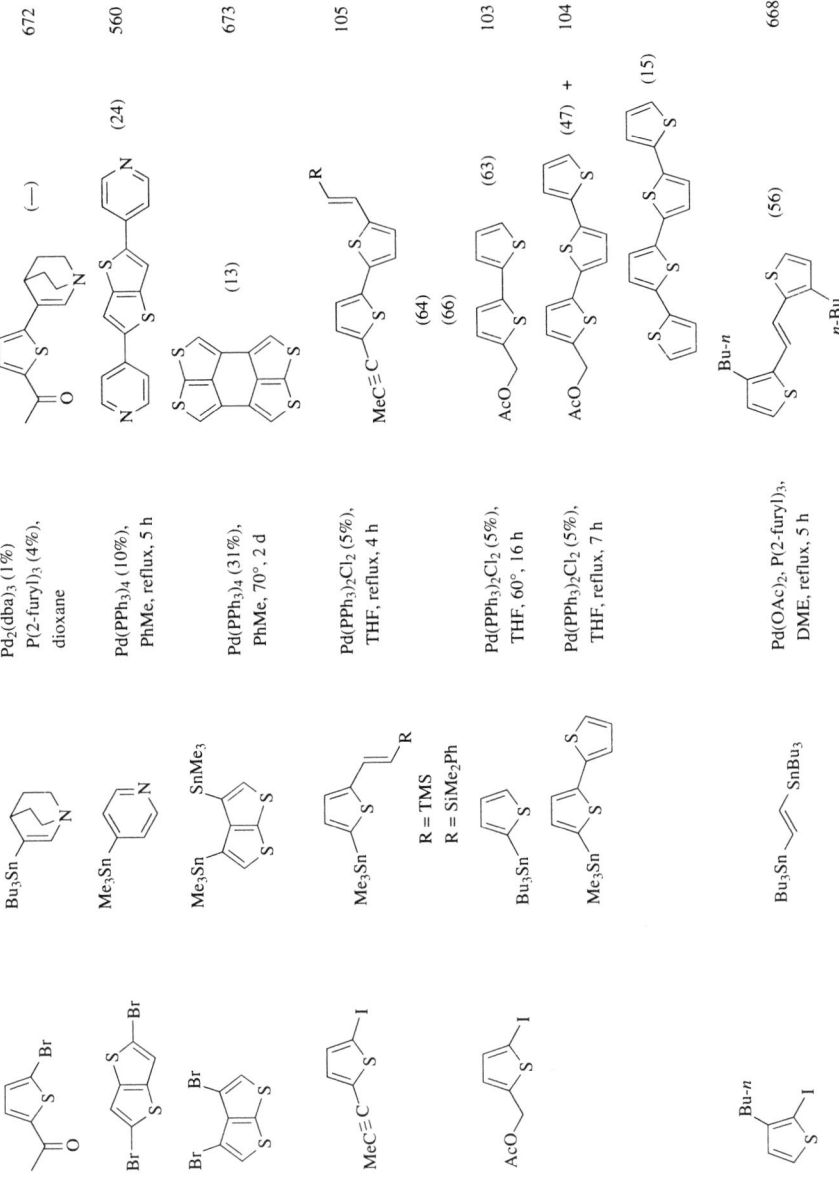

Bu$_3$Sn (substrate with acetyl thiophene, Br)	Pd$_2$(dba)$_3$ (1%) P(2-furyl)$_3$ (4%), dioxane	(—)	672
Me$_3$Sn-pyridine	Pd(PPh$_3$)$_4$ (10%), PhMe, reflux, 5 h	(24)	560
Me$_3$Sn-thienothiophene (SnMe$_3$)	Pd(PPh$_3$)$_4$ (31%), PhMe, 70°, 2 d	(13)	673
Me$_3$Sn-vinyl R (R = TMS, R = SiMe$_2$Ph)	Pd(PPh$_3$)$_2$Cl$_2$ (5%), THF, reflux, 4 h	(64), (66)	105
Bu$_3$Sn-thiophene	Pd(PPh$_3$)$_2$Cl$_2$ (5%), THF, 60°, 16 h	(63)	103
Me$_3$Sn-bithiophene	Pd(PPh$_3$)$_2$Cl$_2$ (5%), THF, reflux, 7 h	(47) + (15)	104
Bu$_3$Sn-vinyl-SnBu$_3$	Pd(OAc)$_2$, P(2-furyl)$_3$, DME, reflux, 5 h	(56)	668

C$_7$

C$_8$

307

TABLE VII. DIRECT CROSS-COUPLING OF THIOPHENE AND BENZOTHIOPHENE ELECTROPHILES (*Continued*)

Substrate	Stannane	Conditions	Product(s) and Yield(s) (%)	Refs.
C9	Me3Sn— (CO2Et)	Pd(PPh3)2Cl2, LiCl, THF, reflux	(51)	438
C10	Bu3Sn— (benzodioxole)	Pd(PPh3)2Cl2 (4%), LiCl, DMF, 80°, 3.5 h	(77)	674
	Me3Sn— (bithiophene)	Pd(PPh3)2Cl2 (4%), THF, reflux, 8 h	(59)	104
	Bu3Sn—OTHP	Pd(PPh3)2Cl2 (5%), THF, 60°, 16 h	(58)	103
C11	THPO Me3Sn— OTHP	Pd(PPh3)4 (1%), PhMe, reflux, 12 h	(70)	106
	Me3Sn— OTHP	Pd(PPh3)4 (1%), PhMe, reflux, 12 h	(50)	106
	Me3SnSnMe3	Pd(PPh3)4, PhMe, reflux	(68)	584

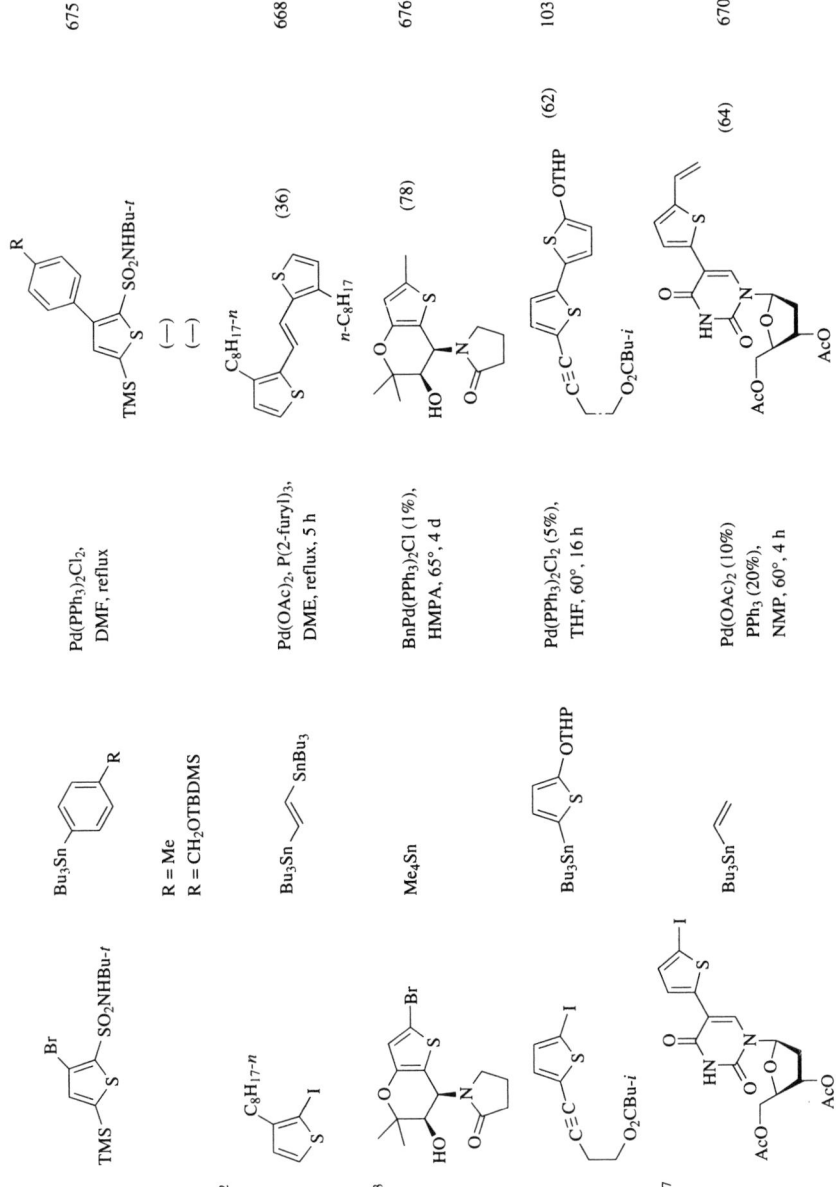

675

668

676

103

670

Substrate	Stannane	Conditions	Product(s) and Yield(s) (%)	Refs.

C₃₇

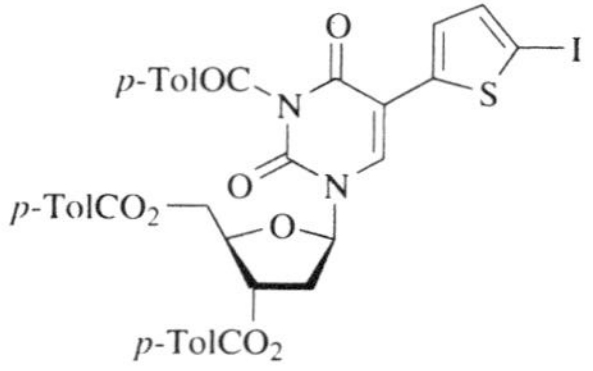

Me₄Sn — Pd(PPh₃)₄, NMP, 60°, 16 h — (65) — 670

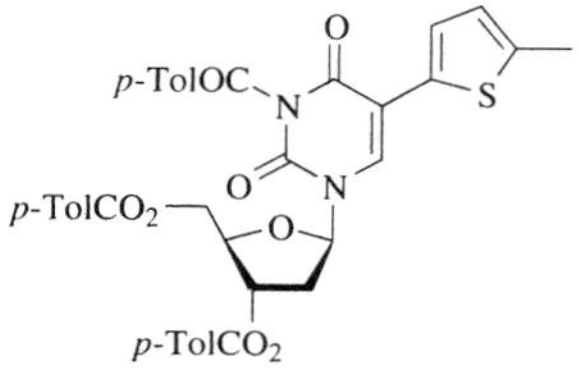

C₄₁

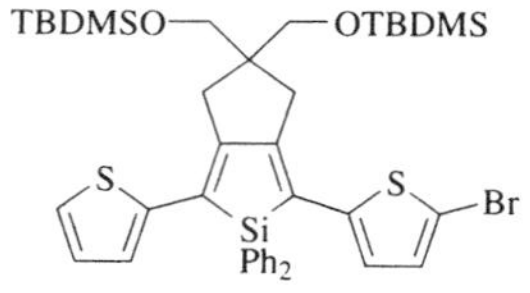

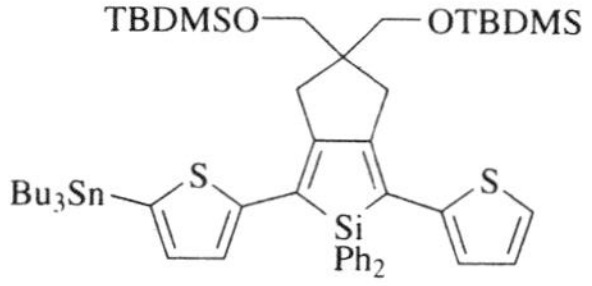

Bu₃Sn — Pd(PPh₃)₄ — (75) — 677

(OTBDMS, TBDMSO, SiPh₂ cyclopenta-fused bithiophene structures)

≡ Bu₃Sn–R–SnBu₃ — SnBu₃ — Pd(PPh₃)₄ — (58) — 677

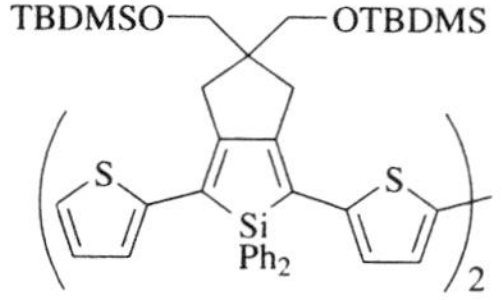

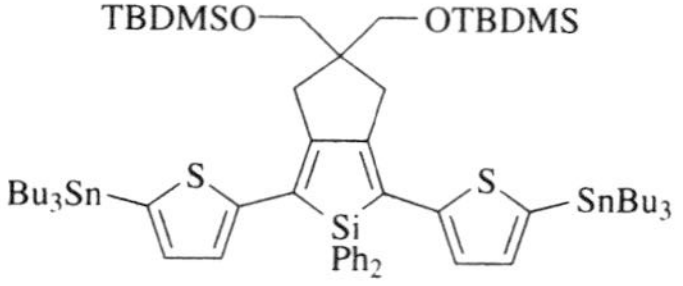

Bu₃Sn (thiophene) — Pd(PPh₃)₄ — (63) — 677

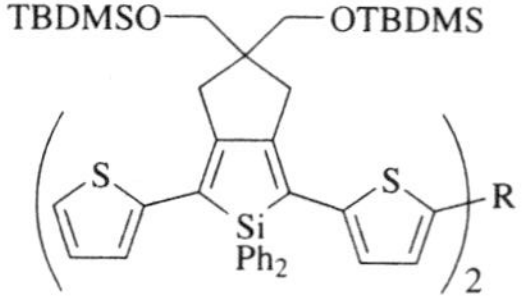

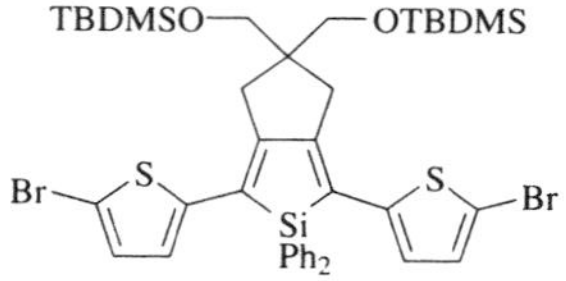

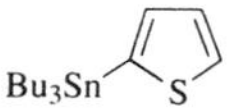

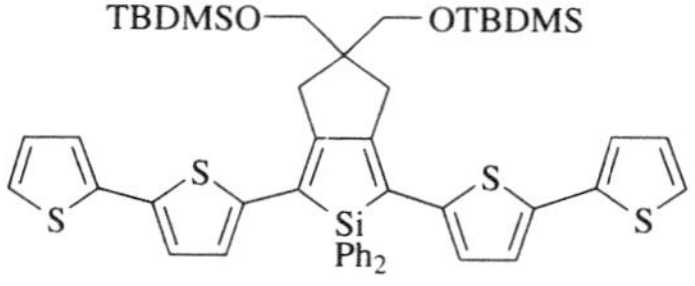

TABLE VIII. DIRECT CROSS-COUPLING OF PYRAN AND BENZOPYRAN ELECTROPHILES

Substrate	Stannane	Conditions	Product(s) and Yield(s) (%)	Refs.
C₆ (structure, OTf)	Me₃SnSnMe₃	Pd(PPh₃)₄ (2.7%) LiCl, THF, reflux, 14 h	(structure, SnMe₃) (71)	678
C₇ (structure, OTf)	Me₃SnSnMe₃	Pd(PPh₃)₄ (2.7%) LiCl, THF, reflux, 14 h	(structure, SnMe₃) (78)	678
C₈ (structure, OTf)	Me₃SnSnMe₃	Pd(PPh₃)₄ (2.7%) LiCl, THF, reflux, 14 h	(structure, SnMe₃) (75)	678
(structure, OTf)	Me₃SnSnMe₃	Pd(PPh₃)₄ (2.7%) LiCl, THF, reflux, 14 h	(structure, SnMe₃) (70)	678
(structure, MeO, OMe, OTf)	Me₃SnSnMe₃	Pd(PPh₃)₄ (2.7%) LiCl, THF, reflux, 14 h	(structure, MeO, OMe, SnMe₃) (69)	678
C₉ (structure, OTf)	Me₃Sn–pyridine	Pd(PPh₃)₄ (2%) LiCl, dioxane, reflux, 5 h	(structure) (89)	679
(structure, OTf)	Bu₃Sn–C₆H₄F	Pd(PPh₃)₄ (2%) LiCl, dioxane, reflux, 6 h	(structure) (76)	679

TABLE VIII. DIRECT CROSS-COUPLING OF PYRAN AND BENZOPYRAN ELECTROPHILES (*Continued*)

Substrate	Stannane	Conditions	Product(s) and Yield(s) (%)	Refs.
		Pd$_2$(dba)$_3$ (2.5%), AsPh$_3$ (10%), DMF, 60°, 2 h	(93)	291
	Bu$_3$Sn (propene) / Bu$_3$SnPh / Bu$_3$SnC≡CPh	Pd(OAc)$_2$, PPh$_3$ (8%), NEt$_3$, 100°, 24 h	$\dfrac{R}{CH=CH_2 \quad (64)}{Ph \quad (71)}{PhC\equiv C \quad (61)}$	327
C$_{10}$	Me$_3$SnPh	Pd(PPh$_3$)$_4$ (2%), LiCl, dioxane, reflux, 18 h	(87)	680
		Pd(PPh$_3$)$_4$ (10%), DMF, 110°	(96)	289
C$_{11}$		Pd$_2$(dba)$_3$·CHCl$_3$, PPh$_3$, LiCl, THF	(83)	681

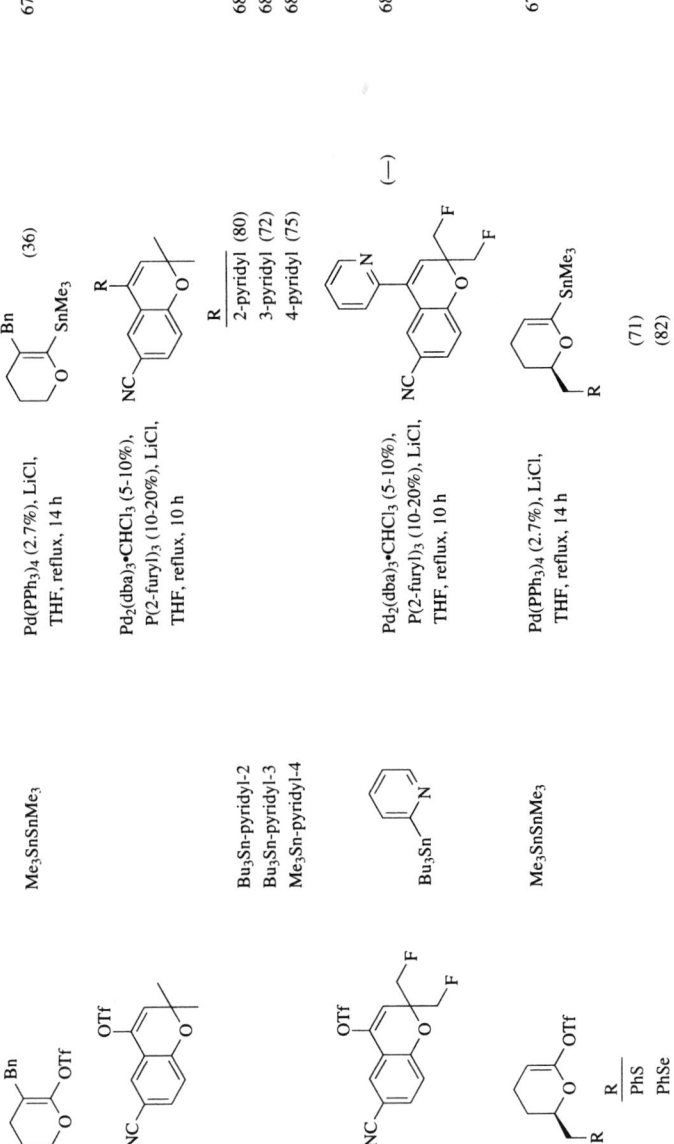

C₁₂

| | Me₃SnSnMe₃ | Pd(PPh₃)₄ (2.7%), LiCl, THF, reflux, 14 h | (36) | 678 |

| | Bu₃Sn-pyridyl-2
Bu₃Sn-pyridyl-3
Me₃Sn-pyridyl-4 | Pd₂(dba)₃•CHCl₃ (5-10%), P(2-furyl)₃ (10-20%), LiCl, THF, reflux, 10 h | R
2-pyridyl (80)
3-pyridyl (72)
4-pyridyl (75) | 683, 682
683
683 |

| | Bu₃Sn | Pd₂(dba)₃•CHCl₃ (5-10%), P(2-furyl)₃ (10-20%), LiCl, THF, reflux, 10 h | (—) | 681 |

| | Me₃SnSnMe₃ | Pd(PPh₃)₄ (2.7%), LiCl, THF, reflux, 14 h | R
PhS (71)
PhSe (82) | 678 |

TABLE VIII. DIRECT CROSS-COUPLING OF PYRAN AND BENZOPYRAN ELECTROPHILES (Continued)

	Substrate	Stannane	Conditions	Product(s) and Yield(s) (%)	Refs.
C₁₄		Bu₃SnR	Pd(PPh₃)₄, LiCl, THF	 R C≡CH (80) CH=CH₂ (100) (E)-CH=CHCO₂Et (80) (E)-CH=CHTMS (100) (E)-CH=CHSnBu₃ (75)	684
		Me₃SnSnMe₃	Pd(PPh₃)₄, LiCl, THF, reflux, 12 h	(69)	685
C₁₅		Sn(CH=CH₂)₄	1. Pd(PPh₃)₂Cl₂ (2%), LiCl, DMF, 100° 2. NaOH (2 N)	(89)	686
		SnEt₄	1. Pd(PPh₃)₂Cl₂ (2%), LiCl, DMF, 100° 2. NaOH (2 N)	(—)	686
		Bu₃Sn(CH=CH₂)	Pd(PPh₃)₄ (5%), THF, 80°, 40 h	(60)	58

314

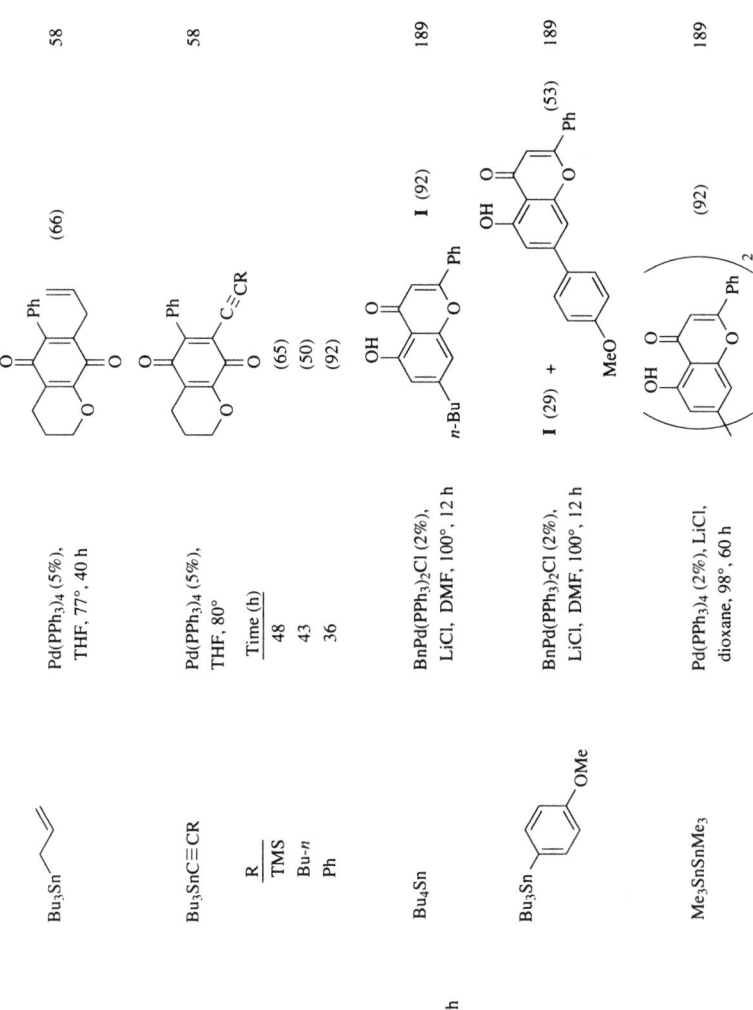

Bu₃Sn⁀	Pd(PPh₃)₄ (5%), THF, 77°, 40 h	(66)	58
Bu₃SnC≡CR	Pd(PPh₃)₄ (5%), THF, 80°		58
R / TMS / Bu-n / Ph	Time (h) / 48 / 43 / 36	(65) / (50) / (92)	
Bu₄Sn	BnPd(PPh₃)₂Cl (2%), LiCl, DMF, 100°, 12 h	I (92)	189
Bu₃Sn—⟨⟩—OMe	BnPd(PPh₃)₂Cl (2%), LiCl, DMF, 100°, 12 h	I (29) + (53)	189
Me₃SnSnMe₃	Pd(PPh₃)₄ (2%), LiCl, dioxane, 98°, 60 h	(92)	189

TABLE VIII. DIRECT CROSS-COUPLING OF PYRAN AND BENZOPYRAN ELECTROPHILES (*Continued*)

Substrate	Stannane	Conditions	Product(s) and Yield(s) (%)	Refs.
C$_{17}$		Pd(PPh$_3$)$_2$Cl$_2$ (15%), PhMe, 115°, 21 h	(25)	687
C$_{18}$		Pd(CH$_3$CN)$_2$Cl$_2$, DMF, 80° or Pd(CH$_3$CN)$_2$Cl$_2$. PPh$_3$, CHCl$_3$, 80°	(0)	688
X = Br X = I		Pd(PPh$_3$)$_2$Cl$_2$ (50%), dioxane, 115°, 3 d Pd(PPh$_3$)$_2$Cl$_2$ (14%), PhMe, 115°, 2 d	(18) (20)	687
C$_{19}$		BnPd(PPh$_3$)$_2$Cl (5%), dioxane, 115°, 4 d	(36)	687

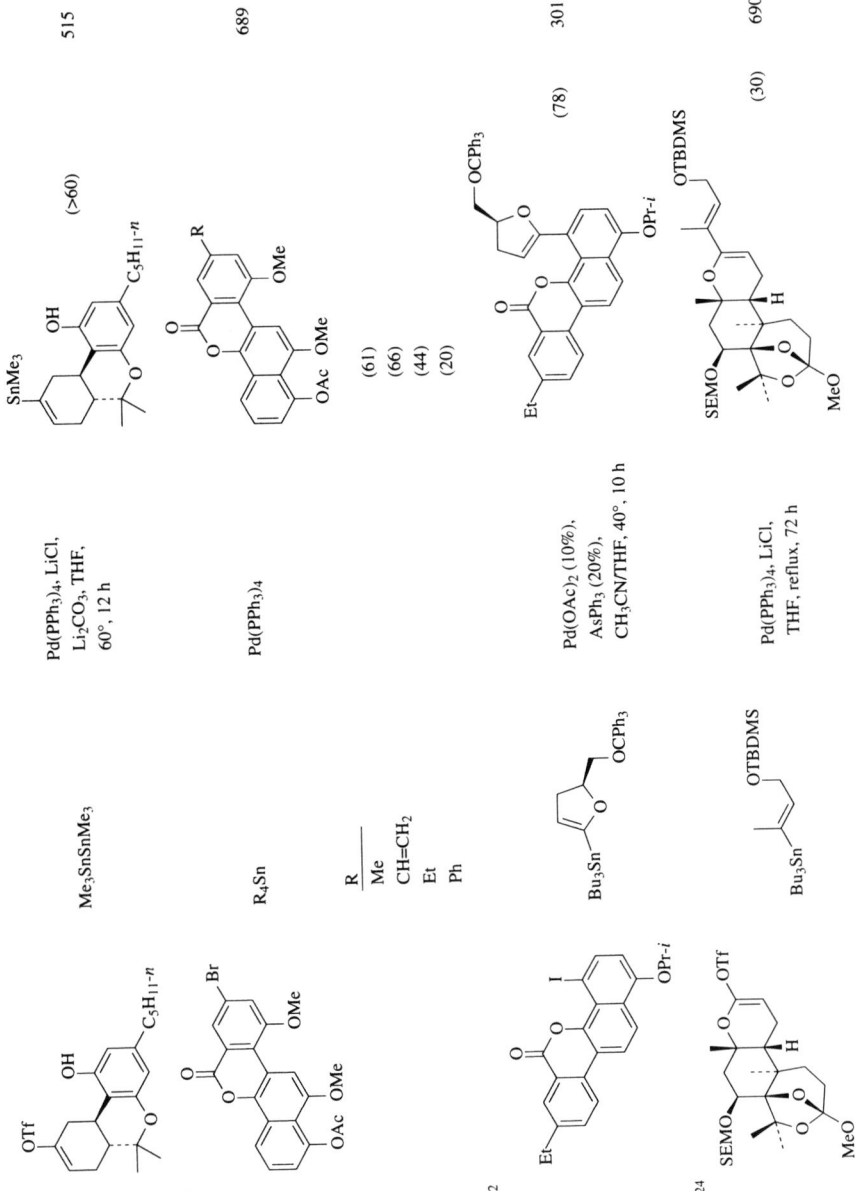

515

(>60)

Me₃SnSnMe₃

Pd(PPh₃)₄, LiCl,
Li₂CO₃, THF,
60°, 12 h

689

R₄Sn

Pd(PPh₃)₄

R	
Me	(61)
CH=CH₂	(66)
Et	(44)
Ph	(20)

301

(78)

Pd(OAc)₂ (10%),
AsPh₃ (20%),
CH₃CN/THF, 40°, 10 h

690

(30)

Pd(PPh₃)₄, LiCl,
THF, reflux, 72 h

C_{20}

C_{21}

C_{22}

C_{24}

Substrate	Stannane	Conditions	Product(s) and Yield(s) (%)	Refs.
C$_{26}$		Pd$_2$(dba)$_3$, P(2-furyl)$_3$, LiCl, THF, 60°	(81)	691
C$_{33}$	Sn$\left(\diagdown\diagup\right)_4$	Pd(PPh$_3$)$_2$Cl$_2$ (10%), THF, reflux, 8 h	(67)	300, 388
	Bu$_3$SnPh	Pd(PPh$_3$)$_2$Cl$_2$ (10%), THF, reflux, 24 h	(20)	300, 388

TABLE IX. DIRECT CROSS-COUPLING OF PYRIDINE ELECTROPHILES

Substrate	Stannane	Conditions	Product(s) and Yield(s) (%)	Refs.
C₅	Bu₃Sn (isoxazole-Me)	Pd(PPh₃)₂Cl₂ (3%), dioxane, reflux, 6 h	(64)	292, 530
	Bu₃Sn OEt	Pd(PPh₃)₂Cl₂ (3%), Et₄NCl, DMF, 80°, 3.5 h	OEt (62)	272, 273
	Me₃Sn (pyridyl)	Pd(PPh₃)₄ (1%), xylene, reflux, 12 h	(77)	93
	Me₃Sn (pyridyl)	Pd(PPh₃)₄ (1%), xylene, reflux, 12 h	(59)	93
	Me₃Sn (oxazoline)	Pd(PPh₃)₄ (5%), C₆H₆, 80°, 24 h	(85)	550, 551
	OHC Bu₃Sn (thienyl)	Pd(PPh₃)₄ (3%), Ag₂O, DMF, 100°, 5 min	CHO (60)	32
	Bu₃Sn (pyrimidinyl-SMe)	Pd(PPh₃)₂Cl₂ (3%), DMF, 80°, 24 h	SMe (72)	651
	Me₃Sn (dithiole)	Pd(PPh₃)₄ (10%), PhMe, reflux, 4 h	(82)	536

319

TABLE IX. DIRECT CROSS-COUPLING OF PYRIDINE ELECTROPHILES (Continued)

Substrate	Stannane	Conditions	Product(s) and Yield(s) (%)	Refs.
		Pd(PPh₃)₄ (2%), C₆H₆, 80°, 20 h	(88)	539
		Pd(PPh₃)₄ (1%), xylene, reflux, 12 h	(79)	93
		Pd(PPh₃)₄ (3%), Ag₂O, DMF, 100°, 0.5 h	(58)	32
		Pd(PPh₃)₄ (10%), PhMe, 110°, 18 h	(0)	371
		Pd(PPh₃)₄ (10%), DMF, 110°	(80)	289
		Pd₂(dba)₃ (2.5%), AsPh₃, (10%), CuI (10%), DMF, 60°, 6 h	(54)	291
		Pd(PPh₃)₄ (10%), DMF, 90°, 8 h	(78)	74

320

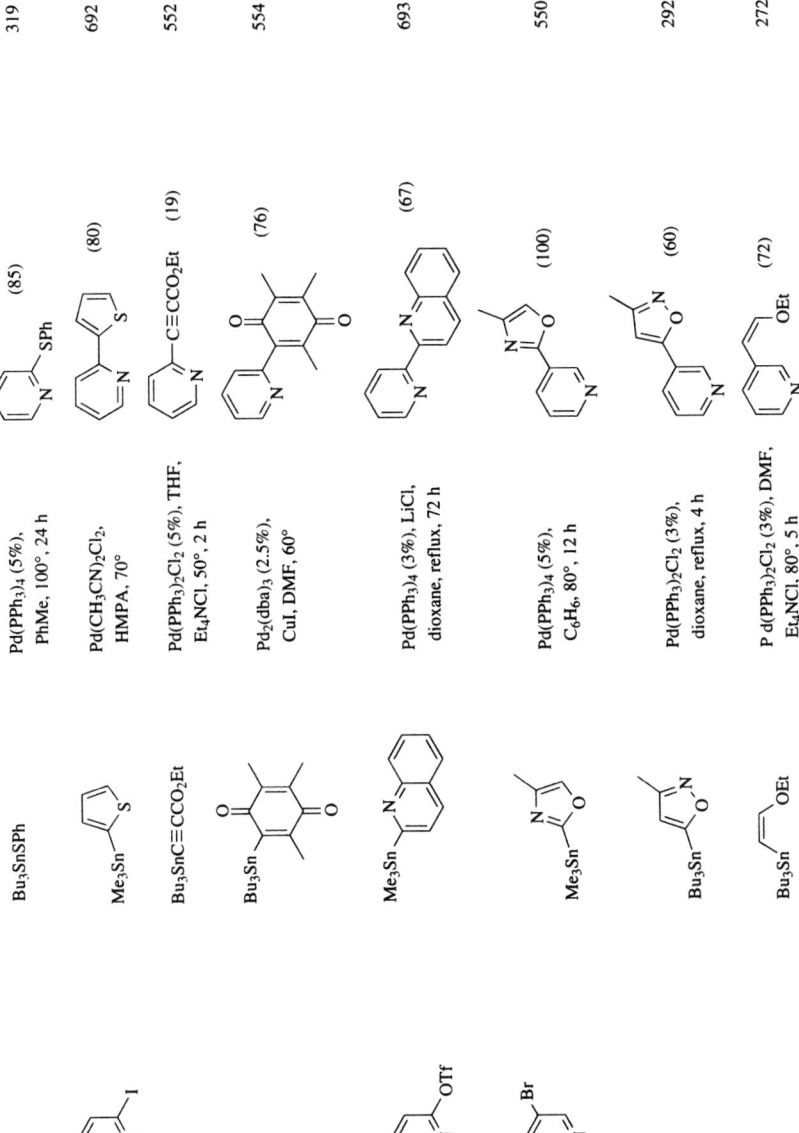

Bu₃SnSPh	Pd(PPh₃)₄ (5%), PhMe, 100°, 24 h	(85) 319
Me₃Sn-thiophene	Pd(CH₃CN)₂Cl₂, HMPA, 70°	(80) 692
Bu₃SnC≡CCO₂Et	Pd(PPh₃)₂Cl₂ (5%), THF, Et₄NCl, 50°, 2 h	(19) 552
Bu₃Sn-quinone	Pd₂(dba)₃ (2.5%), CuI, DMF, 60°	(76) 554
Me₃Sn-quinoline	Pd(PPh₃)₄ (3%), LiCl, dioxane, reflux, 72 h	(67) 693
Me₃Sn-oxazole	Pd(PPh₃)₄ (5%), C₆H₆, 80°, 12 h	(100) 550
Bu₃Sn-isoxazole	Pd(PPh₃)₂Cl₂ (3%), dioxane, reflux, 4 h	(60) 292, 530
Bu₃Sn-OEt	Pd(PPh₃)₂Cl₂ (3%), DMF, Et₄NCl, 80°, 5 h	(72) 272, 273

TABLE IX. DIRECT CROSS-COUPLING OF PYRIDINE ELECTROPHILES (*Continued*)

Substrate	Stannane	Conditions	Product(s) and Yield(s) (%)	Refs.
	Me$_3$Sn	Pd(PPh$_3$)$_4$ (1%), xylene, reflux, 12 h	**I** (63)	93
	Bu$_3$Sn	Pd(dppb)Cl$_2$ (5%), DMF, 100°, 24 h	**I** (99)	96
	Me$_3$Sn	Pd(PPh$_3$)$_4$ (1%), xylene, reflux, 12 h	(68)	93
	Me$_3$Sn	Pd(PPh$_3$)$_4$ (1%), xylene, reflux, 12 h	(70)	93
	Me$_3$Sn	Pd(PPh$_3$)$_4$ (5%), C$_6$H$_6$, 80°, 24 h	(82)	550, 551
	Bu$_3$Sn	Pd(PPh$_3$)$_2$Cl$_2$ (3%), THF, reflux, 24 h	SMe (65)	651
	Me$_3$Sn	Pd(PPh$_3$)$_4$ (5%), PhMe, reflux	(63)	437
	Bu$_3$Sn	BnPd(PPh$_3$)$_2$Cl (1.7%), DMF, 70°, 16 h	OBu-*t* (91)	296

322

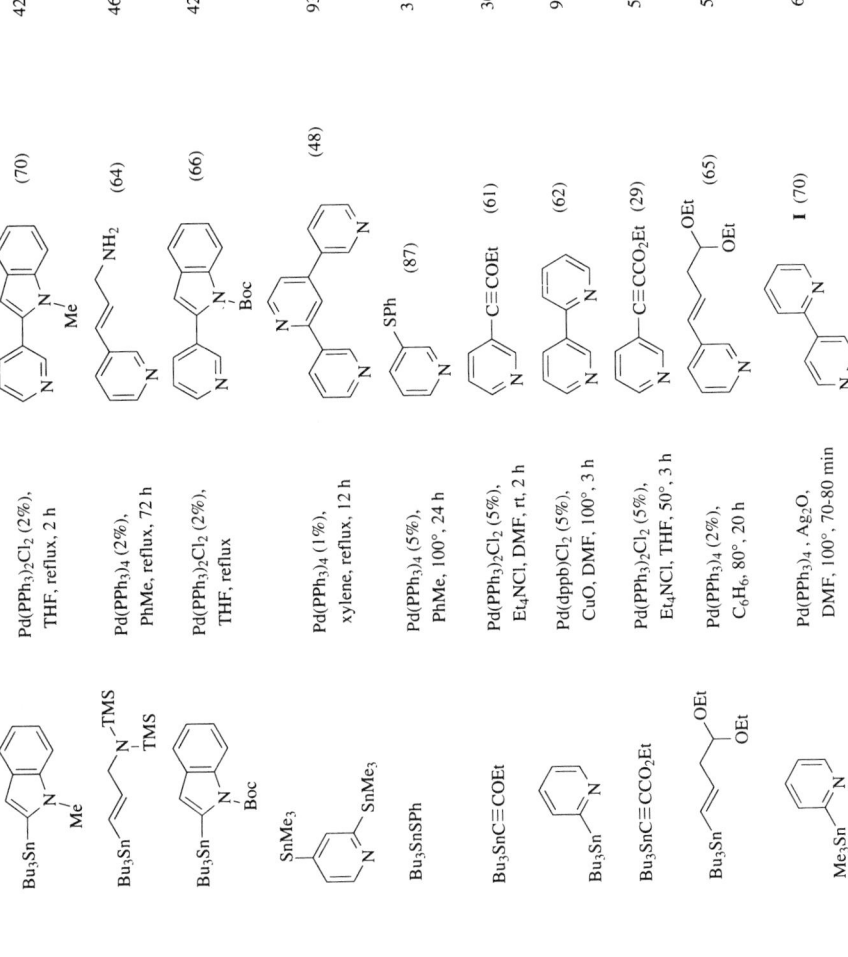

TABLE IX. DIRECT CROSS-COUPLING OF PYRIDINE ELECTROPHILES (Continued)

Substrate	Stannane	Conditions	Product(s) and Yield(s) (%)	Refs.
	Me₃Sn (2-pyridyl)	Pd(PPh₃)₄ (1%), xylene, reflux, 12 h	**I** (65)	93
	Bu₃Sn (2-pyridyl)	Pd(dppb)Cl₂ (5%), CuO, DMF, 100°, 80 min	**I** (75)	96
	Bu₃Sn (3-pyridyl)	Pd(PPh₃)₄ (1%), xylene, reflux, 12 h	(72)	93
	Bu₃Sn (2-SMe-pyrimidin-5-yl)	Pd(PPh₃)₂Cl₂ (3%), DMF, 80°, 10 h	SMe (67)	651
	Cl₃SnPh	PPh(m-C₆H₄SO₃Na)₂, PdCl₂, KOH, 90°	Ph (81)	281
	Bu₃Sn (2-pyridyl)	Pd(dppb)Cl₂ (5%), CuO, DMF, 100°, 4 h	(44)	96
	Me₃Sn (2-thienyl)	PdCl₂, HMPA, 70°	(96)	692
	Bu₃Sn (thiazol-2-yl)	Pd(PPh₃)₂Cl₂ (10%), THF, reflux, 20 h	(52)	667
	Bu₃Sn (2-SEM-indol-2-yl)	Pd(PPh₃)₄ (10%), DMF, 110°	(92)	289

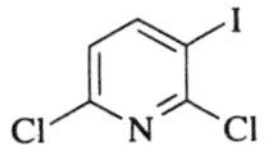

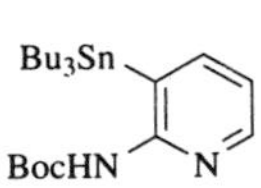

Pd(PPh$_3$)$_4$, CuI, dioxane, reflux — (81) — 582

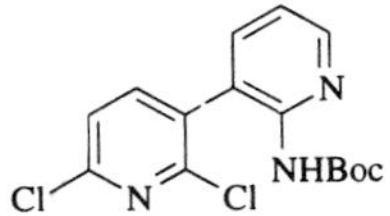

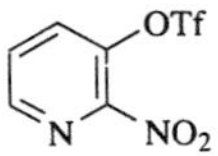

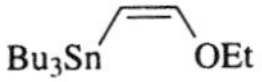

Pd(PPh$_3$)$_2$Cl$_2$ (3%), Et$_4$NCl, CH$_3$CN, reflux, 1 h — (74) + (14) — 95

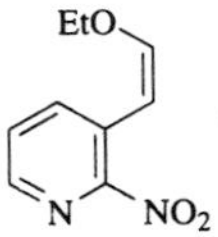

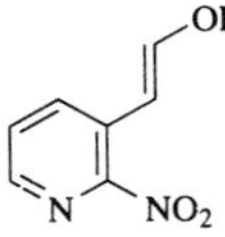

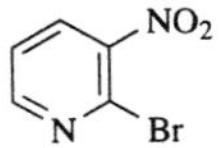

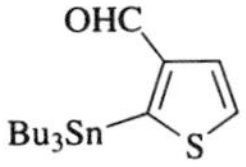

Pd(PPh$_3$)$_4$ (5%), CuO, DMF, 100°, 40 min — (69) — 694

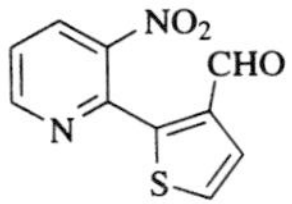

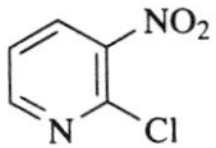

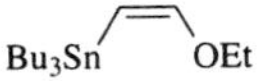

Pd(PPh$_3$)$_2$Cl$_2$ (3%), Et$_4$NCl, CH$_3$CN, reflux, 1.5 h — (96) — 95

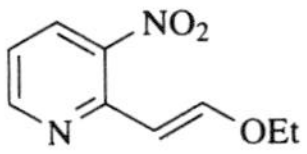

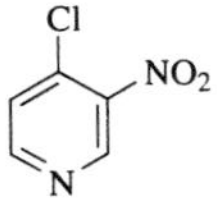

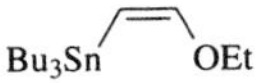

Pd(PPh$_3$)$_2$Cl$_2$ (3%), Et$_4$NCl, CH$_3$CN, reflux, 1.5 h — (36) + (47) — 95

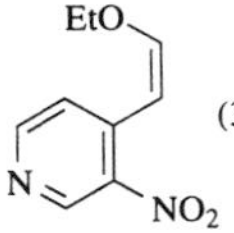

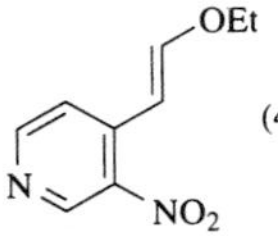

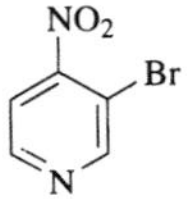

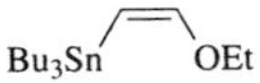

Pd(PPh$_3$)$_2$Cl$_2$ (3%), Et$_4$NCl, CH$_3$CN, reflux, 3 h — (78) + (16) — 95

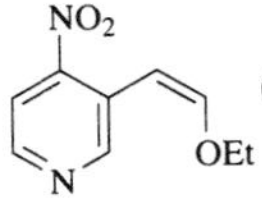

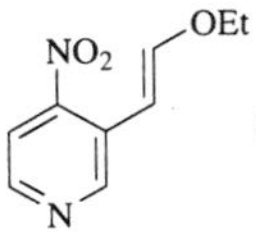

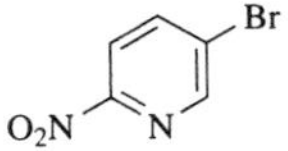

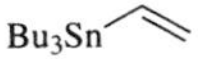

Pd(CH$_3$CN)$_2$Cl$_2$ (1%), DMF, THF, 70°, 10 min — (89) — 695

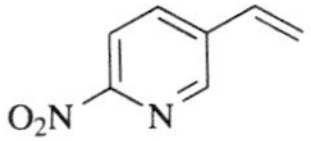

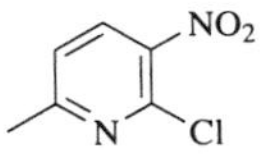

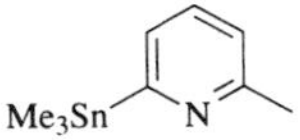

Pd(PPh$_3$)$_2$Cl$_2$ (5%), THF, reflux, 28 h — (58) — 696

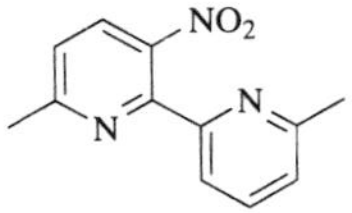

C$_6$

325

TABLE IX. DIRECT CROSS-COUPLING OF PYRIDINE ELECTROPHILES (*Continued*)

Substrate	Stannane	Conditions	Product(s) and Yield(s) (%)	Refs.
		Pd(PPh$_3$)$_2$Cl$_2$ (20%), CuI, DMF, 100°, 18 h	(8)	585
		Pd(PPh$_3$)$_4$ (1%), xylene, reflux, 12 h	(81)	697
		Pd(PPh$_3$)$_4$	(—)	698
		Pd(PPh$_3$)$_4$ (1%), xylene, reflux, 12 h	(26)	697
		Pd(PPh$_3$)$_4$ (1%), xylene, reflux, 12 h	(32)	697
		Pd(PPh$_3$)$_4$ (1%), xylene, reflux, 12 h	(17)	697

326

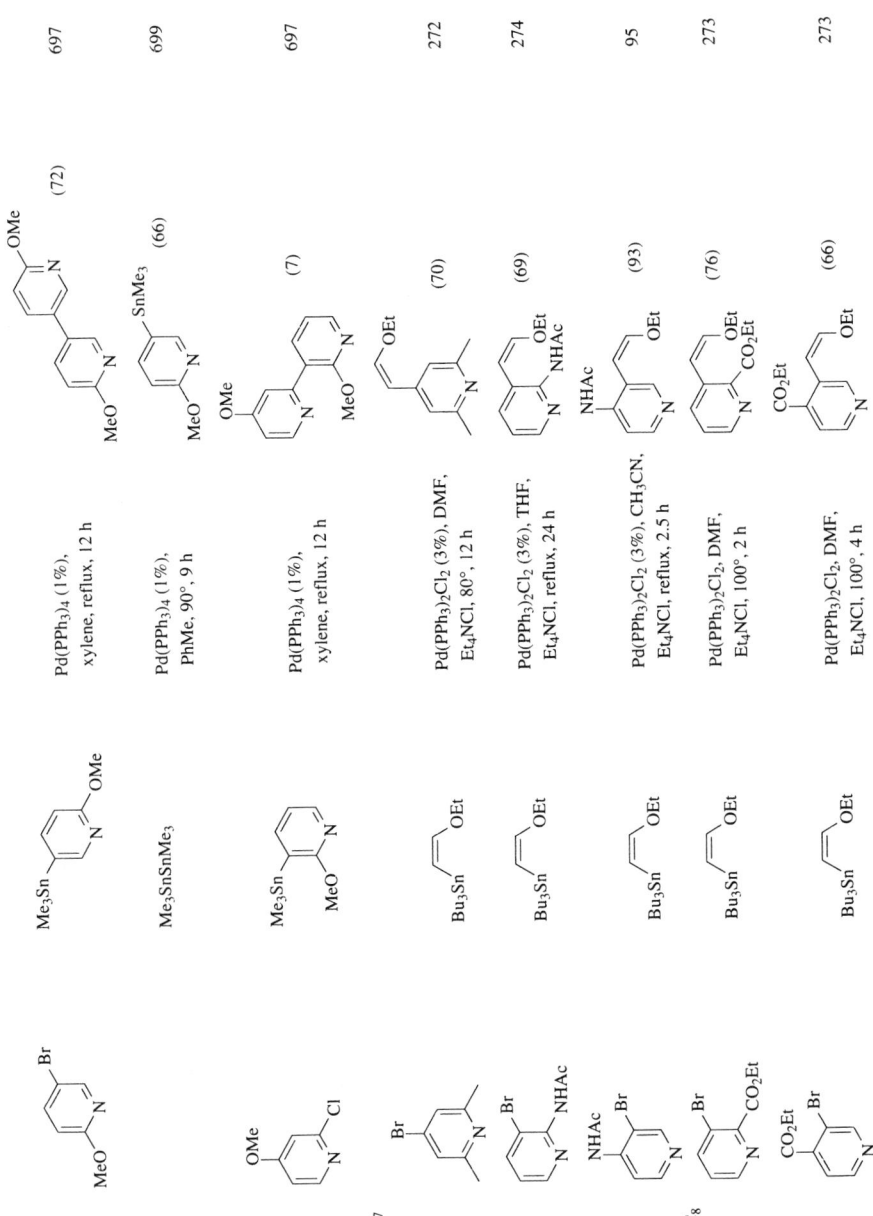

			697
			699
			697
			272
			274
			95
			273
			273

Pd(PPh$_3$)$_4$ (1%), xylene, reflux, 12 h

Pd(PPh$_3$)$_4$ (1%), PhMe, 90°, 9 h

Pd(PPh$_3$)$_4$ (1%), xylene, reflux, 12 h

Pd(PPh$_3$)$_2$Cl$_2$ (3%), DMF, Et$_4$NCl, 80°, 12 h

Pd(PPh$_3$)$_2$Cl$_2$ (3%), THF, Et$_4$NCl, reflux, 24 h

Pd(PPh$_3$)$_2$Cl$_2$ (3%), CH$_3$CN, Et$_4$NCl, reflux, 2.5 h

Pd(PPh$_3$)$_2$Cl$_2$, DMF, Et$_4$NCl, 100°, 2 h

Pd(PPh$_3$)$_2$Cl$_2$, DMF, Et$_4$NCl, 100°, 4 h

C$_7$

C$_8$

TABLE IX. DIRECT CROSS-COUPLING OF PYRIDINE ELECTROPHILES (*Continued*)

	Substrate	Stannane	Conditions	Product(s) and Yield(s) (%)	Refs.
C$_9$	EtO$_2$C— pyridine —Cl	Me$_3$Sn—pyridine(N)	Pd(PPh$_3$)$_2$Cl$_2$ (5%), THF, reflux, 20 h	EtO$_2$C—pyridine—pyridine (54)	700
	CO$_2$Et, pyridine-Br, Me	Bu$_3$Sn—C$_6$H$_4$—OMe	Pd(PPh$_3$)$_4$ (5%), THF, reflux	CO$_2$Et, OMe—C$_6$H$_4$—pyridine (94)	94
C$_{12}$	Br, MeO—pyridine—pyridine—OMe (N)	anthracene-SnMe$_3$	Pd(PPh$_3$)$_4$	OMe, MeO—pyridine—pyridine—anthracene (—)	698
C$_{14}$	Br, Bn—pyridine—NHAc	Ph—C$_6$H$_4$—SnMe$_3$ (Me$_3$Sn)	Pd(PPh$_3$)$_2$Cl$_2$ (3%), PhMe, 120°, 15 h	Bn, Ph—C$_6$H$_3$—(Ac)HN NH(Ac)—pyridine, Bn (60)	606
C$_{15}$	Br—pyridine—F, CH(CH$_3$)-indole	Bu$_3$Sn—CH=CH$_2$	Pd(PPh$_3$)$_2$Cl$_2$ (2%), PhMe, reflux	vinyl—pyridine—F, CH(CH$_3$)-indole (75)	701
		Bu$_3$Sn—C(OEt)=CH$_2$	Pd(PPh$_3$)$_2$Cl$_2$ (2%), PhMe, reflux	EtO—C(=CH$_2$)—pyridine—F, CH(CH$_3$)-indole (95)	701

C16

Reagent	Conditions	Product (yield)	Ref.
Me3Sn— / Me3Sn—SnMe3 (aryl stannane)	Pd(PPh3)2Cl2 (1%), PhMe, 100°, 2 h	(41)	606
Me4Sn	Pd(PPh3)2Cl2 (3%), LiCl, DMF, 130°	(88)	702, 703
Bu3Sn— (vinyl)	Pd(PPh3)2Cl2 (3%), Et3N, DMF, 90°, 4 h	(86)	704
Me3Sn— (quinolinyl)	Pd(PPh3)4 (3%), LiCl, dioxane, reflux, 36 h	(66)	693
Bu3Sn— (thienyl)	Pd(PPh3)4 (3%), PhMe, reflux, 48 h	(89)	556
Me3Sn— (pyridyl)	Pd(PPh3)4 (3%), PhMe, reflux, 48 h	(92)	556

TABLE IX. DIRECT CROSS-COUPLING OF PYRIDINE ELECTROPHILES (*Continued*)

Substrate	Stannane	Conditions	Product(s) and Yield(s) (%)	Refs.
C₁₇				
		Pd(PPh₃)₄ (3%), PhMe, reflux, 48 h	(51)	556
		Pd(PPh₃)₄ (3%), PhMe, reflux, 48 h	(78)	556
		Pd(PPh₃)₂Cl₂ (6%), PhMe, 110°, 16 h	(—)	606
		Pd(PPh₃)₄ (3%), LiCl, dioxane, reflux, 20 h	(65)	693
		1. Pd(PPh₃)₂Cl₂ (5%), PhMe, reflux, 4 h 2. HCl (conc.)	(68)	705

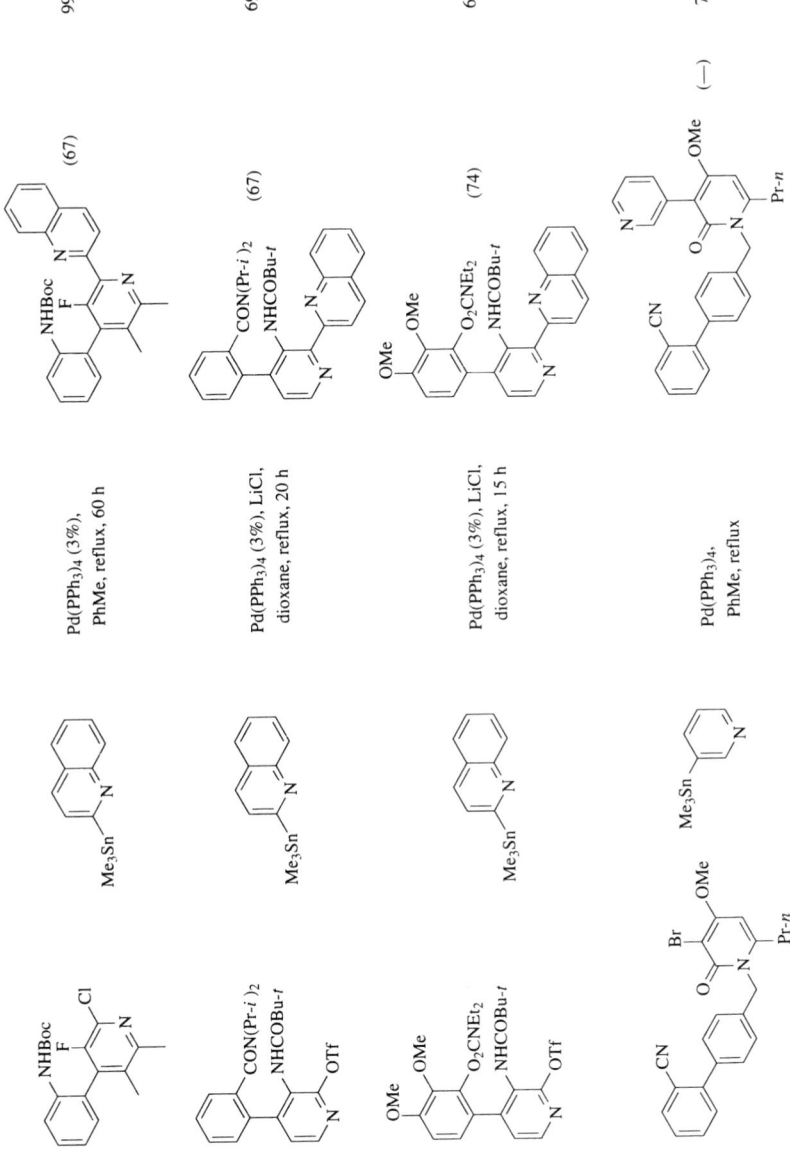

TABLE IX. DIRECT CROSS-COUPLING OF PYRIDINE ELECTROPHILES (*Continued*)

Substrate	Stannane	Conditions	Product(s) and Yield(s) (%)	Refs.
C$_{24}$		Pd(PPh$_3$)$_4$ (10%)	(89)	107

TABLE X. DIRECT CROSS-COUPLING OF PYRIMIDINE ELECTROPHILES

Substrate	Stannane	Conditions	Product(s) and Yield(s) (%)	Refs.
C₄				
(2-chloropyrimidine)	Bu₃SnSPh	Pd(PPh₃)₄ (5%), PhMe, 100°, 24 h	(95)	319
(2-triflyloxypyrimidine)	Bu₃Sn—CH=CH₂	Pd(PPh₃)₄ (3%), LiCl, dioxane, 80°	(68)	196
	Bu₃Sn—CH=CH—Ph	Pd(PPh₃)₄ (3%), LiCl, dioxane, 80°	(57)	196
	Bu₃Sn—thiophene	Pd(PPh₃)₄ (3%), LiCl, dioxane, 80°	(67)	196
(5-bromopyrimidine)	Bu₃Sn—C(OTMS)=CH₂	Pd(PPh₃)₄ (3%), PhMe, 100°, 24 h	(67)	457
	Bu₃SnSPh	Pd(PPh₃)₄ (5%), PhMe, 100°, 24 h	(94)	319
(5-iodopyrimidine)	Bu₃Sn—quinone	Pd₂(dba)₃ (2.5%), CuI, DMF, 60°	(61)	554
	Bu₃Sn—(OMe/Bu-n quinone)	Pd₂(dba)₃ (2.5%), CuI, DMF, 60°	(69)	554

TABLE X. DIRECT CROSS-COUPLING OF PYRIMIDINE ELECTROPHILES (*Continued*)

Substrate	Stannane	Conditions	Product(s) and Yield(s) (%)	Refs.
		Pd$_2$(dba)$_3$ (2.5%), CuI, DMF, 60°	(49)	554
	Bu$_3$Sn	Pd(PPh$_3$)$_2$Cl$_2$, Cl(CH$_2$)$_2$Cl, 80°, 6 h	(61)	137
	Bu$_3$SnPh	Pd(PPh$_3$)$_2$Cl$_2$ (2%), DMF, 80°, 60 h	(73)	141
	Bu$_3$Sn	Pd(PPh$_3$)$_2$Cl$_2$ (2%), DMF, 80°, 6 h	(73)	141
	Bu$_3$SnPh	Pd(PPh$_3$)$_2$Cl$_2$ (2%), DMF, 70°, 7 h	(77)	141
	Bu$_3$Sn	Pd(PPh$_3$)$_2$Cl$_2$ (2%), DMF, 80°, 10 h	(60)	141
	Bu$_3$Sn	Pd(PPh$_3$)$_4$ (5%), Cl(CH$_2$)$_2$Cl, 70°, 24 h	(94)	137
	Bu$_3$Sn	Pd(0), Cl(CH$_2$)$_2$Cl, 70°, 24 h	(83)	137

334

Substrate	Reagent	Conditions	Product (%)	Ref.
2,5-dichloropyrimidine	Bu$_3$Sn–CH=CH–Ph	Pd(PPh$_3$)$_2$Cl$_2$ (2%), DMF, 100°, 7 h	(70)	141
5-bromo-2,4-dichloropyrimidine	Bu$_3$Sn–CH=CH–Ph	Pd(PPh$_3$)$_2$Cl$_2$ (2%), DMF, 70°, 6 h	(73)	141
2,4,5-trichloropyrimidine	Bu$_3$SnPh	Pd(PPh$_3$)$_2$Cl$_2$ (2%), DMF, 70°, 3 h	(69)	141
2,4,5-trichloropyrimidine	Bu$_3$Sn–CH=CH–Ph	Pd(PPh$_3$)$_2$Cl$_2$ (2%), DMF, 80°, 10 h	(71)	141
(same, pyrimidine)	Bu$_3$SnPh	Pd(PPh$_3$)$_2$Cl$_2$ (2%), DMF, 100°, 10 h	(63)	141

C$_5$

Substrate	Reagent	Conditions	Product (%)	Ref.
2-chloro-5-methylpyrimidine	Bu$_3$Sn–CH=CH–Ph	Pd(PPh$_3$)$_2$Cl$_2$ (2%), DMF, 100°, 15 h	(68)	141
4-chloro-2-(methylthio)pyrimidine	Bu$_3$Sn–CH=CH$_2$	Pd(PPh$_3$)$_4$, Cl(CH$_2$)$_2$Cl, 70°, 48 h	(92)	137
4-chloro-2-(methylthio)pyrimidine	Bu$_3$Sn–CH$_2$–OMe	Pd(PPh$_3$)$_2$Cl$_2$ (2%), DMF, 90°, 24 h	(56)	140
4-chloro-2-(methylthio)pyrimidine	Bu$_3$SnPh	Pd(PPh$_3$)$_2$Cl$_2$ (2%), DMF, 100°, 8 h	(60)	141
4-chloro-2-(methylthio)pyrimidine	Bu$_3$Sn–CH=CH–Ph	Pd(PPh$_3$)$_2$Cl$_2$ (2%), DMF, 100°, 1.3 h	(85)	141

335

TABLE X. DIRECT CROSS-COUPLING OF PYRIMIDINE ELECTROPHILES (Continued)

Substrate	Stannane	Conditions	Product(s) and Yield(s) (%)	Refs.
5-Br-4-Cl-2-MeS-pyrimidine	Bu₃Sn–CH₂–OR; R = TBDMS, SiMe₂Thex, TBDPS	Pd(PPh₃)₂Cl₂ (2%), DMF, 90°, 24 h	pyrimidine–CH₂OR (53), (32), (64)	140
4,6-dichloro-2-MeS-pyrimidine	Bu₃Sn–CH=CH–Ph	Pd(PPh₃)₂Cl₂ (2%), DMF, 50°, 27 h	Br, CH=CH–Ph, MeS-pyrimidine (66)	141
	Bu₃SnPh	Pd(PPh₃)₂Cl₂ (2%), DMF, 100°, 5 h	Cl, Ph, MeS-pyrimidine (66)	141
	Bu₃Sn–CH=CH–Ph	Pd(PPh₃)₂Cl₂ (2%), DMF, 100°, 3.5 h	Cl, CH=CH–Ph, MeS-pyrimidine (66)	141
4-I-2-MeS-pyrimidine	Me₃Sn-pyrimidine-SMe	Pd(PPh₃)₄ (0.7%), PhMe, reflux, 20 h	bipyrimidine–SMe (38)	140
5-Br-4-I-2-MeS-pyrimidine	R₃SnSnR₃	Pd(PPh₃)₂(OAc)₂ (3%), Bu₄NCl, THF, rt, 6 h	SnR₃, MeS-pyrimidine: R = Me (54), R = Bu (46)	140
	Bu₃Sn-allyl	Pd[P(OPr-i)₃]₄, Cl(CH₂)₂Cl, 70°, 24 h	Br, vinyl, MeS-pyrimidine (64)	137
4-Cl-5-I-2-MeS-pyrimidine	Me₄Sn	Pd(PPh₃)₂Cl₂ (2%), DMF, 70°, 48 h	Cl, Me, MeS-pyrimidine (78)	135

336

Organostannane	Conditions	Product (%)	Refs.
Bu₃Sn (vinyl)	Pd(PPh₃)₂Cl₂ (2%), THF, reflux, 4 h	(71)	135
Bu₃Sn (allyl)	Pd(PPh₃)₂Cl₂ (2%), DMF, 100°, 2 h	(69)	135
Bu₃Sn (cis-propenyl)	Pd(PPh₃)₂Cl₂ (2%), THF, reflux, 17 h	(90)	135
Bu₄Sn	Pd(PPh₃)₂Cl₂ (2%), DMF, 120°, 14 h	(62)	135
Bu₃Sn (2-thienyl)	Pd(PPh₃)₂Cl₂ (2%), THF, reflux, 7 h	(90)	135
Bu₃SnPh	Pd(PPh₃)₂Cl₂ (2%), THF, reflux, 20 h	(62)	135
Bu₃SnBn	Pd(PPh₃)₂Cl₂ (2%), DMF, 100°, 15 h	(60)	135
Bu₃Sn–CH=CH–Ph	Pd(PPh₃)₂Cl₂ (2%), THF, reflux, 6 h	(90)	135
Bu₃Sn–CH=CH–OTHP	Pd(PPh₃)₂Cl₂ (2%), THF, reflux, 33 h	(66)	135
Me₃Sn-pyrimidine (SMe)	Pd(PPh₃)₄ (0.7%), PhMe, reflux, 20 h	(56)	140

TABLE X. DIRECT CROSS-COUPLING OF PYRIMIDINE ELECTROPHILES (*Continued*)

Substrate	Stannane	Conditions	Product(s) and Yield(s) (%)	Refs.
MeS–pyrimidine–OTf	Bu₃Sn–CH=CH₂	Pd(PPh₃)₄ (3%), LiCl, dioxane, 80°	(vinyl product) (60)	196
MeS–pyrimidine–OTf	Bu₃Sn–CH=CH–Ph	Pd(PPh₃)₄ (3%), LiCl, dioxane, 80°	(styryl product) (68)	196
OTf / MeS / OTf pyrimidine	thienyl–SnBu₃	Pd(PPh₃)₄ (3%), LiCl, dioxane, 80°	(dithienyl SMe product) (73)	196
Br–pyrimidine–SO₂Me	Bu₃Sn–CH=CH₂	Pd(PPh₃)₄, Cl(CH₂)₂Cl, 70°, 2 h	(vinyl product) (90)	137
Br–pyrimidine–SO₂Me	Bu₃Sn–CH=CH–CH₃	Pd(PPh₃)₄, Cl(CH₂)₂Cl, 70°, 24 h	(propenyl product) (86)	137
Cl–pyrimidine–SO₂Me	Bu₃Sn–CH=CH₂	Pd(PPh₃)₄, Cl(CH₂)₂Cl, 70°, 48 h	(vinyl product) (49)	137
	Bu₄Sn	Pd(PPh₃)₂Cl₂ (2%), DMF, 100°, 5.5 h	(Bu-*n* product) (56)	141
	Bu₃SnPh	Pd(PPh₃)₂Cl₂ (2%), DMF, 100°, 2.5 h	(Ph product) (65)	141
	Bu₃Sn–CH=CH–Ph	Pd(PPh₃)₂Cl₂ (2%), DMF, 100°, 3 h	(styryl product) (65)	141

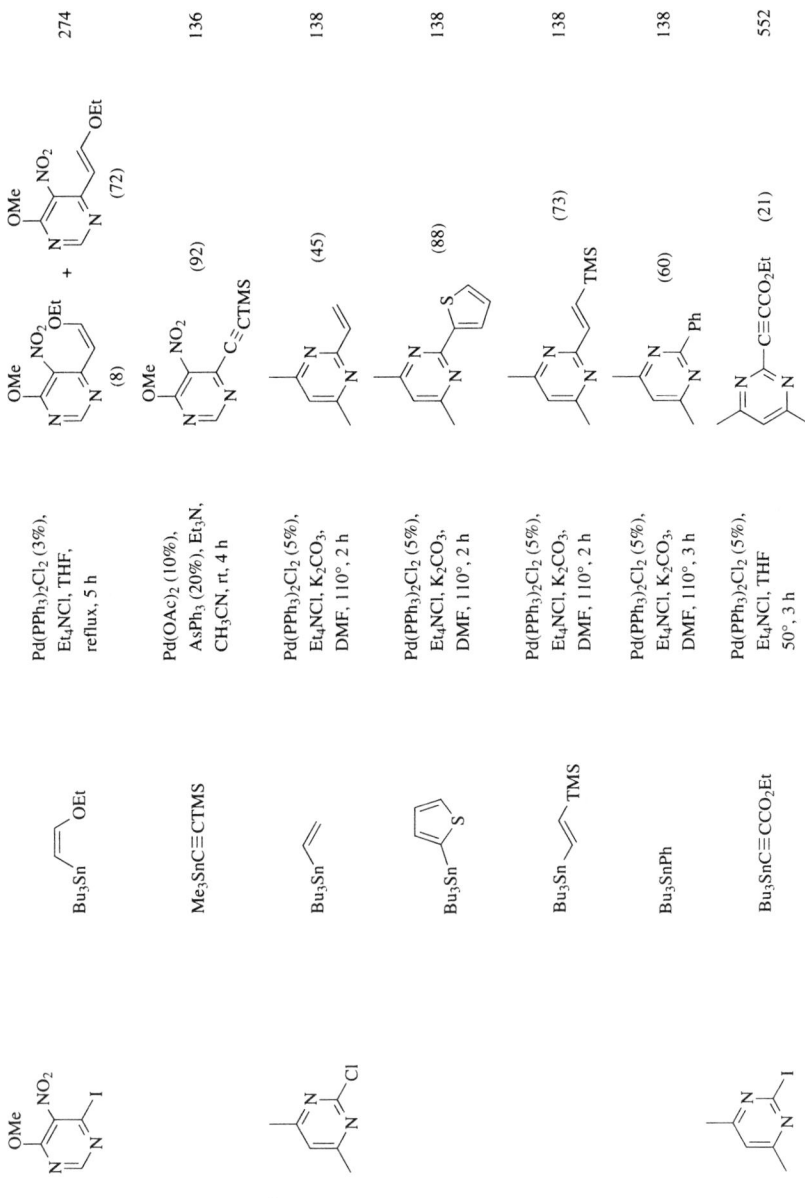

274

136

138

138

138

138

552

TABLE X. DIRECT CROSS-COUPLING OF PYRIMIDINE ELECTROPHILES (*Continued*)

Substrate	Stannane	Conditions	Product(s) and Yield(s) (%)	Refs.
	Bu_3Sn ⟍⟍	Pd(PPh$_3$)$_2$Cl$_2$ (5%), Et$_4$NCl, K$_2$CO$_3$, DMF, 110°, 2 h	(52)	138
	Bu_3Sn (thienyl, S)	Pd(PPh$_3$)$_2$Cl$_2$ (5%), Et$_4$NCl, K$_2$CO$_3$, DMF, 110°, 2 h	(88)	138
	Bu_3Sn ⟍⟍ TMS	Pd(PPh$_3$)$_2$Cl$_2$ (5%), Et$_4$NCl, K$_2$CO$_3$, DMF, 110°, 2 h	TMS (78)	138
	Bu_3SnPh	Pd(PPh$_3$)$_2$Cl$_2$ (5%), Et$_4$NCl, K$_2$CO$_3$, DMF, 110°, 3 h	Ph (71)	138
	Bu_3SnPh	Pd(PPh$_3$)$_2$Cl$_2$ (5%), Et$_4$NCl, K$_2$CO$_3$, DMF, 110°, 8 h	Ph (71)	138
	Bu_3Sn ⟍	Pd(PPh$_3$)$_2$Cl$_2$ (5%), Et$_4$NCl, K$_2$CO$_3$, DMF, 110°, 2 h	(67)	138
	Bu_3Sn (thienyl, S)	Pd(PPh$_3$)$_2$Cl$_2$ (5%), Et$_4$NCl, K$_2$CO$_3$, DMF, 110°, 2 h	(95)	138
	Bu_3Sn ⟍⟍ TMS	Pd(PPh$_3$)$_2$Cl$_2$ (5%), Et$_4$NCl, K$_2$CO$_3$, DMF, 110°, 2 h	TMS (73)	138

340

Organotin	Conditions	Product (yield)	Refs.
Bu$_3$SnPh	Pd(PPh$_3$)$_2$Cl$_2$ (5%), Et$_4$NCl, K$_2$CO$_3$, DMF, 110°, 4 h	[4-Ph-2,?-dimethylpyrimidine] (92)	138
Bu$_3$SnPh	Pd(PPh$_3$)$_2$Cl$_2$ (5%), Et$_4$NCl, K$_2$CO$_3$, DMF, 110°, 4 h	[Ph, Cl pyrimidine] (7) + [Ph, Ph pyrimidine] (50)	138
Bu$_3$SnPh	Pd(PPh$_3$)$_2$Cl$_2$ (5%), Et$_4$NCl, K$_2$CO$_3$, DMF, 110°, 6 h	[Cl, Ph pyrimidine] (50)	138
Bu$_3$Sn—CH=CH$_2$	Pd(PPh$_3$)$_2$Cl$_2$ (5%), Et$_4$NCl, K$_2$CO$_3$, DMF, 110°, 5 h	[vinyl, Cl pyrimidine] (45)	138, 139
Bu$_3$Sn—(2-thienyl)	Pd(PPh$_3$)$_2$Cl$_2$ (5%), Et$_4$NCl, K$_2$CO$_3$, DMF, 110°, 4 h	[thienyl, Cl pyrimidine] (60)	138, 139
Bu$_3$Sn—CH=CH—CO$_2$Et	Pd(PPh$_3$)$_2$Cl$_2$	[CH=CH—CO$_2$Et, Cl pyrimidine] (—)	139
Bu$_3$Sn—CH=CH—TMS	Pd(PPh$_3$)$_2$Cl$_2$ (5%), Et$_4$NCl, K$_2$CO$_3$, DMF, 110°, 5 h	[CH=CH—TMS, Cl pyrimidine] (69)	138, 139
Bu$_3$SnPh	Pd(PPh$_3$)$_2$Cl$_2$ (5%), Et$_4$NCl, K$_2$CO$_3$, DMF, 110°, 7 h	[Ph, Cl pyrimidine] (73) + [Ph, Ph pyrimidine] (20)	138, 139

Substrates (pyrimidines): [Br, Cl]; [Cl, Cl]; [I, Cl]

341

TABLE X. DIRECT CROSS-COUPLING OF PYRIMIDINE ELECTROPHILES (Continued)

Substrate	Stannane	Conditions	Product(s) and Yield(s) (%)	Refs.
(pyrimidine with I, N$_3$, dimethyl)	Bu$_3$Sn–CH=CH–Ph	Pd(PPh$_3$)$_2$Cl$_2$	(—)	139
	Bu$_3$Sn–CH=CH$_2$	PdCl$_2$, Et$_4$NCl, DMF, 110°, 2 h	(11)	139
	Bu$_3$Sn–(2-thienyl)	PdCl$_2$, Et$_4$NCl, DMF, 110°, 2 h	(45)	139
	Bu$_3$Sn–CH=CH–TMS	PdCl$_2$, Et$_4$NCl, DMF, 110°, 2 h	(24)	139
	Bu$_3$SnPh	PdCl$_2$, Et$_4$NCl, DMF, 110°, 3 h	(65)	139
(OMe, NHAc pyrimidine with I)	Bu$_3$Sn–CH=CH–OEt	Pd(PPh$_3$)$_2$Cl$_2$ (4%), Et$_4$NCl, CH$_3$CN, reflux, 33 h	(37)	274
(Br, NHAc, MeO pyrimidine)	Bu$_3$Sn–CH=CH–OEt	Pd(PPh$_3$)$_2$Cl$_2$ (3%), Et$_4$NCl, CH$_3$CN, reflux, 3 h	(71)	274

342

Conditions	Product (yield)	Refs.
Pd(PPh₃)₂Cl₂ (3%), THF, Et₄NCl, reflux, 4 h / 24 h	(44) / (73) + (50) / (21)	274
Pd(OAc)₂ (10%), Bu₄NCl, NaHCO₃, NEt₃, H₂O/EtOH	(83)	707
Pd(OAc)₂ (10%), AsPh₃ (20%), CH₃CN, 40°, 8 h	(65)	301
Pd(OAc)₂ (10%), AsPh₃ (20%), CH₃CN, 40°, 14 h	(54)	301
Pd(OAc)₂ (10%), AsPh₃ (20%), CH₃CN, 60°, 0.5 h	(88)	301
Pd(OAc)₂ (10%), AsPh₃ (20%), CH₃CN, THF, 40°, 10 h	(66)	301

TABLE X. DIRECT CROSS-COUPLING OF PYRIMIDINE ELECTROPHILES (*Continued*)

Substrate	Stannane	Conditions	Product(s) and Yield(s) (%)	Refs.
pyrimidine (OMe, NO_2, I, MeO)	Bu_3Sn–CH=CH–OEt	$Pd(PPh_3)_2Cl_2$ (3%), THF, Et_4NCl, reflux 2 h / 15 h	pyrimidine product (OMe, NO_2, OEt, MeO) (32) (11) + pyrimidine product (OMe, NO_2, CH=CH–OEt, MeO) (51) (73)	274
Bu_3Sn–CH=CH$_2$		$Pd(PPh_3)_4$ (3%), LiCl, dioxane, 80°	vinyl pyrimidine (MeS) (60)	196
thiophene–Bu_3Sn		$Pd(PPh_3)_4$ (3%), LiCl, dioxane, 80°	thienyl pyrimidine (MeS) (88)	196
pyrimidine (Br, NHAc)	Bu_3Sn–CH=CH–OEt	$Pd(PPh_3)_2Cl_2$ (3%), Et_4NCl, DMF, 140°, 1 h	pyrimidine product (OEt, NHAc) (95)	274
pyrimidine (OMe, NHAc, I)	Bu_3Sn–CH=CH–OEt	$Pd(PPh_3)_2Cl_2$ (4%), Et_4NCl, CH_3CN, reflux, 31 h	pyrimidine product (NHAc, CH=CH–OEt, OMe) (75)	274
pyrimidine (OMe, Br, NHAc, MeO)	Bu_3Sn–CH=CH–OEt	$Pd(PPh_3)_2Cl_2$ (3%), Et_4NCl, CH_3CN, reflux, 24 h	pyrimidine product (OEt, NHAc, OMe, MeO) (81)	274

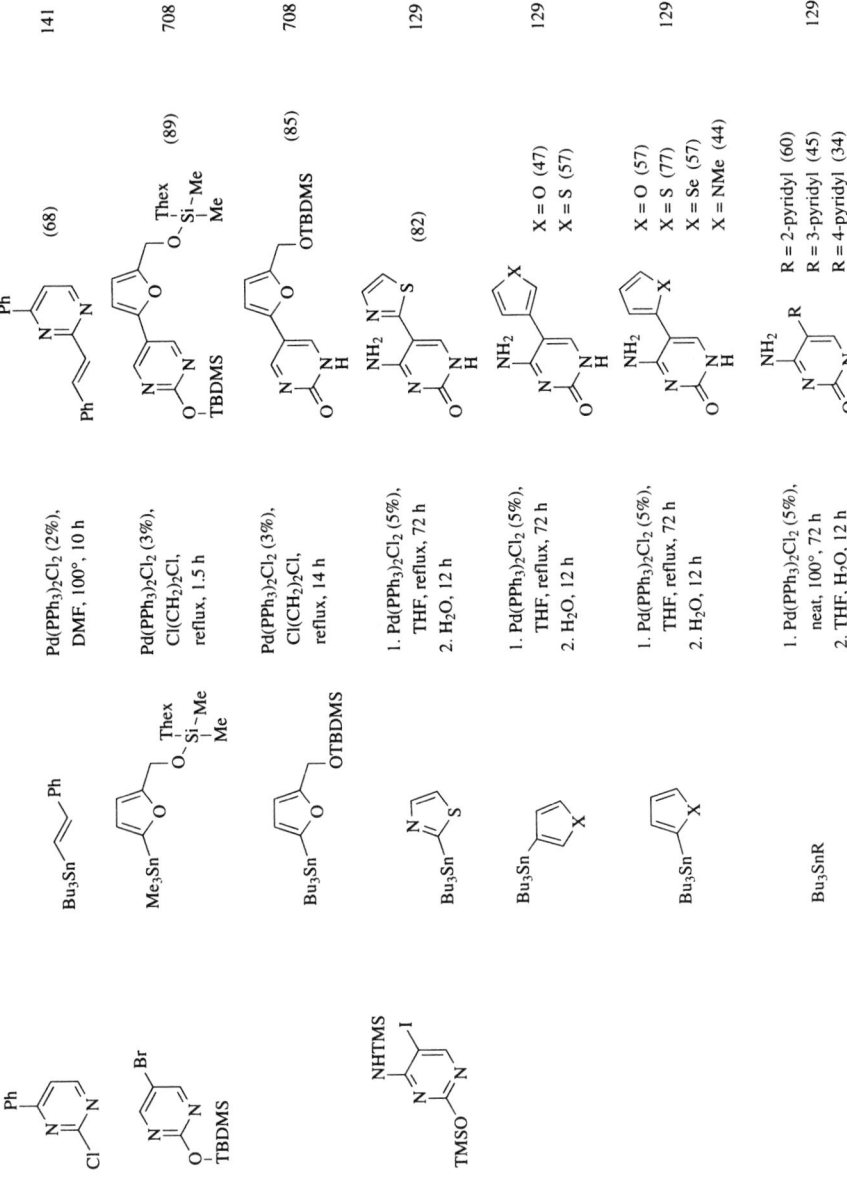

C_{10} 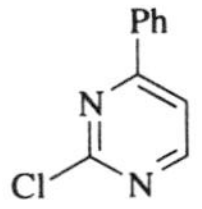	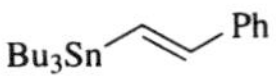	$Pd(PPh_3)_2Cl_2$ (2%), DMF, 100°, 10 h	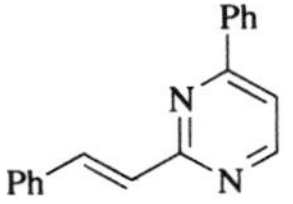 (68)	141
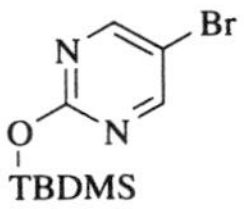	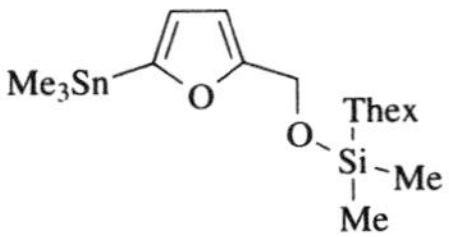	$Pd(PPh_3)_2Cl_2$ (3%), $Cl(CH_2)_2Cl$, reflux, 1.5 h	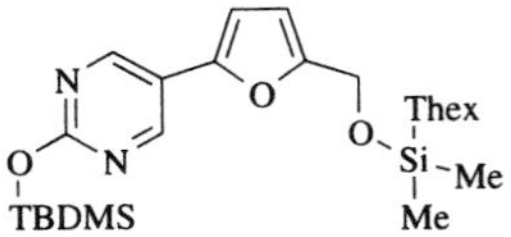 (89)	708
	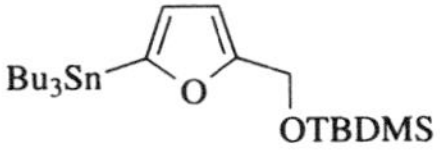	$Pd(PPh_3)_2Cl_2$ (3%), $Cl(CH_2)_2Cl$, reflux, 14 h	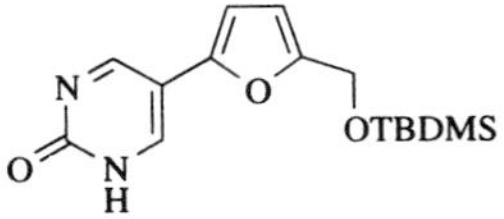 (85)	708
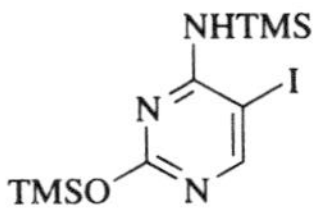	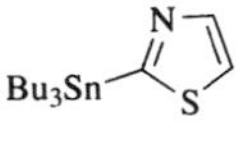	1. $Pd(PPh_3)_2Cl_2$ (5%), THF, reflux, 72 h 2. H_2O, 12 h	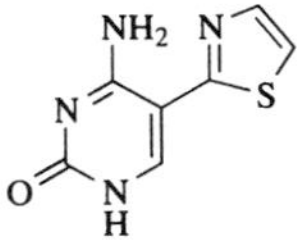 (82)	129
	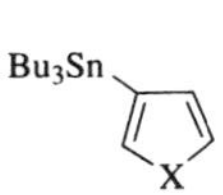	1. $Pd(PPh_3)_2Cl_2$ (5%), THF, reflux, 72 h 2. H_2O, 12 h	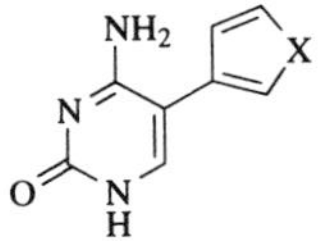 X = O (47) X = S (57)	129
	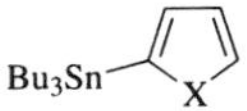	1. $Pd(PPh_3)_2Cl_2$ (5%), THF, reflux, 72 h 2. H_2O, 12 h	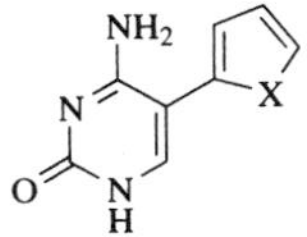 X = O (57) X = S (77) X = Se (57) X = NMe (44)	129
	Bu_3SnR	1. $Pd(PPh_3)_2Cl_2$ (5%), neat, 100°, 72 h 2. THF, H_2O, 12 h	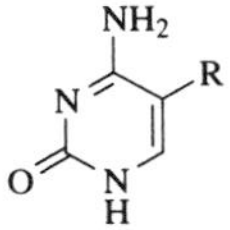 R = 2-pyridyl (60) R = 3-pyridyl (45) R = 4-pyridyl (34)	129

TABLE X. DIRECT CROSS-COUPLING OF PYRIMIDINE ELECTROPHILES (*Continued*)

Substrate	Stannane	Conditions	Product(s) and Yield(s) (%)	Refs.
(5-iodo-4-OTMS-2-TMSO-pyrimidine)	Bu₃SnPh	1. Pd(PPh₃)₂Cl₂ (5%), THF, reflux, 72 h 2. H₂O, 12 h	(62)	129
	Bu₃Sn—(thiophene)—Ph	1. Pd(PPh₃)₂Cl₂ (5%), THF, reflux, 72 h 2. H₂O, 12 h	(82)	129
	Bu₃Sn—(thiazole)	1. Pd(PPh₃)₂Cl₂ (5%), THF, reflux 2. H₂O	(34)	135
	Bu₃Sn—(X)	1. Pd(PPh₃)₂Cl₂ (5%), THF, reflux 2. H₂O	X = S (37) X = Se (39) X = NMe (44)	135
	Bu₃SnR	1. Pd(PPh₃)₂Cl₂ (5%), THF, reflux 2. H₂O	R = 2-pyridyl (28) R = 3-p yridyl (42)	135
(5-bromo-2-PhS-pyrimidine)	Bu₃Sn—(CH=CH)—Ph	Pd(PPh₃)₂Cl₂ (2%), DMF, 100°, 8 h	(69)	141
C₁₁ (6-iodo-2-Ph-4-methyl-pyrimidine)	Bu₃SnC≡CCO₂Et	Pd(PPh₃)₂Cl₂ (5%), Et₄NCl, THF, 50°, 4 h	(38)	552

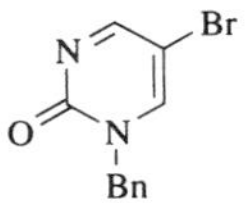

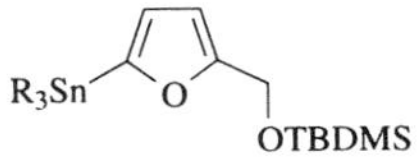

R = Me (98)		Pd(PPh₃)₂Cl₂ (3%), THF, reflux, 1 h	708
R = Bu (72)			
(48)		Pd(PPh₃)₄ (2%), THF, reflux, 2 h	709
(54)		Pd(PPh₃)₂Cl₂ (2%), DMF, 130°, 15 h	141
(70)		Pd(PPh₃)₂Cl₂ (2%), DMF, 80°, 50 h	141
(61)		Pd(PPh₃)₄ (2%), THF, reflux, 2 h	709
(39)		Pd(PPh₃)₂Cl₂ (5%), THF, reflux	135
(82)		Pd(PPh₃)₂Cl₂ (2%), DMF, 100°, 15 h	141

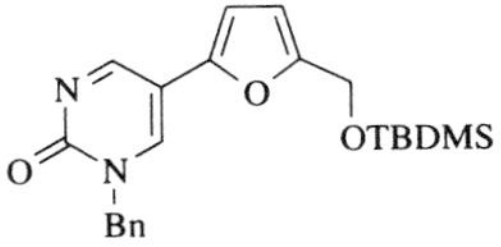

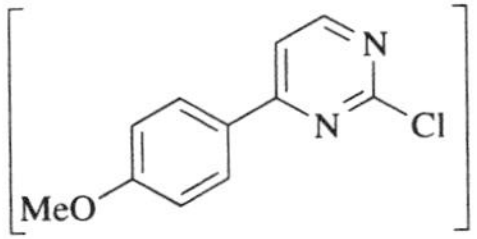

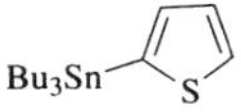

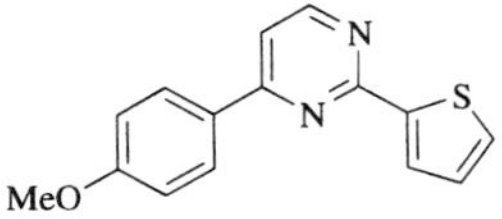

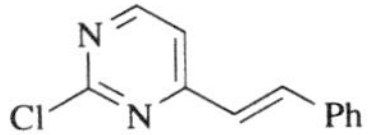

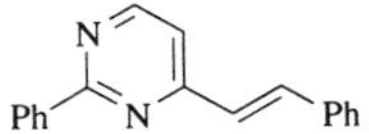

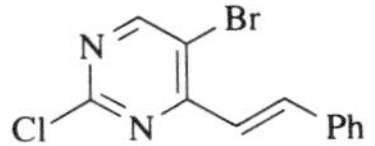

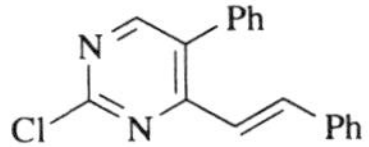

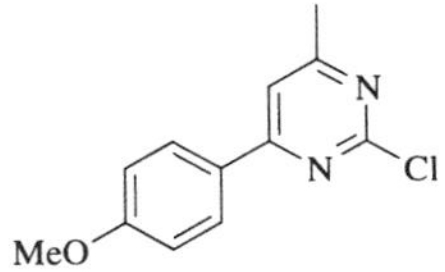

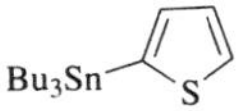

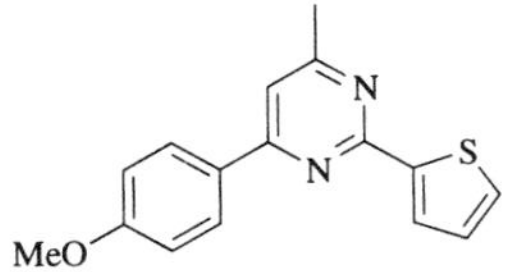

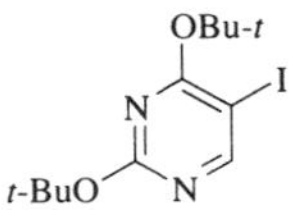

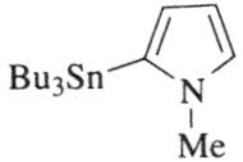

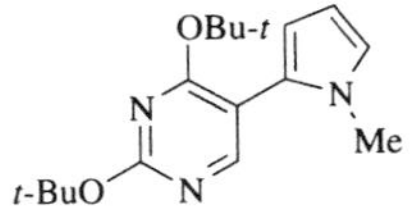

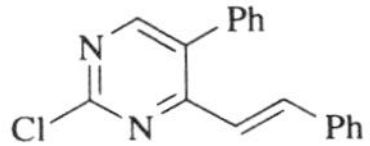

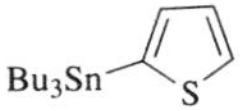

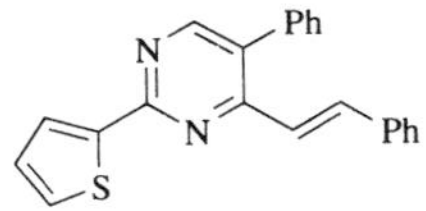

R₃Sn

Bu₃SnPh

Bu₃SnPh

Bu₃Sn

Bu₃Sn

Bu₃Sn

OTBDMS

C₁₂

C₁₈

TABLE X. DIRECT CROSS-COUPLING OF PYRIMIDINE ELECTROPHILES (*Continued*)

Substrate	Stannane	Conditions	Product(s) and Yield(s) (%)	Refs.
		Pd(PPh$_3$)$_4$ (20%), PhMe, reflux, 24 h	(—)	127
		Pd(PPh$_3$)$_4$ (20%), PhMe, reflux, 24 h	(—)	127

348

TABLE XI. DIRECT CROSS-COUPLING OF QUINOLINE AND ISOQUINOLINE ELECTROPHILES

Substrate	Stannane	Conditions	Product(s) and Yield(s) (%)	Refs.
C$_9$				
2-chloroquinoline	Me$_3$Sn–(4,4-dimethyloxazolin-2-yl)	Pd(PPh$_3$)$_4$ (5%), C$_6$H$_6$, 80°, 12 h	2-(4,4-dimethyloxazolin-2-yl)quinoline (75)	550, 551
2-(trifluoromethylsulfonyloxy)quinoline	Bu$_3$SnR, R = H; CH=CH$_2$; 2-thienyl; Ph; C≡CPh; (E)-CH=CHPh	Pd(PPh$_3$)$_2$Cl$_2$ (5%), LiCl, dioxane, 90°, 24 h	2-R-quinoline, R = H (89); CH=CH$_2$ (74); 2-thienyl (71); Ph (88); C≡CPh (65); (E)-CH=CHPh (69)	194
3-bromoquinoline	Me$_3$Sn–(3-thienyl)	Pd(PPh$_3$)$_4$ (5%), DMF, reflux, 16 h	3-(3-thienyl)quinoline (74)	699
	Me$_3$Sn–CH=CH–Ph	Pd(PPh$_3$)$_4$ (5%), DMF, reflux, 16 h	3-(2-phenylvinyl)quinoline (33)	699
	Me$_3$Sn–(4,4-dimethyloxazolin-2-yl)	Pd(PPh$_3$)$_4$ (5%), C$_6$H$_6$, 80°, 48 h	3-(4,4-dimethyloxazolin-2-yl)quinoline (92)	550, 551
	Me$_3$Sn–(2-pyridyl)	Pd(PPh$_3$)$_4$ (1%), xylene, reflux	3-(2-pyridyl)quinoline (79)	93

349

Substrate	Stannane	Conditions	Product(s) and Yield(s) (%)	Refs.
(6,7-dichloro-3-quinolinyl triflate)	Me₃Sn-quinoline	Pd(PPh₃)₄ (15%), C₆H₆, reflux, 2 h	(74)	284
(6,7-dichloro-3-quinolinyl triflate)	Me₃Sn-(3-thienyl)	Pd(PPh₃)₄ (5%), LiCl, dioxane, reflux, 16 h	(75)	699
(5-fluoro-3-quinolinyl triflate)	Me₃Sn-(3-thienyl)	Pd(PPh₃)₄ (5%), LiCl, dioxane, reflux, 16 h	(92)	699
(7-fluoro-3-quinolinyl triflate)	Me₃Sn-(3-thienyl)	Pd(PPh₃)₄ (5%), LiCl, dioxane, reflux, 16 h	(92)	699
(6,7-difluoro-3-quinolinyl triflate)	Me₃Sn-(3-thienyl)	Pd(PPh₃)₄ (5%), LiCl, dioxane, reflux, 16 h	(86)	699
(8-quinolinyl triflate)	Bu₃SnR R CH=CH₂ C≡CPh (*E*)-CH=CHPh	Pd(PPh₃)₄ (5%), LiCl, dioxane 90°, 24 h 90°, 24 h 90°, 24 h	**I** (68) (43) (60)	194
	Me₃SnPh	Pd(PPh₃)₄ (2%), LiCl, dioxane, 98°, 82 h	**I**, R = Ph, (61)	189

Substrate	Reagent	Conditions	Product (yield)	Refs.
(fluorophenylsulfonyloxy quinoline)	Me$_3$SnSnMe$_3$	Pd(PPh$_3$)$_4$ (2%), LiCl, dioxane, 98°, 75 h	(quinoline SnMe$_3$) (67)	189
(3-bromoisoquinoline)	Bu$_3$Sn-CH=CH$_2$	Pd(OAc)$_2$ (5%), LiCl, dppp (5%), DMF, 90°	(8-vinylquinoline) (50)	202
	(oxazoline-Sn Me$_3$)	Pd(PPh$_3$)$_4$ (5%), C$_6$H$_6$, 80°, 12 h	(oxazolinylisoquinoline) (92)	550
(4-bromoisoquinoline)	Bu$_3$Sn-allyl	Pd(PPh$_3$)$_4$ (5%), PhMe, 110°	(4-allylisoquinoline) (85)	710
	(oxazoline-Bu$_3$Sn)	Pd(PPh$_3$)$_4$ (5%), C$_6$H$_6$, 80°, 12 h	(oxazolinylisoquinoline) (92)	551
(1-triflyloxyisoquinoline)	Bu$_3$SnR	Pd(PPh$_3$)$_4$ (5%), LiCl, dioxane, 90°, 24 h ZnCl$_2$	(1-R-isoquinoline)	194

R
CH=CH$_2$	(62)
Ph	(58)
C≡CPh	(70)
(E)-CH=CHPh	(69)

351

TABLE XI. DIRECT CROSS-COUPLING OF QUINOLINE AND ISOQUINOLINE ELECTROPHILES (*Continued*)

Substrate	Stannane	Conditions	Product(s) and Yield(s) (%)	Refs.
C₁₀	Bu₃Sn (F)	Pd(PPh₃)₂Cl₂, LiCl, THF, 65°, 18 h	(61)	264
	Bu₃Sn (OTBDMS)	Pd(PPh₃)₄ (15%), C₆H₆, reflux, 2 h	OTBDMS (85)	98
	Me₃Sn (S)	Pd(PPh₃)₄ (5%), LiCl, dioxane, reflux, 16 h	(99)	699
	Me₃Sn (S)	Pd(PPh₃)₄ (5%), LiCl, dioxane, reflux, 16 h	(34)	699
C₁₁	Me₃Sn (S)	Pd(PPh₃)₄ (5%), LiCl, dioxane, reflux, 16 h	(86)	699
	Me₃Sn (S)	Pd(PPh₃)₄ (5%), LiCl, dioxane, reflux, 16 h	(80)	699

352

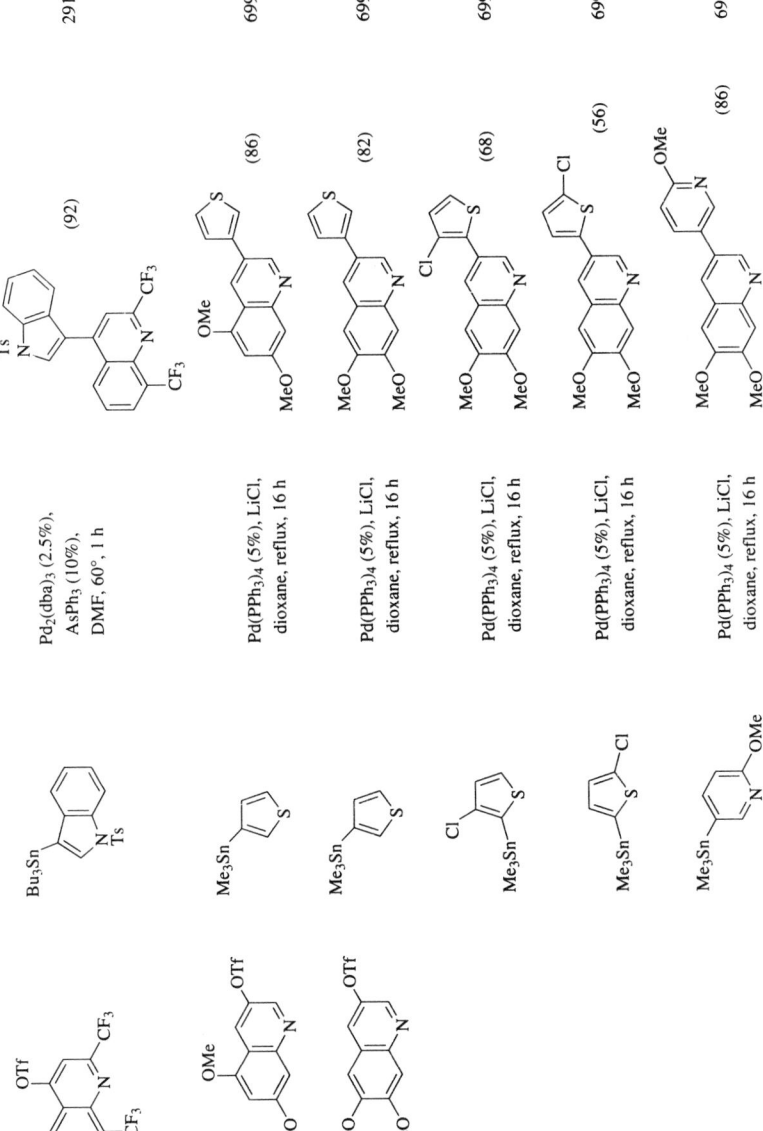

291

699

699

699

699

699

(92)

(86)

(82)

(68)

(56)

(86)

Pd₂(dba)₃ (2.5%),
AsPh₃ (10%),
DMF, 60°, 1 h

Pd(PPh₃)₄ (5%), LiCl,
dioxane, reflux, 16 h

Pd(PPh₃)₄ (5%), LiCl,
dioxane, reflux, 16 h

Pd(PPh₃)₄ (5%), LiCl,
dioxane, reflux, 16 h

Pd(PPh₃)₄ (5%), LiCl,
dioxane, reflux, 16 h

Pd(PPh₃)₄ (5%), LiCl,
dioxane, reflux, 16 h

Substrate	Stannane	Conditions	Product(s) and Yield(s) (%)	Refs.
(OMe OTf quinoline, OMe)	Me₃Sn-aryl (R¹, R²) R¹ R² Cl Cl F F H NO₂ OMe H H OMe NO₂ OMe H CO₂Et CH=CH-CH=CH H Ph	Pd(PPh₃)₄ (5%), LiCl, dioxane, reflux, 16 h	MeO / MeO quinoline with R², R¹ aryl (83) (90) (97) (91) (74) (99) (93) (62) (66)	699
	Me₃Sn-CH=CH-Ph	Pd(PPh₃)₄ (5%), LiCl, dioxane, reflux, 16 h	MeO / MeO quinoline-CH=CH-Ph (91)	699
	Me₃Sn-(3,4,5-tri-OMe-phenyl)	Pd(PPh₃)₄ (5%), LiCl, dioxane, reflux, 16 h	MeO / MeO quinoline with (OMe, OMe, OMe) aryl (51)	699
	Me₃Sn-(2-RHN-phenyl)	Pd(PPh₃)₄ (2%), LiCl, dioxane, 100°	OMe quinoline-NHR (OMe) R = CF₃CO, 16 h, (71) R = Boc, 5-7 h, (87)	711

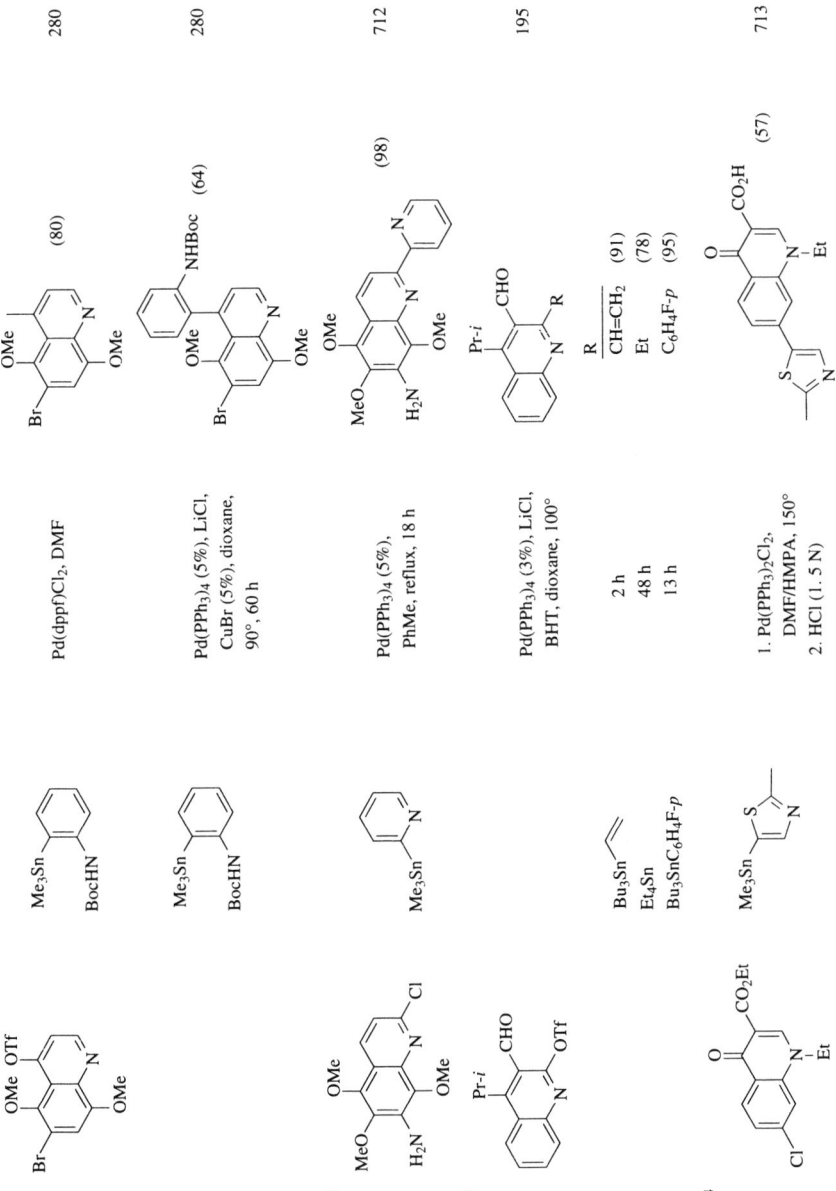

Substrate	Stannane	Conditions	Product(s) and Yield(s) (%)	Refs.
C$_{15}$	Bu$_3$SnR	Pd(PPh$_3$)$_2$Cl$_2$ (2%), LiCl, BHT, THF, 65°, 20 h		
	R: CH=CH$_2$		(44)	257
	(NHBoc)		(31)	257
	(NAc)		(48)	97, 257
	Me$_3$Sn—	Pd(PPh$_3$)$_2$Cl$_2$ (4%), MeO(CH$_2$)$_2$OH, 140°, 20 h	(25)	567
	Bu$_3$Sn—	Pd(PPh$_3$)$_2$Cl$_2$ (3.5%), DMF, 165°, 20 min, 145°, 1 h	(7)	567
	Bu$_3$Sn—	Pd(PPh$_3$)$_2$Cl$_2$ (5%), MeO(CH$_2$)$_2$OH, 140°, 7 h	(43)	567

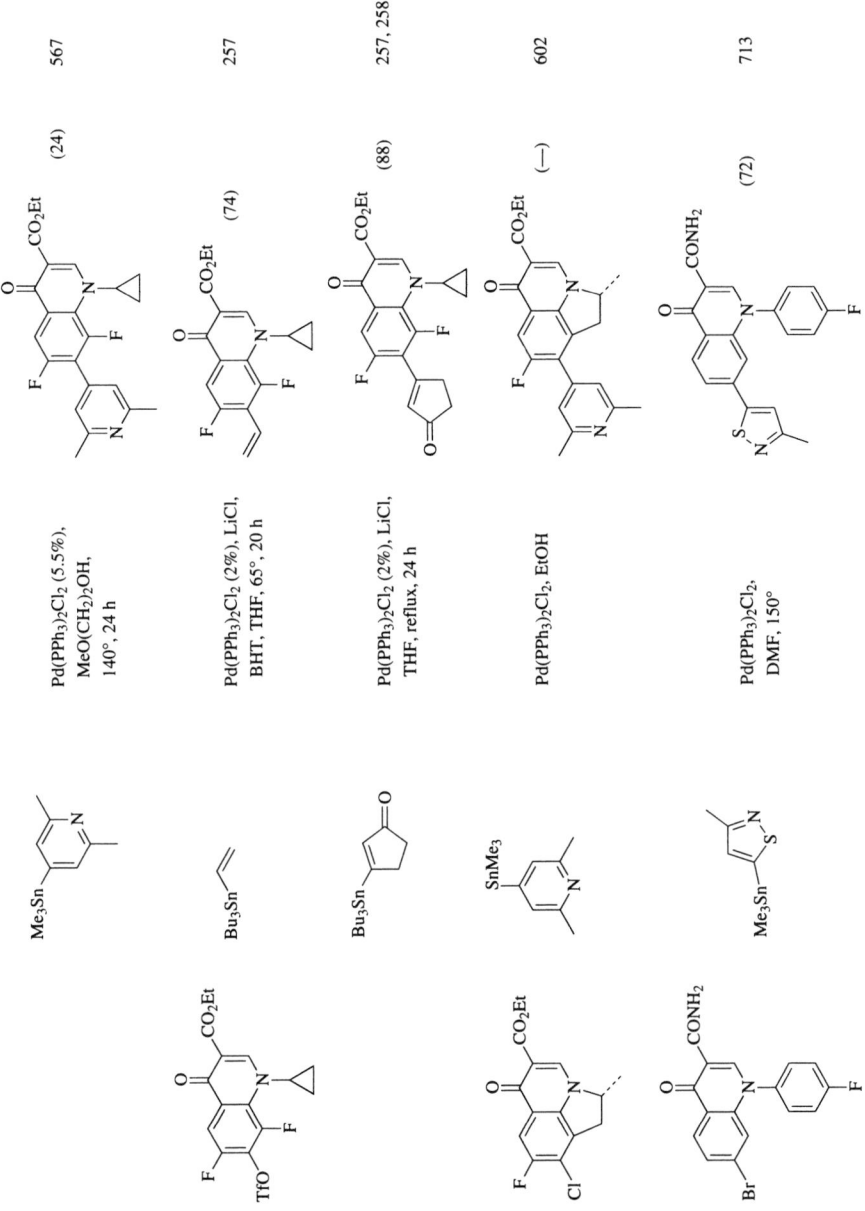

357

TABLE XI. DIRECT CROSS-COUPLING OF QUINOLINE AND ISOQUINOLINE ELECTROPHILES (*Continued*)

Substrate	Stannane	Conditions	Product(s) and Yield(s) (%)	Refs.
(4-oxoquinoline-3-CONH2, N-(4-F-phenyl), 7-Cl)	Bu_3SnPh	$Pd(PPh_3)_2Cl_2$, DMF, 150°, 1 h	(4-oxoquinoline-3-CONH2, N-(4-F-phenyl), 7-Ph) (56)	713
(4-oxoquinoline-3-CONH2, N-(3-F-phenyl), 7-Cl)	SnMe₃–pyridine	1. $Pd(PPh_3)_2Cl_2$, DMF, 150° 2. MeSO₃H, MeOH	(4-oxoquinoline-3-CONH2, N-(3-F-phenyl), 7-(pyridinium), $MeSO_3^-$) (80)	713
(4-oxo-2-OH-quinoline-3-CONH2, N-(4-F-phenyl), 7-Br)	SnMe₃–pyridine	$Pd(PPh_3)_2Cl_2$, DMF, 150°	(4-oxo-2-OH-quinoline-3-CONH2, N-(4-F-phenyl), 7-(pyridinyl)) (—)	713
C₂₅ (tetrahydroisoquinoline, CO₂Me, N-CO-CHPh-Ph, TfO)	Me_3SnPh	$Pd(PPh_3)_4$ (10%), LiCl, dioxane, reflux	(tetrahydroisoquinoline, CO₂Me, N-CO-CHPh-Ph, Ph) (45)	714

358

Substrate	Stannane	Conditions	Product(s) and Yield(s) (%)	Refs.
C₃ (2-bromothiazole)	Me₃Sn–(thiazol-2-yl)	Pd(PPh₃)₄ (5%), C₆H₆, reflux, 48 h	(83)	108
	Me₃Sn–(thiazol-5-yl)	Pd(PPh₃)₄ (5%), C₆H₆, reflux, 48 h	(95)	108
	Me₃Sn–(thiazol-4-yl)	Pd(PPh₃)₄ (5%), C₆H₆, reflux, 48 h	(96)	108
	Bu₃Sn–(pyridin-2-yl)	Pd(dppb)Cl₂ (5%), CuO, DMF, 100°, 0.5h	(81)	96, 669
	Bu₃Sn–(pyridin-3-yl)	Pd(PPh₃)₄, Ag₂O, DMF, 100°, 80–100 min	(25)	669
	Me₃Sn–(oxazoline)	Pd(PPh₃)₄ (5%), C₆H₆, 80°, 24 h	(85)	550
	Bu₃Sn–(2-SMe-pyrimidin-5-yl)	Pd(PPh₃)₂Cl₂ (3%), Cl(CH₂)₂Cl, reflux, 24 h	(68)	651
	Me₃Sn–(thiazol-4-yl)	Pd(PPh₃)₄ (5%), C₆H₆, reflux, 48 h	(79)	108
(4-bromothiazole)	Me₃Sn–(thiazol-5-yl)	Pd(PPh₃)₄ (5%), C₆H₆, reflux, 48 h	(72)	108

359

TABLE XII. DIRECT CROSS-COUPLING OF MISCELLANEOUS HETEROCYCLIC ELECTROPHILES (Continued)

Substrate	Stannane	Conditions	Product(s) and Yield(s) (%)	Refs.
		Pd(PPh$_3$)$_4$ (5%), C$_6$H$_6$, reflux, 48 h	(75)	108
		Pd(PPh$_3$)$_4$ (5%), C$_6$H$_6$, reflux, 48 h	(92)	108
		Pd(PPh$_3$)$_4$ (5%), C$_6$H$_6$, reflux, 48 h	(75)	108
		Pd(PPh$_3$)$_4$ (5%), C$_6$H$_6$, reflux, 48 h	(80)	108
		Pd(PPh$_3$)$_2$Cl$_2$ (10%), THF, reflux, 20 h	(52)	667
		Pd(PPh$_3$)$_2$Cl$_2$ (10%), THF, reflux, 20 h	(25)	667
		Pd(PPh$_3$)$_4$ (5%), C$_6$H$_6$, reflux	TMS (76)	108
		Pd(PPh$_3$)$_4$ (5%), C$_6$H$_6$, reflux, 48 h	TMS (45)	108
C$_4$		Pd(PPh$_3$)$_4$, Ag$_2$O, DMF, 100°, 20 h	(tr)	669

Reagent	Conditions	Product	(Yield)	Refs.
Bu₃Sn–CH=CH₂	Pd₂(dba)₃, AsPh₃, NMP, 40°	5-vinyluracil	(95)	11
Bu₃Sn–C(F)=CH₂	Pd(PPh₃)₄, DMF, 100°, 2 h		(45)	264
Bu₃Sn–CH=CH–CH₃	1. Pd₂(dba)₃ (1%), P(2-furyl)₃ (4%), NMP, rt, 16 h 2. 50°, 5 h		(70)	128
Bu₃Sn–(2-thienyl)	Pd(PPh₃)₂Cl₂ (5%), DMF, reflux		(37)	135, 669
Bu₃Sn–(2-selenophenyl)	Pd(PPh₃)₂Cl₂ (5%), DMF, reflux		(45)	135
Bu₃Sn–(2-pyridyl)	Pd(PPh₃)₂Cl₂, Ag₂O, DMF, 100°		(68)	669, 135
Bu₃SnR	Pd₂(dba)₃ (1%), P(2-furyl)₃ (4%), NMP	95°, 16 h, R = p-MeOC₆H₄, (38); rt, 24 h, R = C≡CPh, (61); rt, 16 h, R = (E)-CH=CHPh, (92)		128

TABLE XII. DIRECT CROSS-COUPLING OF MISCELLANEOUS HETEROCYCLIC ELECTROPHILES (Continued)

Substrate	Stannane	Conditions	Product(s) and Yield(s) (%)	Refs.
	Bu₃Sn–(1-methylindol-2-yl)	Pd(PPh₃)₂Cl₂ (2%), THF, reflux, 20 h	(60)	425
	Bu₃Sn–CH=CH–S(O)Ph	Pd₂(dba)₃, P(2-furyl)₃, NMP, rt, 16 h	(78)	715
	Bu₃Sn–CH=CH–CH(OH)C₈H₁₇-n	Pd₂(dba)₃, P(2-furyl)₃, NMP, 40°, 2 d	(65)	447
	Bu₃Sn–(1-SEM-indol-2-yl)	Pd(PPh₃)₂Cl₂ (2%), THF, reflux	(60)	425
C₅ (6-chloropurine)	Bu₃Sn–C(OEt)=CH₂	Pd(PPh₃)₂Cl₂ (3%), DMF, 80°, 20 h	(58)	133
	Bu₃Sn–(2-thienyl)	Pd(PPh₃)₂Cl₂ (3%), DMF, 90°, 20 h	(81)	133

Bu₃SnR

R	
Ph	(81)
Bn	(36)
(E)-CH=CHPh	(84)

Pd(PPh₃)₂Cl₂ (3%), DMF, 20 h

100°
130°
80°

133

Bu₃SnR

R	
2-thienyl	(76)
Ph	(73)
(E)-CH=CHPh	(75)

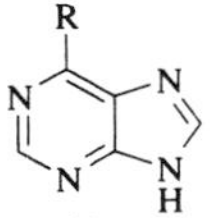

Pd(PPh₃)₂Cl₂ (3%), DMF, 45 h

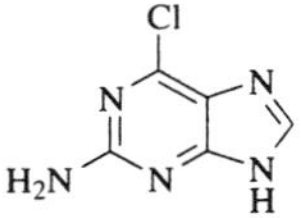

90°
100°
80°

133

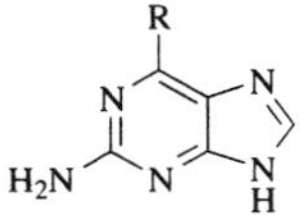

Bu₃Sn

Pd(PPh₃)₂Cl₂ (2%), PhMe, 90°, 18 h

X = O (85)
X = S (84)

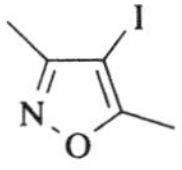

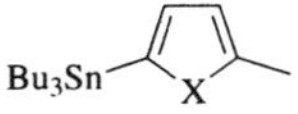

114

Bu₃SnR

R	
3-Me-5-thienyl	(84)
Ph	(58)
m-MeOC₆H₄	(61)

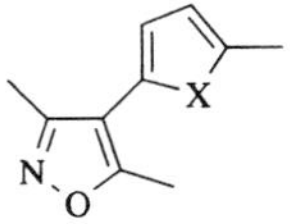

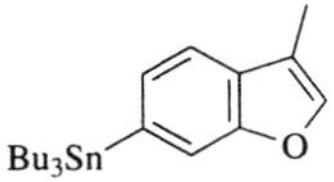

Pd(PPh₃)₂Cl₂ (2%), PhMe, 90°, 18 h

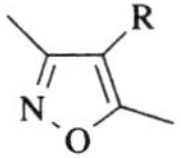

114

Bu₃Sn

Pd(PPh₃)₂Cl₂ (2%), PhMe, 90°, 18 h

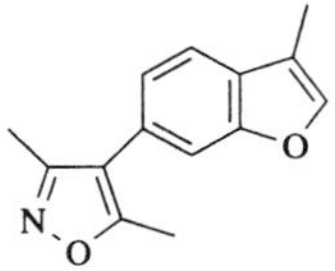

(61)

114

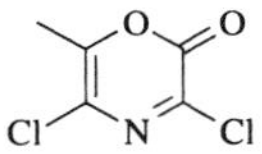

Et₄Sn

Pd(PPh₃)₄ (0.1%), PhMe, reflux, 8 d

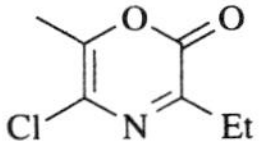

(80)

716

Substrate	Stannane	Conditions	Product(s) and Yield(s) (%)	Refs.
		Pd$_2$(dba)$_3$ (2.5%), CuI, DMF, 60°	(48)	554
		Pd(dba)$_2$, PPh$_3$, PhMe, reflux, 5 h	(50)	116
C$_6$		Pd(PPh$_3$)$_4$ (10%), DMF, 110°	(92)	289
		Pd(PPh$_3$)$_4$ (10%), DMF, 110°, 2 h	(68)	74
	Bu$_3$SnH	Pd(PPh$_3$)$_4$ (2.5%), PhMe, 110°, 1 h	(70)	116
		Pd(dba)$_2$ (5%), P(2-furyl)$_3$ (10%), THF, 70°, 4 h	(60)	116

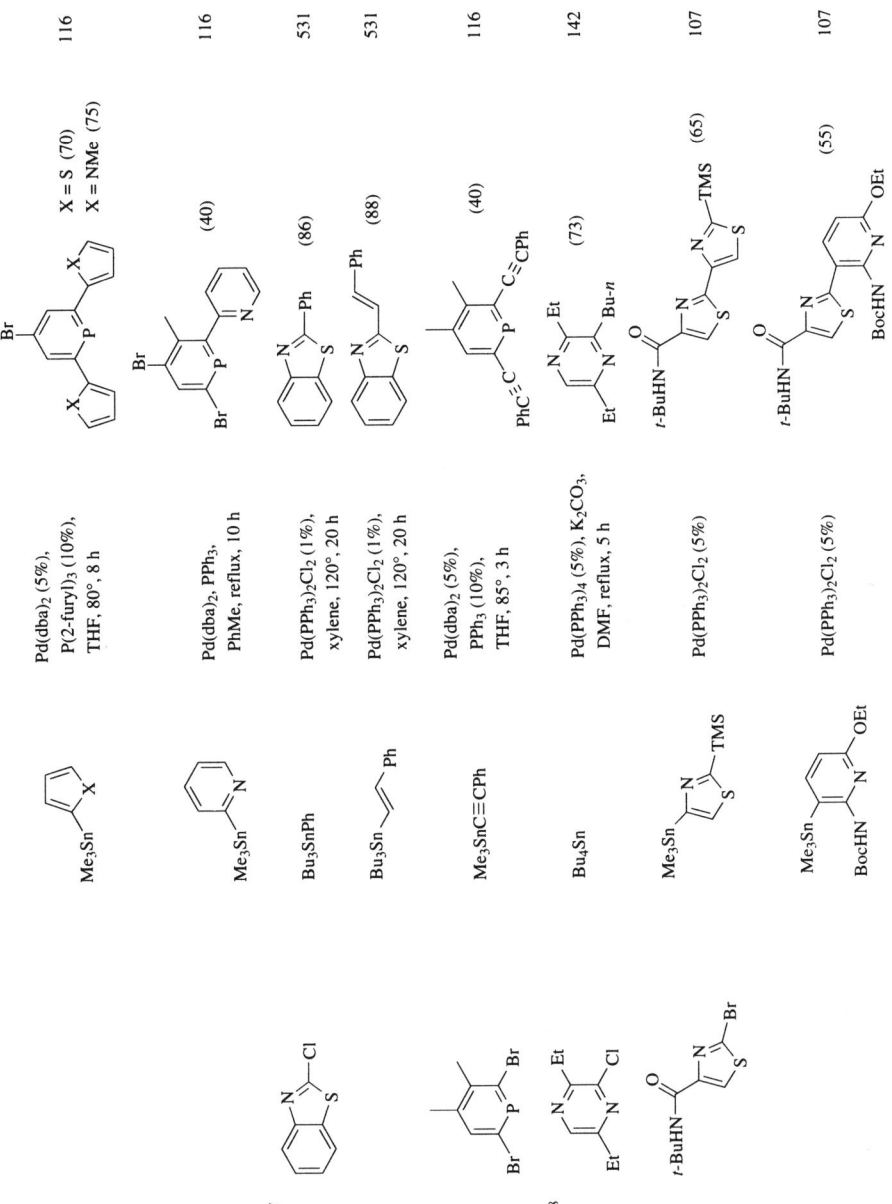

Substrate	Stannane	Conditions	Product(s) and Yield(s) (%)	Refs.
C$_9$	Bu$_3$SnR	Pd(PPh$_3$)$_2$Cl$_2$ (5%), THF, reflux, 24 h		123
	R: CH=CH$_2$ (47); CH$_2$CH=CH$_2$ (55); CH=C(Me)$_2$ (57); Bu (42)			
	Me$_3$Sn—X	Pd(PPh$_3$)$_2$Cl$_2$ (7.6%), dioxane		717, 123
	X: O; S	70°, 2 h (89); 95°, 1 h (77)	(89); (77)	
	Bu$_3$Sn—N–Me	Pd(PPh$_3$)$_2$Cl$_2$ (5%), THF, reflux, 24 h	(60)	123

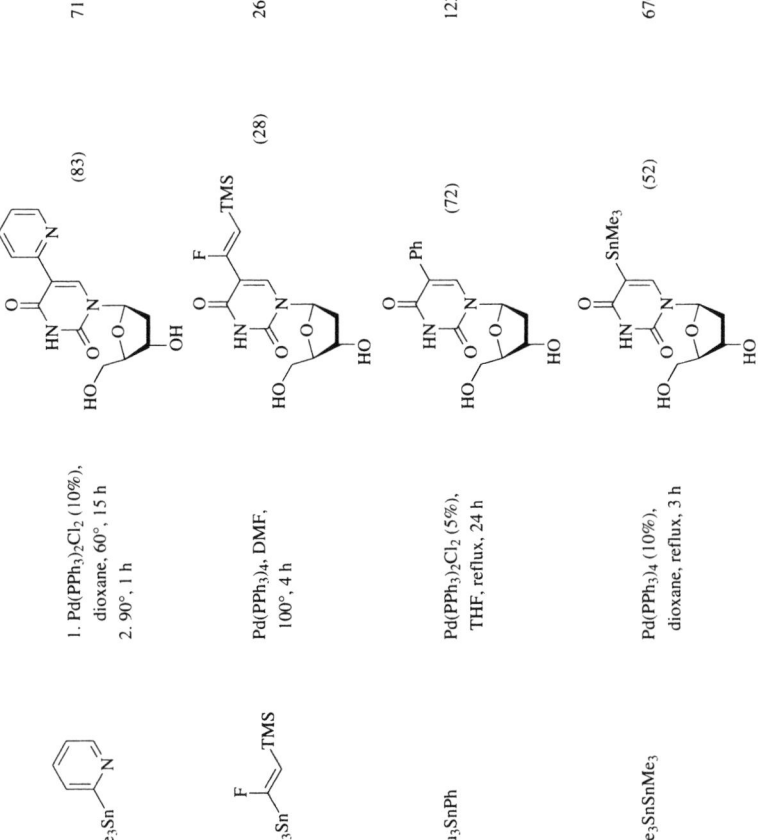

718

263

123

670

TABLE XII. DIRECT CROSS-COUPLING OF MISCELLANEOUS HETEROCYCLIC ELECTROPHILES (*Continued*)

Substrate	Stannane	Conditions	Product(s) and Yield(s) (%)	Refs.
	Bu₃Sn—X	Pd(PPh₃)₂Cl₂ or Pd(PPh₃)₄	X = O (—) X = S (—)	717
		Pd(PPh₃)₄ (20%), PhMe, reflux, 24 h	(0)	127
C₁₀	Bu₃SnR	Pd(PPh₃)₄, LiCl, THF, 65°	R = H (66) R = CH₂CH=CH₂ (66)	719
	Bu₃SnR R ――― Bu [C₅H₁₁-n] [C₆H₄Cl-p] [C₆H₄Me-o] [C₆H₄Me-m] [C₆H₄Me-p] [C₆H₄OMe-p] [C₈H₁₇-n]	Pd(PPh₃)₄ (5%), K₂CO₃, DMF, reflux, 5 h	(42) (40) (67) (10) (59) (80) (65) (66)	142

[Structure: pyridine N-oxide with Pr-i, Cl, i-Pr substituents]

$[Bu_3SnC_5H_{11}\text{-}n]$ — $Pd(OAc)_2$ (5%), K_2CO_3, DMF, reflux, 18 h — [product: pyridine N-oxide with Pr-i, $C_5H_{11}\text{-}n$, i-Pr] (37) — 142

$[Bu_3SnR]$ — $Pd(PPh_3)_4$ (5%), K_2CO_3, DMF, reflux — [product: pyridine N-oxide with Pr-i, R, i-Pr] — 142

R		
$C_6H_4Cl\text{-}p$	5 h	(55)
$C_6H_4Me\text{-}p$	5 h	(81)
$C_6H_4OMe\text{-}p$	2 h	(66)

[Structure: pyridine N-oxide with Pr-i, Cl, i-Pr substituents, N-oxide on other side]

$[Bu_3SnC_8H_{17}\text{-}n]$ — $Pd(OAc)_2$ (5%), K_2CO_3, DMF, reflux, 18 h — [product: pyridine with Pr-i, $C_8H_{17}\text{-}n$, i-Pr, N-oxide] (43) — 142

$[Bu_3SnR]$ — $Pd(PPh_3)_4$ (5%), K_2CO_3, DMF, reflux — [product: pyridine with Pr-i, R, i-Pr, N-oxide] — 142

R		
$C_6H_4Cl\text{-}p$	7 h	(29)
$C_6H_4Me\text{-}p$	5 h	(50)
$C_6H_4OMe\text{-}p$	2 h	(52)

[Structure: Bn–N–S(O2)=N ring with X substituent]

Bu_3SnR — $Pd(PPh_3)_2Cl_2$ (5%), $Cl(CH_2)_2Cl$, 50°, 20 h — [product: Bn–N–S(O2)=N ring with R] — 720

X	R	
Br	$CH=CH_2$	(55)
Br	2-thienyl	(56)
Br	Ph	(45)
Br	$C≡CPh$	(30)
Br	$(E)\text{-}CH=CH=CHPh$	(80)
I	2-thienyl	(68)

369

TABLE XII. DIRECT CROSS-COUPLING OF MISCELLANEOUS HETEROCYCLIC ELECTROPHILES (*Continued*)

Substrate	Stannane	Conditions	Product(s) and Yield(s) (%)	Refs.
	Et₄Sn	Pd(PPh₃)₄, DMF, reflux, 5 h	(48)	721
	Et₄Sn	Pd(PPh₃)₄, DMF, reflux, 5 h	(45)	721
		Pd(OAc)₂ (10%), AsPh₃ (20%), CH₃CN, THF, 40°, 16 h	(81)	301
		Pd(PPh₃)₄ (5%), P(o-Tol)₃, DMF, 100°, 18 h	(50)	120
	R₄Sn 	Pd(PPh₃)₄ (10%), NMP, 110°	(72) (65)	130, 722

For the last row:

R	
Me	2 h
CH=CH₂	14 h

370

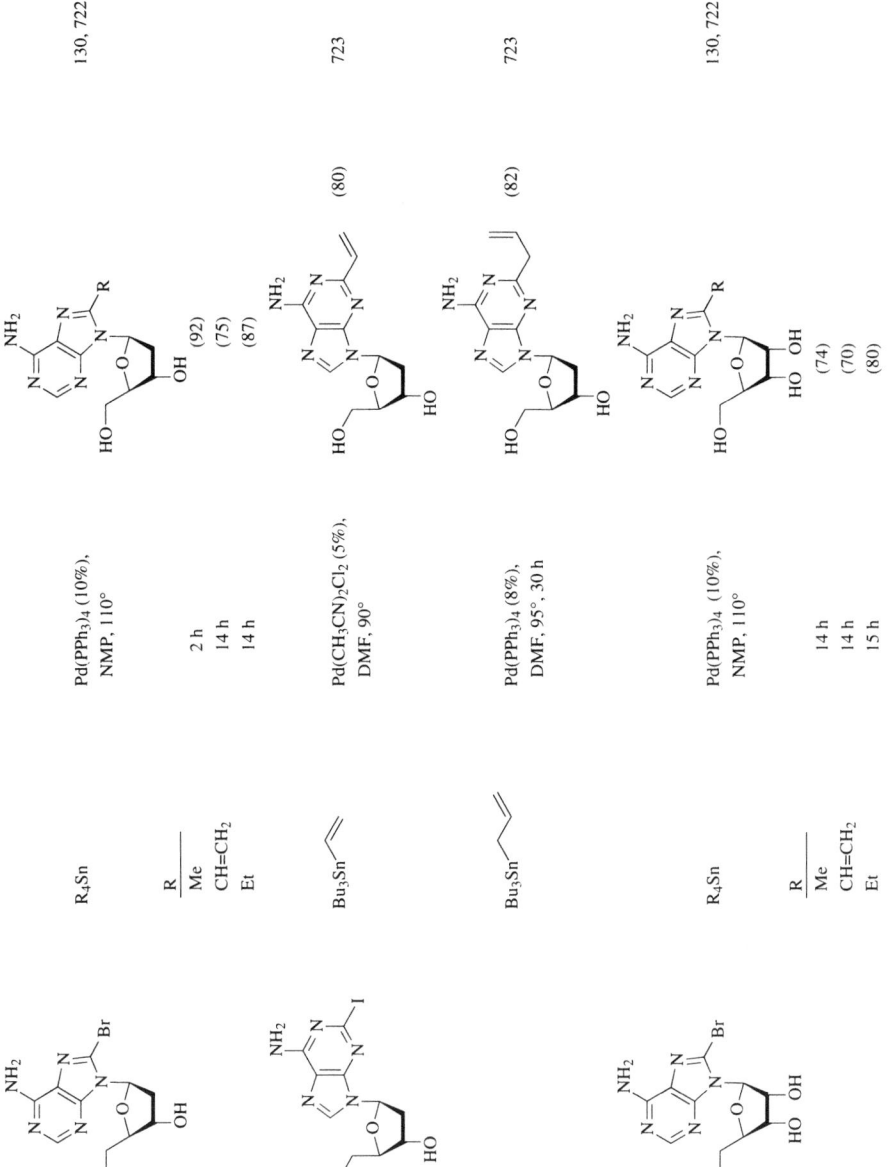

R$_4$Sn

R	
Me	(92)
CH=CH$_2$	(75)
Et	(87)

Pd(PPh$_3$)$_4$ (10%), NMP, 110°

2 h
14 h
14 h

130, 722

Bu$_3$Sn

Pd(CH$_3$CN)$_2$Cl$_2$ (5%), DMF, 90°

(80)

723

Bu$_3$Sn

Pd(PPh$_3$)$_4$ (8%), DMF, 95°, 30 h

(82)

723

R$_4$Sn

R	
Me	(74)
CH=CH$_2$	(70)
Et	(80)

Pd(PPh$_3$)$_4$ (10%), NMP, 110°

14 h
14 h
15 h

130, 722

TABLE XII. DIRECT CROSS-COUPLING OF MISCELLANEOUS HETEROCYCLIC ELECTROPHILES (Continued)

Substrate	Stannane	Conditions	Product(s) and Yield(s) (%)	Refs.
C_{11}	Bu₃SnR	Pd(CH₃CN)₂Cl₂, DMF	R = CN (86) R = CH=CH₂ (84)	119
	Bu₃SnR $\dfrac{R}{\text{CH=CH}_2}$ C(OEt)=CH₂ 2-pyridyl C≡CPh	Pd(PPh₃)₄ (5%), PhMe, 110°, 4 h	**I** (80) (69) (69) (75)	724
	Me₃SnPh	Pd(PPh₃)₄ (5%), PhMe, 110°, 4 h	**I, R = Ph**, (63)	724
	Me₃SnPh	Pd(PPh₃)₂Cl₂ (10%), PhMe, reflux	(58)	111
	Bu₃Sn⟍	Pd(PPh₃)₄ (5%), BHT, PhMe, reflux, 1.5 h	(39)	72
	OEt Bu₃Sn⟍	1. Pd(PPh₃)₂Cl₂ (5%), PhMe, reflux, 17 h 2. HCl, rt, 2 h	(83)	726

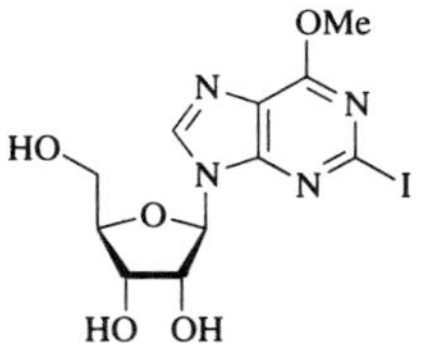

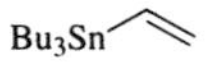

Bu$_3$Sn—CH=CH$_2$	Pd(CH$_3$CN)$_2$Cl$_2$ (10%), PhMe, reflux	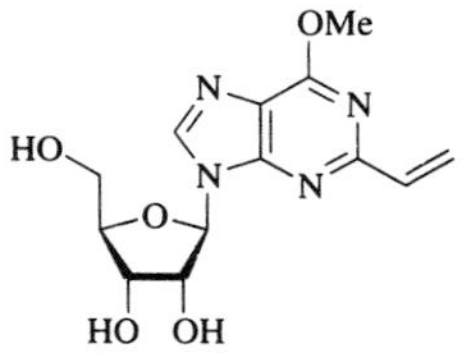 (>90)	118, 120
Bu$_3$Sn—CH$_2$CH=CH$_2$	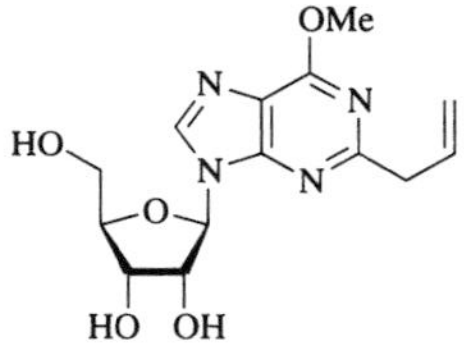 Pd(CH$_3$CN)$_2$Cl$_2$ (5%), DMF, 90°, 6 h	(76)	120
Bu$_3$Sn—CH$_2$CH=CHCH$_3$	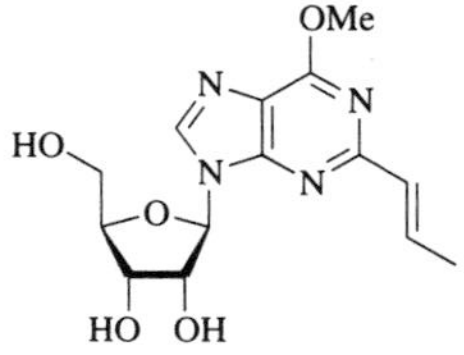 Pd(CH$_3$CN)$_2$Cl$_2$ (5%), DMF, 105°, 24 h	(63)	120

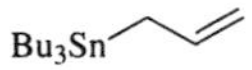

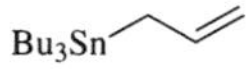

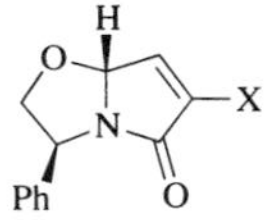

Bu$_3$SnR

Pd(OAc)$_2$ (5%), PPh$_3$ (10%), CuI (10%), NMP, 100°, 30 min

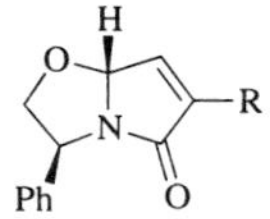

R	
CH=CH$_2$	(63)
2-furyl	(42)
Ph	(57)
CH=CH$_2$	(55)
2-furyl	(54)
Ph	(57)

727

C$_{12}$

X	
Br	
Br	
Br	
I	
I	
I	

TABLE XII. DIRECT CROSS-COUPLING OF MISCELLANEOUS HETEROCYCLIC ELECTROPHILES (*Continued*)

Substrate	Stannane	Conditions	Product(s) and Yield(s) (%)	Refs.
	Bu₃SnC≡CTMS	Pd(PPh₃)₄ (5%), PhMe, 80°, 4 h	(40)	728
	Bu₃SnC≡C–C≡CTMS	Pd(PPh₃)₄ (3%), PhMe, 50°, 18 h	(10)	728
	Bu₃Sn	1. Pd(PPh₃)₂Cl₂ (4%), BHT, DMF, 50°, 5.5 h 2. rt, 64 h	(30)	729
	Bu₃SnR R Bu-*n* 2-thienyl Ph Bn	Pd(PPh₃)₂Cl₂ (5%), DMF reflux, 3 h 100°, 16 h 110°, 4 h reflux, 4 h	(65) (90) (93) (62)	134

374

C13

Substrate	Reagent	Conditions	Product	Yield (%)	Refs.
6-chloro-9-benzylpurine	Bu₃SnR	Pd(PPh₃)₂Cl₂ (5%), DMF	6-R-9-benzylpurine		134
		R			
		CH=CH₂	reflux, 3.5 h	(87)	
		Bu-n	reflux, 21 h	(18)	
		C(OEt)=CH₂	100°, 4.5 h	(81)	
		2-thienyl	100°, 16 h	(87)	
		Ph	110°, 7 h	(75)	
		Bn	reflux, 18 h	(48)	
		(E)-CH=CHPh	130°, 24 h	(76)	
acridone (X)	Bu₃SnSnBu₃	PhPd(PPh₃)₂I (1%), Bu₄NBr, HMPA, 120°, 19 h	SnBu₃-acridone	X = Br (56), X = I (30)	613
bromophenanthridinone	Bu₃SnSnBu₃	PhPd(PPh₃)₂I (1%), Bu₄NBr, HMPA, 110°, 10 h	SnBu₃-phenanthridinone (10)		613
pyrrolooxazolone (X, Ph)	Bu₃SnR	Pd(OAc)₂ (5%), PPh₃ (10%), CuI (10%), NMP, 100°, 30 min			727

X	R	Yield
Br	CH=CH₂	(100)
Br	2-furyl	(70)
Br	Ph	(75)
I	CH=CH₂	(95)
I	2-furyl	(55)
I	Ph	(72)

TABLE XII. DIRECT CROSS-COUPLING OF MISCELLANEOUS HETEROCYCLIC ELECTROPHILES (Continued)

Substrate	Stannane	Conditions	Product(s) and Yield(s) (%)	Refs.
	Bu₃Sn—CH=CH₂ (Bu$_3$Sn, vinyl)	Pd(PPh$_3$)$_4$ (3–4%), LiCl, THF, 65°	(60)	200, 730
	Bu$_3$Sn—CH=CH—CO$_2$Et	Pd(PPh$_3$)$_4$ (3–4%), LiCl, THF, 65°	(78)	200, 730
	Bu$_3$Sn—CH=CH₂	Pd(PPh$_3$)$_2$Cl$_2$ (2%), BHT, THF, reflux, 20 h	(70)	729
	Me$_3$SnR	Pd(PPh$_3$)$_2$Cl$_2$ (5%), THF, reflux		731
	R			
	2-thiazolyl	15 h	(50)	
	2-thienyl	15 h	(44)	
	3-Me-2-thienyl	15 h	(28)	
	5-Ph-2-thienyl	15 h	(56)	
	3-n-hexyl-2-thienyl	72 h	(24)	

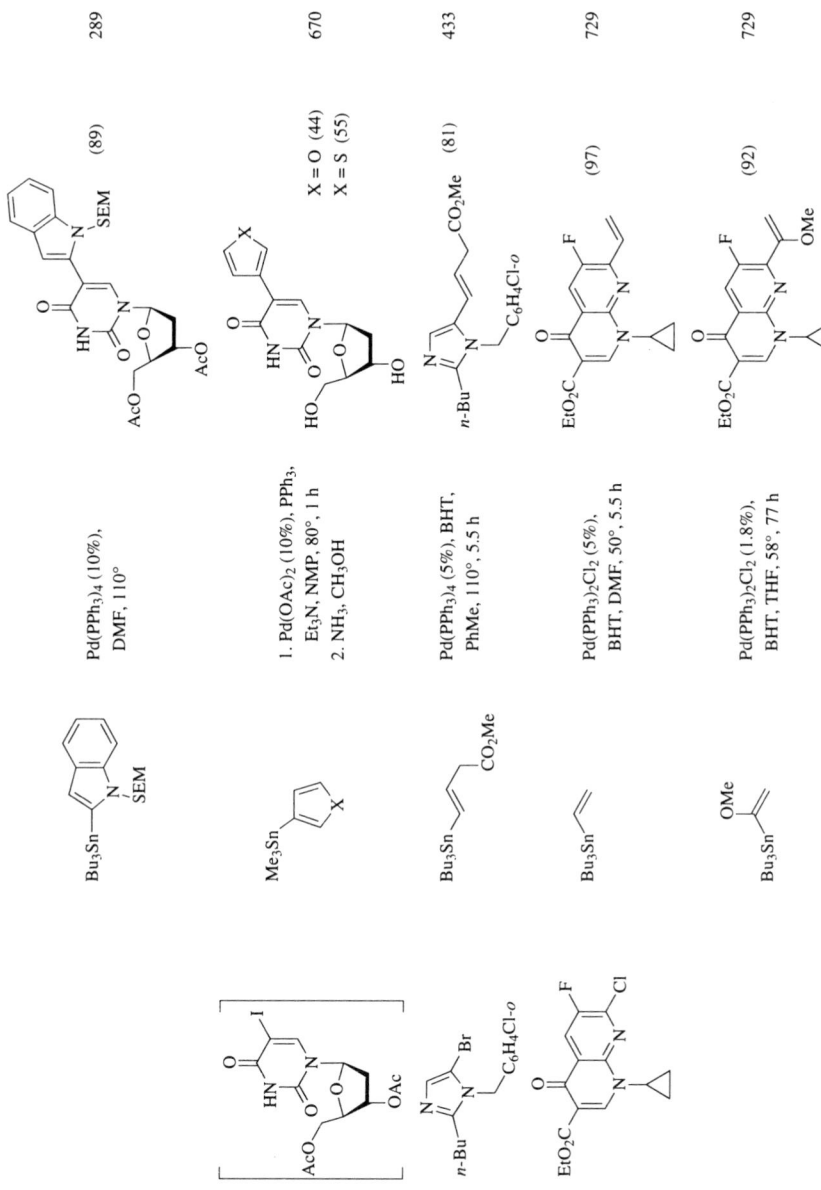

289

(89)

670

X = O (44)
X = S (55)

433

(81)

729

(97)

729

(92)

Pd(PPh$_3$)$_4$ (10%),
DMF, 110°

1. Pd(OAc)$_2$ (10%), PPh$_3$,
Et$_3$N, NMP, 80°, 1 h
2. NH$_3$, CH$_3$OH

Pd(PPh$_3$)$_4$ (5%), BHT,
PhMe, 110°, 5.5 h

Pd(PPh$_3$)$_2$Cl$_2$ (5%),
BHT, DMF, 50°, 5.5 h

Pd(PPh$_3$)$_2$Cl$_2$ (1.8%),
BHT, THF, 58°, 77 h

C$_{14}$

Substrate	Stannane	Conditions	Product(s) and Yield(s) (%)	Refs.
	Bu$_3$Sn–(cyclopent-2-enone)	Pd(PPh$_3$)$_2$Cl$_2$ (2%), BHT, THF, 65°, 28 h	(88)	729
	Bu$_3$Sn–(1,2,3,6-tetrahydropyridine, NAc)	Pd(PPh$_3$)$_2$Cl$_2$ (9%), BHT, dioxane, reflux, 3 h	(54)	729
	Bu$_3$Sn–(cyclopentene–NHBoc)	Pd(PPh$_3$)$_4$ (2%), BHT, dioxane, reflux, 40 h	(48)	729
	Bu$_3$Sn–(cyclohexene–NHBoc)	Pd(PPh$_3$)$_2$Cl$_2$ (2%), DMF, reflux, 24 h	(76)	258, 729
	Bu$_3$Sn–(cyclopentene–OTBDMS)	Pd(PPh$_3$)$_2$Cl$_2$ (2.2%), BHT, THF, 65°, 26 h	(67)	729

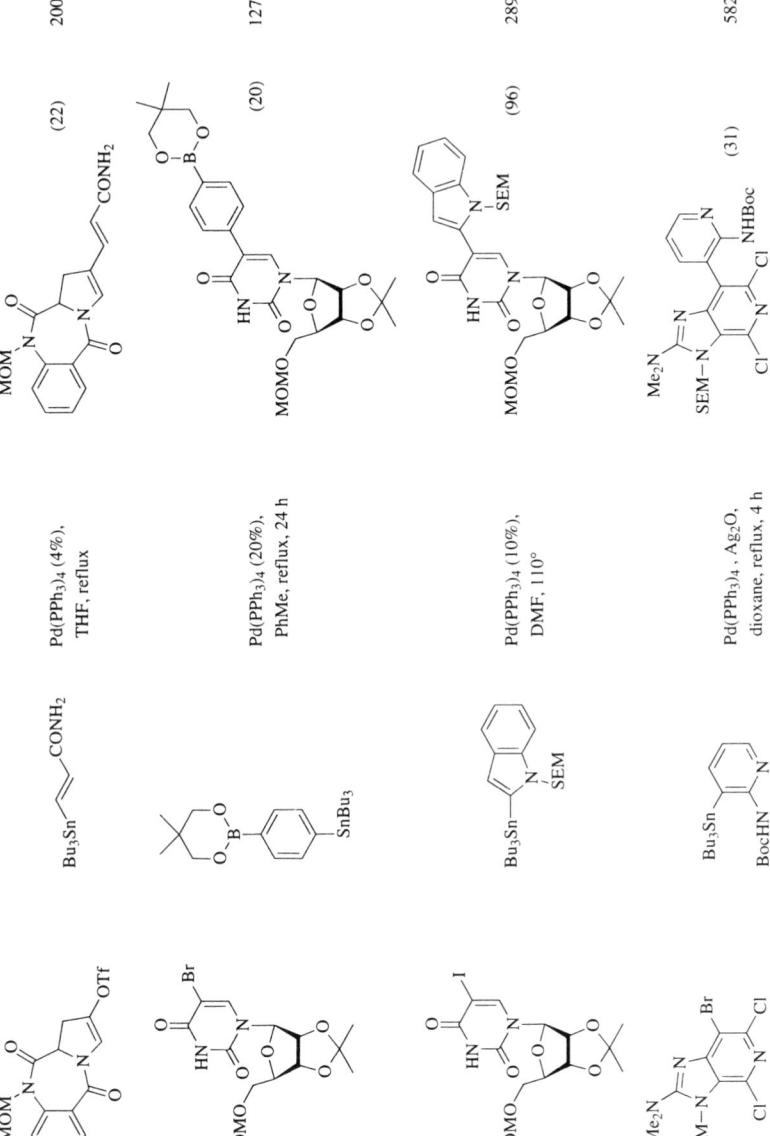

(22) 200

Pd(PPh₃)₄ (4%),
THF, reflux

(20) 127

Pd(PPh₃)₄ (20%),
PhMe, reflux, 24 h

(96) 289

Pd(PPh₃)₄ (10%),
DMF, 110°

(31) 582

Pd(PPh₃)₄ , Ag₂O,
dioxane, reflux, 4 h

TABLE XII. DIRECT CROSS-COUPLING OF MISCELLANEOUS HETEROCYCLIC ELECTROPHILES (*Continued*)

Substrate	Stannane	Conditions	Product(s) and Yield(s) (%)	Refs.
C$_{15}$				
	Bu$_3$Sn⟋CO$_2$Me	Pd(PPh$_3$)$_4$ (4%), THF, reflux	(70)	200
	Bu$_3$Sn⟋	Pd$_2$(dba)$_3$ (1%), P(2-furyl)$_3$ (4%), NMP, rt, 72 h	(91)	128
	Bu$_3$Sn⟋OH	Pd(CH$_3$CN)$_2$Cl$_2$, CH$_3$CN, 100°	(68)	112
	Bu$_3$SnC≡CTMS	Pd(PhCN)$_2$Cl$_2$	(77)	113

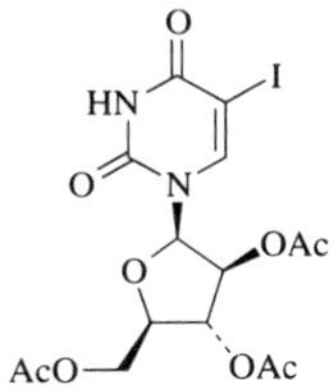

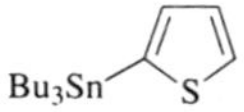

(33)

Bu₃Sn-thiophene

Pd(PPh₃)₄ (4%), PhMe, reflux, 22 h

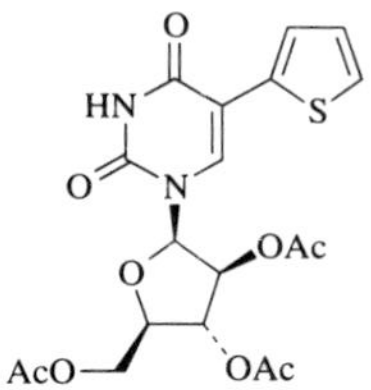

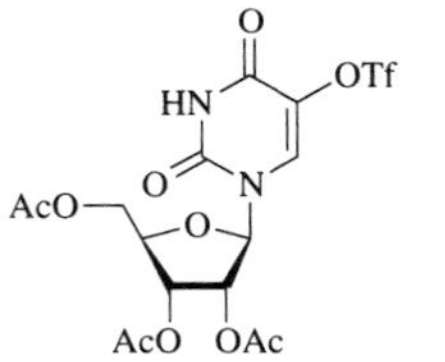

Bu₃SnR

Pd(PPh₃)₄ (5%), LiCl, BHT, dioxane, 100°

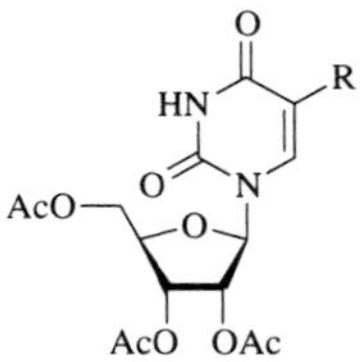

R		
CH=CH₂	4 h	(87)
C(Me)=CH₂	5 h	(86)
(E)-CH=CHCO₂Et	1 h	(92)
C(CO₂Et)=CH₂	15 h	(46)
(E)-CH=CHTMS	4 h	(73)
(E)-CH=CHPh	8 h	(75)
(E)-CH=CHCH₂OSiMe₂Thex	5 h	(92)

Me₃SnR

R		
Ph	20 h	(64)
C₆H₄F-4	8 h	(89)
C₆H₃F₂-3,5	5 h	(81)
C₆H₄CF₃-4	3 h	(91)
C₆H₄OMe-4	28 h	(55)
C₆H₃(CF₃)₂-3,5	4 h	(85)

TABLE XII. DIRECT CROSS-COUPLING OF MISCELLANEOUS HETEROCYCLIC ELECTROPHILES (*Continued*)

Substrate	Stannane	Conditions	Product(s) and Yield(s) (%)	Refs.
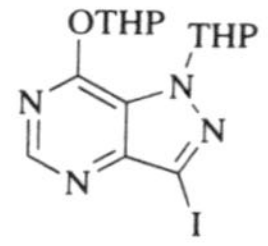	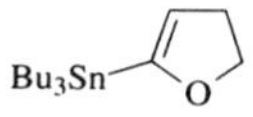	Pd(OAc)$_2$ (10%), AsPh$_3$ (20%), CH$_3$CN, 25°, 8 h	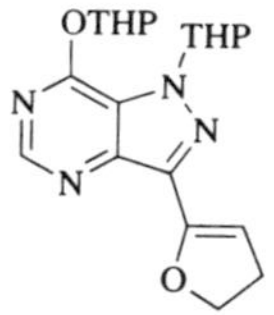 (85)	301
	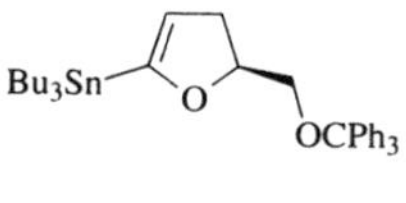	Pd(OAc)$_2$ (10%), AsPh$_3$ (20%), THF, CH$_3$CN, 25°, 12 h	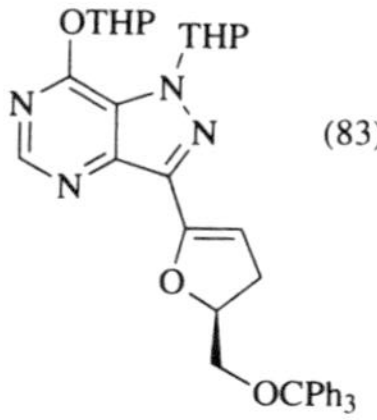 (83)	301
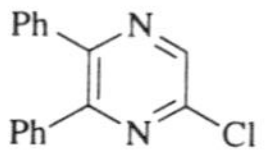	Bu$_4$Sn	Pd(PPh$_3$)$_4$ (5%), K$_2$CO$_3$, DMF, reflux, 5 h	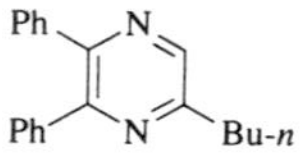 (82)	142
	[Bu$_3$SnR]	Pd(PPh$_3$)$_4$ (5%), K$_2$CO$_3$, DMF, reflux	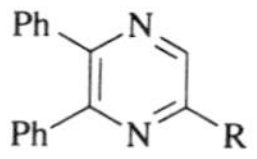	142

C$_{16}$

[Bu$_3$SnR]

R	
C$_5$H$_{11}$-n	5 h (80)
C$_6$H$_4$Cl-p	2 h (52)
C$_6$H$_4$Me-o	5 h (53)
C$_6$H$_4$Me-m	2 h (87)
C$_6$H$_4$Me-p	2 h (88)
C$_6$H$_4$OMe-p	2 h (87)
C$_8$H$_{17}$-n	2 h (83)

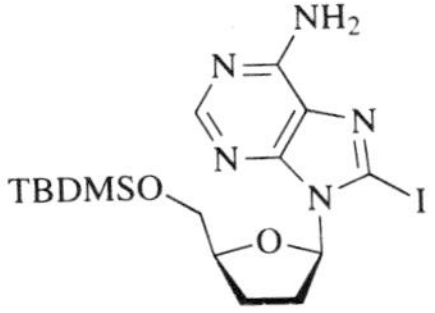

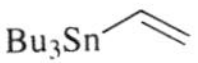

Pd(PPh₃)₄ (5%), DMF, 95°, 45 min — (90) — 131

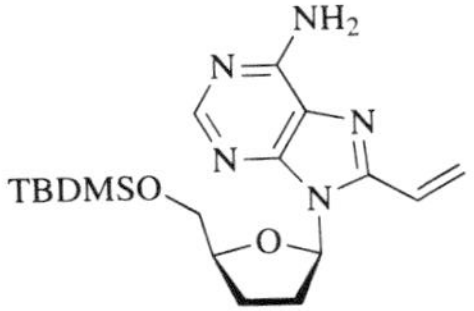

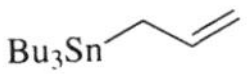

Pd(PPh₃)₄ (5%), HMPA, 145°, 45 min — (75) — 131

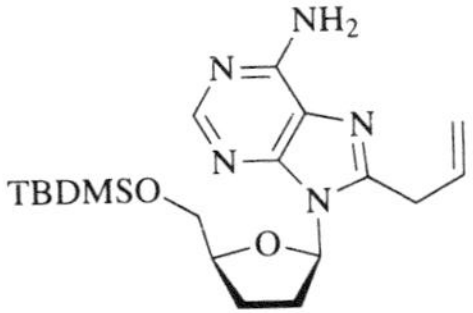

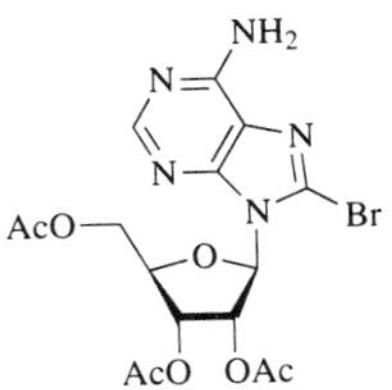

1. Pd(OAc)₂, AsPh₃, NEt₃, NMP, 80°, 12 h
2. NH₄OH, CH₃OH, dioxane
R = 2-thienyl (49)
R = Ph (33) — 732

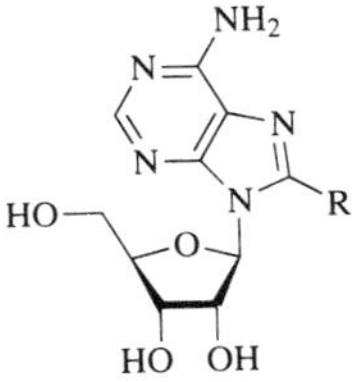

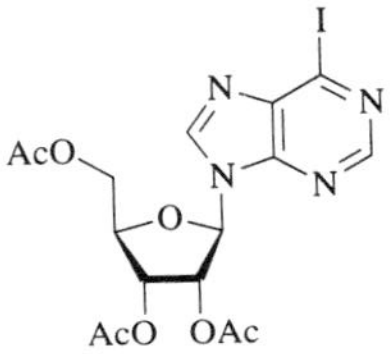

Pd(PPh₃)₄ (10%), NMP
85°, 20 h
80°, 1 h
(53)
(88) — 722

Bu₃Sn

Bu₃Sn

Bu₃Sn-thienyl-2
Ph₄Sn

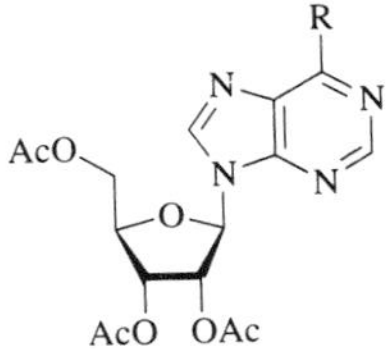

R₄Sn

R
———
Me
CH₂=CH

TABLE XII. DIRECT CROSS-COUPLING OF MISCELLANEOUS HETEROCYCLIC ELECTROPHILES (*Continued*)

Substrate	Stannane	Conditions	Product(s) and Yield(s) (%)	Refs.
(adenine nucleoside with 6-NH$_2$, 2-I, AcO, AcO, OAc sugar) C$_{17}$	R$_4$Sn	Pd(PPh$_3$)$_4$ (10%), NMP	(adenine nucleoside product, 6-NH$_2$, 2-R)	722
			R	
		80°, 2 h	Me (80)	
		80°, 4 h	CH$_2$=CH (95)	
		140°, 5 h	Et (60)	
		110°, 2 h	CH$_2$=CHCH$_2$ (45)	
(quinolone, F, EtO$_2$C, N-2,4-difluorophenyl)	Bu$_3$Sn—(tetrahydropyridine)NAc	Pd(PPh$_3$)$_2$Cl$_2$ (2%), BHT, DMF, reflux, 7 h	(quinolone product with NAc tetrahydropyridine) (83)	729
	Bu$_3$Sn—(cyclopentene)NHBoc	Pd(PPh$_3$)$_4$ (5%), BHT, dioxane, reflux, 55 h or Pd(PPh$_3$)$_2$Cl$_2$, DMF	(quinolone product with NHBoc cyclopentene) (42)	97, 729
(Ph, Me imidazole, 2-Br, N-Bn)	Me$_3$SnPh	Pd(PPh$_3$)$_2$Cl$_2$ (1.7%), PhMe, reflux	(Ph imidazole product, 2-Ph, N-Bn) (60)	111

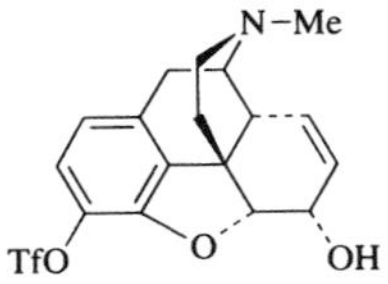

733

734

733

198

(88)

(50)

(77)

R	
2-furyl	(69)
2-thienyl	(41)
Ph	(73)
C_6H_4CN-p	(70)
C_6H_4CHO-p	(73)
$C_6H_4CH_2OH$-p	(67)
C_6H_4OMe-p	(64)
C_6H_4OMe-o	(82)
C_6H_4Ac-p	(77)

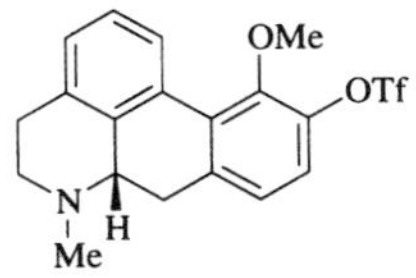

Pd(PPh$_3$)$_4$ (5%), LiCl, DMF, 100°, 18 h

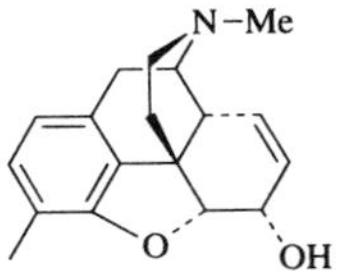

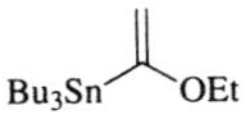

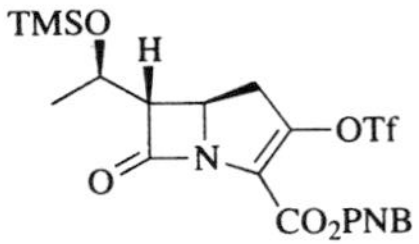

1. Pd(PPh$_3$)$_2$Cl$_2$ (12%), PPh$_3$ (60%), LiCl, BHT, DMF, 120°, 3 h
2. HCl (1 M), THF

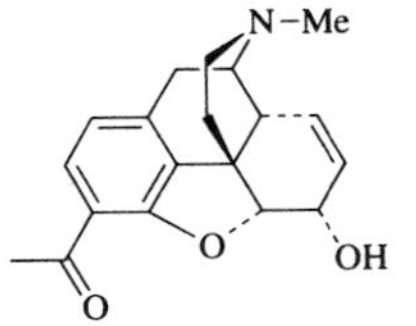

Pd(PPh$_3$)$_4$ (3%), LiCl, DMF, 100°, 16 h

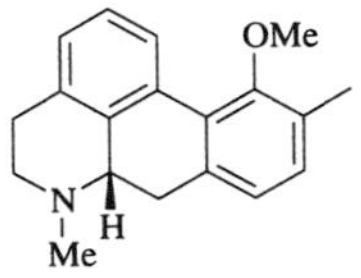

Pd$_2$(dba)$_3$ (2%), P(2,4,6-trimethoxyphenyl)$_3$ (8%), ZnCl$_2$, NMP, rt

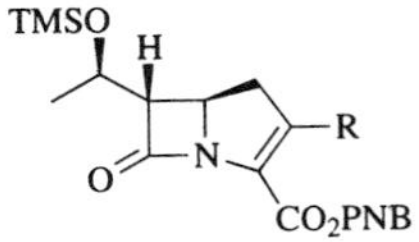

Me$_4$Sn

Bu$_3$Sn—OEt

Me$_4$Sn

Me$_3$SnR

C$_{18}$

385

TABLE XII. DIRECT CROSS-COUPLING OF MISCELLANEOUS HETEROCYCLIC ELECTROPHILES (*Continued*)

Substrate	Stannane	Conditions	Product(s) and Yield(s) (%)	Refs.
	Bu₃SnR	Pd(PPh₃)₄ (5%), PhMe, 100°, 24 h	R / CN (70) / SEt (71) / SPh (89)	319
	R₄Sn	Pd(PPh₃)₂Cl₂ (10%), dioxane, 110°	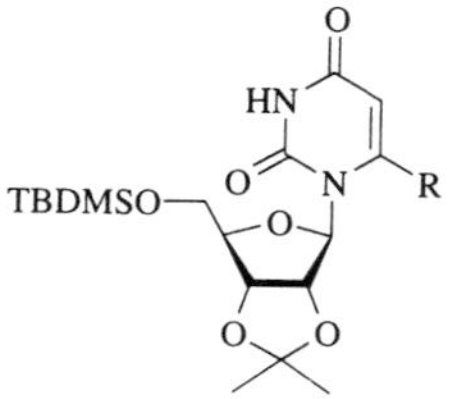 R = Me, 3 h, (>95) / R = Ph, 12 h, (95)	132
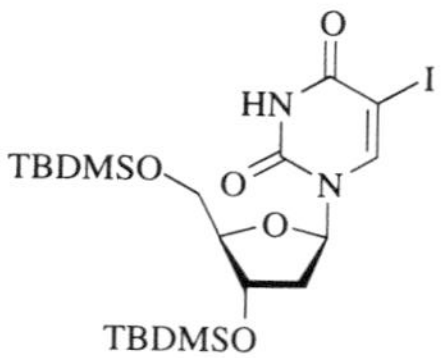	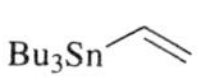 Bu₃Sn⌇	Pd₂(dba)₃ (1%), P(2-furyl)₃ (4%), NMP, rt, 14 h	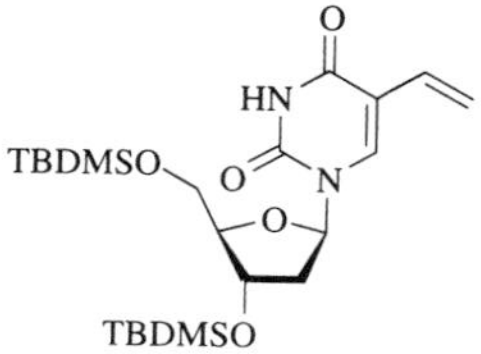 (72)	128
	Me₃SnR	Pd(PPh₃)₄ (10%), dioxane, 75°, 24 h	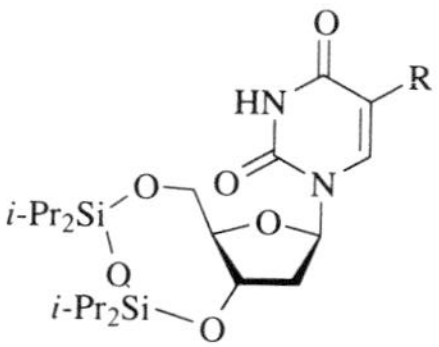	122

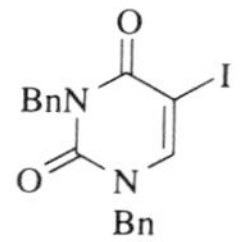

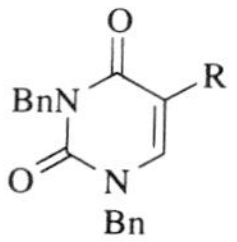

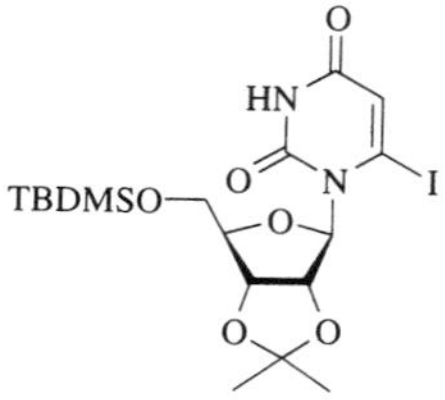

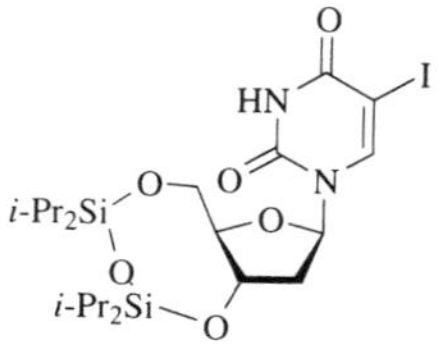

C₂₁

386

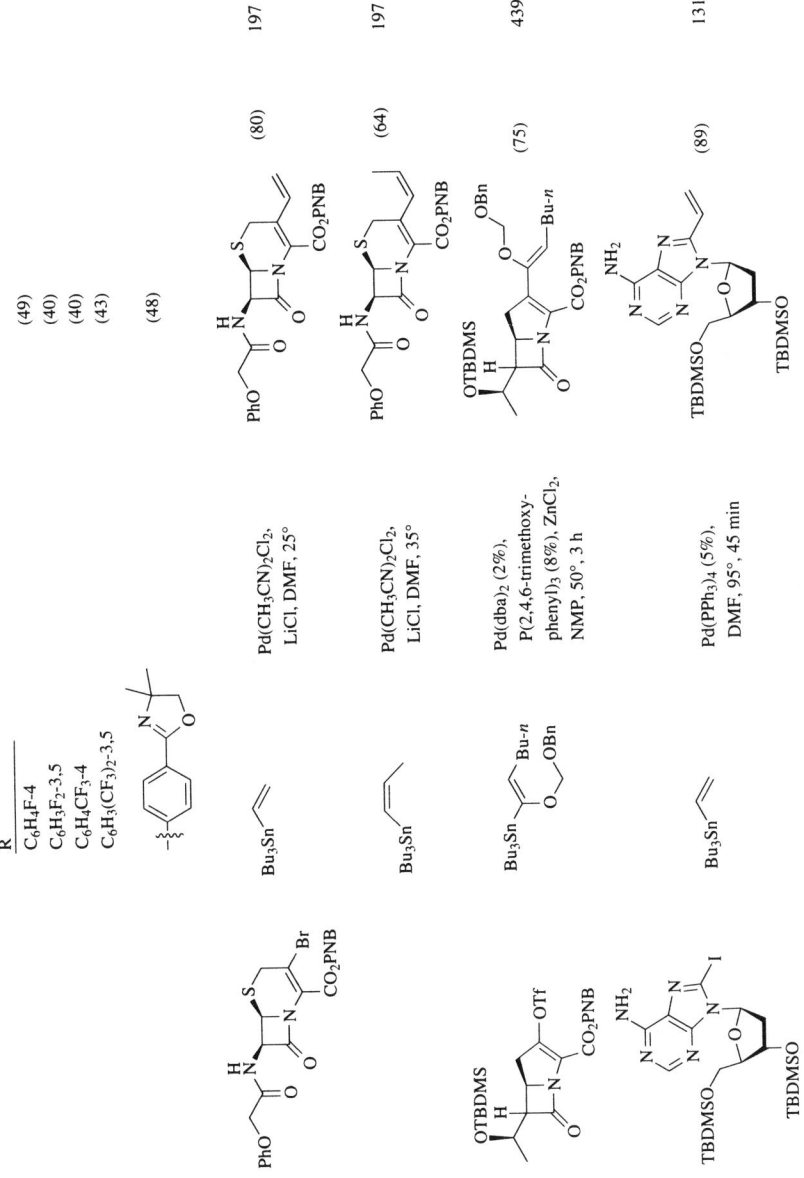

TABLE XII. DIRECT CROSS-COUPLING OF MISCELLANEOUS HETEROCYCLIC ELECTROPHILES (*Continued*)

Substrate	Stannane	Conditions	Product(s) and Yield(s) (%)	Refs.
	Bu₃Sn⌁	Pd(PPh₃)₄ (5%), HMPA, 145°, 45 min	(89)	131
	Bu₃SnCN	Pd(PPh₃)₄, DMF, reflux	(91)	723
	Bu₃Sn⌁	Pd(CH₃CN)₂Cl₂ (5%), DMF, 90°, 45 min	(69)	723
	Bu₃SnCN	1. Pd(PPh₃)₄, DMF, reflux 2. Bu₄NF	(87)	723

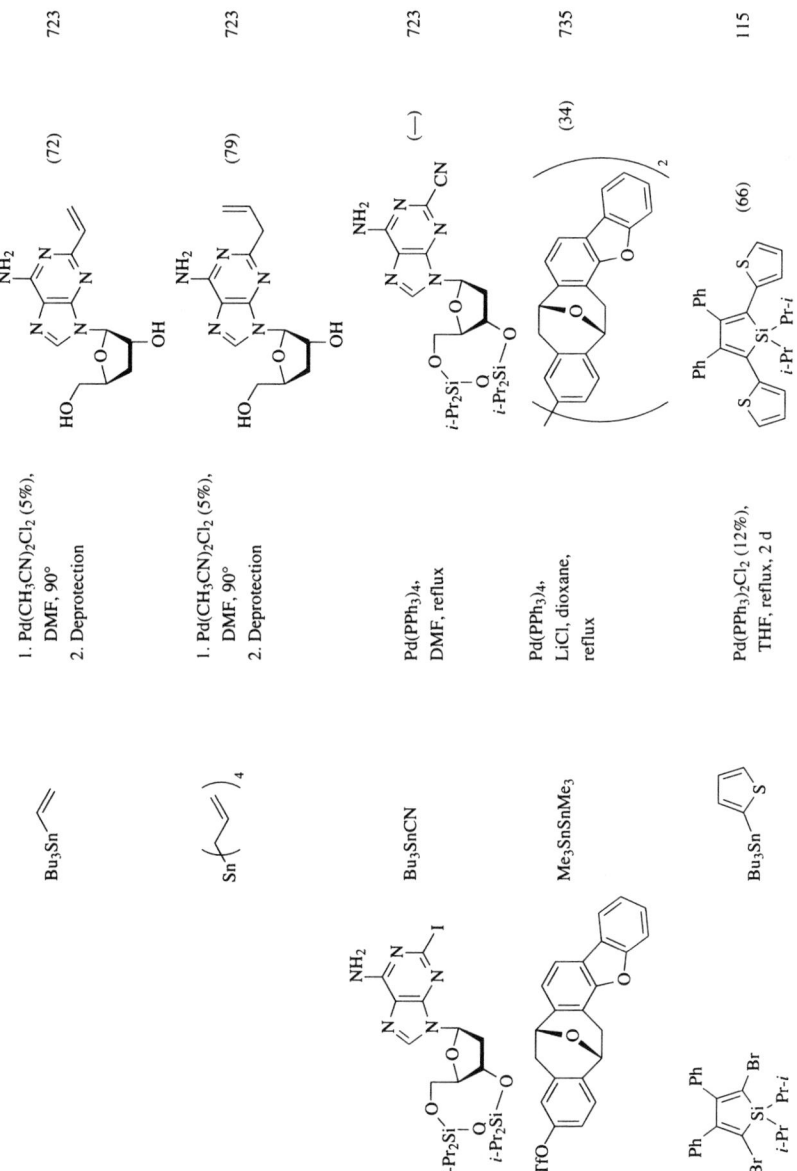

Substrate	Stannane	Conditions	Product(s) and Yield(s) (%)	Refs.
C₂₃ (cephem-type triflate, CO₂PNB ester)	Me₃SnC≡CPh	Pd(PPh₃)₂Cl₂ (5%), THF, reflux, 12 h	(100)	115
	Bu₃Sn–OTBDMS	Pd(CH₃CN)₂Cl₂, LiCl, DMF, 65°	(OTBDMS product) (25)	197
	Bu₃Sn–OBz	Pd(CH₃CN)₂Cl₂, LiCl, DMF, 60°	(OBz product) (28) + (17)	197
	Bu₃Sn (OEt vinyl)	Pd(CH₃CN)₂Cl₂, LiCl, DMF, 25°	(81)	197
(thia-cephem triflate, CO₂PMB ester)	Bu₃Sn (vinyl)	Pd(OAc)₂ (10%), NMP or CH₂Cl₂, rt, 3 min	(55)	451
	Bu₃Sn (propenyl)	Pd(OAc)₂ (10%), NMP or CH₂Cl₂, rt, 20–180 min	**I** (>95)	451

390

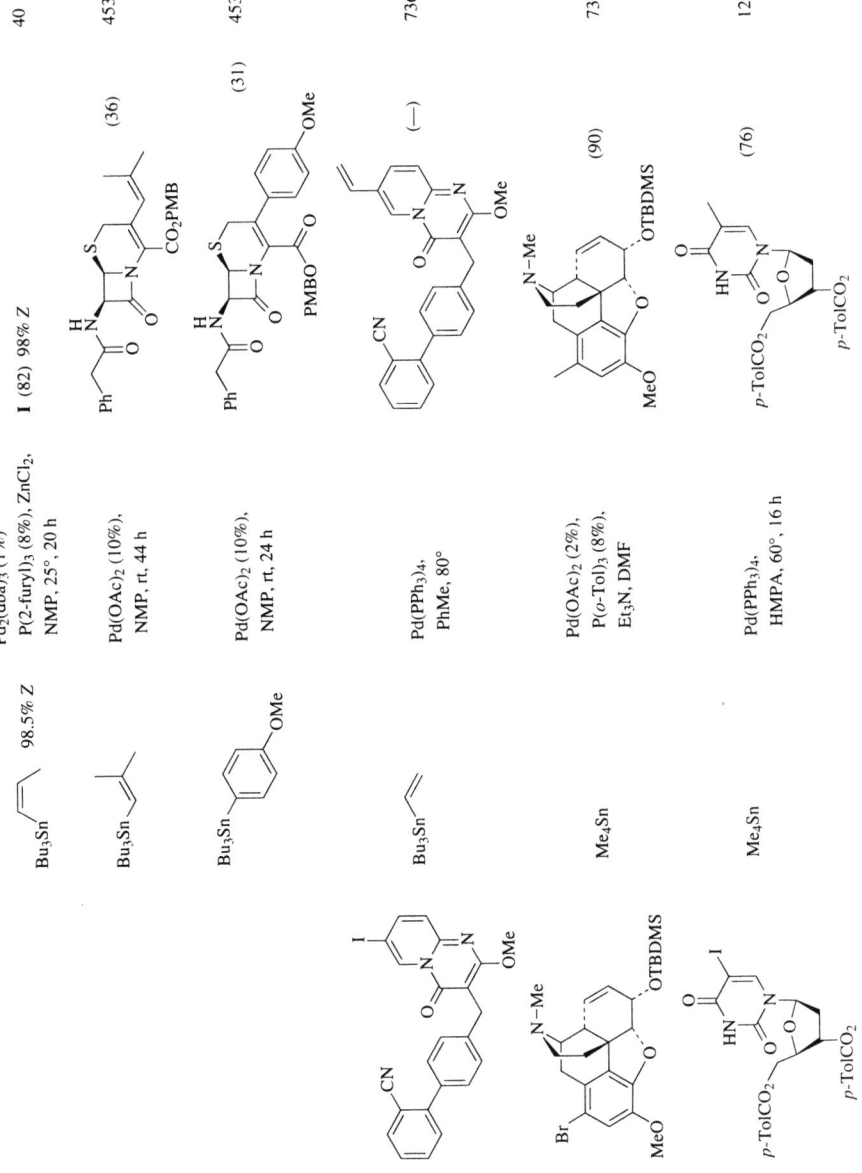

				40
				453
				453
				736
				737
				125

Bu₃Sn — 98.5% Z   Pd₂(dba)₃ (1%), P(2-furyl)₃ (8%), ZnCl₂, NMP, 25°, 20 h   I (82) 98% Z

Bu₃Sn —   Pd(OAc)₂ (10%), NMP, rt, 44 h   (36)

Bu₃Sn—OMe   Pd(OAc)₂ (10%), NMP, rt, 24 h   (31)

Bu₃Sn—   Pd(PPh₃)₄, PhMe, 80°   (—)

Me₄Sn   Pd(OAc)₂ (2%), P(o-Tol)₃ (8%), Et₃N, DMF   (90)

Me₄Sn   Pd(PPh₃)₄, HMPA, 60°, 16 h   (76)

C₂₄

C₂₅

391

TABLE XII. DIRECT CROSS-COUPLING OF MISCELLANEOUS HETEROCYCLIC ELECTROPHILES (*Continued*)

Substrate	Stannane	Conditions	Product(s) and Yield(s) (%)	Refs.
	$(CH_2=CH)_4Sn$	Pd(PPh$_3$)$_4$, HMPA, 60°, 16 h	**I** (80)	125
	Bu$_3$Sn	Pd$_2$(dba)$_3$ (1%), P(2-furyl)$_3$ (8%), NMP, rt, 72 h	**I** (76)	128
	Bu$_3$Sn	Pd(PPh$_3$)$_2$Cl$_2$ (10%), CH$_3$CN, 20°, 6 h	**I** (86)	121
	Bu$_3$Sn (thiazole)	1. Pd(PPh$_3$)$_2$Cl$_2$ (10%), dioxane, 90°, 20 h 2. K$_2$CO$_3$, CH$_3$OH	(81)	718
	Me$_3$Sn (furan)	Pd(PPh$_3$)$_4$, PhMe, reflux, 7 h	(95)	717
	Me$_3$Sn (thiophene)	Pd(PPh$_3$)$_2$Cl$_2$, THF, reflux	(87)	717

392

Reactant	Conditions	Product (%)	Refs.
Bu₃Sn–CH=CH–CO₂Et	Pd(PPh₃)₂Cl₂ (10%), CH₃CN, 50°, 20 h	5-(CH=CH–CO₂Et)uracil (57)	121
Bu₃Sn–CH=CH–TMS	Pd(PPh₃)₂Cl₂ (10%), CH₃CN, 60°, 16 h	5-(CH=CH–TMS)uracil (82)	121
Ph₄Sn	Pd(PPh₃)₄, HMPA, 60°, 3 d	5-Ph-uracil (35)	125
Bu₃Sn–CH=CH–Ph	Pd(PPh₃)₂Cl₂ (10%), CH₃CN, 50°, 16 h	5-(CH=CH–Ph)uracil (81)	121
Bu₃Sn–CH=CH–CH₂OTHP	Pd(PPh₃)₂Cl₂ (10%), CH₃CN, 60°, 16 h	5-(CH=CH–CH₂OTHP)uracil (82)	121
Me₃SnSnMe₃	Pd(PPh₃)₄ (10%), EtOAc, reflux, 24 h	5-(SnMe₃)uracil (72)	670

TABLE XII. DIRECT CROSS-COUPLING OF MISCELLANEOUS HETEROCYCLIC ELECTROPHILES (Continued)

Substrate	Stannane	Conditions	Product(s) and Yield(s) (%)	Refs.
(5-iodo uracil deoxyribonucleoside, p-TolCO$_2$, O_2CTol-p)	Bu_3Sn—(2-thienyl/furyl X)	Pd(PPh$_3$)$_2$Cl$_2$, THF	$X = O$ (—) $X = S$ (—)	717
(BocHN cephem, OTf, CO$_2$CHPh$_2$)	Me$_4$Sn	Pd$_2$(dba)$_3$ (1%), P(2-furyl)$_3$ (4%), NMP, 25°, 16 h	(85)	40
	Bu_3Sn—CH=CH— (98.5% Z)	Pd$_2$(dba)$_3$ (1%), P(2-furyl)$_3$ (4%), NMP, ZnCl$_2$, 25°, 16 h	(90) 98% Z	40
	Bu_3Sn—CH$_2$CH=CH$_2$	Pd$_2$(dba)$_3$ (1%), P(2-furyl)$_3$ (4%), NMP, 50°, 40 h	(48) + (16) + (12)	40
	Bu_4Sn	Pd$_2$(dba)$_3$ (1%), P(2-furyl)$_3$ (4%), NMP, 50°, 7 d	(16)	40

394

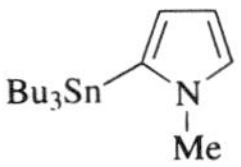

40

(89)

Me

BocHN

Pd$_2$(dba)$_3$ (1%),
P(2-furyl)$_3$ (4%),
NMP, 25°, 1 h

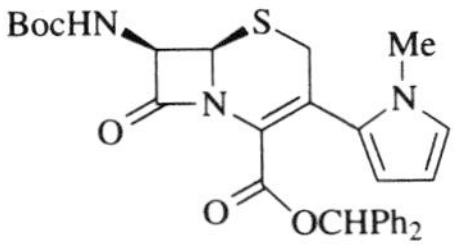

Bu$_3$Sn

N
Me

40

(17) +

Bu-t

BocHN

OCHPh$_2$

Pd$_2$(dba)$_3$ (1%),
P(2-furyl)$_3$ (4%),
NMP, 25°, 16 h

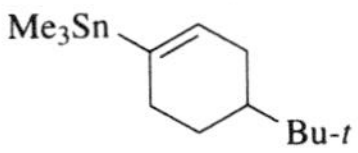

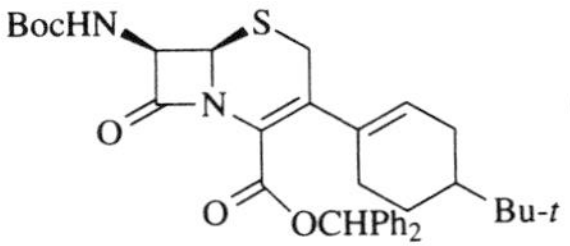

Me$_3$Sn

Bu-t

(74)

CO$_2$CHPh$_2$

BocHN

(73) 40

CO$_2$CHPh$_2$

BocHN

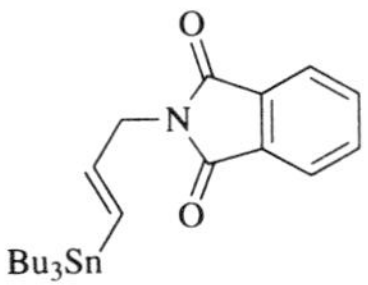

Pd$_2$(dba)$_3$ (1%),
P(2-furyl)$_3$ (4%), ZnCl$_2$,
NMP, 25°, 20 h

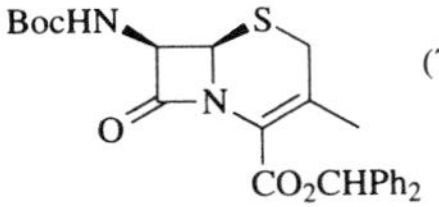

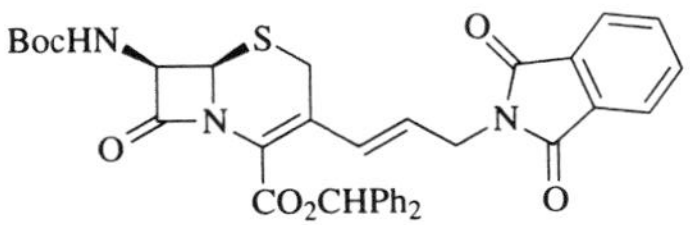

Bu$_3$Sn

CO$_2$Me

738

(80)

CO$_2$Me

N
Bu-n

CN

Pd(0), CuI, DMF

I

N
Bu-n

O

CN

C$_{26}$

739

(67)

CHO

H

N

N

i-PrCOHN

N

O

i-PrCO$_2$

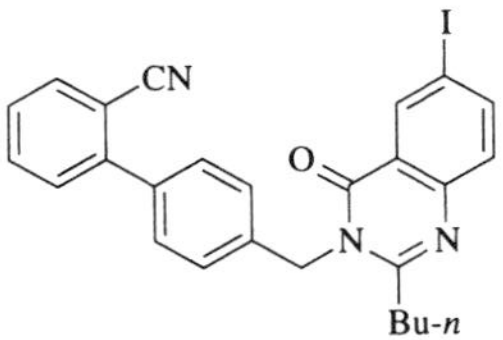

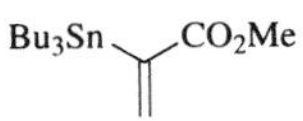

i-PrCO$_2$ O$_2$CPr-i

Pd(PPh$_3$)$_4$

Bu$_3$Sn

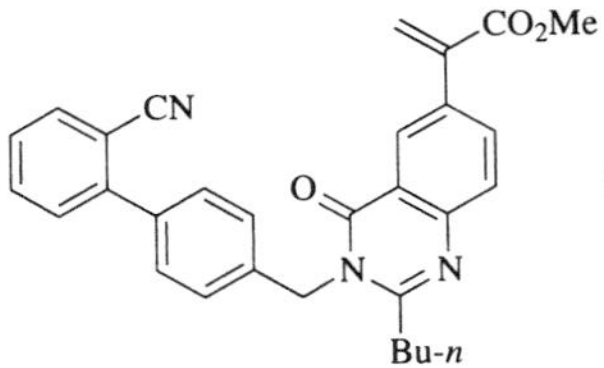

CHO

H

N

N

i-PrCOHN

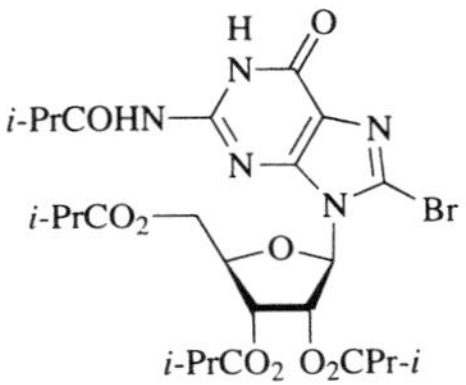

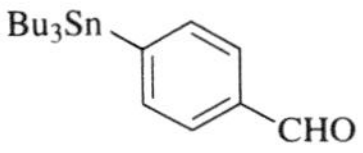

N

Br

O

i-PrCO$_2$

i-PrCO$_2$ O$_2$CPr-i

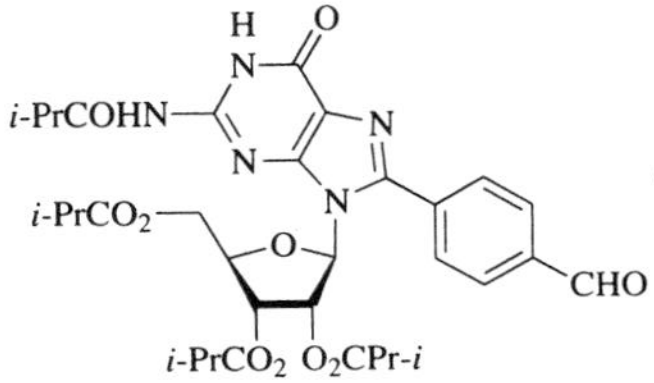

Substrate	Stannane	Conditions	Product(s) and Yield(s) (%)	Refs.
C$_{27}$				
		Pd(PPh$_3$)$_4$ (20%), PhMe, reflux, 24 h	(79)	127
		Pd(PPh$_3$)$_4$, DMF, 100°, 15 h	(88)	740
C$_{28}$				
		1. Pd(CH$_3$CN)$_2$Cl$_2$ (5%), P(o-Tol)$_3$, PhMe 2. Et$_4$NF, CH$_3$CN	(63)	120
		Pd(CH$_3$CN)$_2$Cl$_2$ (6%), P(o-Tol)$_3$, PhMe, reflux, 3 h	(90)	741

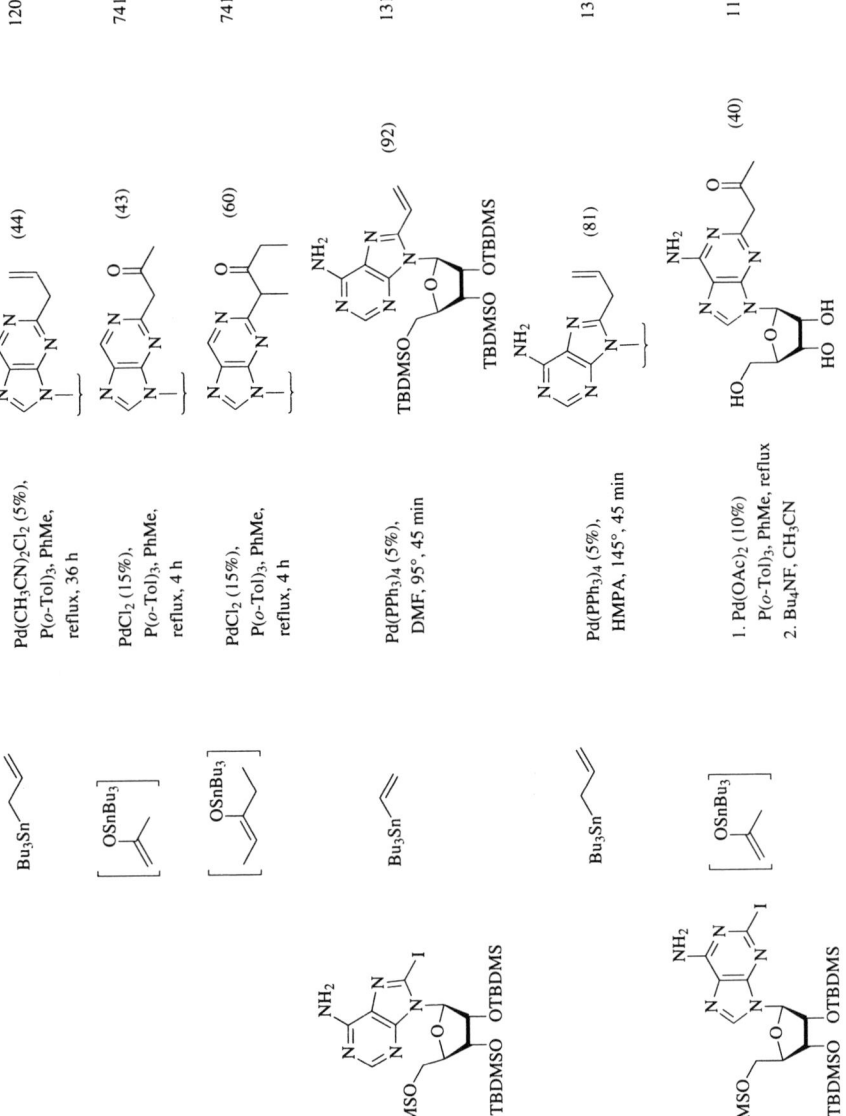

Substrate	Stannane	Conditions	Product(s) and Yield(s) (%)	Refs.
(cephem with OTf, CO₂CHPh₂, BnC(O)NH)	Bu₃SnR R: H, CH=CH₂, CF=CF₂, C≡CMe, CH=CMe₂, C(OEt)=CH₂, C₆H₄OMe-*p*	Pd₂(dba)₃ (1%), P(2-furyl)₃ (4%), ZnCl₂, THF, 65°, 1 h NMP, 25°, 1 h THF, 25°, 2 h NMP, 25°, 16 h NMP, 25°, 19 h NMP, 25°, 19 h NMP, 50°, 6 h	(image: cephem with R, CO₂CHPh₂, Bn–C(O)NH) (68) (79) (55) (50) (66) (52) (57)	40, 742
	(dihydropyran stannane) Bu₃Sn	Pd₂(dba)₃ (1%), P(2-furyl)₃ (4%), ZnCl₂, NMP, rt, 17 h	(61)	743
(cephem with OSO₂F, CO₂CHPh₂, PhO–CH₂C(O)NH)	Bu₃SnR R: CH=CH₂, (Z)-CH=CHMe, CH=CMe₂	Pd(OAc)₂ (10%), NMP, rt or CH₂Cl₂ 5 min 16 h	(image: cephem with R, CO₂CHPh₂, PhO–CH₂C(O)NH) (85) (>98) (47)	744
	Bu₃Sn–C(OEt)=CH₂	1. Pd(OAc)₂ (10%), P(2-furyl)₃, NMP 2. H₃O⁺	(58)	744

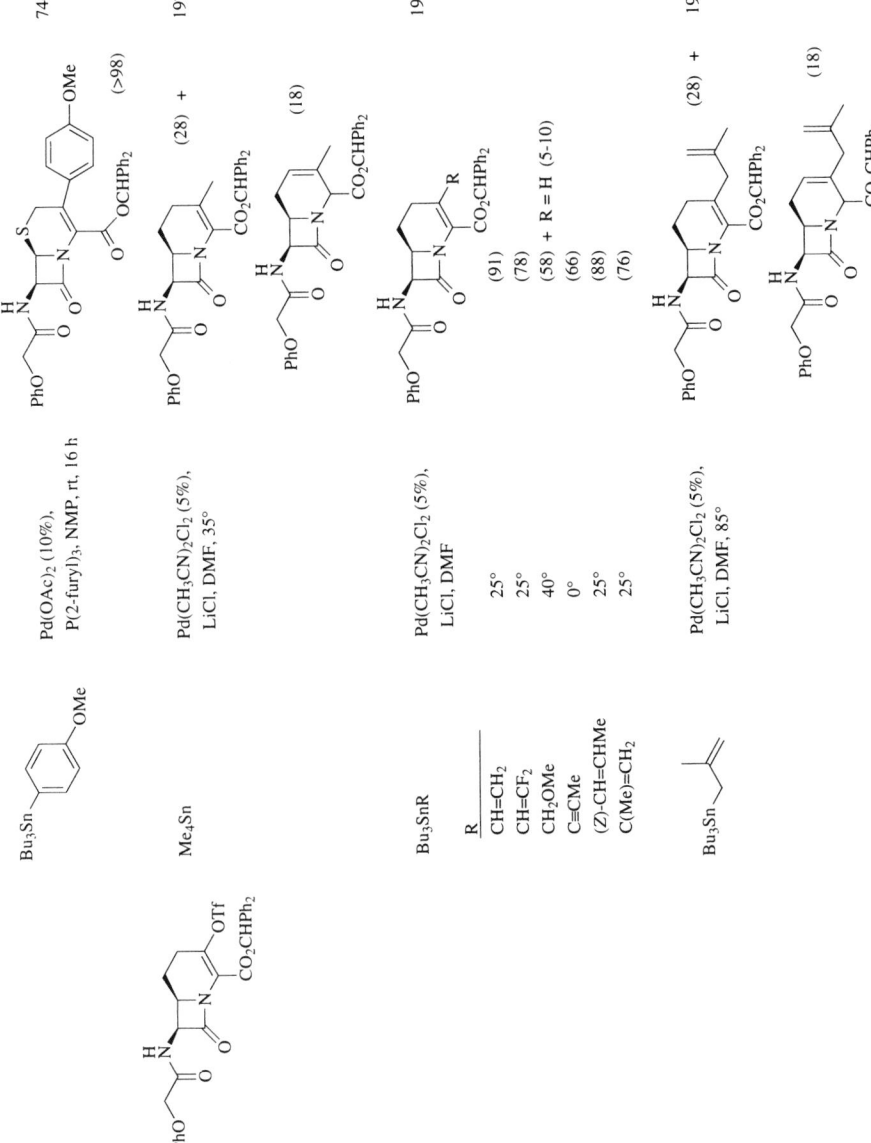

744

(>98)

197

(28) +

(18)

197

R		
CH=CH₂		(91)
CH=CF₂		(78)
CH₂OMe		(58) + R = H (5-10)
C≡CMe		(66)
(Z)-CH=CHMe		(88)
C(Me)=CH₂		(76)

(28) +

197

(18)

Pd(OAc)₂ (10%),
P(2-furyl)₃, NMP, rt, 16 h

Pd(CH₃CN)₂Cl₂ (5%),
LiCl, DMF, 35°

Pd(CH₃CN)₂Cl₂ (5%),
LiCl, DMF

25°
25°
40°
0°
25°
25°

Pd(CH₃CN)₂Cl₂ (5%),
LiCl, DMF, 85°

Bu₃Sn

Me₄Sn

Bu₃SnR

Bu₃Sn

C₂₉

TABLE XII. DIRECT CROSS-COUPLING OF MISCELLANEOUS HETEROCYCLIC ELECTROPHILES (Continued)

Substrate	Stannane	Conditions	Product(s) and Yield(s) (%)	Refs.
	Bu₃Sn⁀OBz	Pd(CH₃CN)₂Cl₂ (5%), LiCl, DMF, 60°	(39) + (5-10)	197
	Bu₃Sn⁀	Pd(CH₃CN)₂Cl₂ (10%), PhMe, reflux	(>90)	117, 118, 120
	Bu₃Sn⁀CN	Pd(OAc)₂ (10%), P(o-Tol)₃, PhMe, reflux	(55)	117, 118
	[OSnBu₃]	Pd(OAc)₂ (10%), P(o-Tol)₃, PhMe, reflux, 6 h	(73)	117, 118
	[OSnBu₃]	Pd(OAc)₂ (10%), P(o-Tol)₃ (20%), PhMe, reflux, 9 h	(77)	117

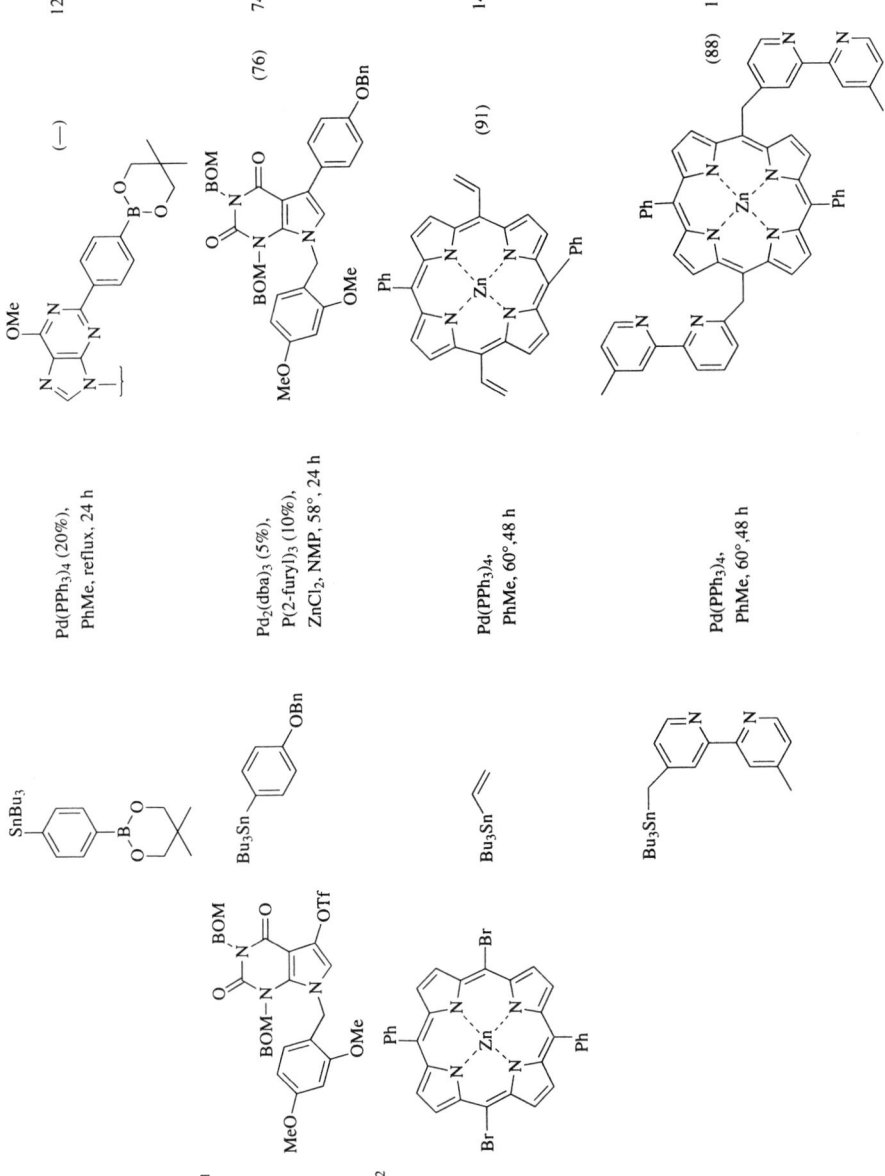

127   (—)

740   (76)

143   (91)

143   (88)

Pd(PPh₃)₄ (20%),
PhMe, reflux, 24 h

Pd₂(dba)₃ (5%),
P(2-furyl)₃ (10%),
ZnCl₂, NMP, 58°, 24 h

Pd(PPh₃)₄,
PhMe, 60°, 48 h

Pd(PPh₃)₄,
PhMe, 60°, 48 h

C₃₁

C₃₂

Substrate	Stannane	Conditions	Product(s) and Yield(s) (%)	Refs.
(porphyrin with Br, Br, MeO$_2$C, MeO$_2$C)	Bu$_3$Sn (vinyl)	Pd(PPh$_3$)$_4$ (60%), BHT, PhMe, reflux, 2 h	(85)	745
C$_{33}$ (nucleoside with I, *p*-TolCO, *p*-TolCO$_2$, *p*-TolCO$_2$)	Me$_4$Sn	1. Pd(PPh$_3$)$_4$, HMPA, 60° 2. NH$_3$	(76)	125
	(CH$_2$=CH)$_4$Sn	1. Pd(PPh$_3$)$_4$, HMPA, 60° 2. NH$_3$	(80)	125
	(CF$_2$=CF)$_4$Sn	1. Pd(OAc)$_2$ (10%), PPh$_3$, NEt$_3$, NMP, rt, 2 h 2. NH$_3$, 48h	(21)	746
	Ph$_4$Sn	1. Pd(PPh$_3$)$_4$, HMPA, 60° 2. NH$_3$	(35)	125

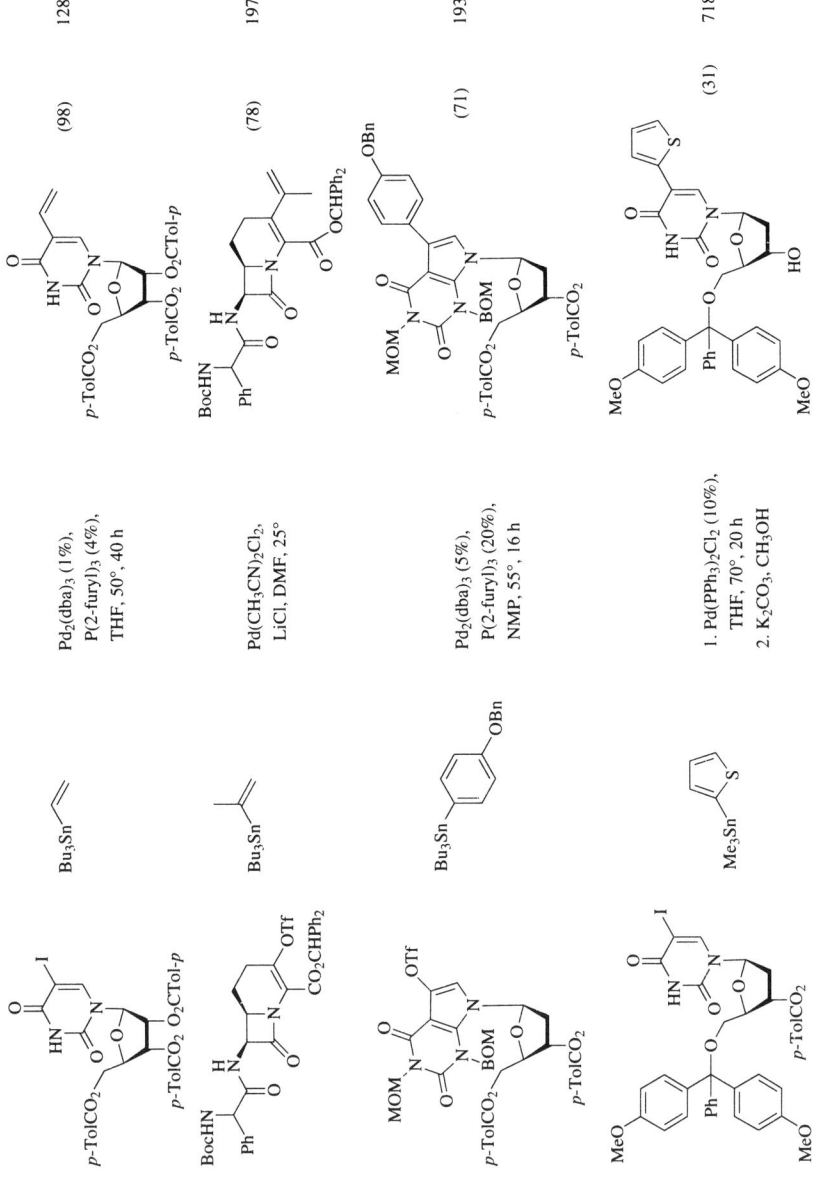

128 (98)

197 (78)

193 (71)

718 (31)

TABLE XII. DIRECT CROSS-COUPLING OF MISCELLANEOUS HETEROCYCLIC ELECTROPHILES (*Continued*)

Substrate	Stannane	Conditions	Product(s) and Yield(s) (%)	Refs.
C$_{45}$				
	Bu$_3$Sn—(furanyl)—X	1. Pd (II) 2. H$_3$O$^+$	X = O (—) X = S (—)	747
	Me$_3$SnR	Pd(PPh$_3$)$_2$Cl$_2$, DMF, 90°		748

$$\frac{R}{\text{2-pyridyl} \quad (70)}{\text{C}_6\text{H}_4\text{Cl-}p \quad (—)}$$

R	
2-pyridyl	(70)
C$_6$H$_4$Cl-p	(—)

404

C_{46}	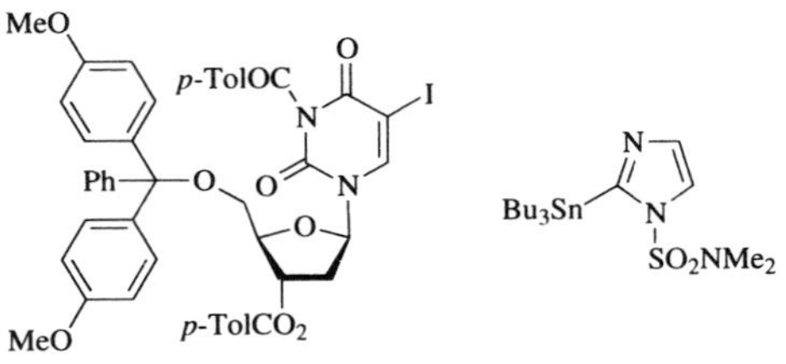	1. $Pd(PPh_3)_2Cl_2$ (10%), Ag_2O, DMF, 100°, 48 h 2. NH_4OH, dioxane, 60°	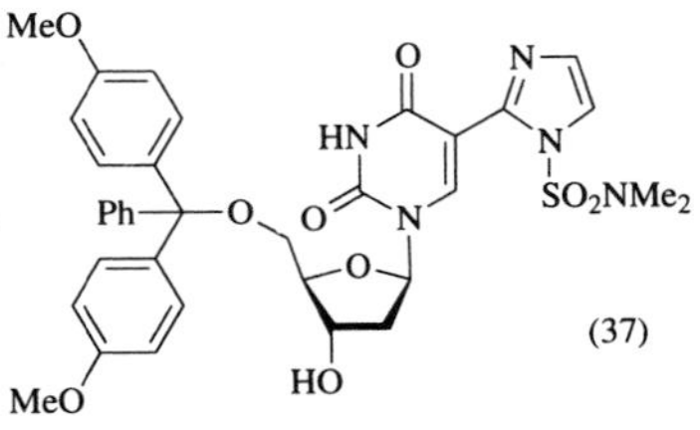 (37)	718

TABLE XIII. DIRECT CROSS-COUPLING OF ACYL CHLORIDES: ALKYL SYSTEMS

Substrate	Stannane	Conditions	Product(s) and Yield(s) (%)	Refs.
C₂	Me₄Sn	Pd(PPh₃)₄ (1%), C₆H₆, 140°, 5 h	(53)	1
	Bu₃Sn	Rh(PPh₃)₃Cl (2%), CH₂Cl₂, 40°, 3 h	(51)	2
	Me₃Sn (OMe)	BnPd(PPh₃)₂Cl (1%), C₆H₆, reflux, 1 h	(44) OMe	749
	Me₃Sn (O)	BnPd(PPh₃)₂Cl (1%), C₆H₆, rt, 0.5 h	(70)	749
	Bu₃Sn (O)	Pd(PPh₃)₂Cl₂ (3%), THF, rt, 12 h	(43)	288
	Bu₃Sn (O O)	BnPd(PPh₃)₂Cl (0.08%), C₆H₆, reflux, 3 h	(82)	271
	Me₃Sn (N N SMe)	Pd(PPh₃)₂Cl₂ (7%), THF, reflux, 2 h	(96) SMe	459
	R₃Sn (N N SO₂Me)	Pd(PPh₃)₂Cl₂ (14%), THF, reflux, 6 h	R = Me (71) R = Bu (52) SO₂Me	459

406

Organotin reagent	Conditions	Product (yield %)	Refs.
Me₃Sn–C(=CH₂)TMS	BnPd(PPh₃)₂Cl (1%), C₆H₆, reflux, 24 h	(42)	749
Me₃SnPh	[(η³-C₃H₅)PdCl]₂ (1%), HMPA, 20°, 24 h	**I** (70)	750, 751, 415
Ph₄Sn	BnPd(PPh₃)₂Cl (0.05%), HMPA, 65°	**I** (76)	147, 1
Bu₃Sn– (sulfone-furan)	Pd(PPh₃)₄ (1%), HMPA, 0°, 2 h	(38)	752
Me₃Sn–CH₂C(=CH₂)TMS	BnPd(PPh₃)₂Cl (2%), CHCl₃, 65°, 12 h	(81)	537
Bu₃Sn–CH₂Ph	Rh(PPh₃)₃Cl (1%), C₆H₆, 80°, 12 h	(69)	2
Bu₃SnC≡C–CH(OEt)₂	Pd(PPh₃)₂Cl₂ (1.8%), Cl(CH₂)₂Cl, 84°, 2 h	(31)	149
Bu₃Sn– (furanoyl)	Pd(PPh₃)₄ (5%), dioxane, 100°, 30 h	(100)	148, 753
Bu₃Sn– (furanyl-vinyl)	Pd(PPh₃)₄ (5%), dioxane, 60°, 3 h	(93)	148, 753

TABLE XIII. DIRECT CROSS-COUPLING OF ACYL CHLORIDES: ALKYL SYSTEMS (*Continued*)

Substrate	Stannane	Conditions	Product(s) and Yield(s) (%)	Refs.
(Bu₃Sn-substituted methylenecyclobutanone/dioxolane spiro)		BnPd(PPh₃)₂Cl (1%), CO (15 psi), C₆H₆, 80°	(68)	268
	Me₃Sn—CH₂C(=CH₂)...Fe(CO)₃	BnPd(PPh₃)₂Cl (2.6%), Cl(CH₂)₂Cl, HMPA, 50°, 24 h	(66)	754
	Bu₃SnC≡CPh	Pd(PPh₃)₂Cl₂ (1.8%), Cl(CH₂)₂Cl, 84°, 2 h	CH₃C(O)C≡CPh (55)	149
	Me₃Sn—(indenyl)	Rh(PPh₃)₃Cl (2%), CH₂Cl₂, 60°, 10 h	(30)	755
	Me₃Sn—(quinolin-3-yl)	Pd(PPh₃)₂Cl₂ (5%), C₆H₆, reflux, 8 h	(70)	285
	Me₃Sn—(quinolin-4-yl)	PdCl₂ (5%), C₆H₆, reflux, 4 d	(24)	284
	Bu₃Sn—CH=CH—Ph	Pd(PPh₃)₄ (5%), dioxane, 100°, 30 h	(47)	148, 753
	Bu₃Sn—CH=CH—Ph	Pd(PPh₃)₄ (5%), dioxane, 60°, 3 h	(86)	148

Organostannane	Conditions	Product (yield)	Refs.
Bu₃SnC≡C—OTBDMS	Pd(PPh₃)₂Cl₂ (1.8%), Cl(CH₂)₂Cl, 84°, 2 h	(48)	149
Bu₃Sn—CO₂Me, TMSO, CF₃	Pd(PPh₃)₄ (0.05%), THF, reflux, 8 h	(68)	756
Me₃Sn— (Pr-i)	Rh(PPh₃)₃Cl, CH₂Cl₂, 60°, 48 h	(72)	150, 304
Me₃Sn—	Rh(PPh₃)₃Cl, CH₂Cl₂, 60°, 48 h	(74)	150, 304
Me₃Sn—	Rh(PPh₃)₃Cl, CH₂Cl₂, 60°, 16 h	(62)	150, 304
Bu₃Sn— CO₂Bu-t	BnPd(PPh₃)₂Cl, C₆H₆, rt, 4 h	(79)	543
Me₃Sn— TMS, Ph	BnPd(PPh₃)₂Cl (1%), 80°, 300 h	(62)	256
Bu₃Sn— C₇H₁₅-n, OBz	Pd(PPh₃)₂Cl₂ (4%), CuCN, PhMe, 75°, 16h	(<5)	243
Bu₃Sn—	BnPd(PPh₃)₂Cl (4%), CO (30 psi), C₆H₆, 90°, 9h	(52)	757

TABLE XIII. DIRECT CROSS-COUPLING OF ACYL CHLORIDES: ALKYL SYSTEMS (*Continued*)

Substrate	Stannane	Conditions	Product(s) and Yield(s) (%)	Refs.
		Pd(PPh₃)₂Cl₂	(45)	758
		Pd(CH₃CN)₂Cl₂ (5%), HMPA, 80°	(60)	513
		Pd(CH₃CN)₂Cl₂ (5%), HMPA, 80°	**I** (40)	513
		Pd(CH₃CN)₂Cl₂ (5%), HMPA, 80°	**I** (10)	513
		Pd(CH₃CN)₂Cl₂ (5%), HMPA, 80°	(30)	513
		Pd(PPh₃)₂Cl₂ (5%), THF, 80°, 24 h	(59)	287, 546
	Me₃SnSnMe₃	BnPd(PPh₃)₂Cl or Pd(PPh₃)₄ (5%), THF, reflux, dark, 12 h	(70)	309, 759

410

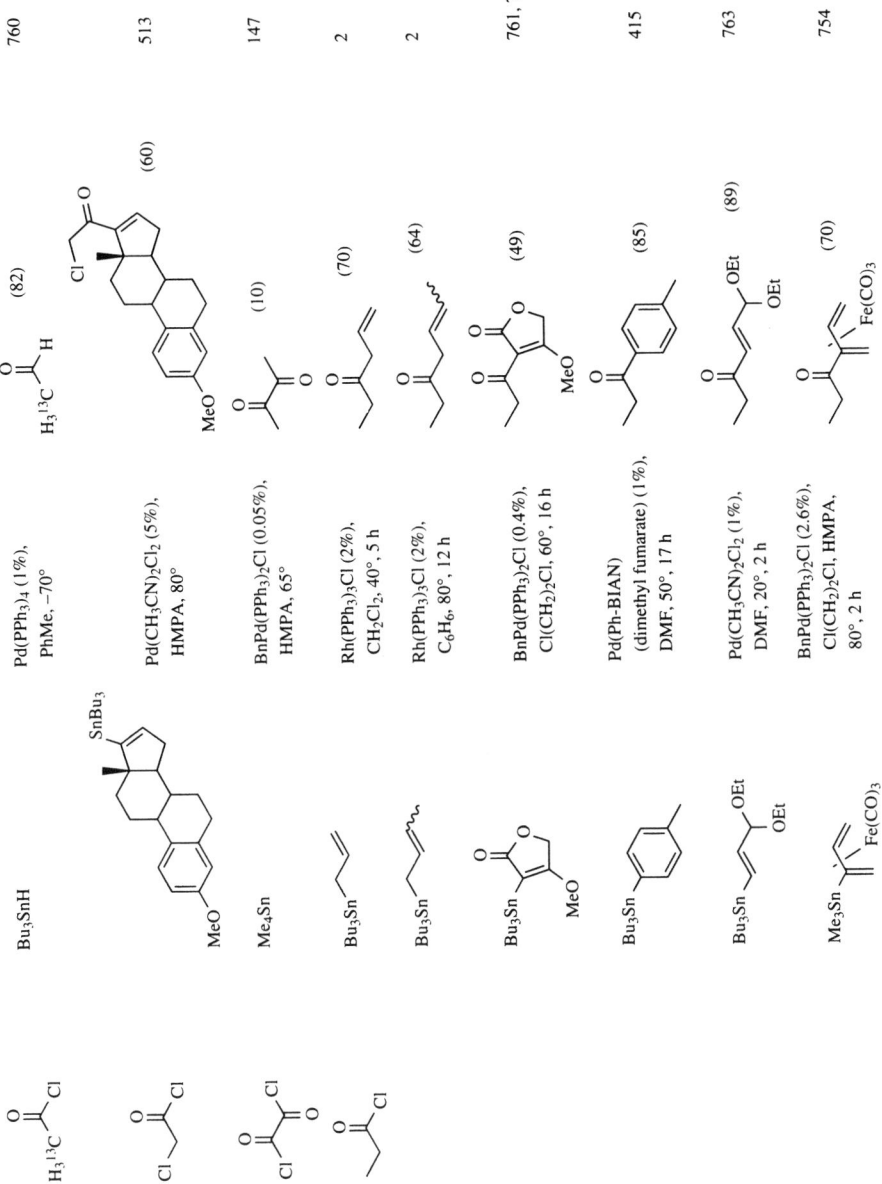

TABLE XIII. DIRECT CROSS-COUPLING OF ACYL CHLORIDES: ALKYL SYSTEMS (*Continued*)

Substrate	Stannane	Conditions	Product(s) and Yield(s) (%)	Refs.
![2-chloropropanoyl chloride substrate] (O=, Cl, Cl)	Me$_3$Sn-quinolin-3-yl	Pd(PPh$_3$)$_2$Cl$_2$ (5%), C$_6$H$_6$, reflux, 8 h	3-propanoylquinoline (76)	284
	Me$_3$Sn-quinolin-4-yl	PdCl$_2$ (5%), C$_6$H$_6$, reflux, 4 d	4-propanoylquinoline (55)	284
	Bu$_3$Sn–CH=CH–CH$_2$OTBDMS	BnPd(PPh$_3$)$_2$Cl, THF, reflux, 9 h	enone–OTBDMS (81)	764
	Bu$_3$Sn–(furan-CH$_2$OTBDMS)	Pd(PPh$_3$)$_2$Cl$_2$ (6.5%), THF, 70°, 20 h	furan–CH$_2$OTBDMS enone (70)	287
	Bu$_3$Sn–C(Ph)=C(CO$_2$Bu-t)	Pd$_2$(dba)$_3$ (0.05%), HMPA, 100°, 2.4 h	Ph, CO$_2$Bu-t enone (30)	247
	Bu$_3$Sn–C(=CH$_2$)–(oxazolidinone N, Ph, Ph)	BnPd(PPh$_3$)$_2$Cl (3.7%), CO (30 psi), C$_6$H$_6$, 90°, 21 h	oxazolidinone product, Ph, Ph (67)	757
	Me$_3$SnSnMe$_3$	BnPd(PPh$_3$)$_2$Cl (5%), THF, reflux, 18 h	acyl–SnMe$_3$ (70)	309, 759
	Bu$_3$Sn–C(Ph)=C(CO$_2$Bu-t)	Pd$_2$(dba)$_3$ (0.05%), CHCl$_3$, 100°, 16 h	Ph, Cl, CO$_2$Bu-t product (22)	247

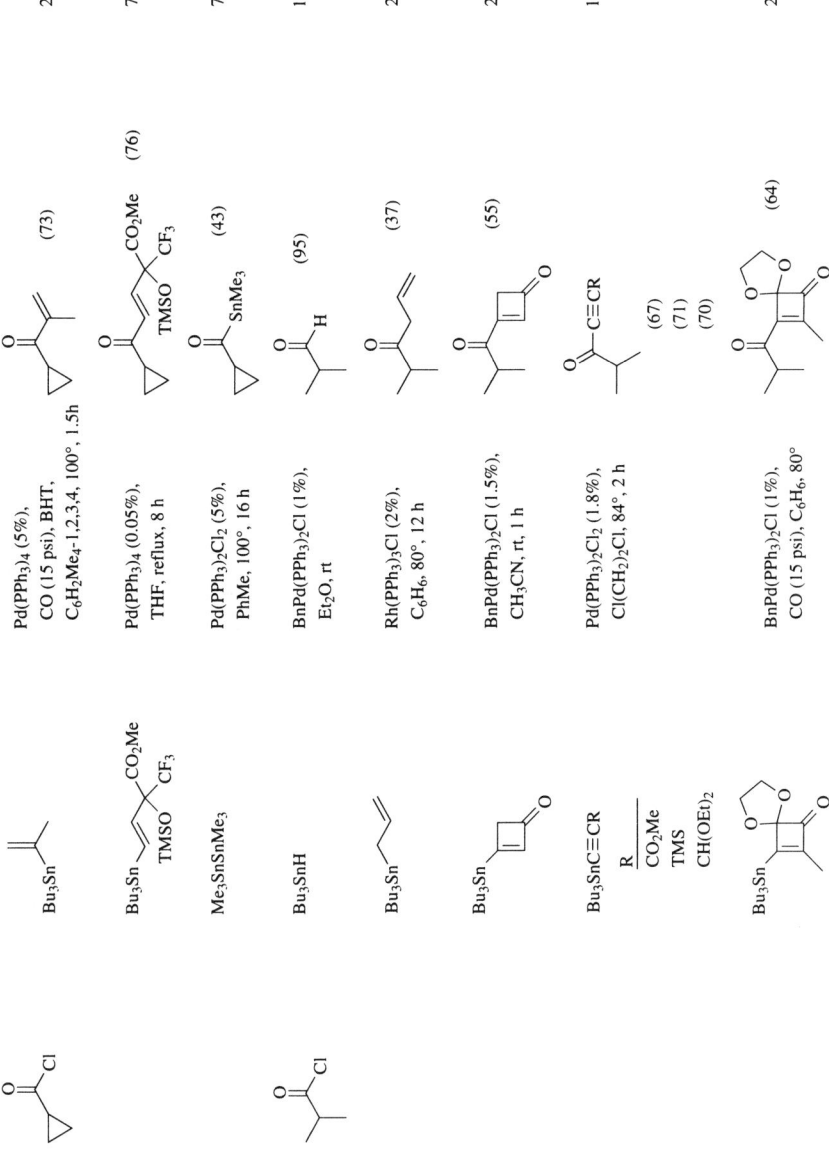

C₄

Pd(PPh₃)₄ (5%), CO (15 psi), BHT, C₆H₂Me₄-1,2,3,4, 100°, 1.5h	(73)		250
Pd(PPh₃)₄ (0.05%), THF, reflux, 8 h	(76)		756
Pd(PPh₃)₂Cl₂ (5%), PhMe, 100°, 16 h	(43)		759
BnPd(PPh₃)₂Cl (1%), Et₂O, rt	(95)		156
Rh(PPh₃)₃Cl (2%), C₆H₆, 80°, 12 h	(37)		2
BnPd(PPh₃)₂Cl (1.5%), CH₃CN, rt, 1 h	(55)		267
Pd(PPh₃)₂Cl₂ (1.8%), Cl(CH₂)₂Cl, 84°, 2 h	R		149
BnPd(PPh₃)₂Cl (1%), CO (15 psi), C₆H₆, 80°	(64)		268

R
CO₂Me   (67)
TMS   (71)
CH(OEt)₂ (70)

TABLE XIII. DIRECT CROSS-COUPLING OF ACYL CHLORIDES: ALKYL SYSTEMS (*Continued*)

Substrate	Stannane	Conditions	Product(s) and Yield(s) (%)	Refs.
	R_3Sn ... $Fe(CO)_3$	BnPd(PPh$_3$)$_2$Cl (2.6%), Cl(CH$_2$)$_2$Cl, HMPA, 80°, 2 h	$Fe(CO)_3$ R = Me, (58) R = Bu, (48)	754
	Bu$_3$Sn ... OEt, OEt	Pd(CH$_3$CN)$_2$Cl$_2$ (1%), DMF, 20°, 2 h	OEt, OEt (83)	763
	Bu$_3$SnC≡CPh	Pd(PPh$_3$)$_2$Cl$_2$ (1.8%), Cl(CH$_2$)$_2$Cl, 84°, 2 h	C≡CPh (77)	149
	Me$_3$Sn– (quinoline)	Pd(PPh$_3$)$_2$Cl$_2$ (5%), C$_6$H$_6$, reflux, 8 h	(82)	284
	Me$_3$Sn– (quinoline)	PdCl$_2$ (5%), C$_6$H$_6$, reflux, 4 d	(28)	284
	Bu$_3$SnC≡C—OTBDMS	Pd(PPh$_3$)$_2$Cl$_2$ (1.8%), Cl(CH$_2$)$_2$Cl, 84°, 2 h	C≡C–OTBDMS (60)	149
	Me–Si(Me)– (furan), Bu$_3$Sn	Pd(CH$_3$CN)$_2$Cl$_2$, THF, reflux, 12 h	(40)	765
EtO–C(O)–C(O)–Cl	OR, Bu$_3$Sn	BnPd(PPh$_3$)$_2$Cl (0.01%), BHT, THF, 10° to rt, 4 h	EtO–...OR R = Me (90) R = Bn (70)	766

414

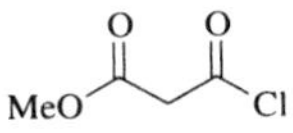

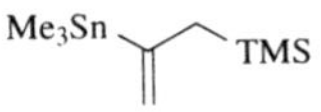

Substrate	Reagent	Conditions	Product (yield)	Ref.
	Me₃Sn∼TMS	BnPd(PPh₃)₂Cl (2%), CHCl₃, 65°, 24 h	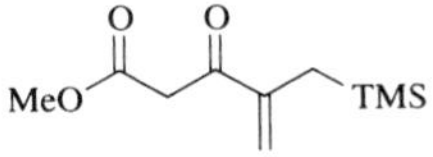 (38)	537
	Bu₃Sn (piperidinyl squarate)	BnPd(PPh₃)₂Cl (2.5%), CuI, THF, 50°, 30 min	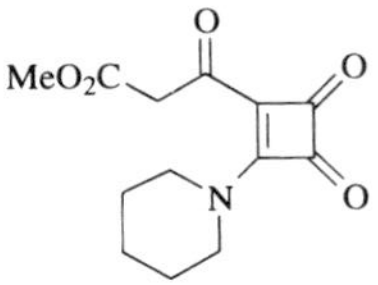 (41)	268
	Me₃Sn∼SnMe₃	BnPd(PPh₃)₂Cl (10%), THF, rt, 2 h	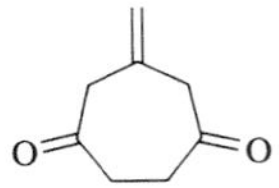 (65)	152
	MeO steroid SnBu₃	Pd(CH₃CN)₂Cl₂ (5%), HMPA, 80°	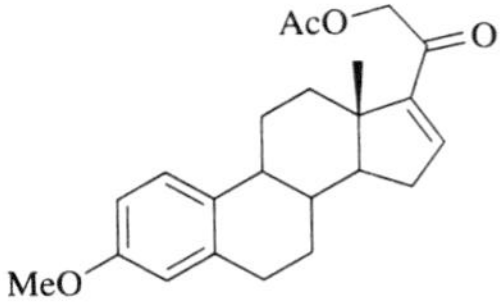 (35)	513
	Bu₃SnH	Pd(PPh₃)₄ (1%), C₆H₆, rt	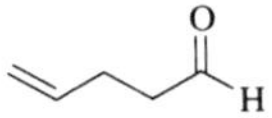 (71)	156
	Bu₃Sn CH(OAc)Ph	Pd(PPh₃)₂Cl₂ (4%), CuCN, PhMe, 75°, 32 h	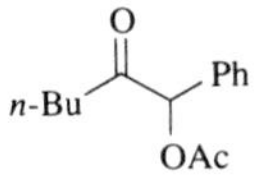 (68)	243
	Bu₃Sn furan SnBu₃	Pd(PPh₃)₂Cl₂ (5%), THF, 80°, 24 h	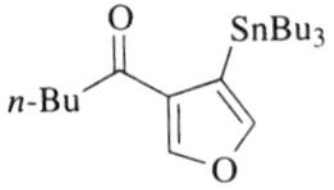 (57)	287, 546
	Bu₃Sn∼	Pd(PPh₃)₄ (5%), CO (15 psi), BHT, C₆H₂Me₄-1,2,3,4, 100°, 1.5h	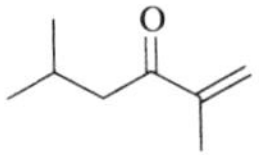 (44)	250

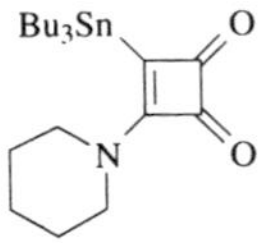

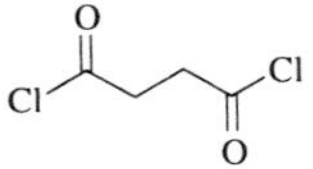

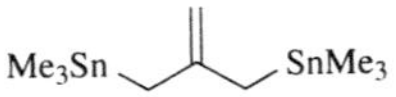

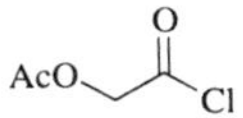

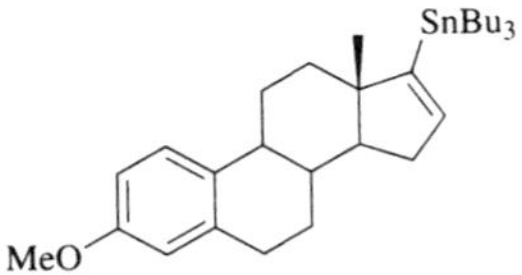

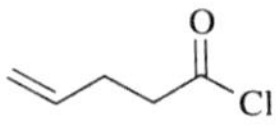

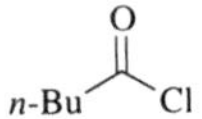

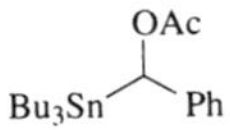

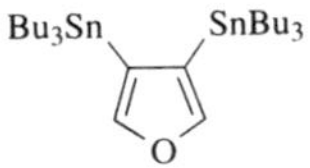

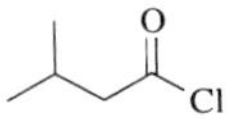

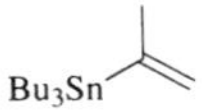

C₅

415

TABLE XIII. DIRECT CROSS-COUPLING OF ACYL CHLORIDES: ALKYL SYSTEMS (*Continued*)

Substrate:

t-Bu–C(=O)–Cl

Stannane	Conditions	Product(s) and Yield(s) (%)	Refs.
Bu_3Sn (azetidinone)	$Pd(PPh_3)_4$	(—)	244
Bu_3Sn (dioxine)	$BnPd(PPh_3)_2Cl$ (0.08%), C_6H_6, reflux, 3 h	(82)	271
Bu_3SnH	$PhCOPd(PPh_3)_2Cl$ (1%), THF, rt	t-Bu–CHO (89)	156
Me_4Sn	$BnPd(PPh_3)_2Cl$ (0.05%), HMPA, 65°	t-Bu–COCH$_3$ (82)	4
Bu_3Sn	$Rh(PPh_3)_3Cl$ (2%), C_6H_6, 80°, 10 h	(72)	2
Bu_3Sn	$Pd(PPh_3)_4$ (5%), CO (15 psi), BHT, $C_6H_2Me_4$-1,2,3,4, 100°, 1.5h	(30)	250
Me_3Sn (OMe)	$BnPd(PPh_3)_2Cl$ (1%), C_6H_6, reflux, 4 h	(79)	749
(polymer, Sn·C≡CPh Bu_2)	$Pd(PPh_3)_4$ (5%), PhMe, 80°, 24 h	(75)	376
Me_3Sn (quinoline)	$Pd(PPh_3)_2Cl_2$ (5%), C_6H_6, reflux, 8 h	(73)	284

	Conditions	Yield	Refs.
(quinoline–Bu₃Sn / product)	PdCl₂ (5%), C₆H₆, reflux, 4 d	(7)	284
	Pd(PPh₃)₄ (5%), dioxane, 100°, 30 h	(87)	148, 753
	BnPd(PPh₃)₂Cl (5%), CO (30 psi), C₆H₆, 90°, 5.5 h	(69)	757
	Pd₂(dba)₃ (2.5%) P(2-furyl)₃ (10%), THF, rt, 6 h	(96)	447
	Pd(PPh₃)₂Cl₂ (5%), PhMe, 100°, 16 h	(80)	759
	BnPd(PPh₃)₂Cl (1%), C₆H₆, reflux, 1 h	(79)	749
	BnPd(PPh₃)₂Cl (0.4%), Cl(CH₂)₂Cl, 60°, 16 h	(47)	761
	BnPd(PPh₃)₂Cl, CO, CHCl₃, 65°, 16 h	(56)	761, 762
	BnPd(PPh₃)₂Cl (2%), CHCl₃, 65°, 48 h	(60)	537

TABLE XIII. DIRECT CROSS-COUPLING OF ACYL CHLORIDES: ALKYL SYSTEMS (*Continued*)

Substrate	Stannane	Conditions	Product(s) and Yield(s) (%)	Refs.
C_6 (4-methylpentanoyl chloride)	$Bu_3Sn\!-\!CH(OBz)\!-\!C_7H_{15}\text{-}n$	Pd(PPh$_3$)$_2$Cl$_2$ (4%), CuCN, PhMe, 75°, 64 h	ketone, OBz, $C_7H_{15}\text{-}n$ (40)	243
(pentanoyl chloride)	$Bu_3Sn\!-\!$furan	Pd(PPh$_3$)$_2$Cl$_2$ (3%), THF, rt, 12 h	furyl ketone (74)	288
(6-bromohexanoyl chloride)	Me$_3$SnR	BnPd(PPh$_3$)$_2$Cl (0.4%), CHCl$_3$, 65°	R ketone, R = Me, 24 h, (45); R = Ph, 2 h, (98)	146
(adipoyl dichloride)	Bu$_3$SnH	Pd(PPh$_3$)$_4$ (1%), THF, rt	dialdehyde (88)	146
	Me$_4$Sn	BnPd(PPh$_3$)$_2$Cl (0.05%), HMPA, 65°	keto aldehyde (90)	147
t-BuO$_2$C–CO–CO–Cl	$Bu_3Sn\!-\!C(=CH_2)CH_2OR$	BnPd(PPh$_3$)$_2$Cl (0.01%), BHT, THF, 10° to rt, 4 h	OR, R = Me (40); R = Bn (50)	766
C_7 MeO$_2$C–acyl chloride	$Bu_3Sn\!-\!CH{=}CH_2$	BnPd(PPh$_3$)$_2$Cl (0.05%), HMPA, 65°	(92)	147
EtO$_2$C–acyl chloride	Me$_3$Sn–C(=CH$_2$)Fe(CO)$_3$	BnPd(PPh$_3$)$_2$Cl (2.6%), Cl(CH$_2$)$_2$Cl, HMPA, 80°, 2 h	EtO$_2$C···Fe(CO)$_3$ (60)	754
(5-methylhex-4-enoyl chloride)	Bu$_3$SnH	Pd(PPh$_3$)$_4$ (1%), C$_6$H$_6$, rt	aldehyde (92)	156

Acyl chloride	Stannane	Conditions	Product	Refs.
![4-pentenoyl chloride]	Bu₃SnC≡CTMS	Pd(PPh₃)₂Cl₂ (1.7%), Cl(CH₂)₂Cl, reflux, 0.5 h	(44)	767
![cyclohexanecarbonyl chloride]	Bu₃SnH	PdCl₂ (1%), PPh₃, C₆H₆, rt	(98)	156
	![isopropenyltin]	Pd(PPh₃)₄ (5%), CO (15 psi), BHT, PhMe, 100°, 2.5 h	(58)	250
	![Me₃Sn OMe vinyl]	BnPd(PPh₃)₂Cl (1%), C₆H₆, reflux, 1 h	(86)	749
	![Me₃Sn pyridine]	Pd(PPh₃)₂Cl₂ (5%), C₆H₆, reflux x, 8 h	(68)	284
	![Me₃Sn methylpyridine]	Pd(PPh₃)₂Cl₂ (5%), C₆H₆, reflux, 8 h	(67)	284
	![Me₃Sn dimethylpyridine]	Pd(PPh₃)₂Cl₂ (5%), C₆H₆, reflux, 8 h	(73)	284
	![Me₃Sn quinoline]	Pd(PPh₃)₂Cl₂ (5%), C₆H₆, reflux, 8 h	(80)	284

TABLE XIII. DIRECT CROSS-COUPLING OF ACYL CHLORIDES: ALKYL SYSTEMS (*Continued*)

Substrate:

$n\text{-}C_6H_{13}$ acyl chloride

Stannane	Conditions	Product(s) and Yield(s) (%)	Refs.
Me_3Sn–quinolin-4-yl	$PdCl_2$ (5%), C_6H_6, reflux, 4 d	cyclohexanecarbonyl-quinoline (50)	284
Me_3Sn–isoquinolin-3-yl	$Pd(PPh_3)_2Cl_2$ (5%), C_6H_6, reflux, 10 h	(73)	284
Me_3Sn–isoquinolin-4-yl	$PdCl_2$ (5%), C_6H_6, reflux, 5 d	(62)	284
Bu_3Sn allyl, $C_5H_{11}\text{-}n$, OAc	$Pd_2(dba)_3$ (2.5%), $P(2\text{-furyl})_3$ (10%), THF, 25°, 16 h	$C_5H_{11}\text{-}n$, OAc (83) $E:Z = 9:1$	447, 768
Bu_3Sn oxazolidinone, Ph, Ph	$BnPd(PPh_3)_2Cl$ (2%), CO (30 psi), C_6H_6, 100°, 3.5 h	Ph, Ph (57)	757
Bu_3Sn, $C_8H_{17}\text{-}n$, OTBDMS	$Pd_2(dba)_3$ (2.5%), $P(2\text{-furyl})_3$ (10%), THF, rt, 6 h	$C_8H_{17}\text{-}n$, OTBDMS (77)	447
Bu_3SnH	$Pd(PPh_3)_4$ (1%), C_6H_6, rt	$n\text{-}C_6H_{13}$ CHO H (77)	156
Bu_3Sn ethyl ketone	$Pd(PPh_3)_2Cl_2$ (1.4%), PhMe, 110°, 24 h	$n\text{-}C_6H_{13}$ ketone (41)	307

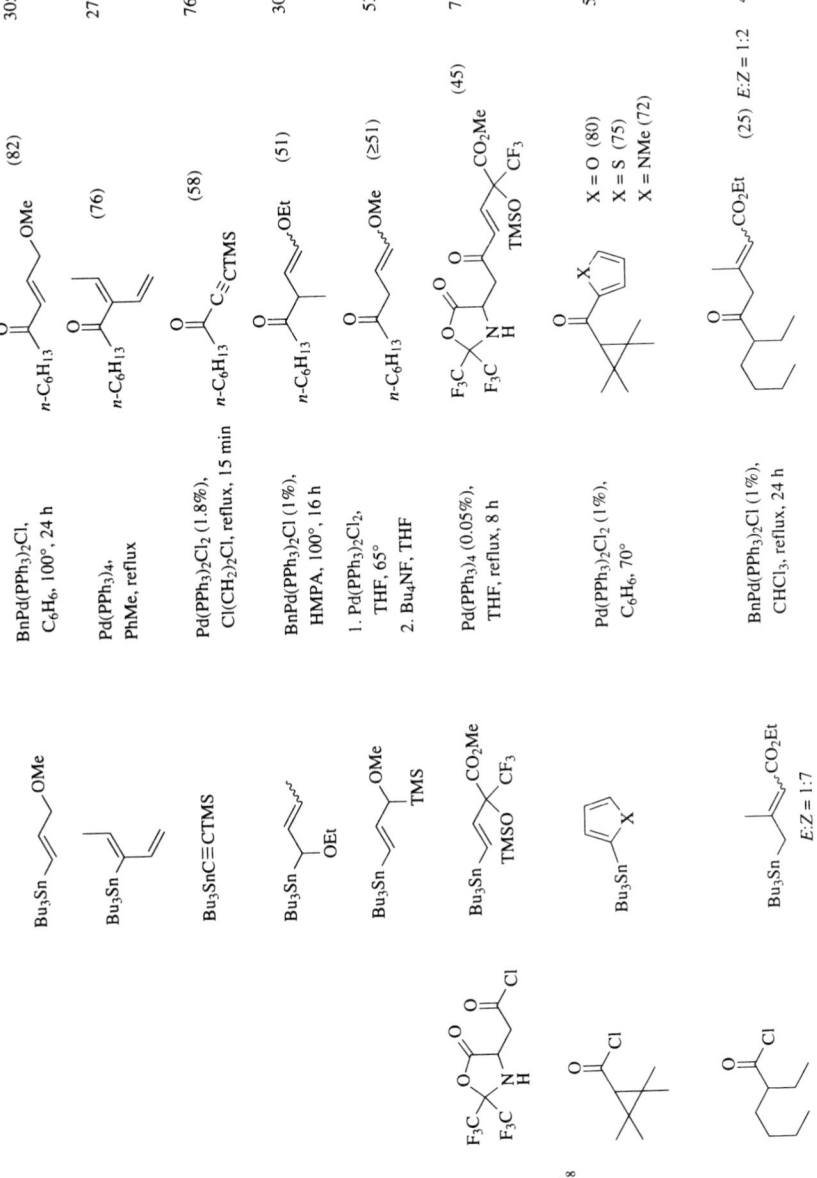

Bu₃Sn⁀OMe	BnPd(PPh₃)₂Cl, C₆H₆, 100°, 24 h	*n*-C₆H₁₃ ⁀ OMe (82)		305
Bu₃Sn⁀	Pd(PPh₃)₄, PhMe, reflux	*n*-C₆H₁₃ ⁀ (76)		277
Bu₃SnC≡CTMS	Pd(PPh₃)₂Cl₂ (1.8%), Cl(CH₂)₂Cl, reflux, 15 min	*n*-C₆H₁₃ ⁀ C≡CTMS (58)		767
Bu₃Sn⁀OEt	BnPd(PPh₃)₂Cl (1%), HMPA, 100°, 16 h	*n*-C₆H₁₃ ⁀ OEt (51)		305
Bu₃Sn⁀OMe TMS	1. Pd(PPh₃)₂Cl₂, THF, 65° 2. Bu₄NF, THF	*n*-C₆H₁₃ ⁀ OMe (≥51)		532
Bu₃Sn⁀CO₂Me TMSO CF₃	Pd(PPh₃)₄ (0.05%), THF, reflux, 8 h	(45)		756
Bu₃Sn⁀X	Pd(PPh₃)₂Cl₂ (1%), C₆H₆, 70°	X = O (80) X = S (75) X = NMe (72)		574
Bu₃Sn⁀CO₂Et *E:Z* = 1:7	BnPd(PPh₃)₂Cl (1%), CHCl₃, reflux, 24 h	CO₂Et (25) *E:Z* = 1:2		427

C₈

421

TABLE XIII. DIRECT CROSS-COUPLING OF ACYL CHLORIDES: ALKYL SYSTEMS (Continued)

Substrate	Stannane	Conditions	Product(s) and Yield(s) (%)	Refs.
$n\text{-}C_7H_{15}\text{-}COCl$	Bu$_3$SnH	Pd(PPh$_3$)$_4$ (1%), C$_6$H$_6$, rt, 2 h	$n\text{-}C_7H_{15}\text{-}CHO$ (77)	769
	Bu$_3$Sn–CH=CH$_2$	BnPd(PPh$_3$)$_2$Cl (1%), C$_6$H$_6$, reflux, 1 h	$n\text{-}C_7H_{15}$ enone (62)	749
	Bu$_3$Sn (isopropenyl)	Pd(PPh$_3$)$_4$ (5%), CO (15 psi), BHT, PhMe, 100°, 2 h	$n\text{-}C_7H_{15}$ enone (59)	250
	Bu$_3$Sn–C(CF$_3$)=CH$_2$	BnPd(PPh$_3$)$_2$Cl (1%), HMPA, 65°, 24 h	$n\text{-}C_7H_{15}$, CF$_3$ enone (78)	262
	Bu$_3$Sn–C(OMe)=CH$_2$	BnPd(PPh$_3$)$_2$Cl (1%), C$_6$H$_6$, reflux, 1 h	$n\text{-}C_7H_{15}$, OMe enone (82)	749
	Bu$_3$Sn–CH$_2$C(CH$_3$)=CH–CO$_2$Et, $E{:}Z = 1{:}7$	BnPd(PPh$_3$)$_2$Cl (1%), CHCl$_3$, reflux, 24 h	$n\text{-}C_7H_{15}$ ...CO$_2$Et (25) $E{:}Z = 1{:}5$	427
	[Et$_3$Sn–C(O)–C$_7$H$_{15}$-n]	[(η^3-C$_3$H$_5$)PdCl]$_2$ (1%), P(OEt)$_3$, CO (120 psi), PhMe, 111°, 2 h	$n\text{-}C_7H_{15}$–C(O)C(O)–C$_7$H$_{15}$-n (78)	308
	Bu$_3$Sn dienoate macrolactone	BnPd(PPh$_3$)$_2$Cl (5%), CO (45 psi), PhMe, 100°, 7 h	macrocycle (38)	770
C$_9$ Ph-CH$_2$CH$_2$-COCl	Me$_4$Sn	BnPd(PPh$_3$)$_2$Cl (0.05%), HMPA, 65°	Ph-CH$_2$CH$_2$-C(O)CH$_3$ (99)	4

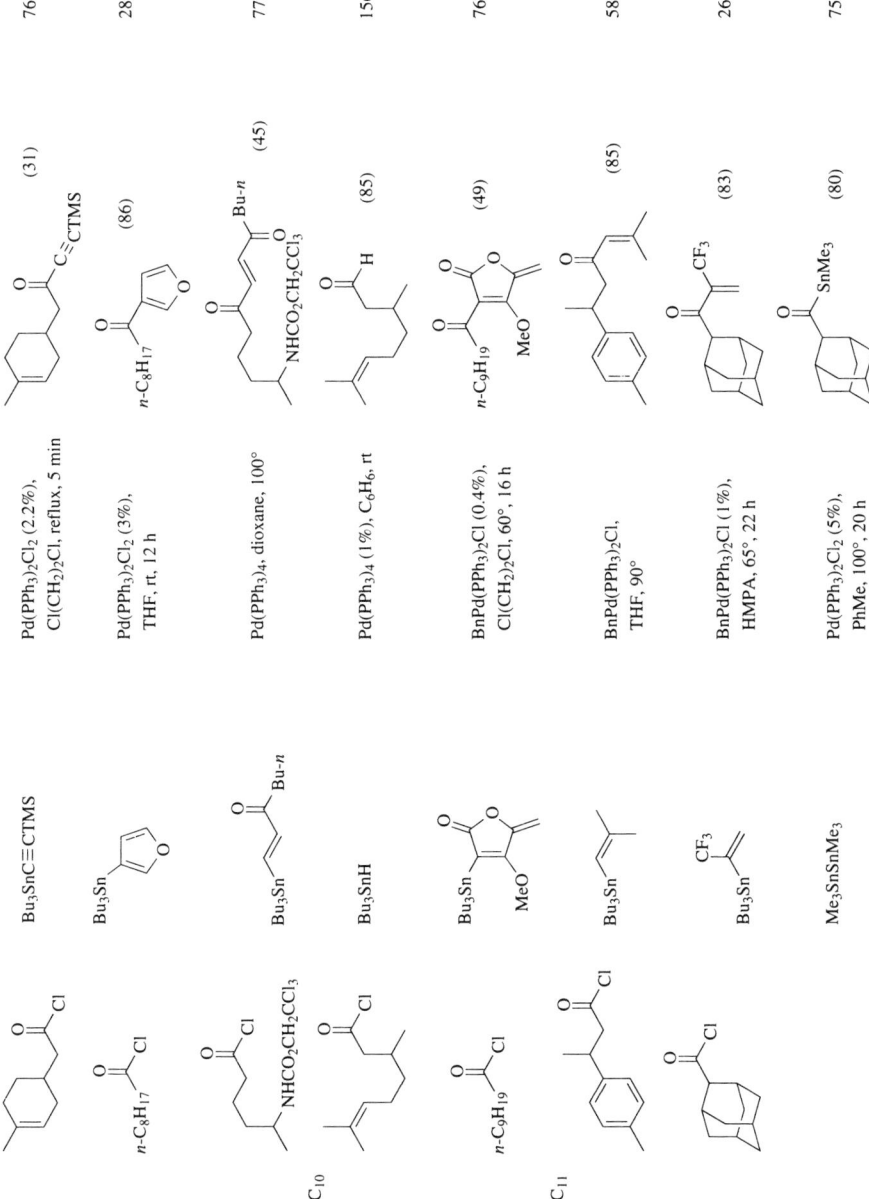

Bu₃SnC≡CTMS	Pd(PPh₃)₂Cl₂ (2.2%), Cl(CH₂)₂Cl, reflux, 5 min	(31)	767
	Pd(PPh₃)₂Cl₂ (3%), THF, rt, 12 h	(86)	288
	Pd(PPh₃)₄, dioxane, 100°	(45)	771
Bu₃SnH	Pd(PPh₃)₄ (1%), C₆H₆, rt	(85)	156
	BnPd(PPh₃)₂Cl (0.4%), Cl(CH₂)₂Cl, 60°, 16 h	(49)	761
	BnPd(PPh₃)₂Cl, THF, 90°	(85)	581
	BnPd(PPh₃)₂Cl (1%), HMPA, 65°, 22 h	(83)	262
Me₃SnSnMe₃	Pd(PPh₃)₂Cl₂ (5%), PhMe, 100°, 20 h	(80)	759

423

Substrate	Stannane	Conditions	Product(s) and Yield(s) (%)	Refs.
	Bu₃Sn (furan)	BnPd(PPh₃)₂Cl (0.63%), HMPA, 65°, 22 h	(95)	772
	Bu₃SnH	Pd(PPh₃)₄ (1%), C₆H₆, rt	(88)	156
	Me₃SnPh	BnPd(PPh₃)₂Cl (0.4%), CHCl₃, 65°, 5 h	(87)	146
	Bu₃Sn (MeO)	BnPd(PPh₃)₂Cl (0.4%), Cl(CH₂)₂Cl, 60°, 16 h	(49)	761, 762
	Me₄Sn	BnPd(PPh₃)₂Cl, CO (15 psi), HMPA, rt, 4 h	(87)	773
	Bu₃Sn	Pd(PPh₃)₄ (5%), dioxane, 100°, 30 h	(45)	148
	Bu₃Sn (N)	Pd(PPh₃)₂Cl₂ (0.7%), C₆H₆, reflux, 12 h	(48)	774

Substrate	Conditions	Product	(Yield)	Refs.
Me₃Sn–C₆H₄(NHBoc)	Pd₂(dba)₃ (0.5%), PhMe, 70°, 4 h	oxazolidinone ketone, N–Cbz, NHBoc-aryl	(79)	775
Me₃Sn–C₆H₃(Cl)(NHBoc)	Pd(CH₃CN)₂Cl₂, PhMe, reflux	oxazolidinone ketone, N–Cbz, 4-Cl-NHBoc-aryl	(62)	775
Me₃Sn–C₆H₂(Cl)₂(NHBoc)	Pd(CH₃CN)₂Cl₂, PhMe, reflux	oxazolidinone ketone, N–Cbz, dichloro-NHBoc-aryl	(53)	775
Bu₃SnH	Pd(PPh₃)₄ (7%), THF, rt	MeO₂C–CH(NHCO₂Bn)–CH₂–CHO	(>77)	776
Bu₃Sn–CH=CH–C(O)CH₃	Pd(PPh₃)₄ (5%), dioxane, 100°, 30 h	enone, NHBz	(37)	148
Bu₃Sn–CH=CH–C(O)Bu-n	Pd(PPh₃)₄ (5%), dioxane, 100°, 30 h	enone, Bu-n, NHBz	(38)	148
Bu₃Sn–CH=CH–C(O)Bu-n	Pd(PPh₃)₄ (5%), dioxane, 60°, 3 h	dienone, Bu-n, NHBz	(50)	148
Bu₃Sn–CH=CH–C(O)Ph	Pd(PPh₃)₄ (5%), dioxane, 100°, 30 h	dienone, Ph, NHBz	(38)	148

Additional substrates (lower left):

MeO₂C–CH(NHCO₂Bn)–CH₂–C(O)Cl

Cl–C(O)–CH₂CH₂CH₂–CH(CH₃)–NHBz

TABLE XIII. DIRECT CROSS-COUPLING OF ACYL CHLORIDES: ALKYL SYSTEMS (Continued)

Substrate	Stannane	Conditions	Product(s) and Yield(s) (%)	Refs.
(bicyclic diketone acyl chloride)	Me$_4$Sn	BnPd(PPh$_3$)$_2$Cl, HMPA, 65°, 3 d	(82)	777
(Me-( )$_5$ diketone acyl chloride)	Bu$_3$Sn–CH=CH$_2$	BnPd(PPh$_3$)$_2$Cl, C$_6$H$_6$, reflux, 3.5 h	(≥54)	778
(oxazolidinone, N–CO$_2$Bn)	Me$_4$Sn	BnPd(PPh$_3$)$_2$Cl (0.5%), HMPA, 65°, 4 h	(58)	779
(oxazolidinone, N–CO$_2$Bn)	R$_4$Sn R: Me / Et / Bu / Bn	BnPd(PPh$_3$)$_2$Cl (0.5%), HMPA, 65°, 4 h	R = Me (74), Et (72), Bu (66), Bn (33)	779
C$_{16}$ (( )$_6$ diketone acyl chloride)	Bu$_3$Sn–CH=CH$_2$	BnPd(PPh$_3$)$_2$Cl (1%), C$_6$H$_6$, reflux, 2.5 h	(≥49)	780
(cyclopentane, t-BuO$_2$C, Et acyl chloride)	Bu$_3$Sn–CH=CH–C≡CTMS	BnPd(PPh$_3$)$_2$Cl (5%), THF, 50°	(76)	781
n-C$_{15}$H$_{31}$–COCl	Bu$_3$SnH	Pd(PPh$_3$)$_4$ (5%), C$_6$H$_6$, rt, 2 h	n-C$_{15}$H$_{31}$–CHO (75)	156, 769

C$_{17-21}$

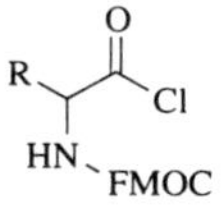

R = H (74), i-Pr (20), s-Bu (25)

Bu$_3$SnH

Pd(PPh$_3$)$_4$, THF, rt

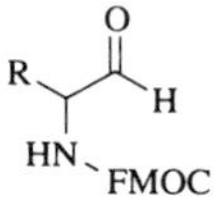

782

C$_{25}$

Bu$_3$Sn—OBn

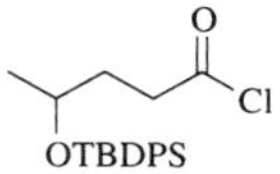

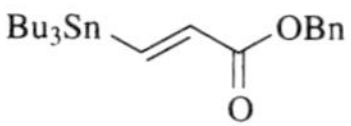

BnPd(PPh$_3$)$_2$Cl (2.4%), CO (15 psi), CHCl$_3$, 65°, 30 h

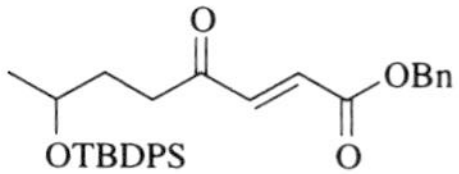

(71)

146, 783

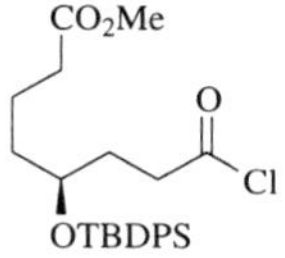

Me$_3$SnC≡CC$_8$H$_{17-n}$

Pd(PPh$_3$)$_2$Cl$_2$ (2%), Cl(CH$_2$)$_2$Cl, reflux, 30 min

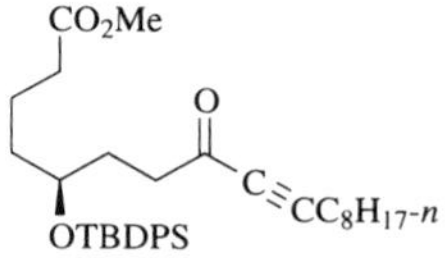

(82)

784

C$_{29}$

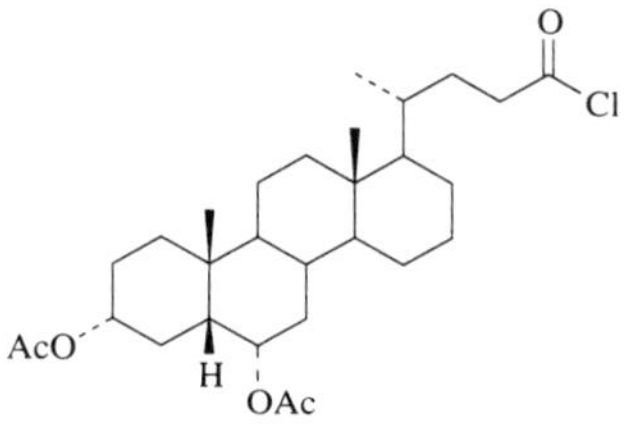

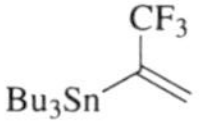

Bu$_3$Sn / CF$_3$

BnPd(PPh$_3$)$_2$Cl (1%), HMPA, 65°, 24 h

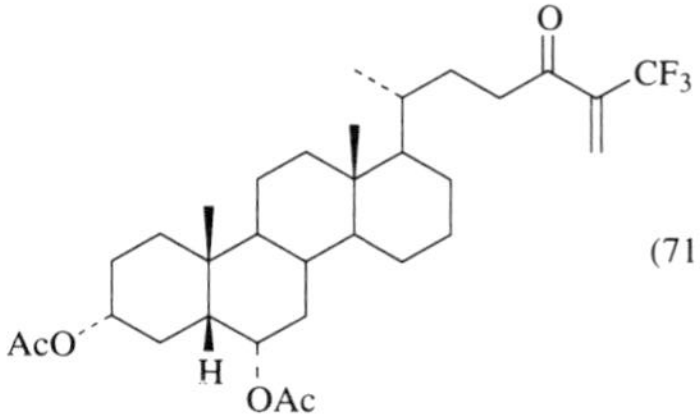

(71)

262

TABLE XIV. DIRECT CROSS-COUPLING OF ACYL CHLORIDES: ARYL SYSTEMS

Substrate	Stannane	Conditions	Product(s) and Yield(s) (%)	Refs.
C₇ PhC(O)Cl	Bu₃SnH	Pd(PPh₃)₂Cl₂ (1%), PPh₃, C₆H₆, rt	PhCHO (95)	156, 769
	Me₄Sn	Pd(Ph-BIAN) (dimethyl fumarate) (1%), DMF, 50°, 16 h	PhCOCH₃ (98)	415, 147, 4, 750, 1
	Et₃SnMe	BnPd(PPh₃)₂Cl (0.45%), HMPA, 65°	PhCOEt + PhCOCH₃ (—) 83:17	27
	Bu₃SnMe	BnPd(PPh₃)₂Cl (0.45%), HMPA, 65°	PhCO-Bu-n + PhCOCH₃ (—) 57:43	27
	Me₃Sn–CH=CH₂	BnPd(PPh₃)₂Cl (0.4%), CHCl₃, 65°, 18 h	I (88) PhCOCH=CH₂	146, 750
	Bu₃Sn–CH=CH₂	Pd(PPh₃)₄ (1%), C₆H₆, 40°, 5 h	I (87)	1, 785, 146
	Sn·CH=CH₂ / Bu₂ polymer	Pd(PPh₃)₄ (5%), PhMe, 40°, 47 h	I (95)	376
	Me₃Sn–OMe	BnPd(PPh₃)₂Cl (0.4%), CHCl₃, 65°, 18 h	PhCOCH₂OMe (48) + PhCOCH₃ (16)	146
	Bu₃Sn–OMe	BnPd(PPh₃)₂Cl (0.4%), CHCl₃, 65°, 18 h	PhCOCH₂OMe (36) + PhCO–Bu-n (14)	146

	Conditions	Product(s)	Refs.
Bu₃Sn (crotyl, cis)	BnPd(PPh₃)₂Cl (0.45%), CHCl₃, 65°	I + II (—) 50:50	27
Bu₃Sn (cis)	Pd₂(dba)₃ (1%), L (8%), THF, 24°; L = PPh₃; L = AsPh₃; L = P(2-furyl)₃	I + II (>90), I:II = 70:30; I + II (97), I:II = 70:30; I + II (>90), I:II = 95:5	11
Bu₃Sn–CH₂CH=CH₂ (allyl), Sn Bu₂ with aryl	Rh(PPh₃)₃Cl (2%), C₆H₆, 80°, 5 h	(86)	2
polymer-supported stannane, (—)ₙ	Pd(PPh₃)₄ (5%), PhMe, 90°, 65 h	(53) + (7) + (5)	376
Bu₃Sn (isopropenyl)	Pd(PPh₃)₄ (5%), CO (15 psi), BHT, PhMe, 100°, 3 h	(75)	250
CF₃ / Bu₃Sn	BnPd(PPh₃)₂Cl (1%), HMPA, 65°, 3 h	(90)	262
Bu₃Sn (aldehyde, H)	Pd(PPh₃)₄ (5%), dioxane, 60°, 3 h	(60)	148
OMe / Me₃Sn	BnPd(PPh₃)₂Cl (1%), C₆H₆, reflux, 1 h	(73)	749

TABLE XIV. DIRECT CROSS-COUPLING OF ACYL CHLORIDES: ARYL SYSTEMS (*Continued*)

Substrate	Stannane	Conditions	Product(s) and Yield(s) (%)	Refs.
	Bu_3Sn —C(=O)CH$_2$CH$_3$	Pd(PPh$_3$)$_2$Cl$_2$ (1%), 100°, 20 h	Ph—C(=O)—C(=O)—Et (43)	529
	polymer-bound allylstannane with OH, Sn Bu$_2$, (benzene ring)$_n$; E:Z = 25:75	Pd(PPh$_3$)$_4$ (5%), PhMe, 40°, 48 h	Ph—C(=O)—CH=CH—CH$_2$OH (74) E:Z = 69:31	376
	Bu$_4$Sn	BnPd(PPh$_3$)$_2$Cl (0.05%), HMPA, 65°	Ph—C(=O)—Bu-n (91)	4, 146, 1, 27
	Bu$_3$Sn—C(CH$_3$)=CH—CH$_3$	BnPd(PPh$_3$)$_2$Cl (0.4%), CHCl$_3$, 65°, 4.5 h	Ph—C(=O)—C(CH$_3$)=CH—CH$_3$ (63) E:Z = 30:70	146, 783
	Bu$_3$Sn—(cyclobutenedione, H$_2$N)	BnPd(PPh$_3$)$_2$Cl (2.5%), CuI, THF, 50°, 20 min	Ph—C(=O)—(cyclobutenedione, H$_2$N) (84)	268
	Bu$_3$Sn—(3-methylisoxazol-5-yl)	Pd(PPh$_3$)$_2$Cl$_2$, dioxane, reflux, 3 h	Ph—C(=O)—(3-methylisoxazol-5-yl) (80)	292, 530
	Bu$_3$Sn—(cyclobutenone, OEt)	BnPd(PPh$_3$)$_2$Cl (1.5%), CH$_3$CN, rt, 2 h	Ph—C(=O)—(cyclobutenone) (81)	267
	Bu$_3$Sn—C(OEt)=CH$_2$	BnPd(PPh$_3$)$_2$Cl (1.5%), C$_6$H$_6$, 100°	Ph—C(=O)—C(OEt)=CH$_2$ (75)	269

430

Organostannane	Conditions	Product	Yield	Refs.
Bu$_3$Sn–CH=CH–C(O)CH$_3$	Pd(PPh$_3$)$_4$ (5%), dioxane, 100°, 30 h	PhC(O)CH$_2$CH$_2$C(O)CH$_3$	(70)	148, 753
Bu$_3$Sn–CH$_2$CH=CH–OMe	BnPd(PPh$_3$)$_2$Cl, CH$_2$Cl$_2$, 65°, 4 h	PhC(O)CH=CHCH$_2$OMe	(71)	305, 532
Bu$_3$Sn–CH$_2$CH=CH–OMe	BnPd(PPh$_3$)$_2$Cl, C$_6$H$_6$, 100°, 48 h	PhC(O)CH$_2$CH=CHOMe	(85)	305
Me$_3$Sn–(2-furyl)	Pd$_2$(dba)$_3$ (1%), p-Tol-BIAN, THF, 65°, 3 h	PhC(O)(2-furyl)	(53)	415
Bu$_3$Sn–(3-furyl)	Pd(PPh$_3$)$_2$Cl$_2$ (3%), THF, rt, 12 h	PhC(O)(3-furyl)	(91)	288, 287
Bu$_3$Sn–CH(CH$_3$)$_2$ C(O)	Pd(PPh$_3$)$_2$Cl$_2$ (1.4%), PhMe, 110°, 24 h	PhC(O)C(O)CH(CH$_3$)$_2$	(59)	307
Bu$_3$Sn–(2,3-dihydro-1,4-dioxin-2-yl)	BnPd(PPh$_3$)$_2$Cl (0.08%), C$_6$H$_6$, reflux, 3 h	PhC(O)(2,3-dihydro-1,4-dioxin-2-yl)	(93)	271
Me$_3$Sn–(2-thienyl)	[(η^3-C$_3$H$_5$)PdCl]$_2$ (1%), HMPA, 20°, 5 min	PhC(O)(2-thienyl)	(87)	750
Bu$_3$SnC≡CPr-n	BnPd(PPh$_3$)$_2$Cl (0.4%), CHCl$_3$, 65°, 23 h	PhC(O)C≡CPr-n	(70)	146
Bu$_3$Sn–C(=CHCH$_3$)CH=CH$_2$	Pd(PPh$_3$)$_4$, PhMe, reflux	PhC(O)C(=CHCH$_3$)CH=CH$_2$	(82)	277

TABLE XIV. DIRECT CROSS-COUPLING OF ACYL CHLORIDES: ARYL SYSTEMS (*Continued*)

Substrate	Stannane	Conditions	Product(s) and Yield(s) (%)	Refs.
	Me₃Sn—(pyridyl)	Pd(PPh₃)₂Cl₂ (5%), C₆H₆, reflux, 8 h	(67)	284
	Me₃Sn—(pyrimidyl)—SMe	Pd(PPh₃)₂Cl₂ (7%), Cl(CH₂)₂Cl, reflux	(97)	459
	R₃Sn—(pyrimidyl)—SO₂Me	Pd(PPh₃)₂Cl₂ (14%), THF, reflux	R = Me, (71) R = Bu, (61)	459
	(polymer, Sn Bu₂, CO₂Me)	Pd(PPh₃)₄ (5%), PhMe, 40°, 44 h	(64)	376
	Bu₃Sn—(furanone, MeO)	BnPd(PPh₃)₂Cl, CO, CHCl₃, 65°, 16 h	(47)	762, 761
	Bu₃SnC≡CTMS	Pd(PPh₃)₂Cl₂ (1.8%), Cl(CH₂)₂Cl, 84°, 2 h	(64)	149
	Bu₃Sn—(F)=TMS	Pd(PPh₃)₂Cl₂, CH₂Cl₂, 82°	(65)	263

432

Organostannane	Conditions	Product	Yield	Refs.
Me₃SnPh	[(η³-C₃H₅)PdCl]₂ (1%), Cl(CH₂)₂Cl, 75°	PhCOPh	(>98)	786, 147, 750, 4, 1, 27
Me₃Sn–C₆H₄–Cl	[(η³-C₃H₅)PdCl]₂ (1%), HMPA, 20°, 10 min	4-Cl-C₆H₄-CO-Ph	(77)	750, 751
Me₃Sn-(2-methylpyridyl)	Pd(PPh₃)₂Cl₂ (5%), C₆H₆, reflux, 8 h	(2-methylpyridyl)CO-Ph	(60)	284
Me₃Sn–C₆H₄–NO₂	[(η³-C₃H₅)PdCl]₂ (1%), HMPA, 20°, 20 min	4-NO₂-C₆H₄-CO-Ph	(97)	750, 751
Bu₃Sn–CH=CH–CH(OEt)CH₃	BnPd(PPh₃)₂Cl (2%), 110–120°, 20 h	PhCO-CH(CH₃)-CH₂-CHO	(77)	534
Bu₃Sn–CH=CH–CH(OEt)CH₃	BnPd(PPh₃)₂Cl (1%), THF, 100°, 16 h	PhCO-CH(CH₃)-CH=CH-OEt	(72)	305
Bu₃Sn-(thiophene-dioxide-furan)	Pd(PPh₃)₄ (1%), HMPA, 0°, 3 h	(sulfolene–furan)CO-Ph	(51)	752
Me₃Sn–CH₂–C(=CH₂)–TMS	BnPd(PPh₃)₂Cl (2%), CHCl₃, 65°, 24 h	PhCO-C(=CH₂)-CH₂-TMS	(73)	537

TABLE XIV. DIRECT CROSS-COUPLING OF ACYL CHLORIDES: ARYL SYSTEMS (Continued)

Substrate	Stannane	Conditions	Product(s) and Yield(s) (%)	Refs.
	Me$_3$Sn—Ph	BnPd(PPh$_3$)$_2$Cl (0.05%), HMPA, 65°	**I** (91)	147, 750, 751
	Me$_3$Sn—Ph	BnPd(PPh$_3$)$_2$Cl (0.45%), CHCl$_3$, 65°	**I** + **II** (—) **II** (—) **I:II** = 17:83	146
	Me$_3$Sn—Ph	BnPd(PPh$_3$)$_2$Cl (0.45%), HMPA, 65°	**I** + **II** (—) **I:II** = 10:90	146
	Me$_3$Sn—Ph	BnPd(PPh$_3$)$_2$Cl (4%), HMPA, 65°	**I** + **II** (—) **I:II** = 60:40	146
	Bu$_3$Sn—Ph	BnPd(PPh$_3$)$_2$Cl (0.45%), CHCl$_3$, 65°	**I** (6) + **III** (34)	146
	Bu$_3$Sn—Ph	BnPd(PPh$_3$)$_2$Cl (0.45%), HMPA, 65°	**I** (9) + **III** (51)	146
	Bu$_3$Sn—Ph	BnPd(PPh$_3$)$_2$Cl (4%), HMPA, 65°	**I** (78) + **III** (6)	146
	(Bn)$_4$Sn	BnPd(PPh$_3$)$_2$Cl (0.05%), HMPA, 65°	**I** (95)	4
	Bu$_3$Sn—CHPh (H, D)	BnPd(PPh$_3$)$_2$Cl (4%), HMPA, 65°, 16 h	(71)	27
	Bu$_3$Sn—(aryl CF$_3$)	BnPd(PPh$_3$)$_2$Cl (0.4%), CHCl$_3$, reflux, 5 h	(64)	146
	Me$_3$Sn—(aryl CN)	[(η^3-C$_3$H$_5$)PdCl]$_2$ (1%), HMPA, 20°, 5 min	(100)	750

434

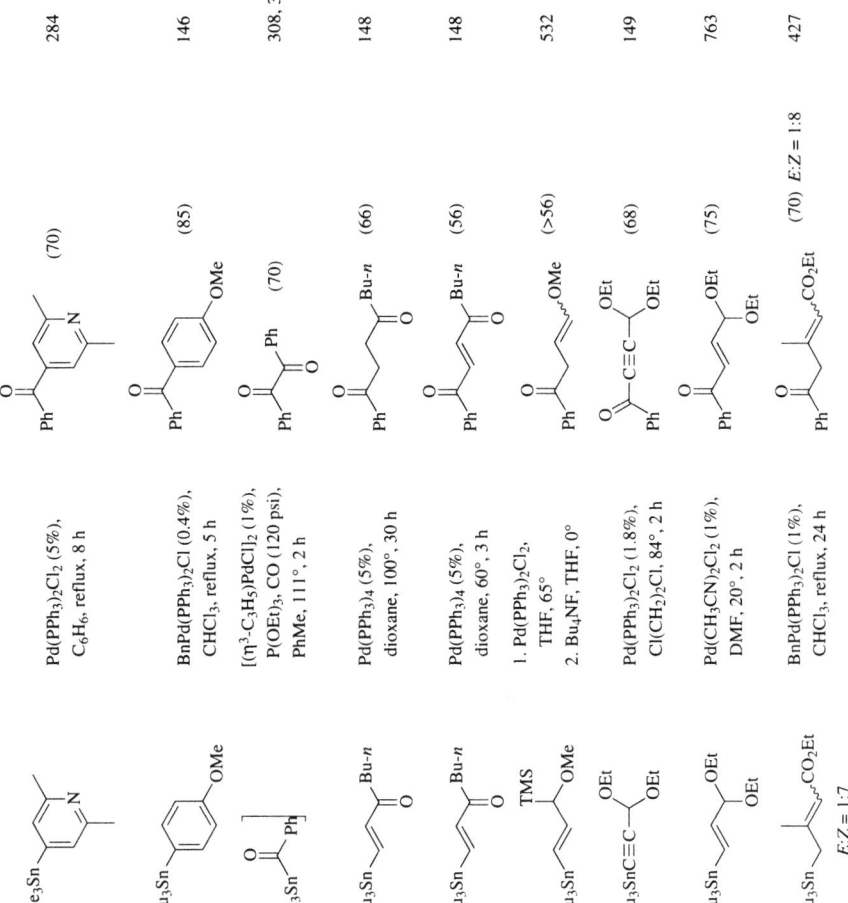

TABLE XIV. DIRECT CROSS-COUPLING OF ACYL CHLORIDES: ARYL SYSTEMS (*Continued*)

Substrate	Conditions	Product(s) and Yield(s) (%)	Refs.
Bu₃Sn ... *i*-PrO	1. BnPd(PPh₃)₂Cl (2.5%), CuI, THF, 50° 2. H₃O⁺	(69)	268
Bu₃Sn ... *i*-PrO	1. BnPd(PPh₃)₂Cl (2.5%), CuI, THF, 50° 2. Piperidine	(79)	268
Bu₃Sn ...	BnPd(PPh₃)₂Cl (1%), CO (15 psi), C₆H₆, 80°	(78)	268
Bu₃Sn ...	Pd(PPh₃)₂Cl₂ (4%), CuCN, PhMe, 75°, 36 h	(50)	243
R₃Sn ... Fe(CO)₃	BnPd(PPh₃)₂Cl (2.6%), Cl(CH₂)₂Cl, HMPA, 80°, 2 h	R = Me (90) R = Bu (43)	754
Bu₃SnC≡CPh	Pd(PPh₃)₂Cl₂ (1.8%), Cl(CH₂)₂Cl, 84°, 2 h	(94)	149, 146
Sn-C≡CPh Bu₂	Pd(PPh₃)₄ (5%), PhMe, 40°, 48 h	(96)	376

Bu₃Sn—CH=CH—Ph	BnPd(PPh₃)₂Cl (0.4%), CHCl₃, 65°, 24 h	$\overset{O}{\underset{Ph}{\parallel}}$—CH=CH—Ph **I** (82)	146
" *E:Z* = 95:5	BnPd(PPh₃)₂Cl (0.5%), CHCl₃, 65°	**I** (—)	27
" *E:Z* = 15:85	"	**I** (—)	27
Bu₃Sn—C(=CH₂)Ph	Pd(PPh₃)₄ (5%), CO (15 psi), BHT, PhMe, 100°, 24 h	(55) PhC(=O)C(=CH₂)Ph	250
Bu₃Sn-(morpholino cyclobutenedione)	BnPd(PPh₃)₂Cl (2.5%), CuI, THF, 50°, 60 min	(73)	268
Bu₃Sn–C₆H₄–CO₂Me	Pd(PPh₃)₂Cl₂ (2%), CHCl₃, 100°, 40 h	(70) CO₂Me	787
Bu₃Sn–CH=CH–CH(OAc)–Pr-n	Pd(PPh₃)₄ (4%), CuCN, PhMe, 75°, 60 h	(57) Pr-n	243
Bu₃Sn–CH=CH–CH=CH–CH(OEt)₂ *E:Z* = 85:15	BnPd(PPh₃)₂Cl (2%), THF, 65°, 17 h	(81) *E:Z* = 85:15 OEt OEt	539
Bu₃Sn–C(=CH₂)OTBDMS	BnPd(PPh₃)₂Cl (1%), PhMe, reflux	(71) OTBDMS	457
Me₃Sn–(quinolin-3-yl)	Pd(PPh₃)₂Cl₂ (5%), C₆H₆, reflux, 8 h	(71)	284

437

TABLE XIV. DIRECT CROSS-COUPLING OF ACYL CHLORIDES: ARYL SYSTEMS (*Continued*)

Substrate	Stannane	Conditions	Product(s) and Yield(s) (%)	Refs.
		PdCl$_2$ (5%), C$_6$H$_6$, reflux, 4 d	(47)	284
		Pd(PPh$_3$)$_2$Cl$_2$ (5%), C$_6$H$_6$, reflux, 10 h	(69)	284
		PdCl$_2$ (5%), C$_6$H$_6$, reflux, 5 d	(49)	284
		Pd(PPh$_3$)$_2$Cl$_2$ (5%), THF, reflux, 3 d	(54)	788
		Pd(PPh$_3$)$_2$Cl$_2$ (5%), THF, reflux, 3 h	(42)	788
		Pd(PPh$_3$)$_2$Cl$_2$ (5%), THF, reflux, 3 h	(0)	788
		BnPd(PPh$_3$)$_2$Cl (2.5%), CuI, THF, 50°, 15 min	(92)	268

Organostannane	Conditions	Product (Yield)	Refs.
Bu₃Sn / NHAc, CO₂Et	Pd₂(dba)₃ (5%), AsPh₃ (40%), THF, rt, 48 h	NHAc, CO₂Et, O, Ph (80)	375
Bu₃Sn / CO₂Et, NHAc	Pd₂(dba)₃ (5%), AsPh₃ (40%), THF, 65°	CO₂Et, NHAc, O, Ph (69)	375
OMOM, Bu₃Sn, Ph	Pd(PPh₃)₂Cl₂ (4%), CuCN, PhMe, 75°, 15 h	Ph, OMOM, O (80)	243
OAc, Bu₃Sn, Ph	Pd(PPh₃)₂Cl₂ (4%), CuCN, PhMe, 75°, 12 h	Ph, OAc, O (78)	243
Bu₃SnC≡C—OTBDMS	Pd(PPh₃)₂Cl₂ (1.8%), Cl(CH₂)₂Cl, 84°, 2 h	OTBDMS, O, Ph (66)	149
Bu₃Sn—OTBDMS	BnPd(PPh₃)₂Cl (0.4%), CHCl₃, 65°, 24 h	OTBDMS, O, Ph (78)	146
R₃Sn—(pyrimidine)—OTBDMS	1. Pd(PPh₃)₂Cl₂ (3.7%), Cl(CH₂)₂Cl, reflux, 20 h 2. AcOH	N, O, N–H, O, Ph R = Me (47) R = Bu (30)	459
Bu₃Sn—CO₂Bn	BnPd(PPh₃)₂Cl (0.4%), CHCl₃, 65°, 20 h	O, Ph, CO₂Bn I (55)	146
Bu₃Sn—CO₂Bn	BnPd(PPh₃)₂Cl (0.5%), THF, 65°	I + O, Ph, CO₂Bn (—) 62:38	27

TABLE XIV. DIRECT CROSS-COUPLING OF ACYL CHLORIDES: ARYL SYSTEMS (*Continued*)

Substrate	Stannane	Conditions	Product(s) and Yield(s) (%)	Refs.
	Bu_3Sn—OBn	Pd(PPh$_3$)$_4$ (5%), C$_6$H$_6$, reflux, 4 h	(62) [structure: Ph—C(O)—CH=CH—CH$_2$—OBn]	789
	Bu_3Sn—C$_7$H$_{15}$-n (OAc)	Pd(PPh$_3$)$_2$Cl$_2$ (4%), CuCN, PhMe, 75°, 18 h	(74) [structure: Ph—C(O)—CH(OAc)—C$_7$H$_{15}$-n]	243
	Bu_3Sn—TMS, Ph	BnPd(PPh$_3$)$_2$Cl (1%), 80°, 18 h	(75) [structure: Ph—C(O)—CH=C(TMS)—Ph]	256
	Bu_3Sn—Ph, OMOM	Pd(PPh$_3$)$_2$Cl$_2$ (4%), CuCN, PhMe, 75°, 38 h	(30) [structure with Ph, OMOM]	243
	Bu_3Sn—Ph, CO$_2$Bu-t	Pd$_2$(dba)$_3$ (0.05%), HMPA, 100°, 1 h	(78) [structure: Ph—C(O)—C(Ph)=CH—CO$_2$Bu-t]	247
	Bu_3Sn—C$_8$H$_{17}$-n, OAc	Pd$_2$(dba)$_3$, P(2-furyl)$_3$, THF, 25°, 16 h	(90) [structure: Ph—C(O)—CH=CH—CH(OAc)—C$_8$H$_{17}$-n]	447, 768
	[lactone structure with BnO, SiEt$_3$]	Pd(PPh$_3$)$_2$Cl$_2$ (4%), CuCN, PhMe, 95°, 18 h	(60) [lactone structure with BnO, C(O)Ph]	242
	Bu_3Sn—SiEt$_3$, Ph	Pd(CH$_3$CN)$_2$Cl$_2$, CHCl$_3$, 60°, 24 h	(65) [structure: Ph—C(O)—CH=C(SiEt$_3$)—Ph]	255

	Conditions	Product	Ref.
Bu₃Sn—CH(C₇H₁₅-n)(O₂CC₆H₄NO₂-p)	Pd(PPh₃)₂Cl₂ (4%), CuCN, PhMe, 75°, 36 h	Ph–CO–CH(C₇H₁₅-n)(O₂CC₆H₄NO₂-p) (50)	243
Bu₃Sn—CH(C₇H₁₅-n)(OBz)	Pd(PPh₃)₂Cl₂ (4%), CuCN, PhMe, 75°, 18 h	(70)	243
Bu₃Sn—CH(OBz)(C₇H₁₅-n), 94% ee	Pd(PPh₃)₂Cl₂ (4%), CuCN, PhMe, 75°, 18 h	(74) 92% ee	243
pyridazine stannane (Ph, Ph)	Pd(PPh₃)₂Cl₂ (5%), Et₄NCl, C₆H₆, reflux, 41 h	(38)	545
phthalimide stannane	Pd(PPh₃)₄ (4%), CuCN, PhMe, 75°, 36 h	(45) + (28)	243
oxazolidinone vinylstannane (Ph)	BnPd(PPh₃)₂Cl (5%), CO (35 psi), C₆H₆, 110°, 24 h	(48)	757
Bu₃Sn–CH=CH–CH(OTBDMS)(C₈H₁₇-n)	Pd₂(dba)₃, P(2-furyl)₃, THF, rt, 6 h	(83)	447
Bu₃Sn–CH=C(CO₂Bu-t)(C₁₅H₃₁-n)	Pd₂(dba)₃ (0.05%), HMPA, 100°, 16 h	(65)	247

TABLE XIV. DIRECT CROSS-COUPLING OF ACYL CHLORIDES: ARYL SYSTEMS (*Continued*)

Substrate	Stannane	Conditions	Product(s) and Yield(s) (%)	Refs.
Me₃Sn substituted indanone-dimethylamide-Boc-pyridyl-thiazoline structure	Bu₃Sn substituted indanone-dimethylamide-Boc-pyridyl-thiazoline structure	[(η³-C₃H₅)PdCl]₂ (11%), proton sponge, THF, reflux, 7 h	Me₂N/Ph indanone product (67)	666
	Bu₃Sn⟶SnBu₃	Pd(PPh₃)₄ (5%), dioxane, 60°, 2 h	Ph—C(O)—CH=CH—SnBu₃ (27)	148
	Bu₃Sn⟶SnBu₃	Pd(PPh₃)₄ (5%), dioxane, 100°, 30 h	Ph—C(O)—CH₂CH₂—C(O)—Ph (50)	148, 753
	Bu₃Sn⟶SnBu₃	Pd(PPh₃)₄ (5%), dioxane, 60°, 3 h	Ph—C(O)—CH=CH—C(O)—Ph (40)	148
	Bu₃Sn furan SnBu₃	Pd(PPh₃)₂Cl₂ (5%), THF, 80°, 24 h or THF, 65°, 8 h	SnBu₃ furan Ph—C(O) product (82)	287, 546
	Bu₃Sn furan SnBu₃	Pd(PPh₃)₂Cl₂ (5%), PhMe, 100°, 32 h	Ph—C(O) furan Ph—C(O) product (20)	287
	Me₃SnSnMe₃	BnPd(PPh₃)₂Cl or Pd(PPh₃)₄ (5%), THF, reflux, 14 h	Ph—C(O)—SnMe₃ (80)	309, 759

442

Substrate	Reagent	Conditions	Product(s) (%)	Refs.
4-bromobenzoyl chloride	Bu$_3$SnH	Pd(PPh$_3$)$_4$ (1%), C$_6$H$_6$, rt	4-bromobenzaldehyde (81)	156
	Me$_4$Sn	BnPd(PPh$_3$)$_2$Cl (0.4%), CHCl$_3$, reflux, 24 h	4'-bromoacetophenone (60)	146
	Me$_4$Sn	BnPd(PPh$_3$)$_2$Cl (0.05%), HMPA, 65°	4'-bromoacetophenone (67) + 4'-methylacetophenone (26)	4
	CH$_2$=C(CF$_3$)SnBu$_3$	BnPd(PPh$_3$)$_2$Cl (1%), HMPA, 65°, 5 h	1-(4-bromophenyl)-2-(trifluoromethyl)prop-2-en-1-one (89)	262
	Bu$_3$SnPh	BnPd(PPh$_3$)$_2$Cl (0.4%), CHCl$_3$, reflux, 2 h	4-bromobenzophenone (89)	146
	Bu$_3$Sn-CH(OEt)CH=CHCH$_3$	BnPd(PPh$_3$)$_2$Cl (1%), THF, 100°, 16 h	enol ether ketone product (69)	305
4-chlorobenzoyl chloride	Me$_4$Sn	BnPd(PPh$_3$)$_2$Cl (0.05%), HMPA, 65°	4'-chloroacetophenone (97)	4
	CH$_2$=C(CF$_3$)SnBu$_3$	BnPd(PPh$_3$)$_2$Cl (1%), HMPA, 65°, 4 h	1-(4-chlorophenyl)-2-(trifluoromethyl)prop-2-en-1-one (93)	262

TABLE XIV. DIRECT CROSS-COUPLING OF ACYL CHLORIDES: ARYL SYSTEMS (*Continued*)

Substrate	Stannane	Conditions	Product(s) and Yield(s) (%)	Refs.
	Bu$_3$Sn–C(=O)–iPr	Pd(PPh$_3$)$_2$Cl$_2$ (1.4%), PhMe, 110°, 24 h	(59)	307
	Bu$_3$Sn—CH=CH—CH$_2$OMe	BnPd(PPh$_3$)$_2$Cl, CH$_2$Cl$_2$, 70°, 8 h	OMe (71)	305
	(furan sulfone stannane)	Pd(PPh$_3$)$_4$ (1%), HMPA, rt, 3 h	(37)	752
	Bu$_3$Sn—CH$_2$—C(=CH$_2$)—TMS	BnPd(PPh$_3$)$_2$Cl (2%), CHCl$_3$, 65°, 24 h	TMS (69)	537
	Et$_3$Sn—C(=O)—C$_6$H$_4$—Cl	[(η^3-C$_3$H$_5$)PdCl]$_2$ (1%), P(OEt)$_3$, CO (120 psi), PhMe, 111°, 2 h	Cl (63)	308, 307
	Bu$_3$Sn—C(Ph)=C—CO$_2$Bu-t	Pd$_2$(dba)$_3$ (0.05%), HMPA, 100°, 45 min	Ph, CO$_2$Bu-t (65)	247
	Me$_3$SnSnMe$_3$	Pd(PPh$_3$)$_2$Cl$_2$ (5%), PhMe, 100°, 15 h	SnMe$_3$ (75)	759

444

F—C(O)Cl (2-fluorobenzoyl chloride)	Me₃SnPh	[(η³-C₃H₅)PdCl]₂ (1%), HMPA, 20°, 10 min	2-F-C₆H₄C(O)Ph (86)	750
F—C₆H₄C(O)Cl (4-fluorobenzoyl chloride)	Bu₃Sn-(indol-3-yl, N-R)	Pd(PPh₃)₂Cl₂ (2%), THF, reflux, 1 h	4-F-C₆H₄C(O)-(indol-3-yl, N-R) R = Me (79) R = Boc (59)	425
NO₂-C₆H₄C(O)Cl (2-nitrobenzoyl chloride)	Me₄Sn	BnPd(PPh₃)₂Cl (0.05%), HMPA, 65°	2-NO₂-C₆H₄C(O)CH₃ (73)	4
O₂N-C₆H₄C(O)Cl (4-nitrobenzoyl chloride)	Bu₃SnH	Pd(PPh₃)₄ (1%), C₆H₆, rt	4-O₂N-C₆H₄CHO (75)	156
	Me₄Sn	[(η³-C₃H₅)PdCl]₂ (1%), HMPA, 20°, 10 min	4-O₂N-C₆H₄C(O)CH₃ (100)	750, 4, 146
	Me₃Sn-CH₂CH=CH₂	BnPd(PPh₃)₂Cl (0.4%), CHCl₃, reflux, 20 min	4-O₂N-C₆H₄C(O)CH=CH₂ (88)	146, 750
	Bu₃Sn-(1,4-dioxen-2-yl)	BnPd(PPh₃)₂Cl (0.08%), C₆H₆, reflux, 3 h	4-O₂N-C₆H₄C(O)-(1,4-dioxen-2-yl) (86)	271
	Me₃Sn-(thien-2-yl)	[(η³-C₃H₅)PdCl]₂ (1%), HMPA, 20°, 2 min	4-O₂N-C₆H₄C(O)-(thien-2-yl) (91)	750

445

TABLE XIV. DIRECT CROSS-COUPLING OF ACYL CHLORIDES: ARYL SYSTEMS (*Continued*)

Substrate	Stannane	Conditions	Product(s) and Yield(s) (%)	Refs.
	Bu₃Sn—CH=CH—C(=O)OEt	BnPd(PPh₃)₂Cl (0.5%), CO (15 psi), CHCl₃, 50°, 12 h	(80)	250
	Bu₃SnC≡CTMS	Pd(PPh₃)₂Cl₂ (1.8%), Cl(CH₂)₂Cl, 84°, 2 h	(51)	149
	Me₃SnPh	BnPd(PPh₃)₂Cl (0.4%), CHCl₃, reflux, 18 h	(97)	146, 750, 751
	Me₃Sn-C₆F₅	[(η³-C₃H₅)PdCl]₂ (1%), HMPA, 20°, 24 h	(32)	750
		Pd(PPh₃)₄ (1%), HMPA, 0°, 0.5 h	(33)	752
	Me₃Sn-C₆H₄-CH₃	Pd(CH₃CN)₂Cl₂ (1%), THF, Et₂O, 20°, 2 h	(70)	367, 590, 750
	Me₃Sn-C₆H₄-OMe	[(η³-C₃H₅)PdCl]₂ (1%), HMPA, 20°, 10 min	(66)	750

446

C$_8$

2-methylbenzoyl chloride (acid chloride substrate)

Organostannane	Product	Conditions	Yield	Ref.
Bu$_3$SnC≡CPh	4-O$_2$N–C$_6$H$_4$–C(O)–C≡CPh	Pd(PPh$_3$)$_2$Cl$_2$ (1.8%), Cl(CH$_2$)$_2$Cl, 84°, 2 h	(57)	149
Bu$_3$Sn–CH(OBz)–C$_7$H$_{15}$-n	4-O$_2$N–C$_6$H$_4$–C(O)–CH(OBz)–C$_7$H$_{15}$-n	Pd(PPh$_3$)$_2$Cl$_2$ (4%), CuCN, PhMe, 75°, 24 h	(40)	243
Bu$_3$Sn–(3-(N-Me)indolyl)	4-O$_2$N–C$_6$H$_4$–C(O)–(3-(N-Me)indolyl)	Pd(PPh$_3$)$_2$Cl$_2$ (2%), THF, reflux, 30 min	(75)	425
Bu$_3$Sn–(glycal, OBn OBn OBn)	4-O$_2$N–C$_6$H$_4$–C(O)–(glycal, OBn OBn CH$_2$OBn)	Pd(CH$_3$CN)$_2$Cl$_2$ (5%), Cl(CH$_2$)$_2$Cl, reflux, 15 min	(71)	423
Bu$_3$SnH	2-Me–C$_6$H$_4$–CHO	Pd(PPh$_3$)$_4$ (1%), C$_6$H$_6$, rt	(92)	156
Bu$_3$Sn–CH(CH$_3$)$_2$	2-Me–C$_6$H$_4$–C(O)–CH(CH$_3$)$_2$	Pd(PPh$_3$)$_2$Cl$_2$ (1.4%), PhMe, 110°, 24 h	(54)	307
Bu$_3$Sn–CH=CH–CH(OEt)–	2-Me–C$_6$H$_4$–C(O)–CH(CH$_3$)–CH=CH–OEt	BnPd(PPh$_3$)$_2$Cl (1%), THF, 100°, 16 h	(52)	305
Me$_3$SnSnMe$_3$	2-Me–C$_6$H$_4$–C(O)–SnMe$_3$	Pd(PPh$_3$)$_2$Cl$_2$ (5%), PhMe, 100°, 16 h	(75)	759

TABLE XIV. DIRECT CROSS-COUPLING OF ACYL CHLORIDES: ARYL SYSTEMS (*Continued*)

Substrate	Stannane	Conditions	Product(s) and Yield(s) (%)	Refs.
4-methylbenzoyl chloride	Bu_3Sn (1-fluorovinyl)	BnPd(PPh₃)₂Cl, THF, 65°, 30 min	(85)	264
	Bu_3Sn (isopropyl ketone vinyl)	Pd(PPh₃)₂Cl₂ (1.4%), PhMe, 110°, 24 h	(63)	307
	Et₃Sn (4-methylbenzoyl)	[(η³-C₃H₅)PdCl]₂ (1%), P(OEt)₃, CO (120 psi), PhMe, 111°, 2 h	(73)	308, 307
4-cyanobenzoyl chloride	Me₃SnSnMe₃	BnPd(PPh₃)₂Cl (5%), THF, reflux, 14 h	SnMe₃ (64)	309
	Me₄Sn	BnPd(PPh₃)₂Cl (0.05%), HMPA, 65°	(99)	4
3-methoxybenzoyl chloride	Bu_3Sn (CF₃ vinyl)	BnPd(PPh₃)₂Cl (1%), HMPA, 65°, 13 h	CF₃ (81)	262
	Bu_3Sn (OEt allyl)	BnPd(PPh₃)₂Cl (1%), THF, 100°, 16 h	OEt (67)	305
4-methoxybenzoyl chloride	Bu₃SnH	Pd(PPh₃)₄ (1%), C₆H₆, rt	H (75)	156

448

Reagent	Conditions	Product (Yield)	Refs.
Me₄Sn	BnPd(PPh₃)₂Cl (0.05%), HMPA, 65°, 25 h	(86)	4
Bu₃Sn (propionyl)	Pd(PPh₃)₂Cl₂ (1.4%), PhMe, 110°, 24 h	(61)	307
Bu₃Sn (aminocyclobutenedione)	BnPd(PPh₃)₂Cl (2.5%), CuI, THF, 50°, 20 min	(98)	268
Bu₃Sn (isobutyryl)	Pd(PPh₃)₂Cl₂ (1.4%), PhMe, 110°, 24 h	(65)	307
Bu₃Sn—CH=CH—CH₂OMe	BnPd(PPh₃)₂Cl, C₆H₆, 100°, 36 h	OMe (78)	305
Bu₃Sn (dioxene)	BnPd(PPh₃)₂Cl (0.08%), C₆H₆, reflux, 3 h	(93)	271
Ph₄Sn	BnPd(PPh₃)₂Cl (0.05%), HMPA, 65°	Ph (84)	4
Bu₃Sn (furan sulfone)	Pd(PPh₃)₄ (1%), HMPA, rt, 3 h	(36)	752

449

TABLE XIV. DIRECT CROSS-COUPLING OF ACYL CHLORIDES: ARYL SYSTEMS (*Continued*)

Substrate	Stannane	Conditions	Product(s) and Yield(s) (%)	Refs.
	Bu₃Sn—CH=CH—CH(OMe)—TMS	1. Pd(PPh₃)₂Cl₂, THF, 65° 2. Bu₄NF, THF, 0°	4-MeO-C₆H₄-CO-CH₂-CH=CH-OMe (>53)	532
	i-PrO squaric acid, Bu₃Sn	1. BnPd(PPh₃)₂Cl (2.5%), CuI, THF, 50° 2. Piperidine	piperidine squarate, 4-MeO-C₆H₄-CO (65)	268
	dioxolane-spiro cyclobutenone, methyl, Bu₃Sn	BnPd(PPh₃)₂Cl (1%), CO (15 psi), C₆H₆, 80°	(92)	268
	[Et₃Sn-CO-C₆H₄-OMe]	[(η³-C₃H₅)PdCl]₂ (1%), POEt₃, CO (120 psi), PhMe, 111°, 2 h	4-MeO-C₆H₄-CO-CO-C₆H₄-OMe (76)	308, 307
	Bu₃Sn cyclobutenedione, N-piperidine	BnPd(PPh₃)₂Cl (2.5%), CuI, THF, 50°, 45 min	(67)	268
	Bu₃Sn—C(Ph)=CH—CO₂Bu-*t*	Pd₂(dba)₃ (0.05%), HMPA, 100°, 15 min	Ph, CO₂Bu-*t*, 4-MeO-C₆H₄-CO (88)	247

450

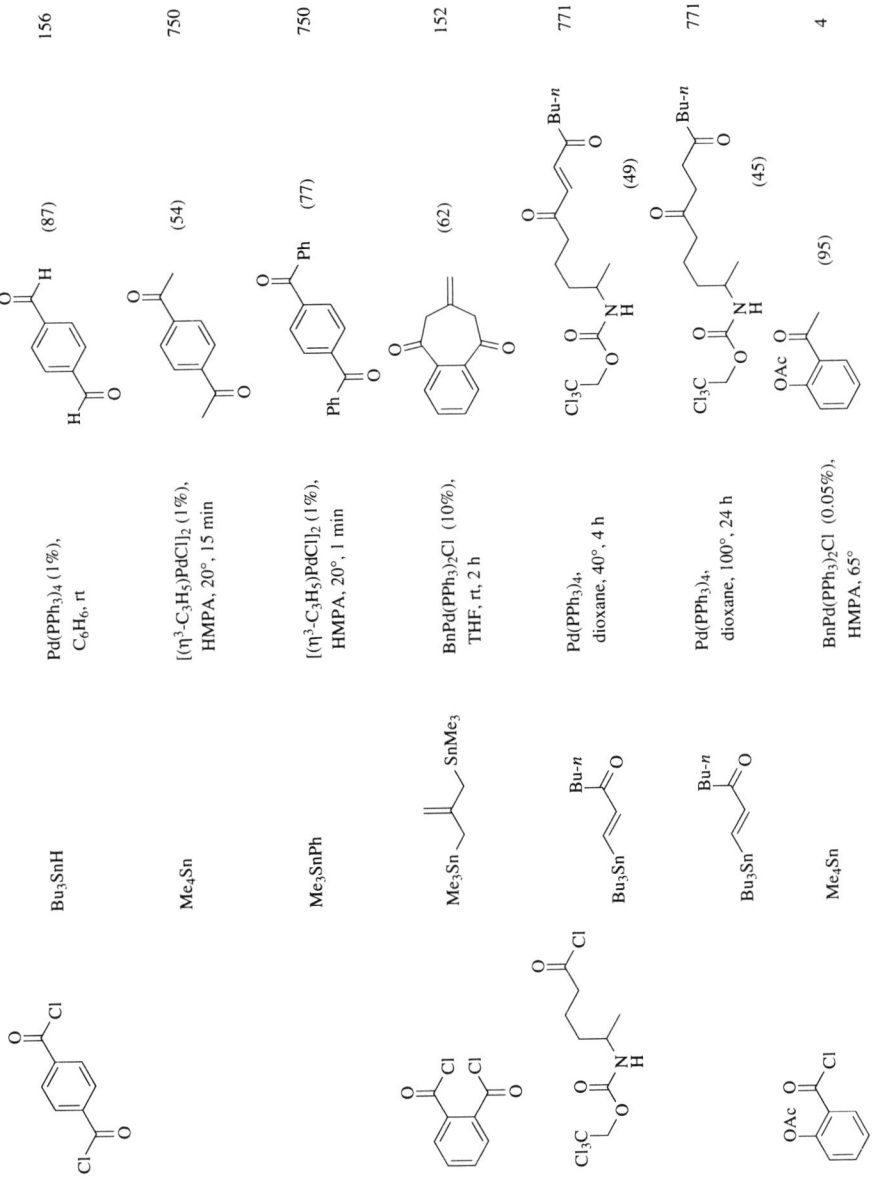

451

Substrate	Stannane	Conditions	Product(s) and Yield(s) (%)	Refs.
	Bu₃Sn (H₂N) squarate	BnPd(PPh₃)₂Cl (2.5%), CuI, BHT, THF, 25°, 2 h	OAc phenyl ketone squarate H₂N (65)	268
	Bu₃Sn (morpholino) squarate	BnPd(PPh₃)₂Cl (2.5%), CuI, BHT, THF, 25°, 2 h	OAc phenyl ketone squarate morpholino (73)	268
	Bu₃Sn (piperidino) squarate	BnPd(PPh₃)₂Cl (2.5%), CuI, BHT, THF, 25°, 3 h	OAc phenyl ketone squarate piperidino (88)	268
C₁₁ naphthalene-1-carbonyl chloride	Bu₃SnH	Pd(PPh₃)₄ (1%), C₆H₆, rt, 2 h	naphthalene-1-carbaldehyde (65)	156, 769
naphthalene-2-carbonyl chloride	Bu₃SnH	Pd(PPh₃)₄ (1%), C₆H₆, rt, 2 h	naphthalene-2-carbaldehyde (85)	156, 769
4-t-Bu-benzoyl chloride	Me₃SnSnMe₃	BnPd(PPh₃)₂Cl (5%), THF, reflux, 45 h	4-t-Bu-C₆H₄-C(O)-SnMe₃ (50)	759

452

Substrate	Organometallic	Conditions	Product (Yield)	Refs.
C14 MEMO—C6H4—COCl	Bu3Sn—(3-piperidinyl-cyclobutene-1,2-dione)	BnPd(PPh3)2Cl (2.5%), CuI, BHT, THF, 25°, 3 h	MEMO—C6H4—CO—(4-piperidinyl-cyclobutene-1,2-dione) (70)	268
C14 Ph-CH2—C6H4—COCl	Me4Sn	Pd(CH3CN)2Cl2, THF, rt, 16 h	Ph-CH2—C6H4—COCH3 (>63)	790
C27 9,9-bis(4-chlorocarbonylphenyl)fluorene	Me4Sn	BnPd(PPh3)2Cl (0.05%), HMPA, 65°	9,9-bis(4-acetylphenyl)fluorene (88)	147

453

TABLE XV. DIRECT CROSS-COUPLING OF ACYL CHLORIDES: BENZYL SYSTEMS

Substrate	Stannane	Conditions	Product(s) and Yield(s) (%)	Refs.
C$_8$ PhCH$_2$COCl	Bu$_3$Sn (isopropenyl)	Pd(PPh$_3$)$_4$ (5%), CO (15 psi), BHT, PhMe, 100°, 5 h	(70)	250
	Bu$_3$Sn (β-lactam)	Pd(PPh$_3$)$_4$	(—)	244
	Me$_3$SnPh	[(η^3-C$_3$H$_5$)PdCl]$_2$ (1%), HMPA, 20°, 10 h	(72)	750
	Bu$_3$Sn (bicyclic)	BnPd(PPh$_3$)$_2$Cl, C$_6$H$_6$, rt, 4 h	(62)	543
C$_{10}$ Ph(MeO)C(CF$_3$)COCl	Bu$_3$Sn (vinyl dioxolane)	BnPd(PPh$_3$)$_2$Cl, CHCl$_3$, 65°	(35)	791
Ph(MeO)C(CF$_3$)COCl	Bu$_3$Sn (vinyl dioxolane)	BnPd(PPh$_3$)$_2$Cl, CHCl$_3$, 65°	(49)	791
MeO$_2$C–C$_6$H$_4$–C(F)(F)COCl	Bu$_3$Sn (tetrahydronaphthalene)	BnPd(PPh$_3$)$_2$Cl, HMPA	(—)	792
C$_{14}$ Ph$_2$CHCOCl	Bu$_3$SnH	Pd(PPh$_3$)$_4$ (1%), C$_6$H$_6$, rt	(75)	156

454

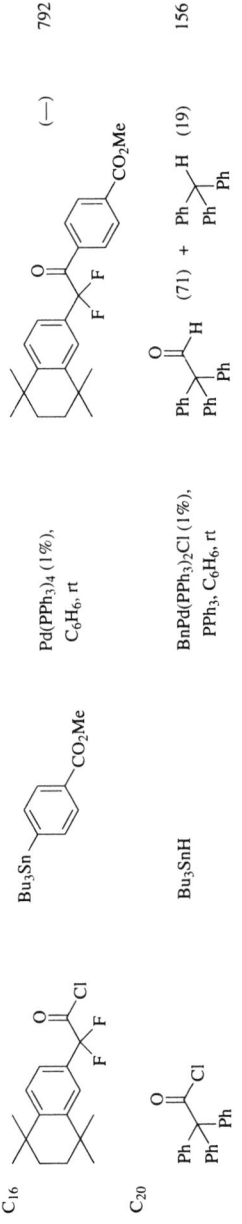

C$_{16}$

C$_{20}$

Bu$_3$Sn—⟨CO$_2$Me⟩

Pd(PPh$_3$)$_4$ (1%), C$_6$H$_6$, rt

(—)   792

Bu$_3$SnH

BnPd(PPh$_3$)$_2$Cl (1%), PPh$_3$, C$_6$H$_6$, rt

(71) + (19)   156

TABLE XVI. DIRECT CROSS-COUPLING OF ACYL CHLORIDES: ALKENYL SYSTEMS

Substrate	Stannane	Conditions	Product(s) and Yield(s) (%)	Refs.
C₃ (acryloyl chloride)	Me₄Sn	BnPd(PPh₃)₂Cl (0.05%), HMPA, 65°	(93)	4
	Bu₃Sn (H₂N squarate)	BnPd(PPh₃)₂Cl (2.5%), CuI, THF, 50°, 15 min	(53)	268
	Bu₃Sn	Pd(PPh₃)₄, dioxane, 100°, 30 h	(86)	148
	Me₃Sn TMS	BnPd(PPh₃)₂Cl (2%), CHCl₃, 65°, 24 h	(75)	537
	Bu₃Sn (N-Me indole)	Pd(PPh₃)₂Cl₂ (2%), THF, reflux, 4 h	(70)	425
	Bu₃Sn (piperidine squarate)	BnPd(PPh₃)₂Cl (2.5%), CuI, THF, 50°, 1 h	(64)	268
	Bu₃Sn Ph	Pd(PPh₃)₄, dioxane, 100°, 30 h	(97)	148
	Bu₃Sn C₁₅H₃₁-n, CO₂Bu-t	Pd₂(dba)₃ (0.05%), hydroquinone, CHCl₃, 25°, 48 h	(55)	247

Substrate	Reagent	Conditions	Product(s) (%)	Refs.
C$_4$				
(crotonyl chloride)	Bu$_3$Sn$\diagup$SnBu$_3$	Pd(PPh$_3$)$_4$, dioxane, 100°, 30 h	(divinyl diketone) (63)	148
	Bu$_3$SnH	Pd(PPh$_3$)$_4$ (1%), C$_6$H$_6$, rt	(85) + (aldehyde) (<5)	156
	Bu$_3$Sn (cyclobutenone)	BnPd(PPh$_3$)$_2$Cl (1.5%), CH$_3$CN, rt, 1.5 h	(53)	267
	Bu$_3$Sn, H$_2$N (cyclobutenedione)	BnPd(PPh$_3$)$_2$Cl (2.5%), CuI, THF, 50°, 15 min	(H$_2$N) (93)	268
	Me$_3$Sn (TMS allyl)	BnPd(PPh$_3$)$_2$Cl (2%), CHCl$_3$, 65°, 24 h	(TMS) (89)	537
	Bu$_3$Sn, OPr-i (cyclobutenedione)	1. BnPd(PPh$_3$)$_2$Cl (2.5%), CuI, THF, 50°, 30 min 2. Piperidine	**I** (68)	268
	Bu$_3$Sn, N (piperidine cyclobutenedione)	BnPd(PPh$_3$)$_2$Cl (2.5%), CuI, THF, 50°, 30 min	**I** (83)	268
C$_4$ (methacryloyl chloride)	Me$_3$Sn (TMS allyl)	BnPd(PPh$_3$)$_2$Cl (2%), CHCl$_3$, 65°, 24 h	(TMS) (75)	537

457

TABLE XVI. DIRECT CROSS-COUPLING OF ACYL CHLORIDES: ALKENYL SYSTEMS (*Continued*)

Substrate	Stannane	Conditions	Product(s) and Yield(s) (%)	Refs.
C5	Bu3Sn–OEt, OEt *E:Z* = 85:15	BnPd(PPh3)2Cl (2%), CHCl3, 65°, 17 h	(80) *E:Z* = 85:15	539
	Bu3Sn	Pd(PPh3)4, dioxane, 100°, 30 h	(86)	148, 753
	Me3Sn–TMS	BnPd(PPh3)2Cl (2%), CHCl3, 65°, 24 h	(74)	537
	Bu3Sn–O	Pd(PPh3)4, dioxane, 100°, 30 h	(63)	753
	Bu3Sn–OEt, OEt	Pd(CH3CN)2Cl2 (1%), DMF, 20°, 2 h	(57)	763
	Bu3Sn–OEt, OEt *E:Z* = 85:15	BnPd(PPh3)2Cl (2%), CHCl3, 65°, 17 h	(85) *E:Z* = 85:15	539
	Bu3Sn–NHBoc	BnPd(PPh3)2Cl (5%)	(71)	465
	Bu3Sn–Ph	Pd(PPh3)4, dioxane, 100°, 30 h	(97)	148, 753
	Me3Sn	Rh(PPh3)3Cl, CH2Cl2, 60°, 48 h	(53)	150, 304

458

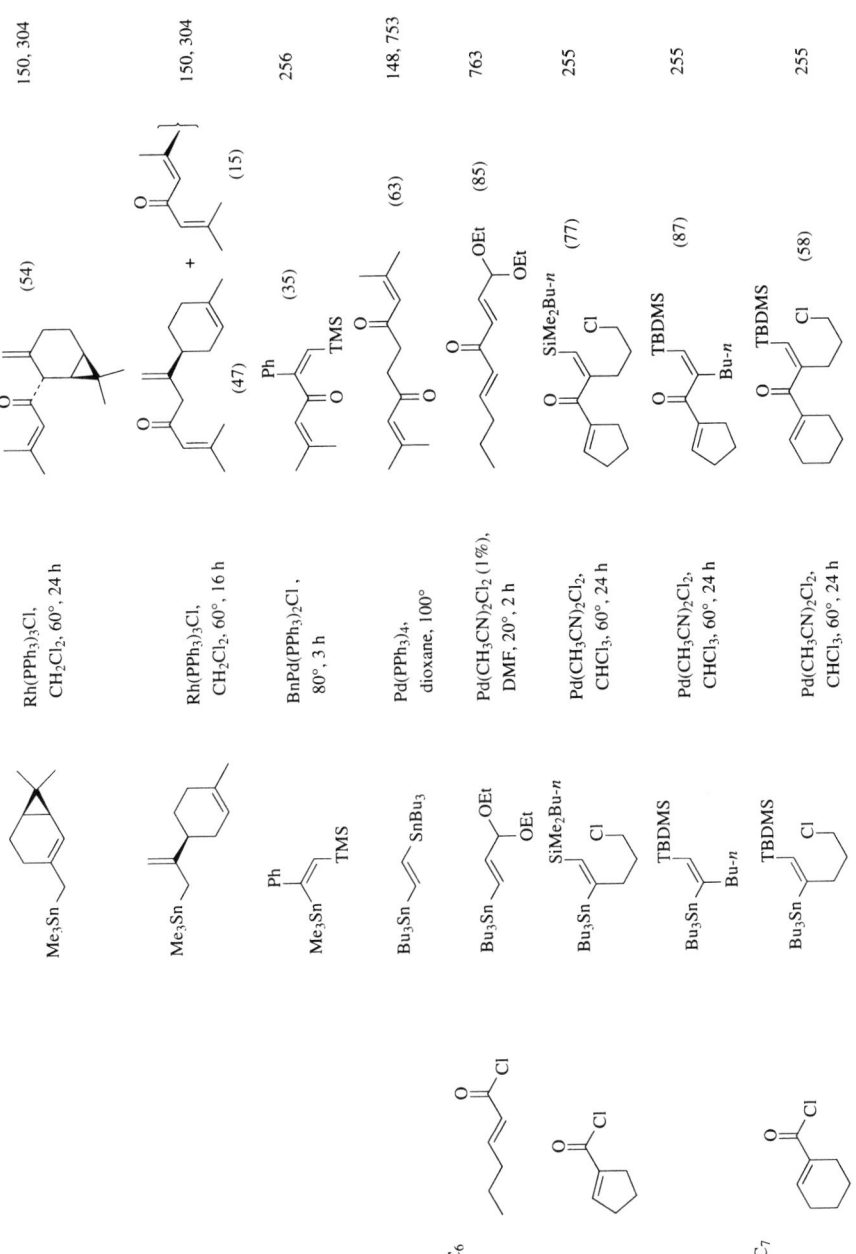

150, 304

150, 304

256

148, 753

763

255

255

255

(54)   (15)

(47)

(35)

(63)

(85)

(77)

(87)

(58)

Rh(PPh₃)₃Cl, CH₂Cl₂, 60°, 24 h

Rh(PPh₃)₃Cl, CH₂Cl₂, 60°, 16 h

BnPd(PPh₃)₂Cl, 80°, 3 h

Pd(PPh₃)₄, dioxane, 100°

Pd(CH₃CN)₂Cl₂ (1%), DMF, 20°, 2 h

Pd(CH₃CN)₂Cl₂, CHCl₃, 60°, 24 h

Pd(CH₃CN)₂Cl₂, CHCl₃, 60°, 24 h

Pd(CH₃CN)₂Cl₂, CHCl₃, 60°, 24 h

C₆

C₇

459

TABLE XVI. DIRECT CROSS-COUPLING OF ACYL CHLORIDES: ALKENYL SYSTEMS (*Continued*)

Substrate	Stannane	Conditions	Product(s) and Yield(s) (%)	Refs.
C₈ (cyclohexylidene acetyl chloride)	Bu₃Sn–CH=C(SiEt₃)Ph	Pd(CH₃CN)₂Cl₂, CHCl₃, 60°, 24 h	(68)	255
C₉ PhC≡C–C(O)Cl	Bu₃Sn–C(CH₃)=CH₂	Pd(AsPh₃)₄	(—)	793
	Me₃SnSnMe₃	BnPd(PPh₃)₂Cl (5%), reflux, 12 h	PhC≡C–C(O)–SnMe₃ (0) + (PhC≡C)₂ (—)	309
(E) Ph–CH=CH–C(O)Cl	Bu₃SnH	Pd(PPh₃)₄ (5%), C₆H₆, rt, 2 h	I (85)	769
	Bu₃SnH	Pd(PPh₃)₄ (1%), C₆H₆, rt	I (73) + (7)	156
	Me₄Sn	BnPd(PPh₃)₂Cl (0.05%), HMPA, 65°	(91)	4
	Bu₃Sn–C(CF₃)=CH₂	BnPd(PPh₃)₂Cl (1%), HMPA, 65°, 4 h	(97)	262
	Bu₃Sn–CH₂CH=CH–C(O)CH₃	Pd(PPh₃)₄, dioxane, 100°, 30 h	(52)	148, 753
	Bu₃Sn–CH₂CH=CH–C(O)CH₃	Pd(PPh₃)₄, dioxane, 60°, 3 h	(36)	148
	Bu₃Sn–(2-furyl/thienyl/pyrryl) X	Pd(PPh₃)₂Cl₂ (1%), C₆H₆, 70°	X = O (84), X = S (86), X = NMe (76)	574

460

R–SnR₃ (organostannane)	Conditions	Product (%)	Refs.

Organostannane	Conditions	Product (%)	Refs.
Bu$_3$Sn-furanone (MeO)	BnPd(PPh$_3$)$_2$Cl, CO, CHCl$_3$, 65°, 16 h or BnPd(PPh$_3$)$_2$Cl (0.4%), Cl(CH$_2$)$_2$Cl, 60°, 16 h	(furanone with MeO and cinnamoyl group) (59)	762, 761
Me$_3$SnPh	[(η^3-C$_3$H$_5$)PdCl]$_2$ (1%), HMPA, 20°, 10 min	(75)	750, 751
Me$_3$Sn–CH$_2$–C(=CH$_2$)–TMS	BnPd(PPh$_3$)$_2$Cl (2%), CHCl$_3$, 65°, 24 h	(73)	537
Me$_3$Sn–C(Ph)=CH–TMS	BnPd(PPh$_3$)$_2$Cl (1%), 80°, 15 min	(85)	256
Bu$_3$Sn–C(Ph)=CH–CO$_2$Bu-t	Pd$_2$(dba)$_3$ (0.05%), hydroquinone, CHCl$_3$, 100°, 2.5 h	(77)	247
Bu$_3$Sn–CH=CH–SnBu$_3$	Pd(PPh$_3$)$_4$, dioxane, 100°, 30 h	(25)	148
Me$_3$SnSnMe$_3$	BnPd(PPh$_3$)$_2$Cl or Pd(PPh$_3$)$_4$ (5%), THF, reflux, 12 h	(80)	309
Bu$_3$Sn–indole (N-Me)	Pd(PPh$_3$)$_2$Cl$_2$ (2%), THF, reflux, 4 h	(60)	425
Bu$_3$Sn–furanone (MeO)	BnPd(PPh$_3$)$_2$Cl (0.4%), Cl(CH$_2$)$_2$Cl, 60°, 16 h	(44)	761

C$_{10}$

(4-chlorocinnamoyl chloride; 2-methyl-3-phenyl-acryloyl chloride)

Substrate	Stannane	Conditions	Product(s) and Yield(s) (%)	Refs.
C$_5$	Me$_4$Sn	BnPd(PPh$_3$)$_2$Cl (0.05%), HMPA, 65°	(91)	4
	CF$_3$ / Bu$_3$Sn	BnPd(PPh$_3$)$_2$Cl (1%), HMPA, 65°, 10 h	(85)	262
	OMe / Me$_3$Sn	BnPd(PPh$_3$)$_2$Cl (1%), C$_6$H$_6$, reflux, 1 h	(83)	749
	Bu$_3$Sn	Pd(PPh$_3$)$_2$Cl$_2$ (3%), THF, rt, 12 h	(95)	288
	Bu$_3$Sn	BnPd(PPh$_3$)$_2$Cl (0.08%), C$_6$H$_6$, reflux, 3 h	(95)	271
	N=SMe / Me$_3$Sn	Pd(PPh$_3$)$_2$Cl$_2$ (7%), THF, reflux, 3 h	(71)	459
	N=SO$_2$Me / R$_3$Sn	Pd(PPh$_3$)$_2$Cl$_2$ (14%), THF, reflux	R = Me (72) R = Bu (62)	459
	Et$_3$Sn	[(η^3-C$_3$H$_5$)PdCl]$_2$ (1%), POEt$_3$, CO (120 psi), PhMe, 111°, 2 h	(41)	308

Bu₃SnCC≡CBu-n not...			

MeO — (48) 761, 762
BnPd(PPh₃)₂Cl (0.4%), Cl(CH₂)₂Cl, 60°, 16 h
or
BnPd(PPh₃)₂Cl, CO, CHCl₃, 65°, 16 h

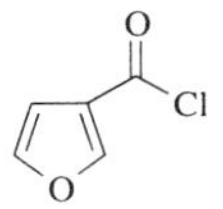

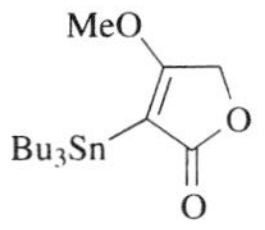

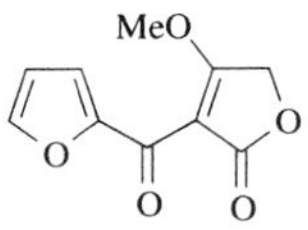

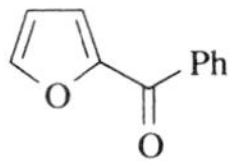

R = H (—)
R = TMS (—)   42
PhCOPd(PPh₃)₂Cl (4%), THF, 40°

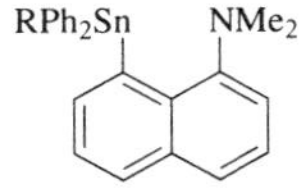

R = Me (—)
R = Ph (—)   42
PhCOPd(PPh₃)₂Cl (4%), THF, 40°

(52)   794
Pd(CH₃CN)₂Cl₂ (5%), PPh₃, CHCl₃, rt, 6 h

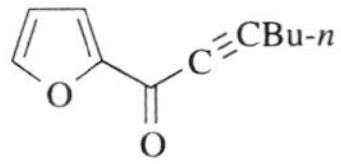

TMS   (67)   537
BnPd(PPh₃)₂Cl (2%), CHCl₃, 65°, 48 h

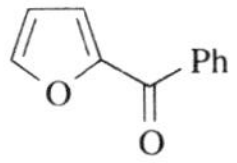

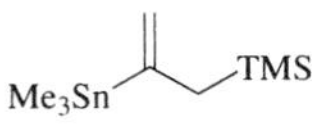

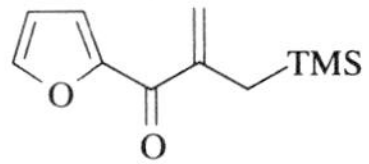

(78)   763
Pd(CH₃CN)₂Cl₂ (1%), DMF, 20°, 2 h

SnMe₃   (80)   309
BnPd(PPh₃)₂Cl (5%), THF, reflux, 15 h

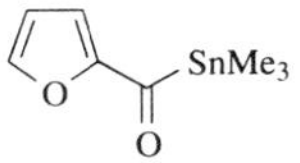

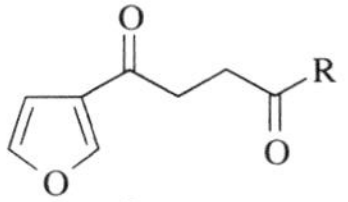

R = Me (70)
R = Ph (64)   148, 753
Pd(PPh₃)₄, dioxane, 100°, 30 h

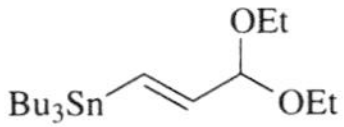

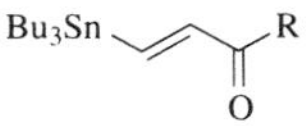

SnBu₃   (35)   148
Pd(PPh₃)₄, dioxane, 60°, 3 h

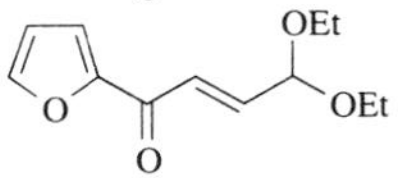

Left column reactants:
MeO / Bu₃Sn

SnPh₂Me / NMe₂ / R

RPh₂Sn / NMe₂

Bu₃SnC≡CBu-n

Me₃Sn / TMS

Bu₃Sn / OEt / OEt

Me₃SnSnMe₃

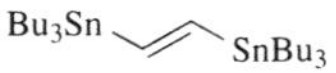

Bu₃Sn / R / O

Bu₃Sn / SnBu₃

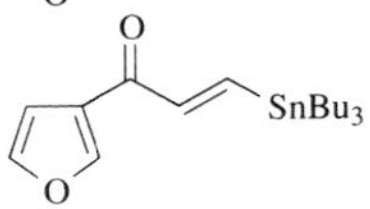

463

Substrate	Stannane	Conditions	Product(s) and Yield(s) (%)	Refs.
C6 (thiophene acyl chloride)	Bu_3Sn—=—$SnBu_3$	$Pd(PPh_3)_4$, dioxane, 100°, 30 h	(55)	148, 753
	Bu_3Sn (furan)	$Pd(PPh_3)_2Cl_2$ (3%), THF, rt, 12 h	(95)	288
	Me_3Sn—TMS	$BnPd(PPh_3)_2Cl$ (2%), $CHCl_3$, 65°, 48 h	TMS (58)	537
	R_3Sn (N—OTBDMS pyrimidine)	1. $Pd(PPh_3)_2Cl_2$ (3.7%), $Cl(CH_2)_2Cl$, reflux, 20 h 2. AcOH	R = Me (36) R = Bu (27)	459
(N-Me pyrrole acyl chloride)	Bu_3Sn—SMe (pyrimidine)	$Pd(PPh_3)_2Cl_2$ (4%), THF, reflux, 6 h or $Pd_2(dba)_3$ (3.7%), $AsPh_3$, THF, rt, 7 h	(72)	151
(nicotinoyl chloride)	Bu_3SnH	$Pd(PPh_3)_4$ (1%), C_6H_6, rt	(90)	156
(MeS-pyrimidine acyl chloride)	Bu_3Sn (N-Me pyrrole)	$Pd(PPh_3)_2Cl_2$ (3%), THF, reflux	(72)	151

Substrate	Stannane/Reagent	Conditions	Product (yield)	Refs.
C9 (acid chloride structure)	Bu3Sn–N–SO2Ph	Pd(PPh3)2Cl2 (3%), THF, reflux	(61)	151
	Bu3Sn–(imidazole)–N–Me	Pd(PPh3)2Cl2 (3%), THF, reflux	(25–31)	151
	Bu3Sn–(furan)–O–SiThex Me Me	Pd(PPh3)2Cl2 (7%), THF, reflux, 30 min	(73)	459
	Bu3Sn–(furan)–Me–O–SiThex Me	Pd(PPh3)2Cl2 (7%), THF, reflux, 30 min	(77)	459
	Bu3Sn–(propenyl)	BnPd(PPh3)2Cl (0.4%), HMPA, 70°, 3 h	(90)	795
C10 (MeO2C...furan acid chloride)	Bu3SnC≡CBu-n	Pd(CH3CN)2Cl2 (5%), PPh3, CH2Cl2, 20°, 24 h	(60)	794
C10 (pyridine acid chloride, CO2Pr-i)	Bu3SnH	PdCl2, PPh3	(60)	796
C11 (pyrrole acid chloride, PhO2S)	Bu3Sn–(pyrimidine)–SMe	Pd(PPh3)2Cl2 (4%), THF, reflux, 6 h	(61)	151

Substrate	Stannane	Conditions	Product(s) and Yield(s) (%)	Refs.
C$_{12}$	R: TMS / CO$_2$Et / Ph	Pd(dppf)Cl$_2$ (5%), CHCl$_3$, rt, 2 d	(53) / (51) / (54) **I**	381
	Me$_3$Sn—CH=CH—Ph	Pd(dppf)Cl$_2$ (5%), CHCl$_3$, rt, 2 d	**I** (94)	381
	E:Z = 66:34	Pd(dppf)Cl$_2$ (5%), CHCl$_3$, rt, 2 d	(25)	381
	E:Z = 83:17	Pd(dppf)Cl$_2$ (5%), CHCl$_3$, rt, 2 d	(41)	381
C$_{13}$	Bu$_3$Sn—C$_6$H$_4$—NO$_2$	[(η^3-C$_3$H$_5$)PdCl]$_2$ (1%), THF, rt, 5 h	(43)	797
C$_{20}$	Bu$_3$Sn—CH=CH$_2$ / Bu$_3$Sn—CH=CH—CO$_2$Et / Bu$_3$Sn—CH=CH—CO$_2$Et / Me$_3$Sn—CH=CH—Ph	Pd(dppf)Cl$_2$ (5%), CHCl$_3$, rt, 2 d	R = CH=CH$_2$ (64) / R = (E)-CH=CHCO$_2$Et (51) / R = (E)-CH=CHCO$_2$Et (64) / R = (E)-CH=CHPh (96)	381

TABLE XVIII. DIRECT CROSS-COUPLING OF CHLOROFORMATES AND CARBAMOYL CHLORIDES

Substrate	Stannane	Conditions	Product(s) and Yield(s) (%)	Refs.
C₂ MeO–C(=O)–Cl	Me₃Sn–C(=CH₂)–TMS	BnPd(PPh₃)₂Cl (2%), CHCl₃, 65°, 48 h	MeO–C(=O)–C(=CH₂)–CH₂–TMS (49)	537
C₃ Me₂N–C(=O)–Cl	Bu₃Sn–CH=CH₂	BnPd(PPh₃)₂Cl (4%), PhMe, 100°, 8 h	Me₂N–C(=O)–CH=CH₂ (71)	157
	Bu₃SnPh	BnPd(PPh₃)₂Cl (4%), PhMe, 100°, 8 h	Me₂N–C(=O)–Ph (72)	157
C₅ EtO–C(=O)–Cl	Bu₃Sn–C₆H₄–NMe₂	BnPd(PPh₃)₂Cl (5%), h hydroquinone, HMPA, 100° 5 h	EtO–C(=O)–C₆H₄–NMe₂ (71)	157
i-BuO–C(=O)–Cl	Bu₃Sn–(furan)–CHO	Pd(PPh₃)₂Cl₂ (0.5%), quinone, PhMe, HMPA, 100°, 3 h	i-BuO–C(=O)–(furan)–CHO (70)	158
	Bu₃Sn–C₆H₄–R R: H / Me / OMe / NMe₂ / CO₂Me	BnPd(PPh₃)₂Cl (5%), hydroquinone, HMPA, 100° 5 h	i-BuO–C(=O)–C₆H₄–R (66) (64) (66) (88) (72)	157
C₈ Ph–N(Me)–C(=O)–Cl	Bu₃Sn–CH=CH₂	BnPd(PPh₃)₂Cl (4%), PhMe, 100°, 8 h	Ph–N(Me)–C(=O)–CH=CH₂ (74)	157

467

TABLE XVIII. DIRECT CROSS-COUPLING OF CHLOROFORMATES AND CARBAMOYL CHLORIDES (*Continued*)

Substrate	Stannane	Conditions	Product(s) and Yield(s) (%)	Refs.
	Bu$_3$Sn (allyl)	BnPd(PPh$_3$)$_2$Cl (4%), PhMe, 100°, 8 h	(18)	157
	Bu$_3$Sn (isopropenyl)	BnPd(PPh$_3$)$_2$Cl (4%), PhMe, 100°, 8 h	(71)	157
	Bu$_3$Sn (2-methyl-1-propenyl)	BnPd(PPh$_3$)$_2$Cl (4%), PhMe, 100°, 8 h	(60)	157
	Bu$_3$Sn–C(OEt)=CH$_2$	BnPd(PPh$_3$)$_2$Cl (4%), PhMe, 100°, 8 h	(65)	157
	Bu$_3$Sn (2-furyl)	BnPd(PPh$_3$)$_2$Cl (4%), PhMe, 100°, 8 h	(68)	157
	Bu$_3$Sn (3-furyl)	BnPd(PPh$_3$)$_2$Cl (4%), PhMe, 100°, 8 h	(73)	157
	Bu$_3$Sn–furyl–CHO	Pd(PPh$_3$)$_2$Cl$_2$ (0.5%), quinone, PhMe, 100°, 1 h	(90)	158
	Bu$_3$Sn–furyl–CHO	Pd(PPh$_3$)$_2$Cl$_2$ (0.5%), quinone, PhMe, 100°, 3 h	(62)	158

C9

Reactant	Conditions	Product(s) and Yield(s) (%)	Refs.

Bu$_3$Sn — (furan, OHC) ; Pd(PPh$_3$)$_2$Cl$_2$ (0.5%), quinone, PhMe, 100°, 4 h ; (70) ; 158

Bu$_3$Sn — (pyrrole, CHO, N–Me) ; Pd(PPh$_3$)$_2$Cl$_2$ (0.5%), quinone, PhMe, 100°, 3 h ; (65) ; 158

Bu$_3$Sn — C$_6$H$_4$R ; BnPd(PPh$_3$)$_2$Cl (4%), PhMe, 100°, 8 h ; 157

R	
H	(81)
Me	(48)
OMe	(57)
NMe$_2$	(50)

Bu$_3$Sn — CH=CH–Ph (E:Z = 80:20) ; BnPd(PPh$_3$)$_2$Cl (4%), PhMe, 100°, 8 h ; (67) ; 157

Bu$_3$SnH ; Pd(PPh$_3$)$_4$ (5%), C$_6$H$_6$, rt, 2 h ; (17) ; 156

Bu$_3$Sn — CH$_2$=C(Me)– ; BnPd(PPh$_3$)$_2$Cl (5%), hydroquinone, PhMe, HMPA, 100°, 5 h ; (47) ; 157

Bu$_3$Sn — CH$_2$=C(Me)CH= ; BnPd(PPh$_3$)$_2$Cl (5%), hydroquinone, PhMe, HMPA, 100°, 5 h ; (70) ; 157

Acid chlorides:
BnO–C(O)Cl

n-C$_8$H$_{17}$O–C(O)Cl

469

TABLE XVIII. DIRECT CROSS-COUPLING OF CHLOROFORMATES AND CARBAMOYL CHLORIDES (*Continued*)

Substrate	Stannane	Conditions	Product(s) and Yield(s) (%)	Refs.
	Bu$_3$Sn–(2-furyl)	BnPd(PPh$_3$)$_2$Cl (5%), hydroquinone, PhMe, HMPA, 100°, 5 h	n-C$_8$H$_{17}$O–C(=O)–(2-furyl) (70)	157
	Bu$_3$Sn–(3-furyl)	BnPd(PPh$_3$)$_2$Cl (5%), hydroquinone, PhMe, HMPA, 100°, 5 h	n-C$_8$H$_{17}$O–C(=O)–(3-furyl) (83)	157
	Bu$_3$Sn–(5-formyl-2-furyl) (CHO)	Pd(PPh$_3$)$_2$Cl$_2$ (0.5%), quinone, PhMe, HMPA, 100°, 3 h	n-C$_8$H$_{17}$O–C(=O)–furyl–CHO (71)	158
	OHC–furyl–Bu$_3$Sn	Pd(PPh$_3$)$_2$Cl$_2$ (0.5%), quinone, PhMe, HMPA, 100°, 3 h	n-C$_8$H$_{17}$O–C(=O)–furyl(CHO) (53)	158
	Bu$_3$Sn–(N-Me-pyrrolyl)–CHO	Pd(PPh$_3$)$_2$Cl$_2$ (0.5%), quinone, PhMe, HMPA, 100°, 3 h	n-C$_8$H$_{17}$O–C(=O)–(N-Me-pyrrolyl)–CHO (70)	158

470

TABLE XIX. INTRAMOLECULAR CROSS-COUPLING OF ACYL CHLORIDES AND CHLOROFORMATES

Substrate	Stannane	Conditions	Product(s) and Yield(s) (%)	Refs.
C₄	Me₃Sn—⧸—SnMe₃	BnPd(PPh₃)₂Cl (10%), THF, rt, 2 h	(65)	152
C₈	Me₃Sn—⧸—SnMe₃	BnPd(PPh₃)₂Cl (10%), THF, rt, 2 h	(62)	152
		BnPd(PPh₃)₂Cl (5%), CO (45 psi), PhMe, 100°, 7 h	(38)	770
C₉		BnPd(PPh₃)₂Cl (5%), CO (45 psi), PhMe, 100°, 14 h	**I** (58)	153, 154
		BnPd(PPh₃)₂Cl (5%), CO (45 psi), PhMe, 100°, 14 h	(15) + **I** (38)	154

TABLE XIX. INTRAMOLECULAR CROSS-COUPLING OF ACYL CHLORIDES AND CHLOROFORMATES (*Continued*)

	Substrate	Stannane	Conditions	Product(s) and Yield(s) (%)	Refs.
C_{10}			BnPd(PPh$_3$)$_2$Cl (5%), CO (45 psi), PhMe, 100°, 14 h	(32) + (30)	153, 154
C_{11}			BnPd(PPh$_3$)$_2$Cl (5%), CO (45 psi), PhMe, 100°, 14 h	(41)	153, 154
C_{12}			BnPd(PPh$_3$)$_2$Cl (4.5%), Cl(CH$_2$)$_2$Cl, 37°, 20 h	C_7H_{15}-n (84)	798
			BnPd(PPh$_3$)$_2$Cl (4.5%), Cl(CH$_2$)$_2$Cl, rt, 92 h	C_6H_{13}-n (49)	798
C_{13}			Pd(PPh$_3$)$_4$ (0.4%), THF, reflux	Bu-n, n-C$_5$H$_{11}$ (77)	155
			BnPd(PPh$_3$)$_2$Cl (5%), CO (45 psi), PhMe, 100°, 14 h	(55)	153, 154

	Substrate	Conditions	Product (Yield %)	Refs.
C_{14}	Bu$_3$Sn— structure, O, Cl, C$_9$H$_{19}$-n	BnPd(PPh$_3$)$_2$Cl (4.5%), Cl(CH$_2$)$_2$Cl, 37°, 20 h	C$_9$H$_{19}$-n lactone (96)	798
C_{15}	Cl, O, (9), SnBu$_3$ structure	BnPd(PPh$_3$)$_2$Cl (5%), CO (45 psi), PhMe, 100°, 14 h	(9) (53)	153, 154
C_{16}	Cl, O, (9), SnBu$_3$ structure	BnPd(PPh$_3$)$_2$Cl (5%), CO (45 psi), PhMe, 100°, 14 h	(9) (70)	153, 154
C_{19}	Cl, O, (13), SnBu$_3$ structure	BnPd(PPh$_3$)$_2$Cl (5%), CO (45 psi), PhMe, 100°, 14 h	(13) (48)	153, 154
C_{21}	C$_6$H$_{11}$, Cl, O, (9), SnBu$_3$ structure	BnPd(PPh$_3$)$_2$Cl (5%), CO (45 psi), PhMe, 100°, 14 h	C$_6$H$_{11}$ (58)	153, 154

TABLE XX. DIRECT CROSS-COUPLING OF ALLYL AND PROPARGYL ELECTROPHILES

Substrate	Stannane	Conditions	Product(s) and Yield(s) (%)	Refs.
C₃ HC≡C–CH₂–Br	(uridine-derived) –SnBu₃	Pd₂(dba)₃ (5%), CuI, P(2-furyl)₃ (20%), THF, 60°, 2 h	(uridine-derived allene product) (47)	170
(allyl / Br)	Bu₃Sn—	[(η³-C₃H₅)PdCl]₂ (1%), HMPA, 20°	(94)	553
	Bu₃Sn—C(=O)—	Pd(PPh₃)₄ (10%), C₆H₆, 60°, 20 h	(44)	529
	Bu₃Sn—	Pd(CH₃CN)₂Cl₂, PPh₃, CHCl₃, 65°	(80)	24
	Bu₃Sn—	BnPd(PPh₃)₂Cl (0.3%), CHCl₃, 65°, 48 h	(20) + (58)	35
	Me₃Sn—(OH)	Pd(CH₃CN)₂Cl₂ (2%), DMF, 70°, 1.5 h	—OH (89)	440
	Me₃Sn—(thiophene)	Pd(PPh₃)₄ (5%), HMPA, 20°, 6 h	(8)	161
	Bu₃Sn—	BnPd(PPh₃)₂Cl (0.3%), CHCl₃, 65°, 48 h	(3) + (48)	35
	Bu₃SnPh	[(η³-C₃H₅)PdCl]₂ (1%), DMF, 70°	Ph (95)	553, 535

Organotin reagent	Conditions	Product (%)	Refs.
Me₃Sn—C₆H₄—Cl	[(η³-C₃H₅)PdCl]₂ (1%), DMF, 70°	(94)	553
Bu₃Sn (methylcyclopentenone)	Pd(dba)₂ (2%), THF, 55°, 5.5 h	(89)	440
HO—/Bu₃Sn/TMS	BnPd(PPh₃)₂Cl (1%), 80°, 240 h	(40)	256
Bu₃Sn—Ph	D₇₁₇-Pd(0), Me₂CO, reflux, 25 h	Ph (92)	535
Me₃Sn—C₆H₄—CH₃	[(η³-C₃H₅)PdCl]₂ (1%), HMPA, 20°	(75)	553
Me₃Sn—C₆H₄—OMe	[(η³-C₃H₅)PdCl]₂ (1%), DMF, 70°	OMe (80)	553
MeO—/Me₃Sn/TMS	BnPd(PPh₃)₂Cl (1%), 80°, 48 h	(55)	256
HO—/Me₃Sn/TMS	BnPd(PPh₃)₂Cl (1%), 80°, 48 h	(32) + (8)	256
Me₃SnC≡CPh	Pd(PPh₃)₄ (5%), HMPA, 20°, 6 h	(3)	161
Bu₃Sn—CH=CH—CH(OH)C₅H₁₁-n	Pd(CH₃CN)₂Cl₂, HMPA, 20°, 5 h	(90)	799

Substrate	Stannane	Conditions	Product(s) and Yield(s) (%)	Refs.
	Me₃Sn, CO₂Et, TMS (allyl)	BnPd(PPh₃)₂Cl (1%), 80°, 36 h	CO₂Et, TMS (allyl) (61)	256
	Bu₃Sn-indole, N-Me	Pd(PPh₃)₂Cl₂ (2%), THF, reflux, 20 h	2-allyl-N-methylindole (70)	425
	Bu₃Sn, CO₂Et, NHAc	Pd₂(dba)₃ (5%), AsPh₃ (40%), THF, 65°	CO₂Et, NHAc, (allyl) (80)	375
	Bu₃Sn-quinone	Pd (0)	allyl-quinone (80)	800
	Bu-*n*, Bu₃Sn, TMS	BnPd(PPh₃)₂Cl (1%), 80°, 55 h	Bu-*n*, TMS (allyl) (61)	256
	Bu-*t*, Me₃Sn, TMS	BnPd(PPh₃)₂Cl (1%), 80°, 450 h	Bu-*t*, TMS (allyl) (43) + *t*-Bu, TMS (5)	256
	Bu₃Sn, OEt, TBDMS	BnPd(PPh₃)₂Cl (1.6%), CuI, DMF, 50°, 7 h	OEt, TBDMS (allyl) (95)	49
	n-C₈H₁₇, Bu₃Sn	Pd(PPh₃)₄ (5%), C₆H₆, reflux	*n*-C₈H₁₇ (allyl) (—)	442
	Ph, R₃Sn, TMS	BnPd(PPh₃)₂Cl (1%), 80°	Ph, TMS (allyl) R = Me, 170 h (51) R = Bu, 45 h (75)	256

476

Stannane	Conditions	Product (%)	Refs.
Me$_3$Sn / TMS, Ph	BnPd(PPh$_3$)$_2$Cl	BrMe$_2$Sn / TMS, Ph (—)	254
Bu$_3$Sn, OEt, TBDMS	BnPd(PPh$_3$)$_2$Cl (1.6%), CuI, DMF, 50°	OEt, TBDMS (97)	49
R, Bu$_3$Sn (quinone)	Pd (0)	R = TMS (84), R = n-Bu (62)	800
PhO, Bu$_3$Sn, OBn	Pd(PPh$_3$)$_4$ (5%), C$_6$H$_6$, reflux, 4 h	OBn (87)	789
PhO, Me$_3$Sn, TMS	BnPd(PPh$_3$)$_2$Cl (1%), 80°, 48 h	TMS (50)	256
SnBu$_3$ (fluorene)	[η^3-C$_3$H$_5$)PdCl]$_2$ (1%), HMPA, 20°	(70)	553
Bu$_3$Sn (indole, SEM)	Pd$_2$(dba)$_3$ (5%), P(2-furyl)$_3$ (20%), THF, 60°	(93)	289, 425
TsHN, Bu$_3$Sn (indole, SEM)	Pd$_2$(dba)$_3$ (10%), P(2-furyl)$_3$ (20%), THF, 65°, 2.5 h	(89)	74

477

TABLE XX. DIRECT CROSS-COUPLING OF ALLYL AND PROPARGYL ELECTROPHILES (*Continued*)

Substrate	Stannane	Conditions	Product(s) and Yield(s) (%)	Refs.
(allyl chloride)	(pyranose: OBn, Bu$_3$Sn, OBn, OBn)	Pd$_2$(dba)$_3$, THF, reflux	(pyranose: OBn, OBn, OBn, allyl) (71–74)	423, 424
	SnMe$_3$, Ph	BnPd(PPh$_3$)$_2$Cl, 70°, 5 d	Ph ... **I** (54)	254
	SnMe$_3$, Ph	BnPd(PPh$_3$)$_2$Cl, THF	**I** + SnMe$_3$ Ph (—)	254
	Me$_3$SnSnMe$_3$	[(η^3-C$_3$H$_5$)PdCl]$_2$ (5%), HMPA, 20°, 10 min	SnMe$_3$ (80)	314, 557
	Bu$_3$SnSnBu$_3$	Pd(PPh$_3$)$_2$Br$_2$ (0.6%), PhMe, 110°, 15 h	SnBu$_3$ (68)	547
	OEt, Bu$_3$Sn	1. Pd(PPh$_3$)$_2$Cl$_2$ (1%), PhMe, 100°, 20 h 2. HCl (5% aq.)	O (43)	269
	(thiophene) Me$_3$Sn–S	Pd(PPh$_3$)$_4$ (5%), HMPA, 20°, 6 h	(thiophene) (11)	161
	Me$_3$Sn–(aryl)–R	Pd(dba)$_2$ (3%), PPh$_3$ (6%), THF, 50°, 24 h	(aryl)–R R = H (71) R = Me (16)	161
	Me$_3$SnC≡CPh	"	C≡CPh (12)	161
	CO$_2$Bu-t, Bu$_3$Sn *E:Z* = 61:39	Pd(dba)$_2$ (3%), PPh$_3$ (6%), THF, 50°, 19 h	CO$_2$Bu-t Ph (56)	248
	Me$_3$SnSnMe$_3$	[(η^3-C$_3$H$_5$)PdCl]$_2$ (5%), HMPA, 20°, 10 min	SnMe$_3$ (83)	314

478

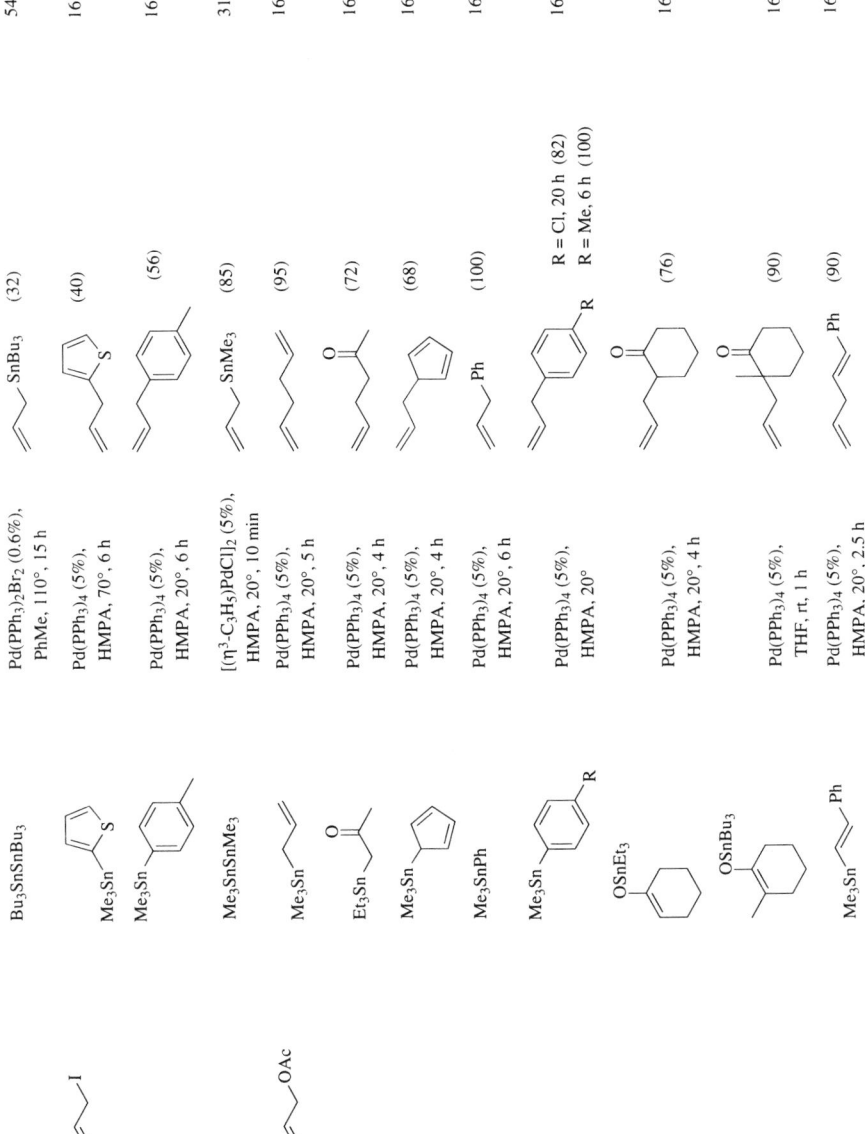

Substrate	Organostannane	Conditions	Product (%)	Refs.
(butenyl iodide)	Bu₃SnSnBu₃	Pd(PPh₃)₂Br₂ (0.6%), PhMe, 110°, 15 h	SnBu₃ (32)	547
	Me₃Sn (thiophene)	Pd(PPh₃)₄ (5%), HMPA, 70°, 6 h	(40)	161
	Me₃Sn (tolyl)	Pd(PPh₃)₄ (5%), HMPA, 20°, 6 h	(56)	161
	Me₃SnSnMe₃	[(η³-C₃H₅)PdCl]₂ (5%), HMPA, 20°, 10 min	SnMe₃ (85)	314
	Me₃Sn (allyl)	Pd(PPh₃)₄ (5%), HMPA, 20°, 5 h	(95)	161
	Et₃Sn (acetonyl)	Pd(PPh₃)₄ (5%), HMPA, 20°, 4 h	(72)	161
	Me₃Sn (cyclopentadienyl)	Pd(PPh₃)₄ (5%), HMPA, 20°, 4 h	(68)	161
	Me₃SnPh	Pd(PPh₃)₄ (5%), HMPA, 20°, 6 h	Ph (100)	161
	Me₃Sn (aryl-R)	Pd(PPh₃)₄ (5%), HMPA, 20°	R (R = Cl, 20 h (82); R = Me, 6 h (100))	161
(butenyl OAc)	OSnEt₃ (cyclohexenyl)	Pd(PPh₃)₄ (5%), HMPA, 20°, 4 h	(76)	161
	OSnBu₃ (methylcyclohexenyl)	Pd(PPh₃)₄ (5%), THF, rt, 1 h	(90)	160
	Me₃Sn (styryl)	Pd(PPh₃)₄ (5%), HMPA, 20°, 2.5 h	Ph (90)	161

TABLE XX. DIRECT CROSS-COUPLING OF ALLYL AND PROPARGYL ELECTROPHILES (*Continued*)

Substrate	Stannane	Conditions	Product(s) and Yield(s) (%)	Refs.
	Me₃Sn–(indene)	Pd(PPh₃)₄ (5%), HMPA, 20°, 4 h	(72)	161
	Me₃Sn–CH(Ph)C(O)OEt	Pd(PPh₃)₄ (5%), HMPA, 20°, 2.5 h	(77)	161
	SnMe₃ (fluorene)	Pd(PPh₃)₄ (5%), HMPA, 20°, 4 h	(97)	161
	Me₃SnSnMe₃	Pd(PPh₃)₄ (5%), HMPA, 20°, 40 min	SnMe₃ (92)	314
Cl, Cl (substrate)	Bu₃SnH	Pd(PPh₃)₄ (6%), THF, rt	**I** + Cl~~~ **II** (>90) 38:62	801
Cl, Cl (substrate)	Bu₃SnH	Pd(PPh₃)₄ (6%), THF, rt	**I** + **II** (>90) 35:65	801
Cl, Cl (substrate)	Bu₃SnH	Pd(PPh₃)₄ (6%), THF, rt	**I** + **II** (>90) 36:64	801
OAc, OAc (substrate)	Bu₃SnH	Pd(PPh₃)₄ (6%), THF, rt	AcO + AcO (>90) 35:65	801
C₄ Br (substrate)	Bu₃Sn	[(η³-C₃H₅)PdCl]₂ (1%), maleic anhydride (5%), THF, 50°, 12 h	(40)	38, 39
	Bu₃Sn–CO₂Et	"	CO₂Et (35)	38, 39

480

Substrate	Reagent	Conditions	Product(s)	Refs.
(E)-crotyl chloride	Me₃SnSnMe₃	[(η³-C₃H₅)PdCl]₂ (5%), HMPA, 20°, 10 min; Me₂CO, 20°, 10 min; DMF, 20°, 10 min	**I** SnMe₃ **II** SnMe₃ — **I** (57) + **II** (16); **I** (58) + **II** (14); **I** (42) + **II** (13)	314
Bu₃Sn (ketone)		Pd(PPh₃)₄ (10%), C₆H₆, 60°, 20 h	(64) + (22)	529
Bu₃Sn (ketone)		Pd(PPh₃)₄ (10%), C₆H₆, 60°, 20 h	(84)	529
Bu₃Sn (oxazoline)		Pd(PPh₃)₂Cl₂ (1%)	(47)	426
Bu₃Sn (ketone)		Pd(PPh₃)₄ (10%), C₆H₆, 60°, 20 h	(54) + (18)	529
3-chloro-1-butene	Me₃SnSnMe₃	[(η³-C₃H₅)PdCl]₂ (5%), HMPA, 20°, 10 min; Me₂CO, 20°, 10 min; DMF, 20°, 10 min	**I** SnMe₃ **II** SnMe₃ — **I** (41); **I** (44) + **II** (12); **I** (43) + **II** (11)	314
OAc / OAc (cis-diol diacetate)	SnBu₂ (dioxastannolane)	Pd(PPh₃)₄, THF, rt	(66)	322
OAc / OAc	SnBu₂	Pd(PPh₃)₄, THF, rt	(77)	322

481

TABLE XX. DIRECT CROSS-COUPLING OF ALLYL AND PROPARGYL ELECTROPHILES (*Continued*)

Substrate	Stannane	Conditions	Product(s) and Yield(s) (%)	Refs.
(epoxide)	Me₃Sn⌒	Pd(CH₃CN)₂Cl₂, DMF, H₂O, rt	HO⌒ (67) + HO⌒ (10) *E:Z* = 10:1	164, 802
	Bu₃Sn⌒TMS	Pd(CH₃CN)₂Cl₂, DMF, H₂O, rt	(82-88) *E:Z* = 7:1 TMS + HO⌒TMS (18-12)	164, 802
	Me₃SnPh	Pd(CH₃CN)₂Cl₂, DMF, H₂O, rt	(73) + HO⌒Ph (10) *E:Z* = 18:1	164, 802
	Bu₃Sn⌒CO₂Me *E:Z* = 4:1	Pd(CH₃CN)₂Cl₂, DMF, H₂O, rt	CO₂Me (74) + CO₂Me (19)	803
	Me₃Sn⌒Ph	Pd(CH₃CN)₂Cl₂, DMF, H₂O, rt	(56) *E:Z* = 13:1 + HO⌒Ph (9) *E:Z* = 13:1	164, 802
	BocHN (aryl stannane) Bu₃Sn	Pd(CH₃CN)₂Cl₂, DMF, H₂O, rt	(56) + (12)	802
Cl₃C⌒Cl	Bu₃SnH	Pd(PPh₃)₄ (6%), THF, rt	Cl₂C=CHCl (>90)	804
Br⌒CN	Bu₃Sn⌒	BnPd(PPh₃)₂Cl, CHCl₃, 65°	CN (65)	24

482

C$_5$

	conditions	product (yield)	ref.
Bu$_3$Sn─ (image)	BnPd(PPh$_3$)$_2$Cl (0.3%), ZnCl$_2$, THF, 65°, 24 h	**I** (43)	35
(image) $\overset{}{\underset{4}{\big\rvert}}$Sn	BnPd(PPh$_3$)$_2$Cl (0.3%), C$_6$H$_6$, 100°, 24 h	**I** (70) + (10)	35
(image) $\overset{}{\underset{4}{\big\rvert}}$Sn	BnPd(PPh$_3$)$_2$Cl (0.3%), ZnCl$_2$, THF, 65°, 48 h	**I** (81)	35
Bu$_3$Sn (ketone image)	Pd(PPh$_3$)$_4$ (10%), C$_6$H$_6$, 60°, 20 h	(54)	529
Me$_3$Sn (OH image)	Pd$_2$(dba)$_3$ (2%), THF, 55°, 3 h	(90)	440
Bu$_3$Sn (oxazoline image)	Pd(PPh$_3$)$_2$Cl$_2$ (1%)	(51)	426
(quinone image) Bu$_3$Sn	Pd(0)	(56), (90), (90), (92)	800

R^1	R^2	R^3
Me	Me	Me
Me	-CH=CHCH=CH-	
TMS	Me	Me
TMS	-CH=CHCH=CH-	

TABLE XX. DIRECT CROSS-COUPLING OF ALLYL AND PROPARGYL ELECTROPHILES (*Continued*)

Substrate	Stannane	Conditions	Product(s) and Yield(s) (%)	Refs.
		Pd$_2$(dba)$_3$ (5%), CuI (20%), P(2-furyl)$_3$ (20%), THF, 60°, 30 min	(84)	170
	Bu$_3$Sn	Pd(CH$_3$CN)$_2$Cl$_2$, PPh$_3$, CHCl$_3$, 65°	(53–81)	24
	Bu$_3$Sn	[(η^3-C$_3$H$_5$)PdCl]$_2$ (2%), maleic anhydride (5%), THF, 25°, 12 h	(38)	38, 39
	Sn $_4$	BnPd(PPh$_3$)$_2$Cl (0.3%), ZnCl$_2$, THF, 65°, 50 h	(16)	35
	Me$_3$SnPh	Pd(dba)$_2$ (5%), LiCl, DMF, 23°, 27 h	**I** (65)	163
I (69)	Me$_3$SnPh	Pd(dba)$_2$ (5%), LiCl, DMF, 23°, 27 h		163
	Me$_3$Sn OMe	Pd(dba)$_2$ (5%), LiCl, DMF, 56°, 47 h	(50)	163
	Me$_3$SnOPh	Pd(PPh$_3$)$_4$ (5%), THF, rt	(42) + (42)	322

484

Substrate	Reagent	Conditions	Product(s) (yield)	Refs.
	Bu₃Sn $\overset{C_5H_{11}\text{-}n}{\underset{OH}{}}$	Pd(CH₃CN)₂Cl₂ (5%), PPh₃ (10%), THF, 50°, 45 min	(75)	365
	Bu₃Sn $\overset{C_5H_{11}\text{-}n}{\underset{OH}{}}$	Pd(CH₃CN)₂Cl₂ (5%), THF, 50°, 40 min	(70) + (15)	365
	Bu₃Sn $\overset{OPh}{\underset{OH}{}}$	Pd(CH₃CN)₂Cl₂ (5%), PPh₃ (10%), THF, 50°, 1 h	(70)	365
	Bu₃Sn $\overset{OPh}{\underset{HO}{}}$	Pd(CH₃CN)₂Cl₂ (5%), PPh₃ (10%), THF, 50°, 45 min	(30)	365
	Bu₃Sn $\overset{OPh}{\underset{OH}{}}$	Pd(CH₃CN)₂Cl₂ (5%), THF, 50°, 40 min	(60) + (10)	365
	PhO $\overset{OTMS}{}$ Bu₃Sn	Pd(CH₃CN)₂Cl₂ (5%), PPh₃ (10%), THF, 50°, 24 h	(60)	365
	Bu₃Sn	Pd(CH₃CN)₂Cl₂, DMF, H₂O, rt	(36) + (31)	164
	Bu₃SnPh	Pd(CH₃CN)₂Cl₂, DMF, H₂O, rt	(40) + (23)	164

485

Substrate	Stannane	Conditions	Product(s) and Yield(s) (%)	Refs.
	Bu₃Sn (MeO₂C ... OTBDMS)	Pd(CH₃CN)₂Cl₂, DMF, H₂O, −10°	(28) + (21)	164
(epoxide substrate)	R₃Sn	Pd(CH₃CN)₂Cl₂, DMF, H₂O, rt	R = Me (77) E:Z = 2:1; R = Bu (80) E:Z = 2:1	164, 802
	Me₃Sn	Pd(CH₃CN)₂Cl₂, DMF, H₂O, rt	(72) E:Z = 1.7:1	164, 802
	Me₃SnPh or Bu₃SnPh	Pd(CH₃CN)₂Cl₂, DMF, H₂O, rt	(83) + (2)	164, 802
	Me₃Sn Bu-*n*	Pd(CH₃CN)₂Cl₂, DMF, H₂O, rt	(63) E:Z = 2:1	164, 802
	Me₃Sn Ph	Pd(CH₃CN)₂Cl₂, DMF, H₂O, rt	(55) + (8) II	164, 802
	Bu₃Sn Ph	Pd(CH₃CN)₂Cl₂, DMF, H₂O, rt	I (56) + II (9) I (E:Z = 2.2-2.6:1)	164, 802
	Me₃Sn (naphthyl)	Pd(CH₃CN)₂Cl₂, DMF, H₂O, rt	(75) E:Z = 2:1	164
	Me₃Sn (Bu-*t* cyclohexenyl)	Pd(CH₃CN)₂Cl₂, DMF, H₂O, rt	(79) E:Z = 1.8-2:1	164, 802

486

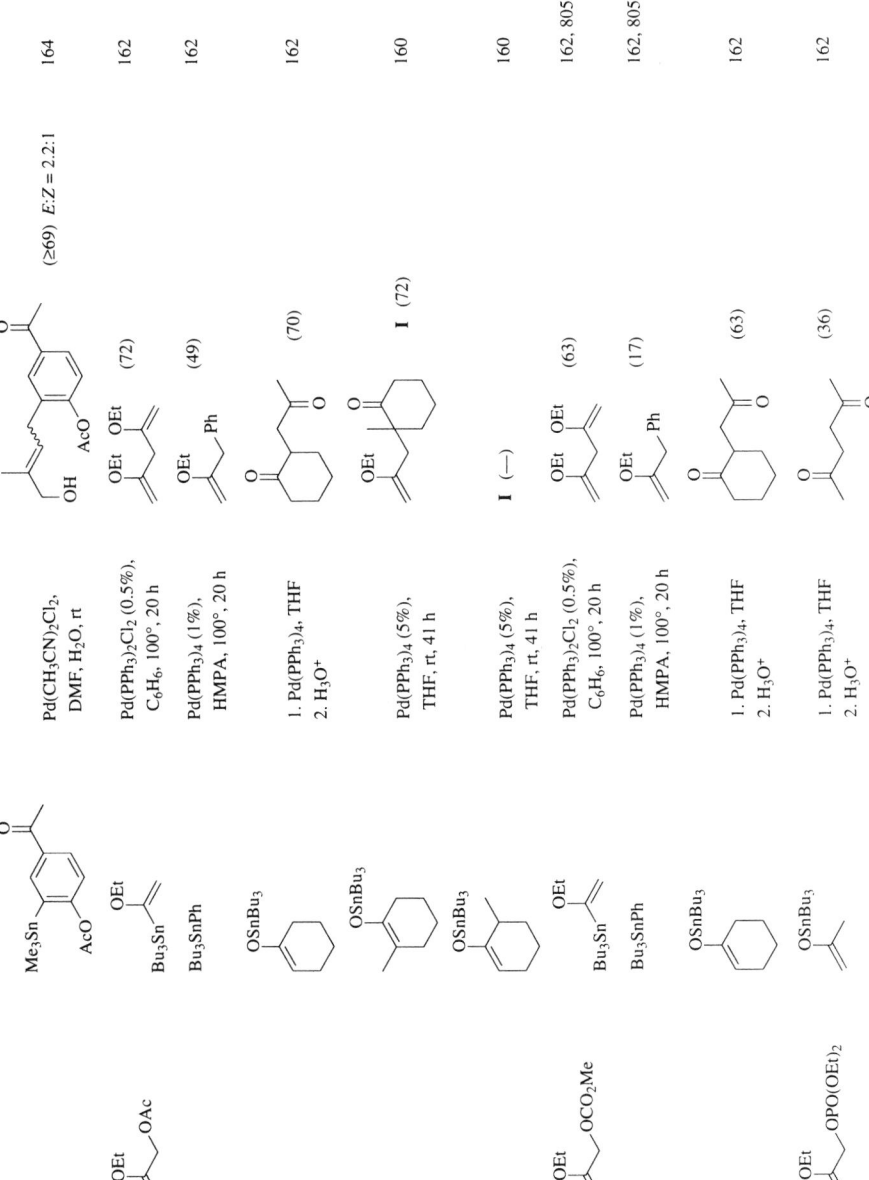

TABLE XX. DIRECT CROSS-COUPLING OF ALLYL AND PROPARGYL ELECTROPHILES (Continued)

Substrate	Stannane	Conditions	Product(s) and Yield(s) (%)	Refs.
	Bu₃Sn–C(OEt)=CH₂	Pd(PPh₃)₂Cl₂ (0.5%), C₆H₆, 100°, 20 h	(OEt)(OEt) bis-methylene product (99)	162, 805
	Bu₃Sn–(2-thienyl)	1. Pd(PPh₃)₂Cl₂, C₆H₆ 2. HCl	thienyl-CH₂C(O)CH₃ (83)	805
	Bu₃Sn–(3-thienyl)	1. Pd(PPh₃)₂Cl₂, C₆H₆ 2. HCl	thienyl-CH₂C(O)CH₃ (72)	805
	Bu₃Sn–(2-pyridyl)	Pd(PPh₃)₂Cl₂, C₆H₆	pyridyl-CH₂C(OEt)=CH₂ (60)	805
	Bu₃Sn–(3-pyridyl)	Pd(PPh₃)₂Cl₂, C₆H₆	pyridyl-CH₂C(OEt)=CH₂ (84)	805
	[OSnBu₃ enol of isopropyl ketone]	1. Pd(PPh₃)₄, THF 2. H₃O⁺	(49)	162
	[OSnBu₃ enol]	1. Pd(PPh₃)₄, THF 2. H₃O⁺	(61)	162
	[OSnBu₃ enol]	1. Pd(PPh₃)₄, THF 2. H₃O⁺	(68)	162
	[OSnBu₃ cyclopentenyl enol]	1. Pd(PPh₃)₄, THF 2. H₃O⁺	(69)	162
	Bu₃SnPh	Pd(PPh₃)₄ (1%), HMPA, 100°, 20 h	(OEt)(Ph) product (74)	162, 805

488

Organostannane	Conditions	Product (Yield)	Ref.
Bu$_3$Sn–C$_6$H$_4$–NO$_2$ (meta)	Pd(PPh$_3$)$_2$Cl$_2$, C$_6$H$_6$	3-NO$_2$-phenyl allyl enol ether (OEt) (34)	805
Bu$_3$Sn–C$_6$H$_4$–NO$_2$ (para)	Pd(PPh$_3$)$_2$Cl$_2$, C$_6$H$_6$	4-NO$_2$-phenyl allyl enol ether (OEt) (48)	805
OSnBu$_3$ (cyclohexenyl)	1. Pd(PPh$_3$)$_4$, THF 2. H$_3$O$^+$	cyclohexanone acetonyl (83)	162
OSnBu$_3$ (Ph)	1. Pd(PPh$_3$)$_4$, THF 2. H$_3$O$^+$	(68)	162
OSnBu$_3$ (Ph, propenyl)	1. Pd(PPh$_3$)$_4$, THF 2. H$_3$O$^+$	(71)	162
Bu$_3$Sn–CH=CH$_2$ + MeO$_2$C–CH=CH–CH$_2$Br	Pd(CH$_3$CN)$_2$Cl$_2$, PPh$_3$, CHCl$_3$, 65°	CO$_2$Me diene (87–100)	24
Bu$_3$Sn–C$_6$H$_4$–Cl (meta)	1. Pd(PPh$_3$)$_2$Cl$_2$, THF, reflux 2. LiOH, H$_2$O, reflux	HO$_2$C–CH=CH–C$_6$H$_4$–Cl (86)	806
Bu$_3$Sn–C$_6$H$_3$(R)(Cl)	1. Pd(PPh$_3$)$_2$Cl$_2$, THF, reflux 2. LiOH, H$_2$O, reflux	HO$_2$C–CH=CH–C$_6$H$_3$(R)(Cl) R = H (74) R = Cl (73)	806
Bu$_3$Sn–C$_6$H$_4$–F (meta)	1. Pd(PPh$_3$)$_2$Cl$_2$, THF, reflux 2. LiOH, H$_2$O, reflux	HO$_2$C–CH=CH–C$_6$H$_4$–F (91)	806
Bu$_3$Sn–C$_6$H$_3$(R)(F)	1. Pd(PPh$_3$)$_2$Cl$_2$, THF, reflux 2. LiOH, H$_2$O, reflux	HO$_2$C–CH=CH–C$_6$H$_3$(R)(F) R = H (72) R = F (71)	806
Bu$_3$Sn–C$_6$H$_4$–CF$_3$	1. Pd(PPh$_3$)$_2$Cl$_2$, THF, reflux 2. LiOH, H$_2$O, reflux	HO$_2$C–CH=CH–C$_6$H$_4$–CF$_3$ (74)	806

489

TABLE XX. DIRECT CROSS-COUPLING OF ALLYL AND PROPARGYL ELECTROPHILES (*Continued*)

Substrate	Stannane	Conditions	Product(s) and Yield(s) (%)	Refs.
MeO$_2$C⌒⌒Cl	Bu$_3$Sn⌒	Pd(CH$_3$CN)$_2$Cl$_2$, PPh$_3$, CHCl$_3$, 65°	⌒⌒⌒CO$_2$Me (74)	24
CO$_2$Me / Cl	Bu$_3$Sn⌒	Pd(CH$_3$CN)$_2$Cl$_2$, PPh$_3$, CHCl$_3$, 65°	CO$_2$Me (94)	24
OMOM / Cl	Bu$_3$Sn–(2-thienyl)	Pd(PPh$_3$)$_2$Cl$_2$, C$_6$H$_6$	(42)	805
	Bu$_3$Sn–(3-thienyl)	Pd(PPh$_3$)$_2$Cl$_2$, C$_6$H$_6$	(55)	805
	Bu$_3$SnPh	Pd(PPh$_3$)$_2$Cl$_2$, THF, 120°	(79)	805
	Bu$_3$Sn–(4-NO$_2$C$_6$H$_4$)	Pd(PPh$_3$)$_2$Cl$_2$, C$_6$H$_6$	(70)	805
	Bu$_3$Sn–(3-NO$_2$C$_6$H$_4$)	Pd(PPh$_3$)$_2$Cl$_2$, C$_6$H$_6$	(71)	805
(cyclohexenyl)Br	Me$_3$SnSnMe$_3$	[(η^3-C$_3$H$_5$)PdCl]$_2$ (5%), HMPA, 20°, 15 min	(cyclohexenyl)SnMe$_3$ (83)	314, 557
	Bu$_3$SnPh	Pd(dba)$_2$ (3%), PPh$_3$ (6%), THF, 50°, 24 h	(cyclohexenyl)Ph (62)	24
	Bu$_3$Sn⌒CO$_2$Bn	Pd(dba)$_2$ (3%), PPh$_3$ (6%), THF, 50°, 24 h	(cyclohexenyl)⌒CO$_2$Bn (72)	24
(cyclohexenyl)OAc	Me$_3$SnSnMe$_3$	[(η^3-C$_3$H$_5$)PdCl]$_2$ (5%), HMPA, 20°, 22 h	(cyclohexenyl)SnMe$_3$ (26)	314

490

Halide	Stannane	Conditions	Product (%)	Refs.
n-Pr–CH=CH–CH₂Br	Bu₃Sn–CH₂C(O)Et	Pd(PPh₃)₄ (10%), C₆H₆, 60°, 20 h	*n*-Pr–CH=CH–...–C(O)Et (75) + *n*-Pr ketone (22)	529
cyclohexene oxide	Me₃SnPh	Pd(CH₃CN)₂Cl₂, DMF, H₂O, rt	Ph–cyclohexenol (58) + Ph–cyclohexenol (20)	164, 802
vinyl-dimethyloxirane	Bu₃Sn–(isopropenyl)	Pd(CH₃CN)₂Cl₂, DMF, H₂O, rt	OH dienes (71) + (4)	164
EtO₂C–...–Br	Me₃Sn–C(=CH₂)CH(OH)CH₃	Pd₂(dba)₃ (2%), THF, 55°, 3 h	EtO₂C–...–OH (79)	440
EtO₂C–...–CH₂Br	Bu₃Sn–(furan-3-yl)	Pd(CH₃CN)₂Cl₂, DMF, 60°, 2 h	EtO₂C–...–furan (60)	287
	Bu₃Sn–(furan)–SnBu₃	Pd(CH₃CN)₂Cl₂, DMF, 70°, 2 h	EtO₂C–...–furan–...–CO₂Et (67)	287
MeO₂C–C(Br)=... *E:Z* = 1:1	Bu₃Sn–...–SnBu₃	Pd(CH₃CN)₂Cl₂, PPh₃, CHCl₃, 80°, 3 h	MeO₂C–...–SnBu₃ (76) *E:Z* = 1:1	468, 469
CO₂Et –...–Br	Bu₃SnPh	Pd(dba)₂ (3%), PPh₃ (6%), THF, 50°, 24 h	CO₂Et –...–Ph (69)	24
NC–CH₂–C₆H₄–CH₂Br	Bu₃Sn–...	Pd(dba)₂ (3%), PPh₃ (6%), THF, 50°, 24 h	EtO₂C–...–C₆H₄–CH₂CN (81)	24, 807
MeO₂C–C(OMe)=...–Br	Bu₃Sn–CH=CH₂	BnPd(PPh₃)₂Cl, CHCl₃, 65°	MeO₂C–C(OMe)=...–CH=CH₂ (56)	24

TABLE XX. DIRECT CROSS-COUPLING OF ALLYL AND PROPARGYL ELECTROPHILES (Continued)

Substrate	Stannane	Conditions	Product(s) and Yield(s) (%)	Refs.
C7 (cyclopentenyl–OAc, R)	Bu_3Sn—	$Pd(dba)_2$ (3%), PPh_3 (6%), THF, 50°, 24 h	MeO_2C ... OMe (86)	24, 807
(OAc, R, Cl)	Bu_3Sn—OH	$Pd(dba)_2$ (3%), PPh_3 (6%), THF, 50°, 24 h	MeO_2C ... OMe ...OH (82)	24, 807
(cyclopentenyl)	OSnBu₃ (2-methylcyclohexenyl)	$Pd(PPh_3)_4$ (5%), THF, rt, 42 h	(cyclohexanone) (24)	160
$OPO(OEt)_2$, Cl	Me_3SnOPh	$Pd(PPh_3)_4$ (5%), THF, rt	I + II ; R OPh	322
$OPO(OEt)_2$ Ph	Bu_3SnPh	$PdCl_2(PPh_3)_2$ (0.5%), HMPA, 120°	$OPO(OEt)_2$ Ph (58)	805
(thiophene, OAc)	Bu_3Sn—(thiophene)	$PdCl_2$ (1.2%), PPh_3, Et_4NOTs, DMF, 50°, 1.16 F/mol, 10 mA	(55)	808
C8 (OAc, R)	Me_3SnOPh	$Pd(PPh_3)_4$ (5%), THF, rt	I + II ; R OPh	322
(Br bicyclic ketone)	Bu_3Sn—C_5H_{11}-n OH	$Pd(CH_3CN)_2Cl_2$ (1%), HMPA, 20°, 20 h	(bicyclic) C_5H_{11}-n OH (72)	799

For the Me_3SnOPh row (322, first):

	R	I	II
	Pr-n	(66)	(16)
	Pr-i	(66)	(6)

For the Me_3SnOPh row (322, second):

	R	I	II
	Bu-n	(64)	(17)
	Bu-i	(77)	(8)

(cyclohexene with Cl, CO₂Me) $trans:cis = 73:27$	Bu₃Sn (vinyl)	BnPd(PPh₃)₂Cl, THF, 40°	(cyclohexene with vinyl, CO₂Me) (70) $trans:cis = 29:71$	24
	Bu₃SnPh	BnPd(PPh₃)₂Cl, THF, 40°	(cyclohexene with Ph, CO₂Me) (57) $trans:cis = 25:75$	24
	Bu₃Sn—CO₂Bn	Pd(dba)₂ (3%), PPh₃ (6%), THF, 50°, 24 h	(cyclohexene with CO₂Bn vinyl, CO₂Me) (87)	24, 807
(cyclohexene with Cl, CO₂Me)	Bu₃Sn (allyl)	[(η³-C₃H₅)PdCl]₂ (5%), L, CD₂Cl₂, rt, 24 h L:	I + II	
		maleic anhydride (20%)	(>95) **I:II** = 90:10	809
		dimethyl fumarate (20%)	(>95) **I:II** = 20:80	809
		cyclooctadiene (20%)	(>95) **I:II** = 4:96	809
		styrene (20%)	(>95) **I:II** = 1:99	809
		—	(>95) **I:II** = 0:100	809
		maleic anhydride, C₆H₆, 48 h	(92) **I:II** = 92:8	25
	Bu₃SnPh	[(η³-C₃H₅)PdCl]₂ (5%), maleic anhydride, C₆H₆, rt, 48 h	I + II I (48) II (80) **I:II** = 98:2	25
	Bu₃SnPh	[(η³-C₃H₅)PdCl]₂ (8%), C₆H₆, 25°, 24 h	I (48)	809
	Bu₃SnPh	[(η³-C₃H₅)PdCl]₂ (8%), maleic anhydride (32%), C₆H₆, rt, 24 h	I (100)	809

TABLE XX. DIRECT CROSS-COUPLING OF ALLYL AND PROPARGYL ELECTROPHILES (*Continued*)

Substrate	Stannane	Conditions	Product(s) and Yield(s) (%)	Refs.
(OAc, CO₂Me cyclohexene)	Me₃SnPh	Pd(dba)₂ (5%), LiCl, DMF, 55°, 19 h	**I** (47)	163
	(OSnBu₃ methylcyclohexene)	Pd(PPh₃)₄ (5%), THF, rt, 24 h	(85)	160
	Bu₃SnOPh	Pd(PPh₃)₄ (5%), THF, rt, 30 min	(53) + (27) [OPh, CO₂Me products]	322
C₉ (cyclooctene epoxide)	Me₃SnPh	Pd(CH₃CN)₂Cl₂, DMF, H₂O, rt	(45)	164
Ph—CH=CH—CH₂—Br	Bu₃Sn—C(OTMS)=CH₂	1. Pd(PPh₃)₄ (3%), PhMe, 100°, 24 h 2. HCl (1 N)	(55)	457
	Bu₃Sn—CH₂—C(CO₂Et)=CH—... (E:Z = 13:87)	Pd(PPh₃)₄ (5%), C₆H₆, reflux, 24 h	(70)	306
	Ph—C(=CH—Bu₃Sn)... TMS	BnPd(PPh₃)₂Cl (1%), 80°, 120 h	(43) + (6)	256
	Me₃SnSnMe₃	[(η³-C₃H₅)PdCl]₂ (5%), Me₂CO, 20°, 10 min	Ph—CH=CH—CH₂—SnMe₃ (97)	314

494

Substrate	Stannane	Conditions	Product (% yield)	Ref.
Ph–CH=CH–CH2–Cl	Bu3Sn–CH=CH2	Pd2(dba)3, L = AsPh3, P(2-furyl)3 or PPh3, NMP, 40°	Ph~~~~ (>95)	11
	Bu3Sn–C(O)Et	Pd(PPh3)4 (10%), C6H6, 60°, 20 h	(36) + Ph (26)	529
	[(η3-C3H5)PdCl]2 (1%), maleic anhydride (5%), THF, 50°, 12 h	Ph (64)	38
	[Bu3Sn–CH2CH=CH–Ph]	PdCl2 (1.2%), PPh3, Et4NOTs, DMF, 50°, 1.14 F/mol, 10 mA	Ph (89)	808
	Bu3Sn-(indole, N–R)	Pd(PPh3)2Cl2 (2%), THF, reflux	R = Me (80), R = Boc (72)	425
	Bu3Sn (CO2Et, NHAc)	Pd2(dba)3 (5%), AsPh3 (40%), THF, 65°	Ph (CO2Et, NHAc) (99)	375
	Bu3Sn (CO2Et, NHAc)	Pd2(dba)3 (5%), AsPh3 (40%), THF, rt	Ph (CO2Et, NHAc) (96)	375
	Bu3SnH	Pd(PPh3)4, THF	Ph (100)	810
	Bu3Sn–CH(H)–C=C=CH2	Pd(PPh3)4, THF, rt	(37) + Ph–C=C=CH (25), Ph H C=C=CH2	165
	Me3Sn~~~	Pd2(dba)3 (5%), LiCl, DMF, 23°, 22 h	Ph (80)	163
	Sn(~~~)4	Pd(PPh3)4 (4%), THF, reflux	Ph (71)	36
Ph–CH=CH–CH2–OAc	Bu3Sn~~~OH	Pd2(dba)3 (5%), LiCl, DMF, 23°, 22 h	Ph~~~OH (69)	163

495

TABLE XX. DIRECT CROSS-COUPLING OF ALLYL AND PROPARGYL ELECTROPHILES (*Continued*)

Substrate	Stannane	Conditions	Product(s) and Yield(s) (%)	Refs.
	Bu₃Sn	Pd(PPh₃)₄ (4%), THF, reflux	(32)	36
	Bu₃Sn	Pd(PPh₃)₄ (4%), THF, reflux	(69)	36
	Bu₃Sn—OEt	1. Pd(dba)₂ (5%), LiCl, DMF, 23°, 20 h 2. HCl (1 N)	(88)	163
	Bu₃Sn	Pd(PPh₃)₄ (4%), THF, reflux	(4)	36
	Me₃SnPh	Pd(dba)₂ (5%), LiCl, DMF, 23°, 69 h	Ph (57)	163, 36
	OSnBu₃	Pd(PPh₃)₄ (5%), THF, rt	(82–89)	160
	OSnBu₃	Pd(PPh₃)₄ (5%), THF, rt	(—)	160
	CO₂Et Bu₃Sn *E:Z* = 13:87	Pd(PPh₃)₄ (5%), C₆H₆, reflux, 24 h	CO₂Et (80)	306
	[Bu₃Sn——Ph]	PdCl₂ (1.2%), PPh₃, Et₄NOTs, DMF, 50°, 1.4 F/mol, 10 mA	(82)	808, 36
	Bu₃Sn———OTBDMS *E:Z* = 1:1	Pd(dba)₂ (5%), LiCl, DMF, 23°, 62 h	OTBDMS (60) *E:Z* = 1:1	163

496

Reactant	Conditions	Product (yield)	Refs.
Bu₃Sn—〔 〕—OTBDMS	Pd(dba)₂ (5%), LiCl, DMF, THF, 55°, 70 h	Ph—〔 〕—OTBDMS (64)	163
Bu₃Sn—〔 〕—C(O)OBn	Pd(dba)₂ (5%), LiCl, DMF, 60°, 72 h	Ph—〔 〕—OBn (70)	163
Bu₃Sn—〔 〕—SnBu₃	1. Pd(PPh₃)₄ (10%), THF, rt, 19 h 2. HCl (1 N)	Ph—〔 〕 (45)	811
Me₃SnOPh	Pd(PPh₃)₄ (4%), THF, rt	Ph—〔 〕—OPh (92)	322
AcO sugar AcO OSnBu₃	Pd(PPh₃)₄ (4%), THF, rt	AcO sugar O—CH₂CH=CHPh (38) β:α = 2:1	322
MeO₂C AcO sugar AcO OSnBu₃	Pd(PPh₃)₄ (4%), THF, rt	MeO₂C AcO sugar O—CH₂CH=CHPh (33) β:α = 1:2	322
Ph sugar Sn–O Bu₂ OMe	Pd(PPh₃)₄ (4%), THF, rt	Ph HO sugar OMe O—CH₂CH=CHPh (47) + Ph HO sugar OMe (34)	322
BnO BnO sugar BnO OSnBu₃	Pd(PPh₃)₄ (4%), THF, rt	BnO BnO sugar BnO O—CH₂CH=CHPh (51) β:α = 1:2	322
Me₃SnSnMe₃	Pd(PPh₃)₄ (5%), HMPA, 20°, 44 h	Ph—〔 〕—SnMe₃ (96)	314
Me₃SnPh	Pd(dba)₂ (5%), LiCl, DMF, 23°, 19 h	Ph—〔 〕 (32)	163

Ph—CH(OAc)—CH=CH₂

TABLE XX. DIRECT CROSS-COUPLING OF ALLYL AND PROPARGYL ELECTROPHILES (*Continued*)

Substrate	Stannane	Conditions	Product(s) and Yield(s) (%)	Refs.
[OAc allylic substrate]	[Bu₃Sn–CH₂CH=CH–Ph]	PdCl₂ (1.2%), PPh₃, Et₄NOTs, DMF, 50°, 1.39 F/mol, 10 mA	Ph—CH=CH—...—Ph (79)	808
[OAc substrate]	OSnBu₃ (2-methylcyclohex-1-enol)	Pd(PPh₃)₄ (5%), THF, rt, 0.5 h	(96)	160
[D, OAc, Bu-i substrate]	Me₃SnOPh	Pd(PPh₃)₄ (5%), THF, rt	OPh product (68) + OPh product (20)	322
[D, OAc, Bu-i substrate]	Me₃SnOPh	Pd(PPh₃)₄ (5%), THF, rt, 3 h	PhO, D, Bu-i (9) + D, Bu-i, OPh (70)	322
[OAc, Br substrate]	OSnBu₃	Pd(PPh₃)₄ (5%), THF, rt, 19 h	(89)	160
[OAc, Br substrate]	OSnBu₃	Pd(PPh₃)₄ (5%), THF, rt, 3 h	(91)	160
[OAc, AcO cyclohexene substrate]	Bu₃SnOPh	Pd(PPh₃)₄ (5%), THF, rt, 20 min	OPh, AcO product (55) + OPh, AcO product (28)	322

498

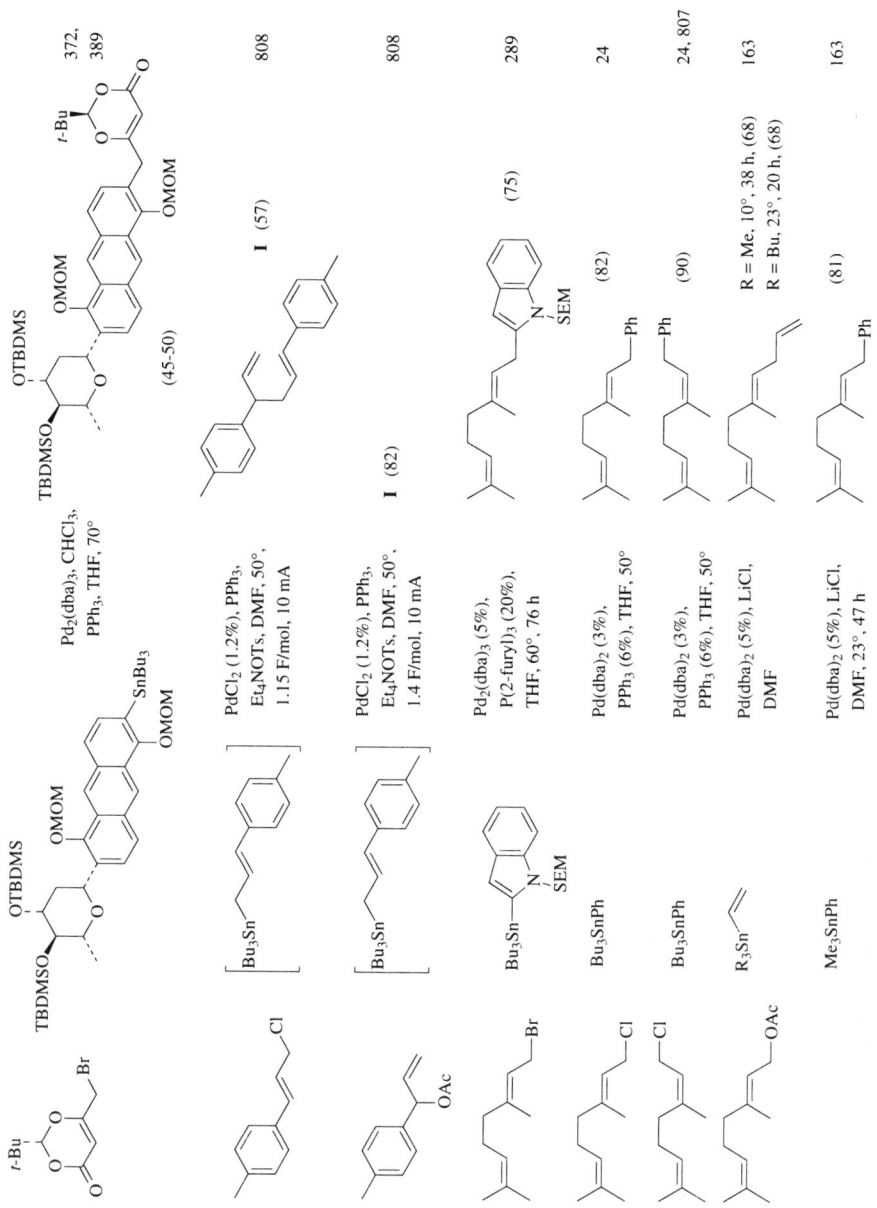

			372, 389
		I (57)	808
		I (82)	808
		(75)	289
		(82)	24
		(90)	24, 807
		R = Me, 10°, 38 h, (68) R = Bu, 23°, 20 h, (68)	163
		(81)	163

C_{10}

TABLE XX. DIRECT CROSS-COUPLING OF ALLYL AND PROPARGYL ELECTROPHILES (*Continued*)

Substrate	Stannane	Conditions	Product(s) and Yield(s) (%)	Refs.
	Me$_3$Sn—C$_6$H$_4$—OAc	Pd(dba)$_2$ (5%), LiCl, DMF, 23°, 22 h	(80)	163
	Bu$_3$Sn—(CH$_2$)—CO$_2$Bn	Pd(dba)$_2$ (5%), LiCl, DMF, 60°, 48 h	(81)	163
geranyl-type OAc	Bu$_3$Sn—CH=CH$_2$	Pd(dba)$_2$ (5%), LiCl, DMF, 23°, 44 h	(40)	163
	Me$_3$SnPh	Pd(dba)$_2$ (5%), LiCl, DMF, 23°, 43 h	(76)	163
	OSnBu$_3$ (2-methylcyclohexenyl)	Pd(PPh$_3$)$_4$ (5%), THF, rt, 42 h	(78)	160
Ph—CH=CH—CH(OAc)—CN	Bu$_3$SnH	Pd(PPh$_3$)$_4$, THF	(99)	810
	Bu$_3$Sn—C≡C—CH$_2$ (H)	Pd(PPh$_3$)$_4$, THF	(19)	165
	Sn(CH$_2$CH=CH$_2$)$_4$	Pd(PPh$_3$)$_4$, THF	(—)	165
	OSnBu$_3$ (2-methylcyclohexenyl)	Pd(PPh$_3$)$_4$, THF	(—)	37

500

Substrate	Reagent	Conditions	Product	Refs.
(Ph, CN structure)	Me₃SnOR	Pd(PPh₃)₄ (5%), THF, rt	Ph CN, OR R = Me (55), R = Ph (89)	322
(Ph–N-morpholine, CN)	Bu₃SnH	Pd(PPh₃)₄, THF, H₂O, NH₄Cl	Ph CN (68)	810
(Cl–C₆H₄, OAc, CN)	Bu₃SnH	Pd(PPh₃)₄, THF, BHT	Cl–C₆H₄ CN (99)	810
(Cl–C₆H₄, CN structure)	Bu₃Sn–C(H)=C=CH₂	Pd(PPh₃)₄, THF, rt	Cl–C₆H₄ CN C≡CH (15)	165
(MeO–C₆H₄, OAc, allyl)	[Bu₃Sn–CH=CH–C₆H₄–OMe]	PdCl₂ (1.2%), PPh₃, Et₄NOTs, DMF, 50°, 1.31 F/mol, 10 mA	MeO–C₆H₄ ... OMe (42)	808
(Ph, epoxide)	Me₃Sn–allyl	Pd(CH₃CN)₂Cl₂, DMF, H₂O, rt	Ph HO (80) E:Z = 9:1	164, 802
(Ph, epoxide)	Me₃SnPh	Pd(CH₃CN)₂Cl₂, DMF, H₂O, rt	Ph HO Ph (75) E:Z = 11:1	164, 802
(Ph, epoxide)	Me₃Sn–CH=CH–Ph	Pd(CH₃CN)₂Cl₂, DMF, H₂O, rt	Ph HO Ph (55) E:Z = 10:1	164, 802
(Ph, epoxide)	Bu₃Sn–allyl	Pd(CH₃CN)₂Cl₂, DMF, H₂O, rt	Ph HO (8) + HO Ph (75)	164, 802

501

TABLE XX. DIRECT CROSS-COUPLING OF ALLYL AND PROPARGYL ELECTROPHILES (*Continued*)

Substrate	Stannane	Conditions	Product(s) and Yield(s) (%)	Refs.
C_{11} (CHO, Cl-substituted structure)	Me₃SnPh	Pd(CH₃CN)₂Cl₂, DMF, H₂O, rt	(structure) (54)	164
	Bu₃Sn—C₆H₄—OMe	Pd(dba)₂ (3%), PPh₃ (6%), THF, 50°, 24 h	CHO / OMe (structure) (87)	164, 802
(OAc, CN tolyl structure)	Bu₃SnPh	Pd(PPh₃)₄, THF, H₂O (1%)	CN (structure) (98)	810
	Bu₃Sn—C(H)=C=CH₂	Pd(PPh₃)₄, THF, rt	CN / C≡CH (structure) (33)	165
(MeO-substituted OAc, CN structure)	Bu₃Sn—C(H)=C=CH₂	Pd(PPh₃)₄, THF, rt	CN / C≡CH (structure) (24)	165
C_{12} (OAc, C≡CMe, Ph structure)	Bu₃Sn—C(H)=C=CH₂	Pd(PPh₃)₄, THF, rt	C≡CH / C≡CMe / Ph (structure) (65) + C≡CH / C≡CMe / Ph (structure) (19) + HC=C=CH₂ / C≡CMe / Ph (structure) (13)	165

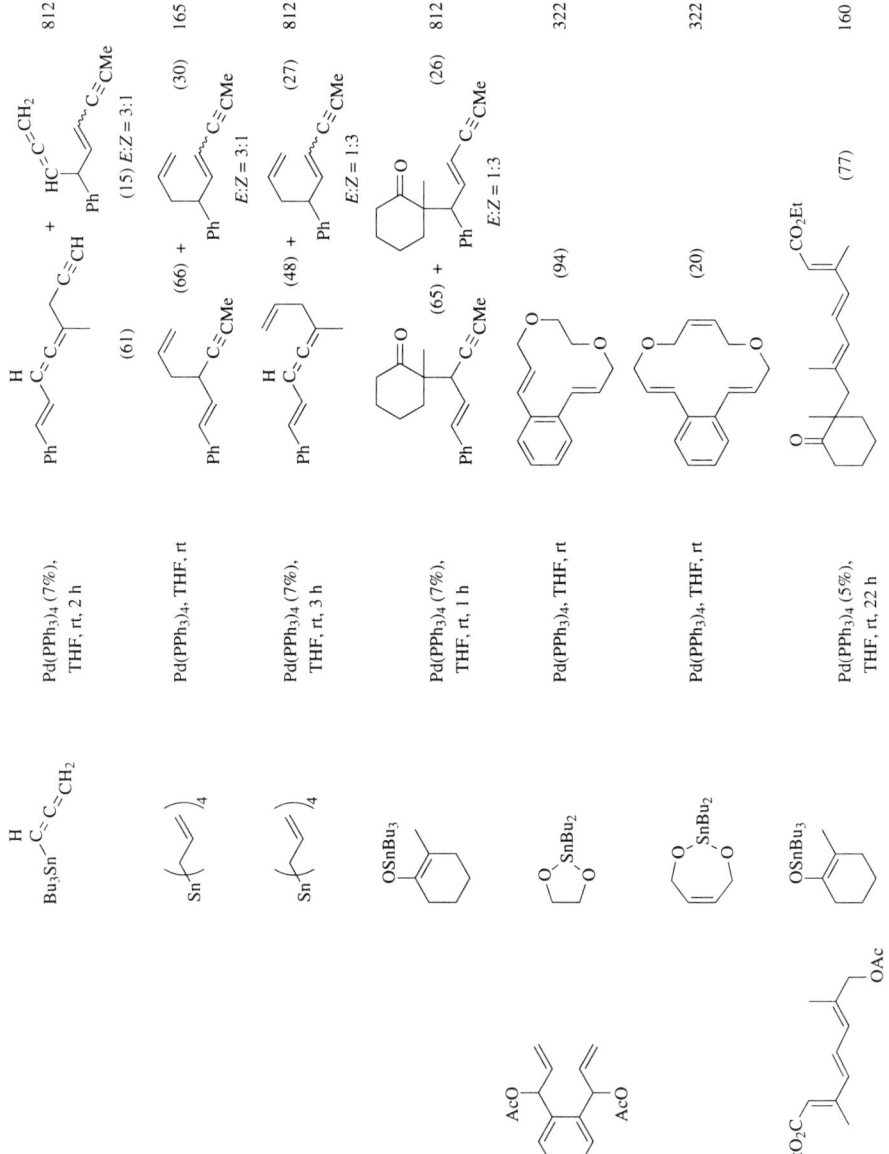

TABLE XX. DIRECT CROSS-COUPLING OF ALLYL AND PROPARGYL ELECTROPHILES (*Continued*)

Substrate	Stannane	Conditions	Product(s) and Yield(s) (%)	Refs.
(C₁₄ substrate with OMe, O, Br, MeO, SPh pyranone)	TBDMSO epoxide Bu₃Sn OMOM	Pd(CH₃CN)₂Cl₂, PPh₃, DME, reflux	TBDMSO epoxide OMOM product (75)	813
CH₂=C(SPh)C(CMe₃)OAc	Bu₃Sn (butenyl)	Pd(PPh₃)₄ (5%), THF, reflux, 0.5 h	SPh diene product (40)	166
	Me₃SnSnMe₃	Pd(PPh₃)₄ (5%), THF, reflux, 0.5 h	Me₃Sn SPh diene product (46)	166
C₁₄ pyridine-CH=CH-CH(OAc)Ph	Me₃SnOMe	Pd(PPh₃)₄ (5%), THF, rt, 3 h	Ph–CH=CH–CH(OMe)–pyridine (16) + Ph–CH(OMe)–CH=CH–pyridine (82)	322
Ph–CH=CH–CH(OAc)–C≡C–TMS	Bu₃Sn–C(H)=C=CH₂	Pd(PPh₃)₄, THF, rt	C≡CH + C≡C–TMS products (52) + (10) + HC≡C–CH₂ (6) + C≡C–TMS product	165
C₁₅ (4-Br-C₆H₄)CH=CH–CH(OAc)Ph	Bu₃SnH	Pd(PPh₃)₄, BHT, THF	(4-Br-C₆H₄)–CH=CH–CH=CH–Ph (92)	810

504

Pd(PPh₃)₄, THF, rt → $Pd(PPh_3)_4$, THF, rt

Starting material	Reagent	Conditions	Products (yield %)	Refs.

Entry 1: $Bu_3SnCH=C=CH_2$ (H) — $Pd(PPh_3)_4$, THF, rt — (42) + (38) — 165

Entry 2: Bu_3SnOR — $Pd(PPh_3)_4$, THF, rt

R	
Me	(45) (45)
Ph	(36) (36)

— 322

Entry 3: Bu_3SnH — $Pd(PPh_3)_4$, BHT, THF — (90) — 810

Entry 4: $Bu_3SnCH=C=CH_2$ (H) — $Pd(PPh_3)_4$, THF, rt — (52) + (41) — 165

Entry 5 (C_{16}): Me_3Sn-arene (OMe, OMe) — $Pd(dba)_2$ (5%), LiCl, DMF, 100°, 31 h — (71) — 163

Entry 6: $Bu_3SnCH=C=CH_2$ (H) — $Pd(PPh_3)_4$, THF, rt — (47) + (47) — 165

TABLE XX. DIRECT CROSS-COUPLING OF ALLYL AND PROPARGYL ELECTROPHILES (Continued)

Substrate	Stannane	Conditions	Product(s) and Yield(s) (%)	Refs.
C_{17}	Bu_3SnH	$Pd(PPh_3)_4$, BHT, THF	(49)	810
	Bu_3Sn	$Pd(PPh_3)_4$ (7%), THF, rt, 2 h	(38) + (19) $E{:}Z = 1{:}3$	812
	$Bu_3Sn-\overset{H}{C}=C=CH_2$	$Pd(PPh_3)_4$, THF, rt	(23) $E{:}Z = 55{:}45$	165
C_{19}	Sn	$BnPd(PPh_3)_2Cl$ (0.3%), $ZnCl_2$, THF, 65°, 50 h	(>95)	35
	HO Bu_3Sn O TMS	$Pd_2(dba)_3$, PPh_3, 50°, 12 h	(66) $E{:}Z = 1{:}1$	814
C_{22}	Bu_3Sn OMOM	$Pd(CH_3CN)_2Cl_2$ (3%), PPh_3 (5%), DME, reflux	(75)	815
C_{24}	Bu_3SnH	$Pd_2(dba)_3$ (1%), P(2-furyl)$_3$ (4%), THF, 25°, 0.5 h	(98)	40

506

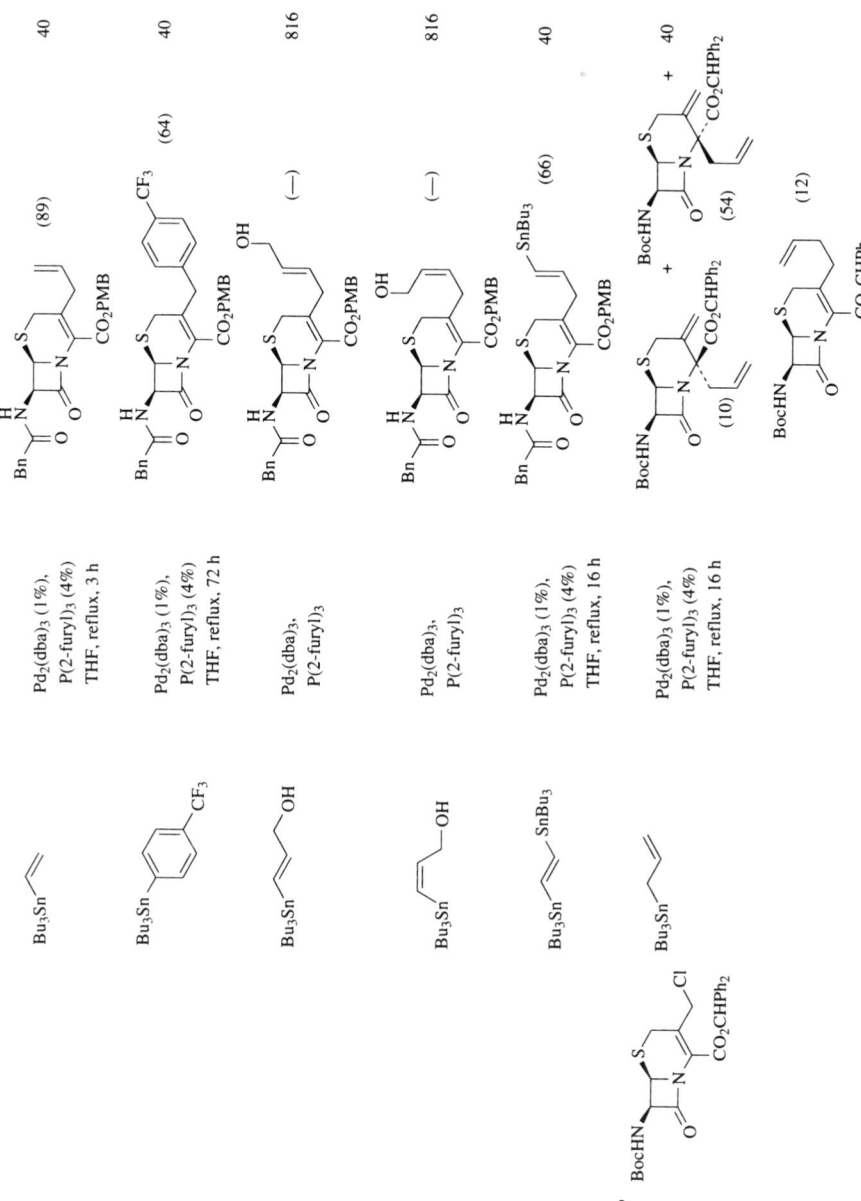

Substrate	Stannane	Conditions	Product(s) and Yield(s) (%)	Refs.
C27		Pd(dba)2, LiCl, DMF	(92)	817
C28	Bu3SnH	Pd(PPh3)4, BHT, THF	(51)	810
C29	Bu3Sn	Pd2(dba)3 (1%), P(2-furyl)3 (4%), THF, reflux, 16 h	(80)	40
	Bu3Sn	Pd2(dba)3 (2%), P(2-furyl)3 (4%), THF, reflux, 3 h	(88)	743
	Bu3Sn	Pd2(dba)3 (1%), P(2-furyl)3 (4%), THF, reflux, 24 h	(81)	40, 818

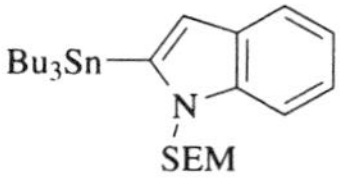

Bu₃Sn — SEM indole stannane

Pd₂(dba)₃ (5%), P(2-furyl)₃ (20%), THF, 65°

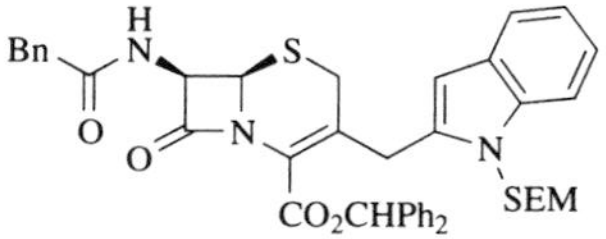

(95) 289

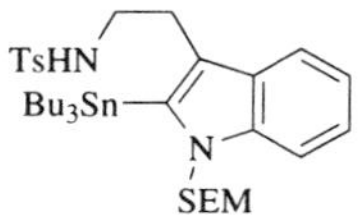

Pd₂(dba)₃ (10%), P(2-furyl)₃ (20%), THF, 65°, 3 h

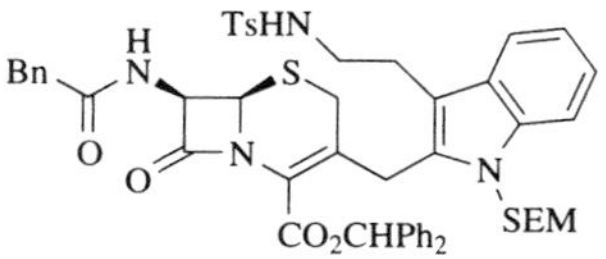

(95) 74

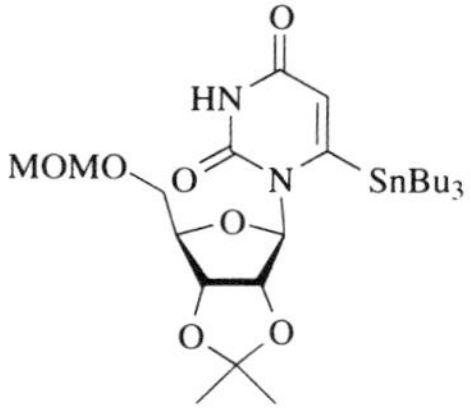

Pd₂(dba)₃ (5%), CuI (20%), P(2-furyl)₃ (20%), THF, 60°, 0.5 h

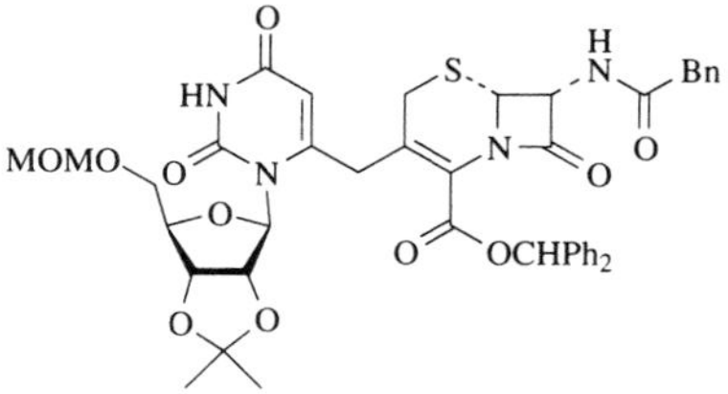

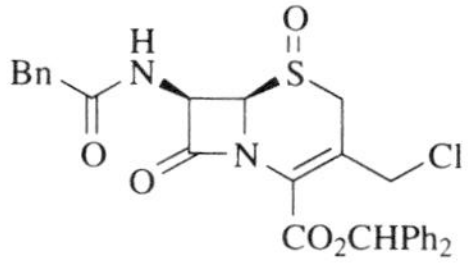

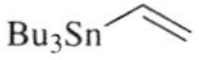

(79) 170

Pd₂(dba)₃ (1%), P(2-furyl)₃ (4%), THF, reflux, 40 min

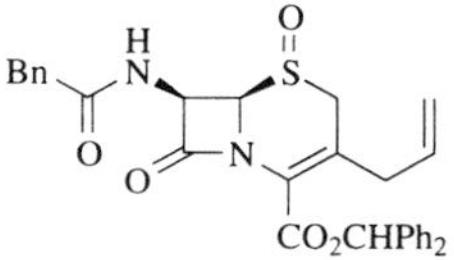

(87)

(40)

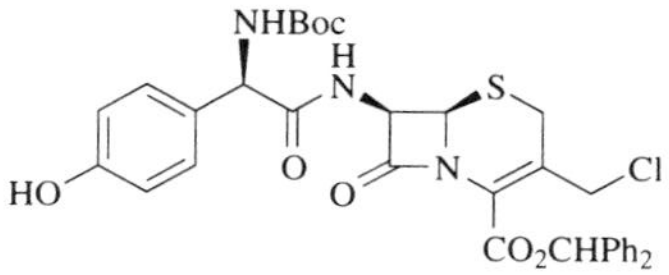

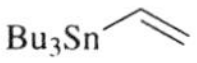

Pd₂(dba)₃ (1%), P(2-furyl)₃ (4%), THF, reflux, 3 h

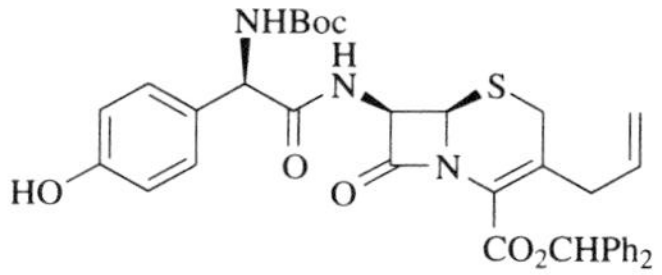

(82) 40, 818

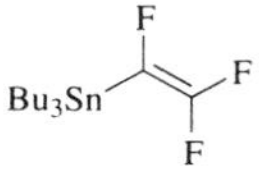

Pd₂(dba)₃ (1%), P(2-furyl)₃ (4%), THF, reflux, 72 h

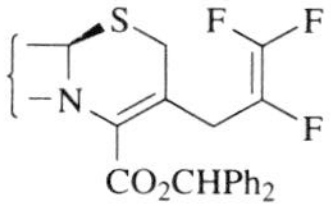

(65) 40, 818

C₃₄

509

Substrate	Stannane	Conditions	Product(s) and Yield(s) (%)	Refs.
	Bu$_3$Sn	Pd$_2$(dba)$_3$ (1%), P(2-furyl)$_3$ (4%), THF, reflux, 16 h	(78)	40, 818
	Bu$_3$Sn	Pd$_2$(dba)$_3$ (1%), P(2-furyl)$_3$ (4%), THF, reflux, 16 h	(9) + (47)	818
	Bu$_3$SnC≡CMe	Pd$_2$(dba)$_3$ (1%), P(2-furyl)$_3$ (4%), THF, reflux, 16 h	(32)	40
	Bu$_3$Sn	Pd$_2$(dba)$_3$ (1%), P(2-furyl)$_3$ (4%), THF, reflux, 72 h	(60)	40, 818
	Bu$_3$Sn OEt	Pd$_2$(dba)$_3$ (1%), P(2-furyl)$_3$ (4%), THF, reflux, 2 h	(71)	40, 818
C$_{40}$	Bu$_3$Sn OH	Pd$_2$(dba)$_3$, P(2-furyl)$_3$	(—)	816
	Bu$_3$Sn OH	Pd$_2$(dba)$_3$, P(2-furyl)$_3$	(—)	816

510

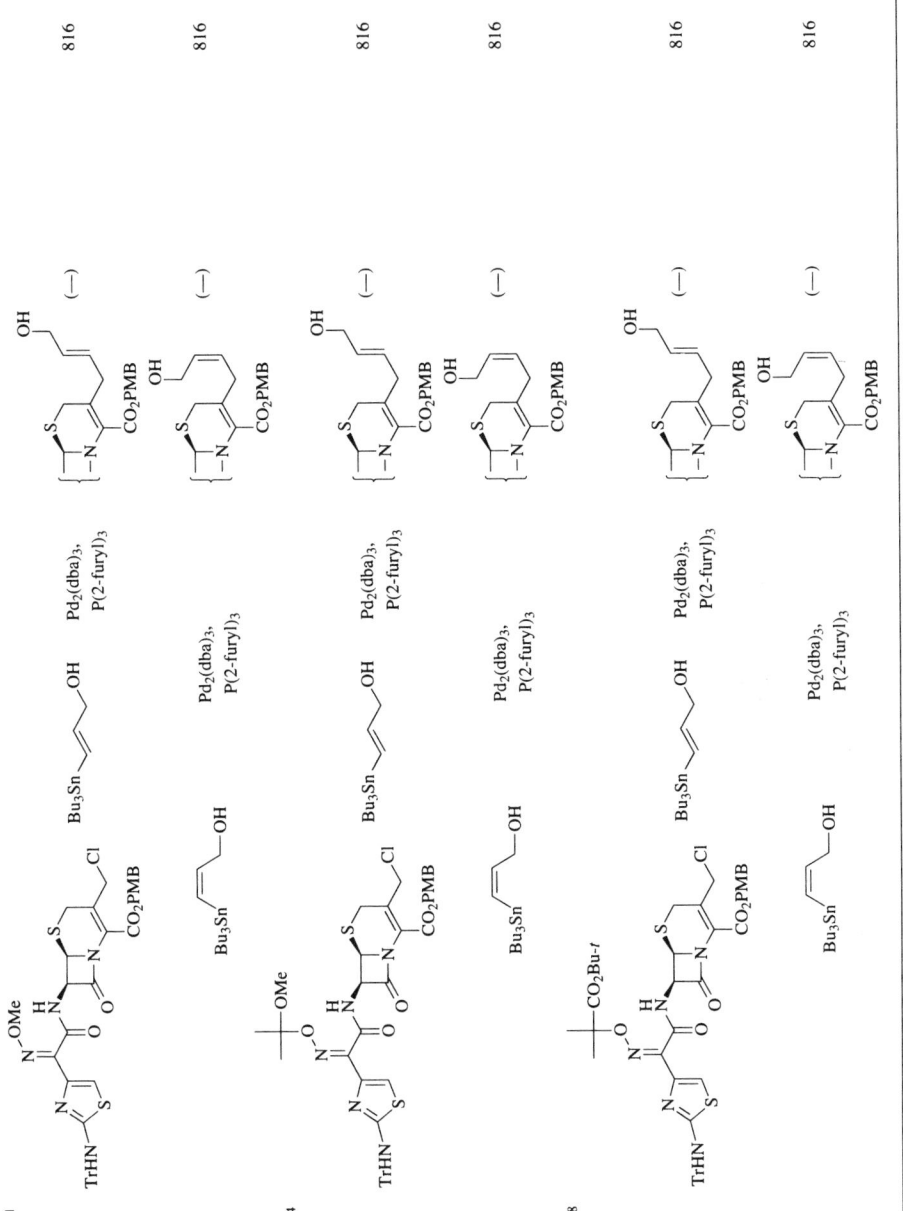

TABLE XXI. DIRECT CROSS-COUPLING OF BENZYL ELECTROPHILES

Substrate	Stannane	Conditions	Product(s) and Yield(s) (%)	Refs.
C₇ Ph–Br	Me₄Sn	Pd(2, 2'-bipyridine), (fumaronitrile) (1.5%) HMPA, 60°, 15 h	Ph⌒ (94)	171, 819, 19
	Bu₃Sn⌒ (allyl)	Pd(PPh₃)₄ (0.7%), HMPA, 65°	Ph⌒ (100)	19
	Bu₄Sn	Pd(PPh₃)₄ (0.7%), HMPA, 65°	Ph⌒Bu-n (42)	19
	Bu₃Sn / SnBu₃ furan	Pd(PPh₃)₄, DMF, 70°, 10 h	(furan with Ph, Ph) (45)	287
	(oxazoline) Bu₃Sn	Pd(PPh₃)₂Cl₂ (1%)	(oxazoline, Ph) (58)	426
	MeSnPh₃	Pd(PPh₃)₄ (0.7%), HMPA, 65°	Ph⌒Ph (95)	19
	Me₃Sn⌒Ph	Pd(PPh₃)₄ (0.7%), HMPA, 65°	Ph⌒⌒Ph (89)	19
	Bu₃Sn–(tolyl)	Pd(Ph-BIAN) (1%), HMPA, 20°, 20 h or DMF, 50°, 20 h	(76)	819
	Bu₃Sn (N-Me indole)	Pd(PPh₃)₂Cl₂ (2%), THF, reflux, 1 h	(70)	425
	Bu₃Sn / CO_2Et NHAc	Pd₂(dba)₃ (5%), AsPh₃ (40%), THF, 65°	CO_2Et NHAc (100)	375

512

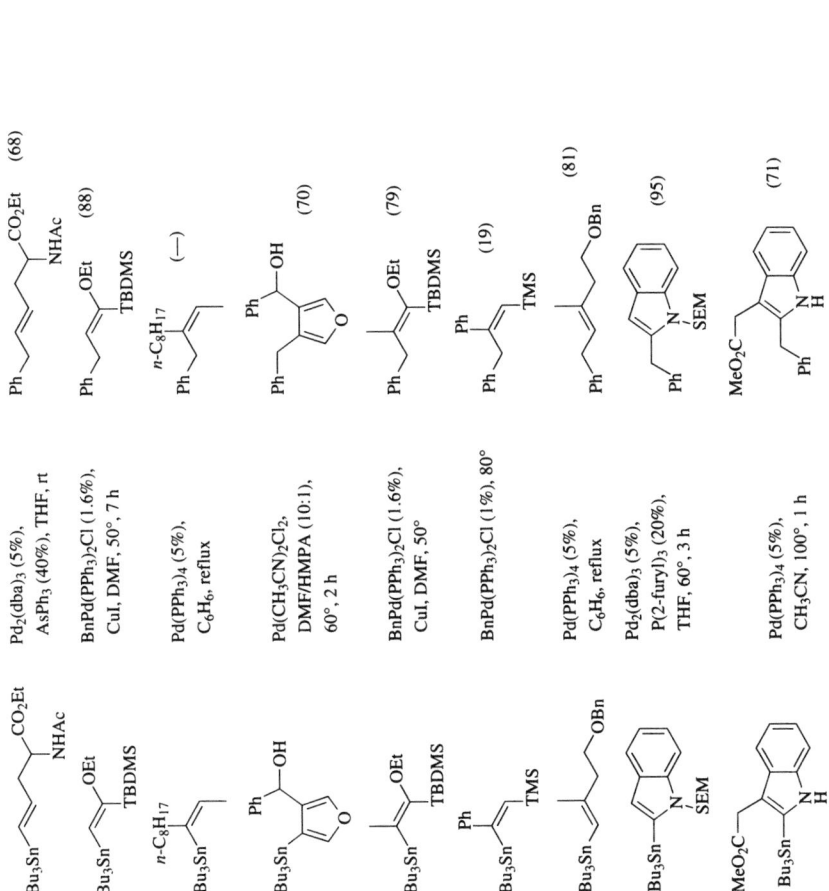

Pd$_2$(dba)$_3$ (5%), AsPh$_3$ (40%), THF, rt   (68)   375

BnPd(PPh$_3$)$_2$Cl (1.6%), CuI, DMF, 50°, 7 h   (88)   49

Pd(PPh$_3$)$_4$ (5%), C$_6$H$_6$, reflux   (—)   442

Pd(CH$_3$CN)$_2$Cl$_2$, DMF/HMPA (10:1), 60°, 2 h   (70)   287

BnPd(PPh$_3$)$_2$Cl (1.6%), CuI, DMF, 50°   (79)   49

BnPd(PPh$_3$)$_2$Cl (1%), 80°   (19)   256

Pd(PPh$_3$)$_4$ (5%), C$_6$H$_6$, reflux   (81)   789

Pd$_2$(dba)$_3$ (5%), P(2-furyl)$_3$ (20%), THF, 60°, 3 h   (95)   289

Pd(PPh$_3$)$_4$ (5%), CH$_3$CN, 100°, 1 h   (71)   290

513

TABLE XXI. DIRECT CROSS-COUPLING OF BENZYL ELECTROPHILES (*Continued*)

Substrate	Stannane	Conditions	Product(s) and Yield(s) (%)	Refs.
Ph—CH₂Cl (shown as Ph⌒Cl at bottom)	TsHN-ethyl-2-benzyl-3-(SnBu₃)-indole (N-SEM)	Pd₂(dba)₃ (10%), P(2-furyl)₃ (20%), THF, 65°, 1.5 h	TsHN-ethyl indole with SEM, benzyl (93)	74
	Bu₃Sn-glycal with OBn, OBn, OBn	Pd(PPh₃)₄ (1%), PhMe, reflux, 1 h	benzyl glycal with OBn, OBn, OBn (74)	424
	Bu₃SnSnBu₃	Pd(PPh₃)₂Br₂ (0.6%), PhMe, 110°, 15 h	Ph⌒SnBu₃ (76)	547
	Bu₃SnSnBu₃	Pd(CH₃CN)₂Cl₂ (1%), HMPA, 25°, 40 min	Ph⌒SnBu₃ (24) + Ph⌒⌒Ph (51)	313
	Me₄Sn	Pd(PPh₃)₄ (0.7%), HMPA, 65°	Ph⌒ (61)	19
	Bu₃Sn-C(O)Et	Pd(PPh₃)₄ (10%), C₆H₆, 100°, 20 h	Ph⌒C(O)Et (30)	529
	Bu₃Sn-C(=CH₂)OEt	1. Pd(PPh₃)₂Cl₂ (1%), PhMe, 100°, 20 h 2. HCl (5%)	Ph⌒C(O)CH₃ (72)	269
	Bu₃Sn-2-(N-Me)indole	Pd(PPh₃)₂Cl₂ (2%), THF, reflux, 1 h	2-benzyl-N-Me indole (—)	425
	Bu₃Sn- / -SnBu₃ furan (3,4-bis)	Pd(PPh₃)₄, DMF, 100°, 12 h	3,4-bis(benzyl)furan (70)	287
	Bu₃SnSnBu₃	Pd(PPh₃)₄ (0.6%), PhMe, 110°, 15 h	Ph⌒SnBu₃ (95)	547

514

Substrate	Reagent	Conditions	Product (%)	Refs.
Ph–C(D)(H)–Br, 61% ee	Me₄Sn	BnPd(PPh₃)₂Cl (1.6%), HMPA, 60°, 20 h	Ph–C(D)(H)–Me (70) 8% ee	19
3-Cl-benzyl–X	R₃SnSnR₃	Pd(PPh₃)₄ or Pd(PPh₃)₂Br₂ (0.6%), PhMe, 110°, 15 h	3-Cl-benzyl–SnR₃ X = Br, R = Me (65); X = Br, R = Bu (51); X = Cl, R = Bu (58)	547
2-(Bu₃Sn)-1-methylindole	(1-methylindol-2-yl)	Pd(PPh₃)₂Cl₂ (2%), THF, reflux, 1 h	(79)	425
4-O₂N-benzyl–Br (C₇₋₈)	Me₃SnSnMe₃	Pd(PPh₃)₄ (0.6%), PhMe, 110°, 15 h	4-R-benzyl–SnMe₃ R = NO₂ (68); R = CN (95)	547
3-R-benzyl–Br	Bu₃SnSnBu₃	Pd(PPh₃)₂Br₂ (0.6%), PhMe, 110°, 15 h	3-R-benzyl–SnBu₃ R = NO₂ (30); R = CN (56)	547
Ph–CH(Me)–Br (C₈)	Me₄Sn	Et₂Pd(2,2′-bipyridine) (1.5%), fumaronitrile, HMPA, 60°, 113 h	(77)	171
	Et₄Sn	Pd(2,2′-bipyridine) (fumaronitrile) (1.5%), HMPA, 60°, 27 h	(37)	171
4-NC-benzyl–Br	Me₃SnSnMe₃	Pd(PPh₃)₄ (0.6%), PhMe, 110°, 15 h	4-NC-benzyl–SnMe₃ (96)	547
3-R-benzyl–Cl	Bu₃SnSnBu₃	Pd(PPh₃)₄ (0.6%), PhMe, 110°, 15 h	3-R-benzyl–SnBu₃ R = m-OMe (87); R = p-OMe (35)	547

515

TABLE XXI. DIRECT CROSS-COUPLING OF BENZYL ELECTROPHILES (*Continued*)

Substrate	Stannane	Conditions	Product(s) and Yield(s) (%)	Refs.
C$_9$	Bu$_3$SnC≡CPh	Pd(PPh$_3$)$_4$ (4%), CCl$_4$, 80°, 20 h	(65)	207
C$_9$	Bu$_3$SnSnBu$_3$	Pd(PPh$_3$)$_2$Br$_2$ (0.6%), PhMe, 110°, 15 h	SnBu$_3$ (56)	547
C$_{13}$	Bu$_3$Sn	BnPd(PPh$_3$)$_2$Cl (0.7%), HMPA, 65°, 6 h	(92)	820
C$_{33}$	Bu$_3$Sn OTHP	Pd(PPh$_3$)$_4$, C$_6$H$_6$, reflux	(56–63)	172, 821
C$_{34}$	Bu$_3$Sn OTHP	Pd(PPh$_3$)$_4$, CHCl$_3$, reflux	(50–60)	822, 823

TABLE XXII. INTRAMOLECULAR CROSS-COUPLING OF ALLYL AND BENZYL ELECTROPHILES

Substrate	Stannane	Conditions	Product(s) and Yield(s) (%)	Refs.
C$_4$		Pd(PPh$_3$)$_4$, THF, rt	(66)	322
		Pd(PPh$_3$)$_4$, THF, rt	(77)	322
	Me$_3$SnSnMe$_3$	Pd(OAc)$_2$ (10%), PPh$_3$, PhMe, 110°, 18h	(90)	645
	Me$_3$SnSnMe$_3$	Pd(OAc)$_2$ (5%), PPh$_3$, 1-hexene, THF, reflux	(83)	824
C$_{12}$		Pd(PPh$_3$)$_4$, THF, rt	(94)	322
		Pd(PPh$_3$)$_4$, THF, rt	(20)	322

TABLE XXII. INTRAMOLECULAR CROSS-COUPLING OF ALLYL AND BENZYL ELECTROPHILES (*Continued*)

Substrate	Stannane	Conditions	Product(s) and Yield(s) (%)	Refs.
C$_{14}$	Me$_3$SnSnMe$_3$	Pd(OAc)$_2$ (5%), PPh$_3$, 1-hexene, THF, reflux	(52)	824
C$_{15}$	SnBu$_3$	Pd$_2$(dba)$_3$, AsPh$_3$, cyclohexane, reflux	(26)	167
	SnBu$_3$	Pd$_2$(dba)$_3$, AsPh$_3$, cyclohexane, reflux	(40)	167
C$_{16}$	Me$_3$SnSnMe$_3$	Pd(OAc)$_2$ (5%), PPh$_3$, 1-hexene, THF, reflux	(15) + (45)	824
	Bu$_3$SnSnBu$_3$	Pd(PPh$_3$)$_2$Cl$_2$ (5%), Li$_2$CO$_3$, Et$_4$NBr, PhMe, reflux	(68)	563

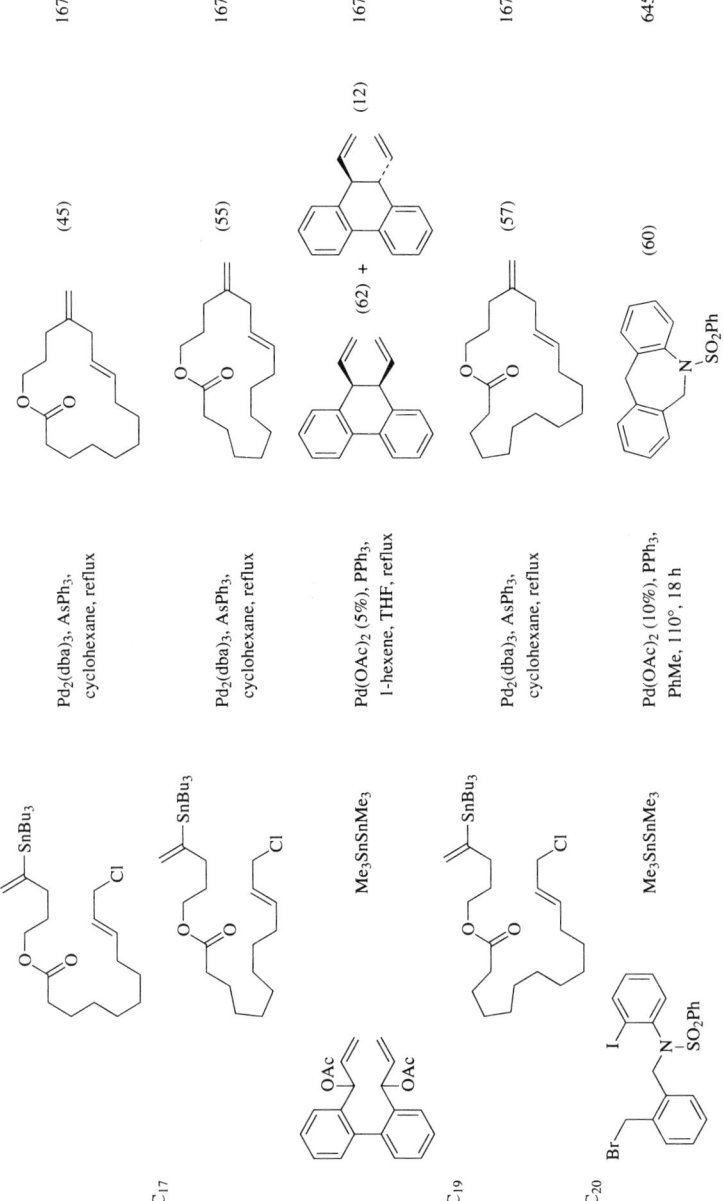

Substrate	Stannane	Conditions	Product(s) and Yield(s) (%)	Refs.
C$_{24}$		Pd$_2$(dba)$_3$, AsPh$_3$, cyclohexane, reflux	(38)	167
	Me$_3$SnSnMe$_3$	Pd(OAc)$_2$ (5%), PPh$_3$, 1-hexene, THF, reflux	(71)	824
C$_{31}$		Pd$_2$(dba)$_3$, P(2-furyl)$_3$ THF, reflux	(89)	74

520

TABLE XXIII. DIRECT CROSS-COUPLING OF ORGANOMETALLIC ELECTROPHILES

Substrate	Stannane	Conditions	Product(s) and Yield(s) (%)	Refs.
C₇	Me₃SnC≡CTMS	Pd₂(dba)₃ (5.5%), AsPh₃ (22%), DMF, 20°, 18 h	(—)	223
	Me₃SnC≡C–C≡CTMS	Pd₂(dba)₃ (5.5%), AsPh₃ (22%), DMF, 20°, 18 h	(44)	223
		Pd₂(dba)₃ (5.5%), AsPh₃ (22%), DMF, 20°, 18 h	(55)	223
	Me₃SnC≡CSnMe₃	Pd₂(dba)₃ (5.5%), AsPh₃ (22%), DMF, 20°, 18 h	(65)	223
	Me₃SnC≡C–C≡CSnMe₃	Pd₂(dba)₃ (5.5%), AsPh₃ (22%), DMF, 20°, 18 h	(67)	223
	Me₃SnC≡CH	Pd₂(dba)₃ (5.5%), AsPh₃ (22%), DMF, 20°, 18 h	(25)	219

521

Substrate	Stannane	Conditions	Product(s) and Yield(s) (%)	Refs.
	$Me_3SnC≡CR$	$Pd_2(dba)_3$ (5.5%), $AsPh_3$ (22%), DMF, 20°, 18 h	(20)	
				219, 220
			R: TMS (20) / (40); Bu-t	219
	$Me_3SnC≡C—C≡CR$	$Pd_2(dba)_3$ (5.5%), $AsPh_3$ (22%), DMF, 20°, 18 h	R: Pr-i (26); TMS (34); Bu-t (57); C_5H_{11}-n (30)	220
	$Me_3SnC≡CR$	$Pd_2(dba)_3$ (5.5%), $AsPh_3$ (22%), DMF, 20°, 18 h	(18)	220
R: C_8H_{17}-n		$Pd_2(dba)_3$ (5.5%), $AsPh_3$ (22%), DMF, 20°, 18 h	(83)	219

522

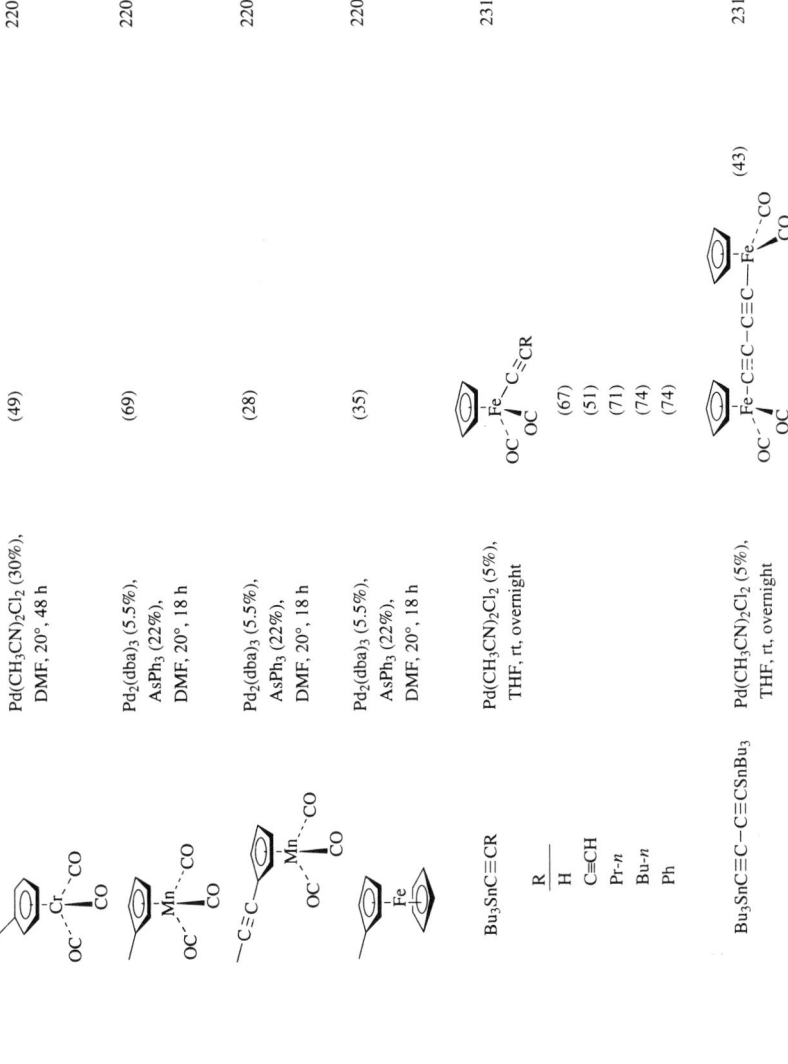

Bu₃SnC≡CR

R	
H	(67)
C≡CH	(51)
Pr-n	(71)
Bu-n	(74)
Ph	(74)

Pd(CH₃CN)₂Cl₂ (5%), THF, rt, overnight — 231

Bu₃SnC≡C–C≡CSnBu₃ — Pd(CH₃CN)₂Cl₂ (5%), THF, rt, overnight — (43) — 231

Pd(CH₃CN)₂Cl₂ (30%), DMF, 20°, 48 h — (49) — 220

Pd₂(dba)₃ (5.5%), AsPh₃ (22%), DMF, 20°, 18 h — (69) — 220

Pd₂(dba)₃ (5.5%), AsPh₃ (22%), DMF, 20°, 18 h — (28) — 220

Pd₂(dba)₃ (5.5%), AsPh₃ (22%), DMF, 20°, 18 h — (35) — 220

TABLE XXIII. DIRECT CROSS-COUPLING OF ORGANOMETALLIC ELECTROPHILES (*Continued*)

Substrate	Stannane	Conditions	Product(s) and Yield(s) (%)	Refs.
C$_8$	Bu$_3$SnC≡CH	Pd(CH$_3$CN)$_2$Cl$_2$ (5.5%), DMF, 25°, 12 h	(86)	825
	Bu$_3$Sn⟍	Pd(CH$_3$CN)$_2$Cl$_2$ (5%), DMF, rt, 2 h	(91)	826
	Bu$_3$SnC≡CSnBu$_3$	Pd(CH$_3$CN)$_2$Cl$_2$ (5.5%), DMF, 25°, 12 h	(72)	226
	Bu$_3$SnC≡CH	Pd(CH$_3$CN)$_2$Cl$_2$ (5.5%), DMF, 25°, 12 h	(92)	825
	Me$_3$SnC≡C−C≡CTMS	Pd$_2$(dba)$_3$ (5.5%), AsPh$_3$ (22%), DMF, 20°, 18 h	(79)	220
	Bu$_3$SnC≡C	Pd(CH$_3$CN)$_2$Cl$_2$ (5.5%), DMF, 25°, 12 h	(79)	825
	Bu$_3$SnC≡C	Pd(CH$_3$CN)$_2$Cl$_2$ (5.5%), DMF, 25°, 12 h	M = Mo (70) M = W (81)	825

524

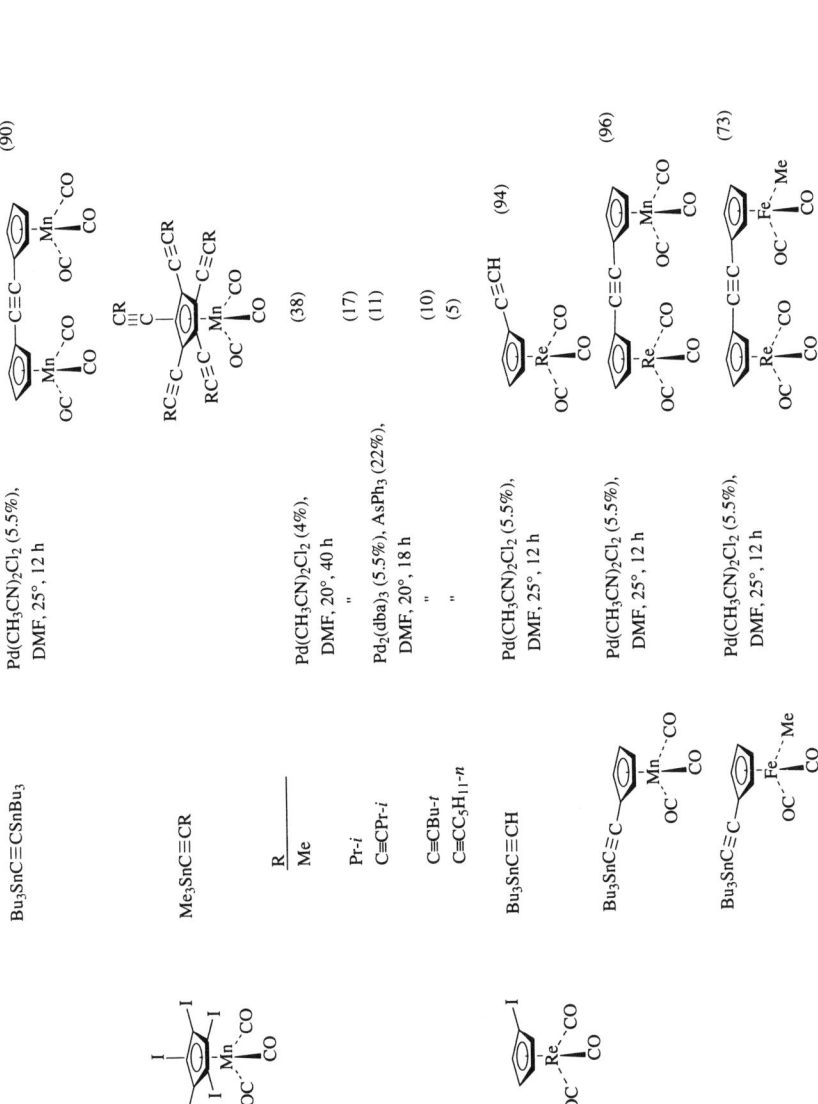

Substrate	Stannane	Conditions	Product(s) and Yield(s) (%)	Refs.
Bu$_3$SnC≡C–[M(Me)(CO)Cp]	Bu$_3$SnC≡C–[M(Me)(CO)(OC)Cp]	Pd(CH$_3$CN)$_2$Cl$_2$ (5.5%), DMF, 25°, 12 h	M = Mo (68) M = W (85)	825
(Cr(CO)$_3$ arene with X)	Bu$_3$SnC≡CSnBu$_3$	Pd(CH$_3$CN)$_2$Cl$_2$ (5.5%), DMF, 25°, 12 h	(78)	226
	Me$_3$Sn–(benzofuran)	Pd(PPh$_3$)$_4$ (4%), THF, reflux, 16 h	X = Cl (80) X = I (64)	392
(Cr(CO)$_3$ arene with Cl, Cl)	Bu$_3$Sn–CH=CH$_2$	[(η^3-C$_3$H$_5$)PdCl]$_2$ (10%), (R)-BINAP (12%), 40°, 18 h	**I** + **II** (80) 75:25 ee = 0%	222, 827
	Bu$_3$Sn–CH=CH$_2$	[(η^3-C$_3$H$_5$)PdCl]$_2$ (10%), (S)-(R)-PPFA (12%), 40°, 18 h	**I** (ee = 0%) + **II** (46) 87:13	222, 827
	Bu$_3$Sn–CH=CH$_2$	[(η^3-C$_3$H$_5$)PdCl]$_2$ (10%), (R)-MeO-MOP (24%), 0°, 18 h	**I** (ee = 0%) + **II** (46) 0:100	222, 827
C$_9$ (Cr(CO)$_3$ arene with Cl, Cl)	Me$_3$SnC≡CPh	Pd(CH$_3$CN)$_2$Cl$_2$ (2%), CH$_2$Cl$_2$, rt, 72 h	PhC≡C–(Cr(CO)$_3$ arene)–C≡CPh (84)	828

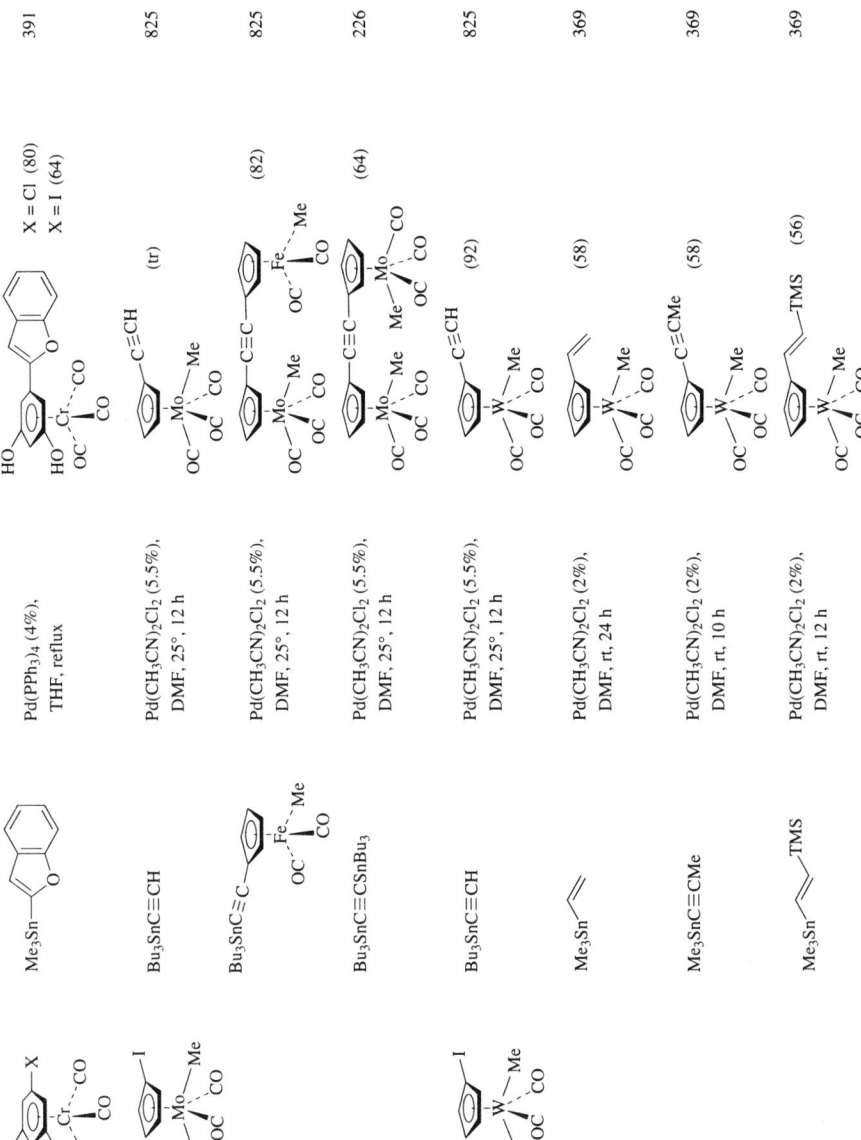

TABLE XXIII. DIRECT CROSS-COUPLING OF ORGANOMETALLIC ELECTROPHILES (*Continued*)

Substrate	Stannane	Conditions	Product(s) and Yield(s) (%)	Refs.
	Bu₃Sn⌒CO₂Me	Pd(CH₃CN)₂Cl₂ (2%), DMF, rt, 12 h	(55)	369
	Bu₃SnPh	Pd(CH₃CN)₂Cl₂ (2%), DMF, rt, 72 h	(32)	369
	Me₃SnC≡CBu-n	Pd(CH₃CN)₂Cl₂ (2%), DMF, rt, 4 h	(56)	369
	Me₃Sn⟨tolyl⟩	Pd(CH₃CN)₂Cl₂ (2%), DMF, rt, 12 h	(major) + (5)	369
	Me₃SnC≡CPh	Pd(CH₃CN)₂Cl₂ (2%), DMF, rt, 12 h	(29)	369
	Bu₃SnC≡C⟨Fe⟩	Pd(CH₃CN)₂Cl₂ (2%), DMF, rt, 12 h	(85)	825
	Bu₃SnC≡C⟨Mo⟩	Pd(CH₃CN)₂Cl₂ (2%), DMF, rt, 12 h	(76)	825

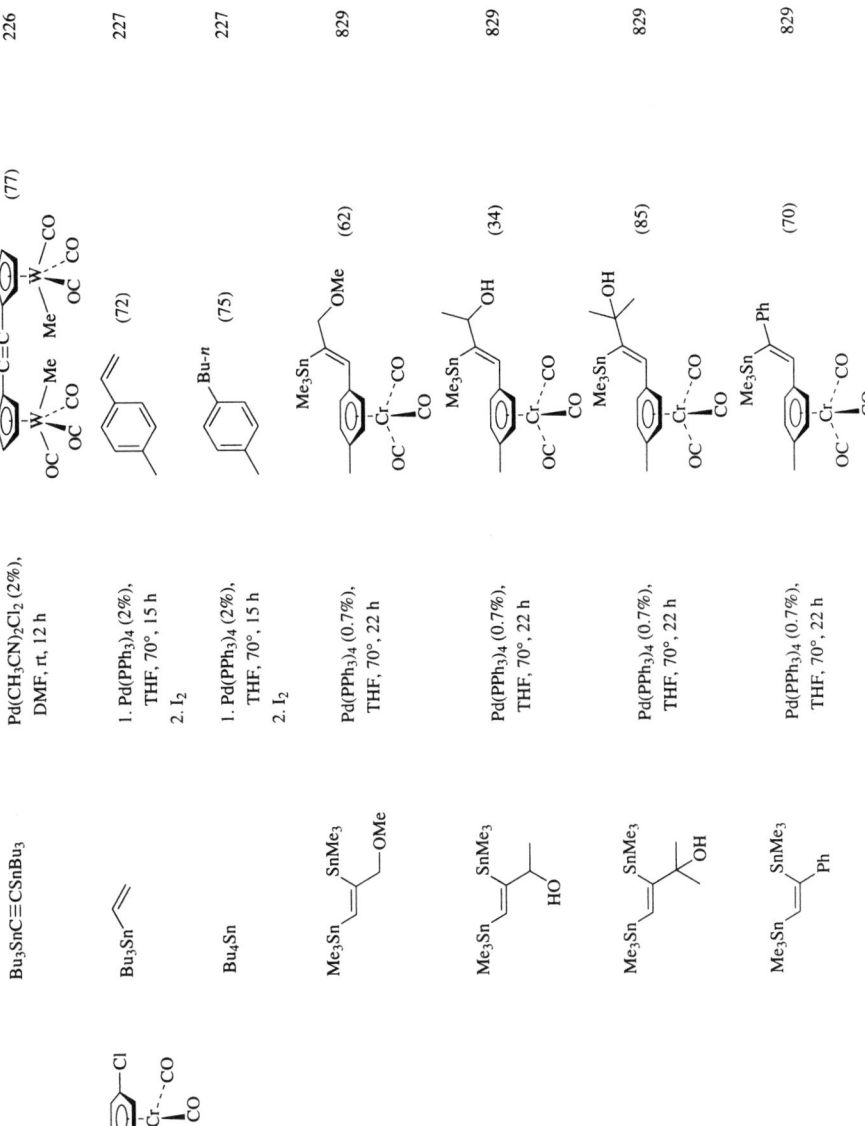

$Bu_3SnC{\equiv}CSnBu_3$	$Pd(CH_3CN)_2Cl_2$ (2%), DMF, rt, 12 h	(77)	226
Bu_3Sn	1. $Pd(PPh_3)_4$ (2%), THF, 70°, 15 h 2. I_2	(72)	227
Bu_4Sn	1. $Pd(PPh_3)_4$ (2%), THF, 70°, 15 h 2. I_2	(75)	227
	$Pd(PPh_3)_4$ (0.7%), THF, 70°, 22 h	(62)	829
	$Pd(PPh_3)_4$ (0.7%), THF, 70°, 22 h	(34)	829
	$Pd(PPh_3)_4$ (0.7%), THF, 70°, 22 h	(85)	829
	$Pd(PPh_3)_4$ (0.7%), THF, 70°, 22 h	(70)	829

C_{10}

TABLE XXIII. DIRECT CROSS-COUPLING OF ORGANOMETALLIC ELECTROPHILES (*Continued*)

Substrate	Stannane	Conditions	Product(s) and Yield(s) (%)	Refs.
(chromium arene complex)	Bu$_4$Sn	1. Pd(PPh$_3$)$_4$ (2%), THF, 70°, 15 h 2. I$_2$	(68)	227
(chromium arene complex)	Bu$_3$Sn—	1. Pd(PPh$_3$)$_4$ (2%), THF, 70°, 15 h 2. I$_2$	(68)	227
(chromium arene complex)	Bu$_4$Sn	1. Pd(PPh$_3$)$_4$ (2%), THF, 70°, 15 h 2. I$_2$	(82)	227
(iron complex, TMS)	Me$_3$SnC≡CSnMe$_3$	Pd$_2$(dba)$_3$ (5.5%), AsPh$_3$ (22%), DMF, 20°, 18 h	(35)	223
(diiodoferrocene)	Me$_3$SnC≡CPh	Pd(PPh$_3$)$_4$ (2%), THF, 75°, 20 h	(35) + (51)	830
Me$_3$SnC≡C—⟨⟩—C≡CSnMe$_3$	Pd(PPh$_3$)$_4$ (2%), THF, 75°, 20 h	(57)	830
C$_{11}$ (chromium arene complex, R)	(benzofuran Me$_3$Sn, R)	Pd(PPh$_3$)$_4$ (4%), THF, reflux, 16 h	R = H (46) R = OTBDPS (54)	391, 392

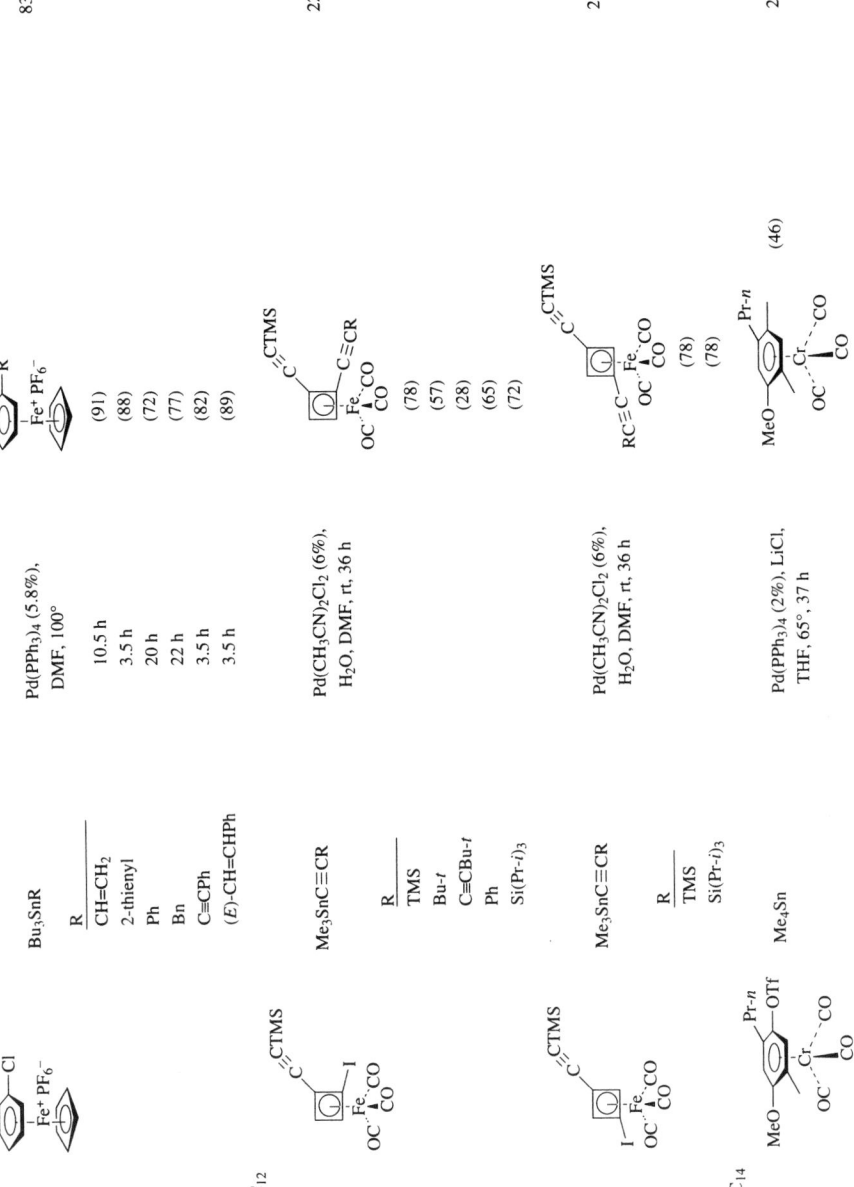

Bu₃SnR — Bu$_3$SnR, Pd(PPh$_3$)$_4$ (5.8%), DMF, 100°

R		
CH=CH$_2$	10.5 h	(91)
2-thienyl	3.5 h	(88)
Ph	20 h	(72)
Bn	22 h	(77)
C≡CPh	3.5 h	(82)
(E)-CH=CHPh	3.5 h	(89)

831

C$_{12}$

Me$_3$SnC≡CR, Pd(CH$_3$CN)$_2$Cl$_2$ (6%), H$_2$O, DMF, rt, 36 h

R	
TMS	(78)
Bu-t	(57)
C≡CBu-t	(28)
Ph	(65)
Si(Pr-i)$_3$	(72)

224

Me$_3$SnC≡CR, Pd(CH$_3$CN)$_2$Cl$_2$ (6%), H$_2$O, DMF, rt, 36 h

R	
TMS	(78)
Si(Pr-i)$_3$	(78)

224

C$_{14}$

Me$_4$Sn, Pd(PPh$_3$)$_4$ (2%), LiCl, THF, 65°, 37 h

(46)

217

531

TABLE XXIII. DIRECT CROSS-COUPLING OF ORGANOMETALLIC ELECTROPHILES (Continued)

Substrate	Stannane	Conditions	Product(s) and Yield(s) (%)	Refs.
	Bu₃Sn⌒	Pd(PPh₃)₄ (2%), LiCl, THF, 65°, 16 h	(91)	217
	Bu₃SnC≡CTMS	Pd(PPh₃)₄ (2%), LiCl, THF, 65°, 37 h	(79)	217
C₁₅	Bu₃SnC≡C–Mn(CO)₂Cp	Pd(CH₃CN)₂Cl₂ (3%), DMF, rt, overnight	(99)	230
C₁₆	Bu₃Sn⌒	Pd(PPh₃)₄ (2%), LiCl, THF, 65°, 24 h	(99)	217
C₁₇	Me₄Sn	Pd(PPh₃)₄ (2%), LiCl, THF, 65°, 132 h	(41)	217
	Bu₃Sn⌒	Pd(PPh₃)₄ (2%), LiCl, THF, 65°, 15 h	(76)	217

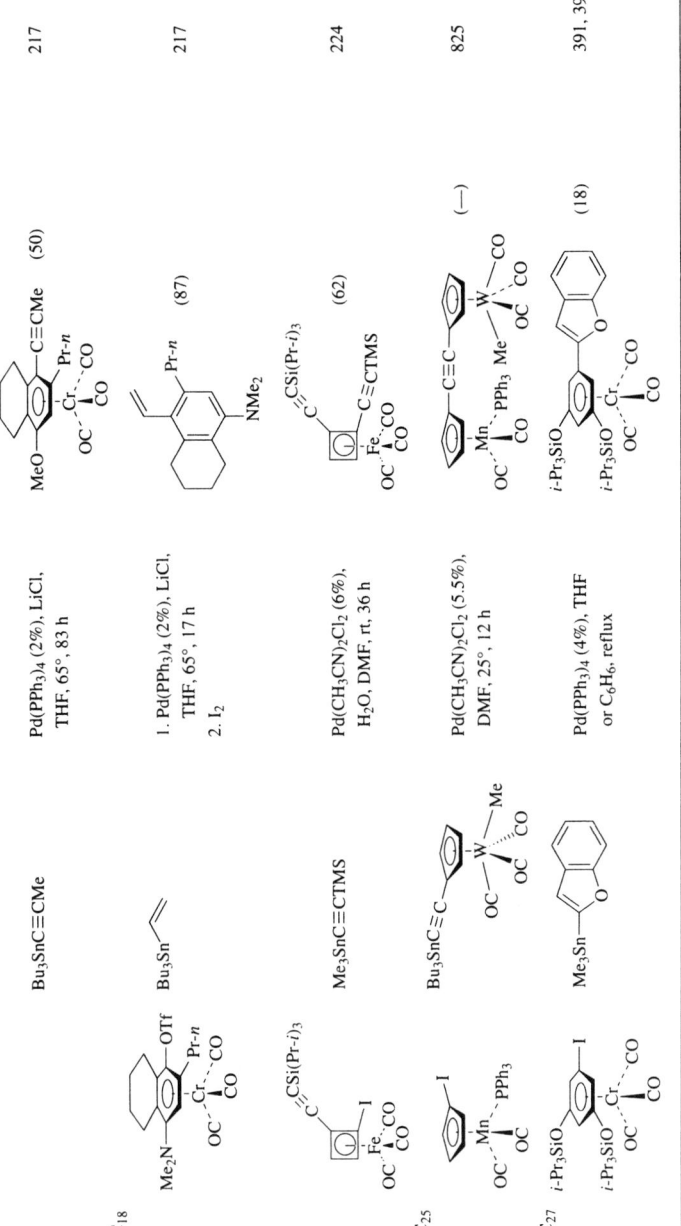

533

TABLE XXIV. DIRECT CROSS-COUPLING OF MISCELLANEOUS ELECTROPHILES

Substrate	Stannane	Conditions	Product(s) and Yield(s) (%)	Refs.
C₁ MeI	Bu₃Sn–C(=O)CH₃	Pd(dba)₂ (0.5%), PPh₃ (2%), C₆H₆, 60°, 20 h	(0) + MeSnBu₃ (26)	529
CF₃I	Bu₃Sn⁀	Pd(PPh₃)₄ (10%), C₆H₆, 80°, 3 h	F₃C⁀ (11)	210
	Bu₃Sn⁀Ph	Pd(PPh₃)₄ (10%), C₆H₆, 80°, 3 h	F₃C⁀Ph (11)	210
O₂N⁀Br	Bu₃SnSnBu₃ + TMSCl	Pd(PPh₃)₄ (10%), tetralin, 80°, 20 h	O₂N⁀TMS (0)	832
C₂ MeSO₂Cl	Bu₃Sn⁀Ph	Pd(PPh₃)₄ (1%), THF, 65–70°, 15 min	Me–S(=O)(=O)⁀Ph (90)	229
F₃C⁀I	Bu₃Sn⁀Ph	Pd(PPh₃)₄ (10%), C₆H₆, 80°, 4 h	F₃C⁀Ph (38)	210
	Bu₃Sn⁀Ph	Pd(PPh₃)₄ (10%), C₆H₆, 80°, 4 h	F₃C⁀Ph (35)	210
C₃ NC⁀Br	Bu₃SnSnBu₃ + TMSCl	Pd(PPh₃)₄ (10%), tetralin, 80°, 20 h	NC⁀TMS (61)	832
O=C(CH₃)CH₂Br	Me₃SnTMS	Pd(PPh₃)₄ (10%), C₆H₆, 80°, 5 h	OTMS (isopropenyl) (67)	833
MeO₂C⁀Br	Bu₃Sn⁀	D₇₁₇-Pd(0) on polymer, Me₂CO, reflux, 25 h	MeO₂C⁀ (91)	535
	Bu₃SnPh	D₇₁₇-Pd(0) on polymer, Me₂CO, reflux, 25 h	MeO₂C⁀Ph (90)	535
C₄ n-C₄F₉I	Bu₃Sn⁀OH	Pd(PPh₃)₄ (10%), C₆H₁₄, 70°, 4 h	n-C₄F₉⁀OH (52)	210
	Bu₃Sn⁀OH	Pd(PPh₃)₄ (10%), C₆H₁₄, 70°, 4 h	n-C₄F₉⁀OH (68)	210

Substrate	Reagent	Conditions	Product (%)	Refs.
(bromoethyl ketone)	(E) or (Z) Bu₃Sn $\diagup$ Ph	Pd(PPh₃)₄ (10%), C₆H₁₄, 70°, 4 h	n-C₄F₉ $\diagup$ Ph (70)	210
(3-bromo-2-butanone)	Me₃SnTMS	Pd(PPh₃)₄ (10%), C₆H₆, 80°, 5 h	OTMS (72)	833, 832
	[isopropenyl OSnBu₃]	Pd(PhCN)₂Cl₂ (5%), C₆H₆, 80°, 20 h	(diketone) (41)	834
MeO₂CC≡CBr	Bu₃SnSnBu₃ + TMSCl	Pd(PPh₃)₄ (10%), tetralin, 80°, 20 h	OTMS (77)	832, 833
	Bu₃Sn $\diagup$ OEt OEt	Pd(PPh₃)₂Cl₂ (5%), hydroquinone, DMF, 20°, 2 h	MeO₂CC≡C— OEt OEt (52)	215
	Bu₃Sn $\diagup$ OEt OEt	Pd(PPh₃)₂Cl₂ (5%), hydroquinone, DMF, 20°, 15 h	MeO₂CC≡C— OEt OEt (73)	215
(α-halo-γ-butyrolactone, X)	[isopropenyl OSnBu₃]	Pd(PPh₃)₂Cl₂ (1%), THF, 50°, 12 h	(acetonyl lactone) X = Br (73) X = I (82)	208
(α-iodo-γ-butyrolactone)	Bu₃Sn $\diagup\!\!\diagup$	Pd(PPh₃)₂Cl₂ (1%), THF, 50°, 12 h	(allyl lactone) (97)	208
EtO₂C $\diagdown$ Br	Bu₂Sn($\diagup\!\!\diagup$)₂	Pd(PPh₃)₂Cl₂ (1%), PhMe, 100°, 6 h	EtO₂C $\diagup\!\!\diagup$ (67)	208
	[isopropenyl OSnBu₃]	Pd(PPh₃)₂Cl₂ (1%), PhMe, 100°, 9 h	EtO₂C (ketone) (41)	208

TABLE XXIV. DIRECT CROSS-COUPLING OF MISCELLANEOUS ELECTROPHILES (*Continued*)

Substrate	Stannane	Conditions	Product(s) and Yield(s) (%)	Refs.
C₅ MeO–O–Cl	Bu₃SnPh	D₇₁₇-Pd(0) on polymer, Me₂CO, reflux, 25 h	EtO₂C–Ph (94)	535
	Bu₃SnSnBu₃ + TMSCl	Pd(PPh₃)₄ (10%), HMPA, 80°, 20 h	EtO₂C–TMS (79)	832
	Bu₃Sn, TMSO, CO₂Et (structure)	Pd(PPh₃)₄ (2.5%), DMF, 100°, 12 h	TMSO, CO₂Et (structure) (60)	441
	[OSnBu₃] (structure)	Pd(PhCN)₂Cl₂ (1%), C₆H₆, 80°, 20 h	(structure) (0)	834
	Bu₃SnSnBu₃ + TMSCl	Pd(PPh₃)₄ (10%), tetralin, 80°, 20 h	OTMS (structure) (75)	832, 833
⊙Br (structure)	Me₃SnTMS	Pd(PPh₃)₄ (10%), C₆H₆, 80°, 5 h	OTMS (structure) (93)	833
⊙Br (structure)	[OSnBu₃] (structure)	Pd(PhCN)₂Cl₂ (1%), C₆H₆, 80°, 20 h	(structure) (70)	834
	[OSnBu₃] (structure)	Pd(PhCN)₂Cl₂ (1%), C₆H₆, 80°, 20 h	(structure) (21)	834
EtO₂C–Br (structure)	Bu₃SnSnBu₃ + TMSCl	Pd(PPh₃)₄ (10%), HMPA, 80°, 20 h	EtO₂C–TMS (structure) (49)	832

Halide	Organostannane	Conditions	Product (%)	Ref.
TMSC≡CBr	(H₂N, Bu₃Sn, SEM indole)	Pd(PPh₃)₄ (0.7%), DMF, 110°, 0.5 h	(95)	74
TMSC≡Cl	(Bu₃Sn cyclopropene, TMS)	Pd(PPh₃)₄ (5%), THF, 20°, 24 h	(0)	422
	(Bu₃Sn indole, SEM)	Pd(PPh₃)₄ (0.7%), DMF, 110°, 1 h	(88)	289
	(MOMO, SnBu₃ uracil)	Pd(PPh₃)₄ (10%), CuI (20%), DMF, 80°, 20 min	(69)	170
C₆				
(prenyl-C≡CCl)	(Bu₃Sn furanone)	Pd(PPh₃)₂Cl₂, DMF	(37)	214
n-BuC≡CBr	(Bu₃Sn–CH=CH–CH(OEt)₂)	Pd(PPh₃)₂Cl₂ (5%), hydroquinone, DMF, 20°, 20 h	(80)	215
	(Bu₃Sn–CH=CH–CH₂–CH(OEt)₂)	Pd(PPh₃)₂Cl₂ (5%), hydroquinone, DMF, 20°, 20 h	(28)	215

TABLE XXIV. DIRECT CROSS-COUPLING OF MISCELLANEOUS ELECTROPHILES (*Continued*)

Substrate	Stannane	Conditions	Product(s) and Yield(s) (%)	Refs.
4-chlorophenylsulfonyl chloride (O=S=O, Cl)	Bu$_3$Sn–CH=CH–Ph	Pd(PPh$_3$)$_4$ (1%), THF, 65–70°, 15 min	sulfone–CH=CH–Ph (Cl) (75)	229
n-C$_6$F$_{13}$I	Bu$_3$Sn–CH$_2$CH=CH$_2$	Pd(PPh$_3$)$_4$ (10%), C$_6$H$_{14}$, rt, 1 h	n-C$_6$F$_{13}$–CH$_2$CH=CH$_2$ (100)	210
	Me$_3$SnC≡CPh	Pd(PPh$_3$)$_4$ (10%), C$_6$H$_{14}$, 70°, 6 h	n-C$_6$F$_{13}$C≡CPh (27)	210
	Me$_3$SnC≡CC$_6$H$_{13}$-n	Pd(PPh$_3$)$_4$ (10%), C$_6$H$_{14}$, 70°, 6 h	n-C$_6$F$_{13}$C≡CC$_6$H$_{13}$-n (55)	210
	Me$_3$SnC≡C(CH$_2$)$_4$OTHP	Pd(PPh$_3$)$_4$ (10%), C$_6$H$_{14}$, 70°, 6 h	n-C$_6$F$_{13}$C≡C(CH$_2$)$_4$OTHP (60)	210
	Me$_3$SnTMS	Pd(PPh$_3$)$_4$ (10%), C$_6$H$_6$, 80°, 5 h	cyclohexenyl OTMS (52)	833
2-chlorocyclohexanone (Cl, t-Bu)	[isopropenyl OSnBu$_3$]	Pd(PhCN)$_2$Cl$_2$ (1%), C$_6$H$_6$, 80°, 20 h	t-Bu diketone (83)	834
	[dimethylvinyl OSnBu$_3$]	Pd(PhCN)$_2$Cl$_2$ (1%), C$_6$H$_6$, 80°, 20 h	t-Bu diketone (35)	834
t-Bu–CO–CH$_2$Br	[cyclohexenyl OSnBu$_3$]	Pd(PhCN)$_2$Cl$_2$ (1%), C$_6$H$_6$, 80°, 20 h	t-Bu ketone (95)	834
	[Bu-t vinyl OSnBu$_3$]	Pd(PhCN)$_2$Cl$_2$ (1%), C$_6$H$_6$, 80°, 20 h	t-Bu…Bu-t diketone (97)	834

C_7

Substrate	Conditions	Product	Refs.
Bu₃SnSnBu₃ + TMSCl	Pd(PPh₃)₄ (10%), tetralin, 80°, 20 h	(OTMS / t-Bu) (81)	832, 833
Bu₃SnC≡CPh	Pd(PPh₃)₄ (4%), CCl₄, 80°, 12 h	TMS—O—C≡CPh (61)	207
Bu₃SnC≡CR	Pd(PPh₃)₂Cl₂ (1%), C₆H₆, 70°, 5 h	RC≡C—C(NPh)—C≡CR	835

R:
- TMS (60)
- C≡CEt (31)
- Bu-n (70)
- C(Me)₂OMe (64)
- Ph (65)
- 1-cyclohexenyl (74)

Substrate	Conditions	Product	Refs.
Bu₃SnC≡C—O—(CH₂)ₙ—O—C≡CSnBu₃	Pd(PPh₃)₂Cl₂ (5%), LiClO₄, PhMe, 50°	PhN=C< ring with (CH₂)ₙ	212

n = 3 (31) n = 8 (28)
n = 4 (42) n = 10 (24)
n = 5 (45)
n = 6 (42)

| Bu₃SnC≡C—O—((CH₂)O)ₙ | Pd(PPh₃)₂Cl₂ (5%), LiClO₄, PhMe, 50° | NPh ring | 212 |

n = 2 (52)
n = 3 (32)

| Bu₃SnC≡C—X—(CH₂)ₙ—X—C≡CSnBu₃ | Pd(PPh₃)₂Cl₂ (5%), LiClO₄, PhMe, 50° | PhN=C< ring with (CH₂)₅ and X | 212 |

X = CH₂ (34)
X = NPr-i (41)

Left column substrates (C₇):

TMS—O—CH₂—Cl

(NPh)C(Cl)—Cl

TABLE XXIV. DIRECT CROSS-COUPLING OF MISCELLANEOUS ELECTROPHILES (*Continued*)

Substrate	Stannane	Conditions	Product(s) and Yield(s) (%)	Refs.
		Pd(PPh₃)₂Cl₂ (5%), LiClO₄, PhMe, 50°	(40)	212
		Pd(PPh₃)₂Cl₂ (5%), LiClO₄, PhMe, 50°	(23)	212
C₆H₁₁N=CCl₂ (Cl, Cl)	Bu₃SnC≡CTMS	Pd(PPh₃)₂Cl₂ (1%), C₆H₆, 70°, 5 h	TMSC≡C—(C₆H₁₁N=)C≡CTMS (35)	835
	Bu₃SnC≡C—(O)₂	Pd(PPh₃)₂Cl₂ (5%), LiClO₄, PhMe, 50°	(17)	212
	Bu₃SnC≡C—(CH₂)₅—C≡CSnBu₃	Pd(PPh₃)₂Cl₂ (5%), LiClO₄, PhMe, 50°	(CH₂)₅ (52)	212
p-ClC₆H₄N=CCl₂ (Cl, Cl)	Bu₃SnC≡C—(O)₂	Pd(PPh₃)₂Cl₂ (5%), LiClO₄, PhMe, 50°	p-ClC₆H₄N= (55)	212

540

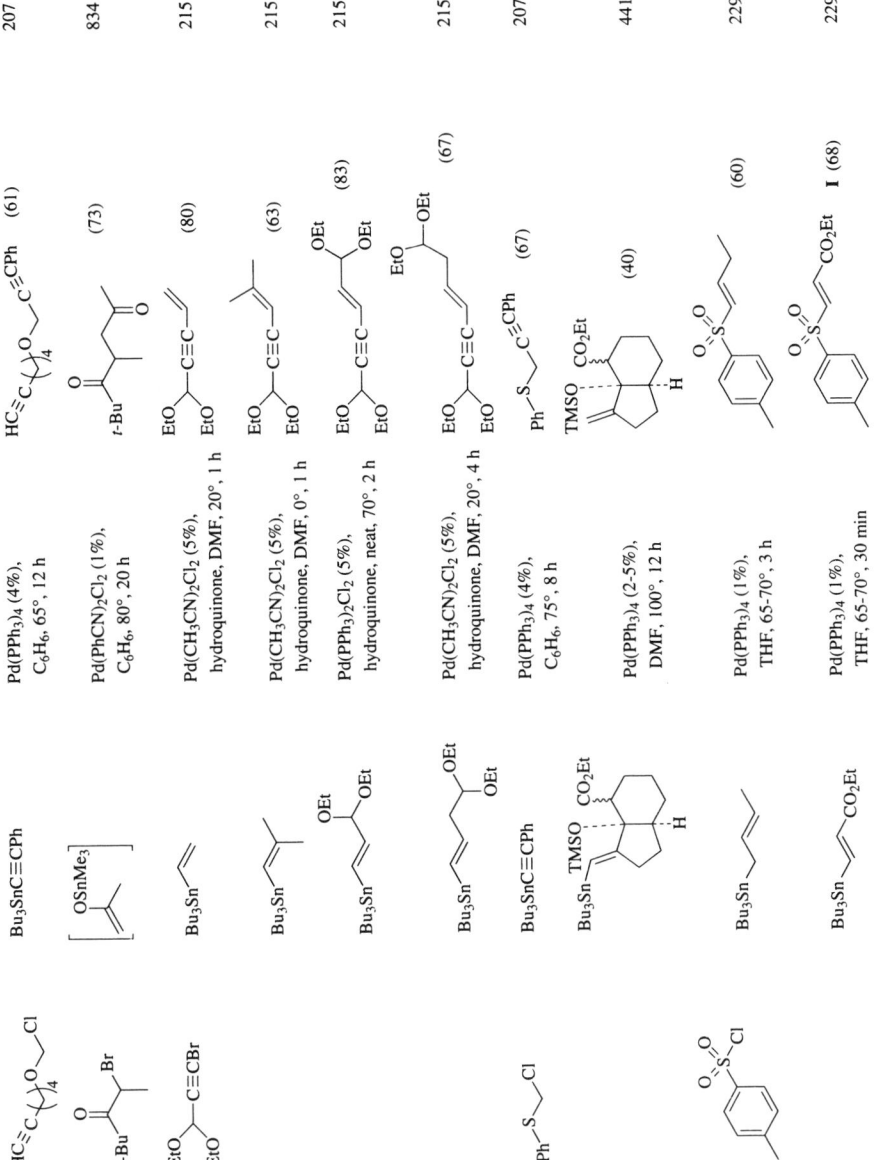

TABLE XXIV. DIRECT CROSS-COUPLING OF MISCELLANEOUS ELECTROPHILES (*Continued*)

Substrate	Stannane	Conditions	Product(s) and Yield(s) (%)	Refs.
MeO–C$_6$H$_4$–SO$_2$Cl	Bu$_3$Sn–CH=CH–CO$_2$Et	Pd(PPh$_3$)$_4$ (1%), THF, 65-70°, 30 min	**I** (64)	229
	Bu$_3$Sn–CH=CH–Ph	Pd(PPh$_3$)$_4$ (1%), THF, 65-70°, 15 min	(sulfone)–CH=CH–Ph (77)	229
	Bu$_3$Sn–CH$_2$–C$_6$H$_4$–CH$_3$	Pd(PPh$_3$)$_4$ (1%), THF, 65-70°, 12 h	(57)	229
	Bu$_3$Sn–CH=CH–CH(OEt)$_2$	Pd(PPh$_3$)$_4$ (2%), C$_6$H$_6$, 80°, 20 h	CH(OEt)$_2$ (85)	763
	Bu$_3$Sn–CH=CH–CH$_2$–CH(OEt)$_2$ *E:Z = 85:15*	Pd(PPh$_3$)$_4$ (2%), C$_6$H$_6$, 80°, 20 h	CH(OEt)$_2$ (85)	539
	Bu$_3$Sn–CH=CH–CH$_2$–OTHP	Pd(PPh$_3$)$_4$ (1%), THF, 65-70°, 30 min	OTHP (90)	229
	Bu$_3$Sn–CH=CH–C$_6$H$_{13}$-*n*	Pd(PPh$_3$)$_4$ (1%), THF, 65-70°, 30 min	C$_6$H$_{13}$-*n* (87)	229
HO$_2$C–C$_6$H$_4$–SO$_2$Cl	Bu$_3$Sn–CH=CH–Ph	Pd(PPh$_3$)$_4$ (1%), THF, 65-70°, 15 min	Ph (75)	229
C$_8$ PhC≡CBr	Bu$_3$Sn–CH=CH–CH(OEt)$_2$	Pd(PPh$_3$)$_2$Cl$_2$ (5%), hydroquinone, DMF, 20°, 20 h	PhC≡C–CH=CH–CH(OEt)$_2$ (36)	215

542

Substrate	Stannane	Conditions	Product (%)	Refs.
PhC≡Cl	Bu₃Sn–CH=CH–CH(OEt)₂	Pd(CH₃CN)₂Cl₂ (5%), hydroquinone, DMF, 20°, 20 h	EtO, OEt, PhC≡C (64)	215
	(dimethylcyclopropene)–Bu₃Sn / TMS	Pd(PPh₃)₄ (5%), THF, 20°, 24 h	(dimethylcyclopropene) PhC≡C / TMS (0)	422
	MOMO-uracil–SnBu₃ (acetonide)	Pd(PPh₃)₄ (10%), CuI (20%), DMF, 80°, 30 min	MOMO-uracil C≡CPh (acetonide) (77)	170
Ph–C(O)–CH₂Br	Bu₃Sn–CH₂CH=CH₂	D₇₁₇-Pd(0) on polymer, Me₂CO, reflux, 25 h	Ph–C(O)–CH₂CH₂CH=CH₂ (86)	535
	Me₃SnTMS	Pd(PPh₃)₄ (10%), C₆H₆, 80°, 5 h	Ph, OTMS (69)	833, 832
PhCH₂O–CH₂Cl	Bu₃SnC≡CPh	Pd(PPh₃)₄ (4%), C₆H₆, 80°, 6 h	PhCH₂O–CH₂–C≡CPh (73)	207
	Bu₃Sn–CH₂CH=CH₂	Pd(PPh₃)₄ (4%), C₆H₆, 80°, 4 h	Ph–O–CH₂CH=CH₂ (75)	207
Ph–CH(OMe)Cl	Bu₃SnC≡CPh	Pd(PPh₃)₄ (4%), CCl₄, 80°, 20 h	Ph–CH(OMe)–C≡CPh (71)	207
n-C₆H₁₃–C(O)–CH₂Br	[isopropenyl–OSnBu₃]	Pd(PhCN)₂Cl₂ (5%), C₆H₆, 80°, 20 h	n-C₆H₁₃ diketone (21)	834

TABLE XXIV. DIRECT CROSS-COUPLING OF MISCELLANEOUS ELECTROPHILES (*Continued*)

Substrate	Stannane	Conditions	Product(s) and Yield(s) (%)	Refs.
EtO—CH(OEt)—CH2—C≡CBr	Bu3Sn—CH=CH2	Pd(CH3CN)2Cl2 (5%), hydroquinone, DMF, 0°, 4 h	EtO—CH(OEt)—CH2—C≡C—CH=CH2 (78)	215
	Bu3Sn—C(CH3)=CH2 (prenyl)	Pd(CH3CN)2Cl2 (5%), hydroquinone, DMF, 0°, 9 h	(44)	215
	Bu3Sn—CH=CH—CH(OEt)2	Pd(PPh3)2Cl2 (5%), hydroquinone, DMF, 20°, 72 h	(57)	215
	Bu3Sn—CH2—CH=CH—CH(OEt)2	Pd(CH3CN)2Cl2 (5%), hydroquinone, DMF, 20°, 96 h	(58)	215
n-C6H13O—CO—CH2Br	Bu3SnPh	D17-Pd(0) on polymer, Me2CO, reflux, 25 h	n-C6H13O—CO—CH2—Ph (84)	535
Et—C(=NPh)—CCl	Bu3SnC≡CPh	Pd(PPh3)2Cl2 (4%), PhEt, 70°, 5 h	Et—C(=NPh)—C≡CPh (69)	211

C9

Substrate	Stannane	Conditions	Product(s) and Yield(s) (%)	Refs.
n-C8H17—O—CH2—X	Bu3Sn—CH=CH2	Pd(PPh3)4 (4%), Cl(CH2)2Cl, 24°	n-C8H17—O—CH2—CH=CH2; X = Br, 1.5 h (84); X = Cl, 10 h (84)	207
n-C8H17—O—CH2—Br	Bu3SnC≡CPh	Pd(PPh3)4 (4%), CH2Cl2, 40°, 8 h	n-C8H17—O—CH2—C≡CPh (55)	207
n-C8H17—O—CH2—Cl	Bu3Sn—C(OEt)=CH2	Pd(PPh3)4 (4%), C6H6, 80°, 12 h	n-C8H17—O—CH2—C(OEt)=CH2 (81)	207
	Bu3SnPh	Pd(PPh3)4 (4%), C6H6, 80°, 2 h	n-C8H17—O—CH2—Ph (86)	207

	Organotin reagent	Conditions	Product (%)	Refs.
C₁₀				
(substrate: CH(OMe)CH₂CH₂Ph, CHCl)	Bu₃SnC≡CPh	Pd(PPh₃)₄ (4%), C₆H₆, 80°, 1.5 h	$n\text{-}C_8H_{17}\text{O}\text{-}CH_2\text{C}{\equiv}\text{CPh}$ (71)	207
	Bu₃SnC≡CPh	Pd(PPh₃)₄ (4%), CCl₄, 65°, 12 h	(65)	207
2-naphthalenesulfonyl chloride	Bu₃Sn⁓ (crotyl)	Pd(PPh₃)₄ (1%), THF, 65-70°, 3 h	(60)	229
	Bu₃Sn⁓OTHP	Pd(PPh₃)₄ (1%), THF, 65-70°, 30 min	OTHP (85)	229
	Bu₃SnC≡CPh	Pd(PPh₃)₄ (1%), THF, 65-70°, 15 min	C≡CPh (0)	229
	Bu₃Sn⁓Ph	Pd(PPh₃)₄ (1%), THF, 65-70°, 15 min	Ph (70)	229
	Bu₃Sn⁓C₆H₁₃-n	Pd(PPh₃)₄ (1%), THF, 65-70°, 30 min	$C_6H_{13}\text{-}n$ (70)	229
C₁₁				
$n\text{-BuN}{=}C(Cl)Ph$	Ph₄Sn	Pd(PPh₃)₂Cl₂ (4%), PhEt, 130°, 63 h	(25)	211
OAc / CHCl / CH₂CH₂Ph	Bu₃SnC≡CPh	Pd(PPh₃)₄ (4%), C₆H₆, 80°, 12 h	(0)	207
C₁₂				
thienyl-C(Cl)=N-(4-tolyl)	Ph₄Sn	Pd(PPh₃)₂Cl₂ (4%), PhEt, 130°, 15 h	(63)	211

545

TABLE XXIV. DIRECT CROSS-COUPLING OF MISCELLANEOUS ELECTROPHILES (*Continued*)

Substrate	Stannane	Conditions	Product(s) and Yield(s) (%)	Refs.
C$_{13}$ Ph $\overset{NPh}{\underset{}{C}}$–Cl	Bu$_3$Sn—	Pd(PPh$_3$)$_2$Cl$_2$ (10%), xylene, 120°, 10 h	(67)	531
	Bu$_3$Sn—	Pd(PPh$_3$)$_2$Cl$_2$ (10%), xylene, 120°, 20 h	(67)	531
	Bu$_3$Sn—	Pd(PPh$_3$)$_2$Cl$_2$ (10%), xylene, 120°, 20 h	**I** (67)	531
	[OSnBu$_3$]	Pd(PPh$_3$)$_2$Cl$_2$ (10%), xylene, 120°, 5 h	(9) + (25)	531
	Bu$_3$SnPh	Pd(PPh$_3$)$_2$Cl$_2$ (10%), xylene, 120°, 60 h	(78)	531
	[Bu$_3$Sn–O—]	Pd(PPh$_3$)$_2$Cl$_2$ (10%), xylene, 120°, 5 h	(74)	531
Ph–S $\overset{NPh}{\underset{}{C}}$–Cl	Bu$_3$SnC≡CR R: TMS (93), Bu-*n* (64), C(Me)$_2$OMe (76), Ph (83), CH$_2$O(CH$_2$)$_3$CH=CH$_2$ (83)	Pd(dppf)Cl$_2$ (1%), C$_6$H$_6$, 80°	(93), (64), (76), (83), (83)	836
	Bu$_3$SnC≡C—⟨ ⟩—C≡CSnBu$_3$	Pd(dppf)Cl$_2$ (1%), C$_6$H$_6$, 80°	(80)	836

546

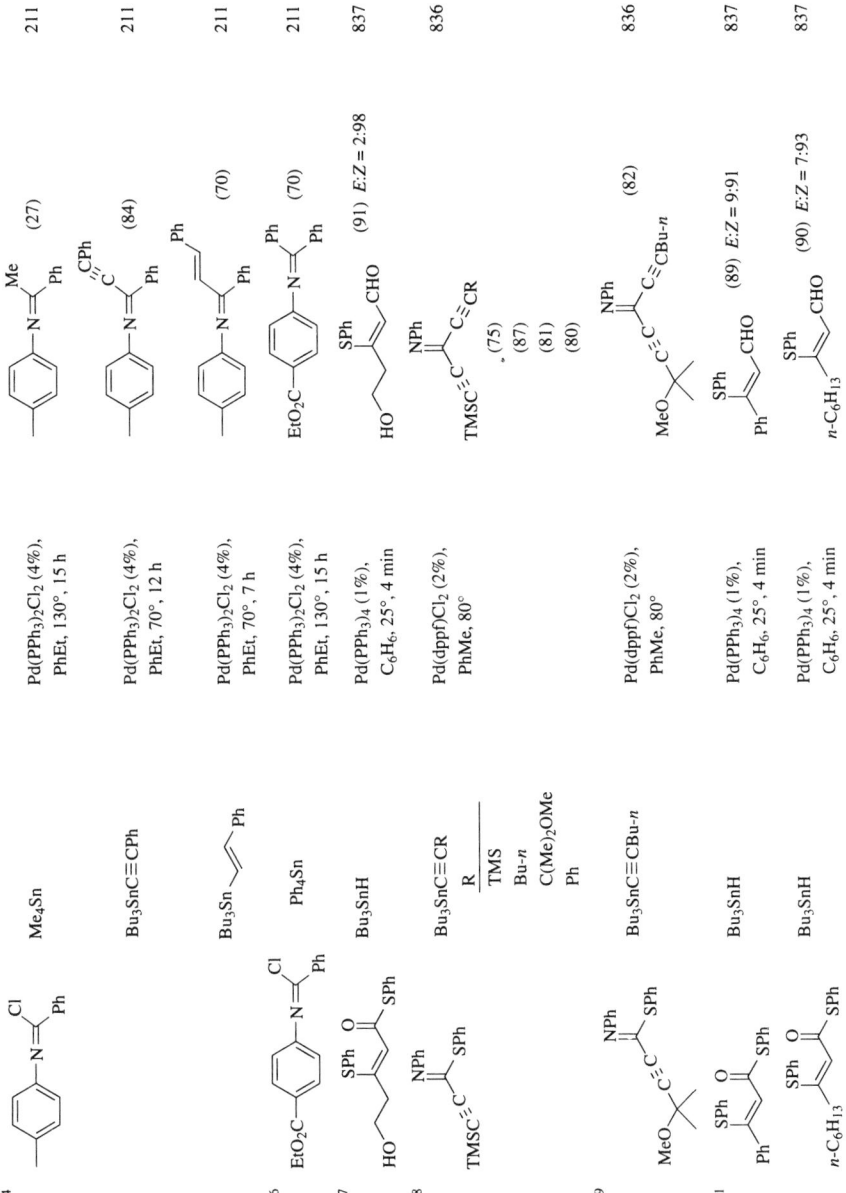

Substrate	Stannane	Conditions	Product(s) and Yield(s) (%)	Refs.
C$_{23}$	Bu$_3$SnH	Pd(PPh$_3$)$_4$ (1%), C$_6$H$_6$, 25°, 4 min	(98) *E:Z* 0:100	837
	Bu$_3$SnH	Pd(PPh$_3$)$_4$ (1%), C$_6$H$_6$, 25°, 4 min	(98) *E:Z* 4:96	837
C$_{31}$		Pd(PPh$_3$)$_4$ (4%), PhMe, 65°, 12 h	(71)	213
		Pd(PPh$_3$)$_4$ (4%), PhMe, 65°, 12 h	(71)	213
		Pd(PPh$_3$)$_4$ (10%), DMF, 60°, 1 h	(—)	216

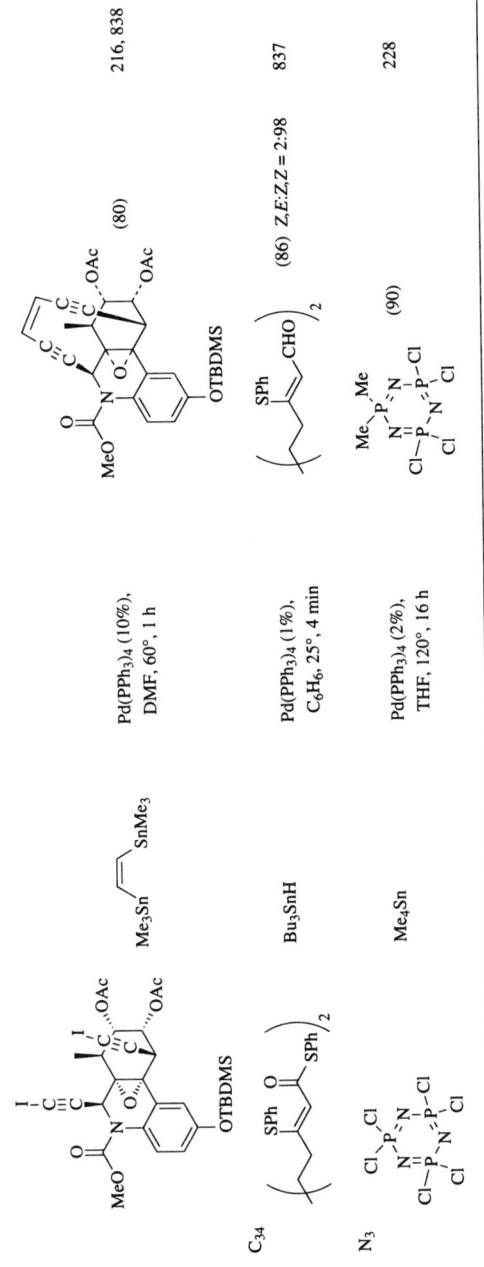

| | | | 216, 838 |
| | Pd(PPh₃)₄ (10%), DMF, 60°, 1 h | | (80) |

Me₃Sn⎯⎯SnMe₃   Pd(PPh₃)₄ (10%), DMF, 60°, 1 h   (80)   216, 838

C₃₄

Bu₃SnH   Pd(PPh₃)₄ (1%), C₆H₆, 25°, 4 min   (86) *Z:E:ZZ* = 2:98   837

N₃

Me₄Sn   Pd(PPh₃)₄ (2%), THF, 120°, 16 h   (90)   228

TABLE XXV. CARBONYLATIVE CROSS-COUPLING OF ALKENYL ELECTROPHILES

Substrate	Stannane	Conditions	Product(s) and Yield(s) (%)	Refs.
C₄ (2-methyl-2-butenyl iodide)	Bu₃Sn–(cyclopentenyl)	CO (50 psi), Pd(PPh₃)₂Cl₂ (2%)^{ya}, THF, 50°	(40)	324
C₅ (1-cyclopentenyl iodide)	Me₃SnH	CO (15 psi), Pd(PPh₃)₄ (4%), THF, 50°, 3.5h	CHO (35)	331
	R₃Sn–CH=CH₂	CO (50 psi), Pd(PPh₃)₂Cl₂ (2%)^{ya}, THF, 45-50°	R = Me, 6 h, (86) R = Bu, 18 h, (70)	324
	Bu₃Sn–CH=CH– (E:Z = 1:6)	CO (50 psi), Pd(PPh₃)₂Cl₂ (2%)^{ya}, THF, 45-50°, 24 h	(63) E:Z = 2.5:1	324
	Bu₃Sn–CH=CMe₂	CO (50 psi), Pd(PPh₃)₂Cl₂ (2%)^{ya}, THF, 50°	(0)	324
	Bu₃SnC≡CPr-n	CO (50 psi), Pd(PPh₃)₂Cl₂ (2%)^{ya}, THF, 45-50°, 7 h	C≡CPr-n (54)	324
	Bu₃Sn–CH=CH–Ph	CO (50 psi), Pd(PPh₃)₂Cl₂ (2%)^{ya}, THF, 45-50°, 8 h	Ph (60)	324
	Bu₃Sn–CH=CH–CO₂Bn	CO (50 psi), Pd(PPh₃)₂Cl₂ (2%)^{ya}, THF, 45-50°, 10 h	CO₂Bn (40)	324
(3-iodo-2-cyclopentenone)	Bu₃SnH	CO (45 psi), Pd(PPh₃)₄ (4%), PhMe, 50°, 3.5 h	(0) + (100)	331

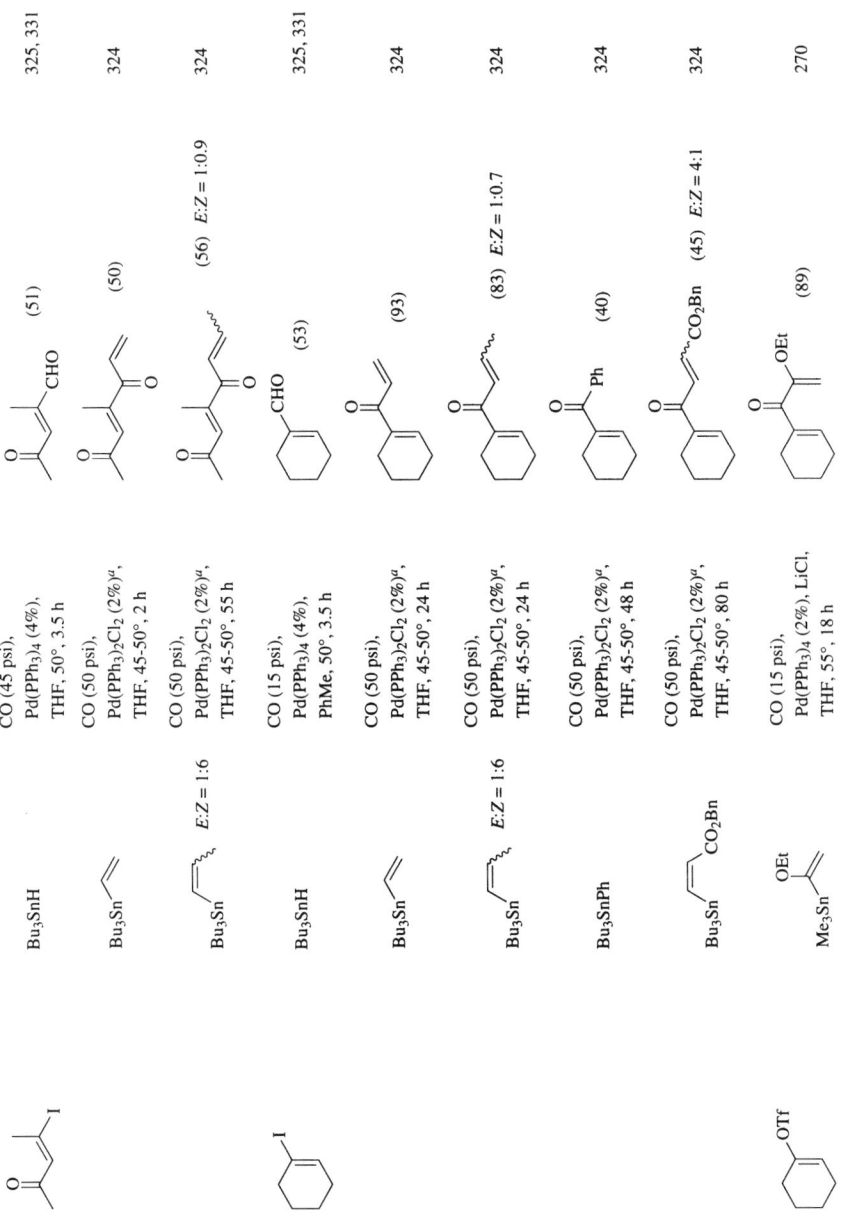

TABLE XXV. CARBONYLATIVE CROSS-COUPLING OF ALKENYL ELECTROPHILES (*Continued*)

Substrate	Stannane	Conditions	Product(s) and Yield(s) (%)	Refs.
n-Bu I	Bu₃SnH	CO (15 psi), Pd(PPh₃)₄ (4%), PhMe, 50°, 3.5 h	n-Bu—CHO (88)	331
n-Bu I	R₃Sn	CO (50 psi), Pd(PPh₃)₂Cl₂ (2%)^u, THF, 35-40°	R = Me, 5 h, (65); R = Bu, 12 h, (70)	324
	Me₃Sn E:Z = 1:2	CO (50 psi), Pd(PPh₃)₂Cl₂ (2%)^u, THF, 45-50°, 44 h	(70)	324
	Bu₃Sn E:Z = 1:6	CO (50 psi), Pd(PPh₃)₂Cl₂ (2%)^u, THF, 45-50°, 48 h	(62)	324
	R₃Sn Ph	CO (50 psi), Pd(PPh₃)₂Cl₂ (2%)^u, THF, 45-50°	Ph R = Me, 18 h, (65); R = Bu, 23 h, (40)	324
	Me₃Sn TMS	CO (15 psi), Pd(PPh₃)₄ (3%), LiCl, THF, 55°	TMS (77)	334
n-Bu OTf	Bu₃SnH	CO (15 psi), Pd(PPh₃)₄ (4%), PhMe, 50°, 3.5 h	**I** (20) + **II** (69)	331
	Bu₃SnH	CO (45 psi), Pd(PPh₃)₄ (4%), PhMe, 50°, 3.5 h	**I** (84) + **II** (18)	325, 331
C₇ I	Bu₃Sn	CO (50 psi), Pd(PPh₃)₂Cl₂ (2%)^u, THF, 65°, 5 h	(65)	324

552

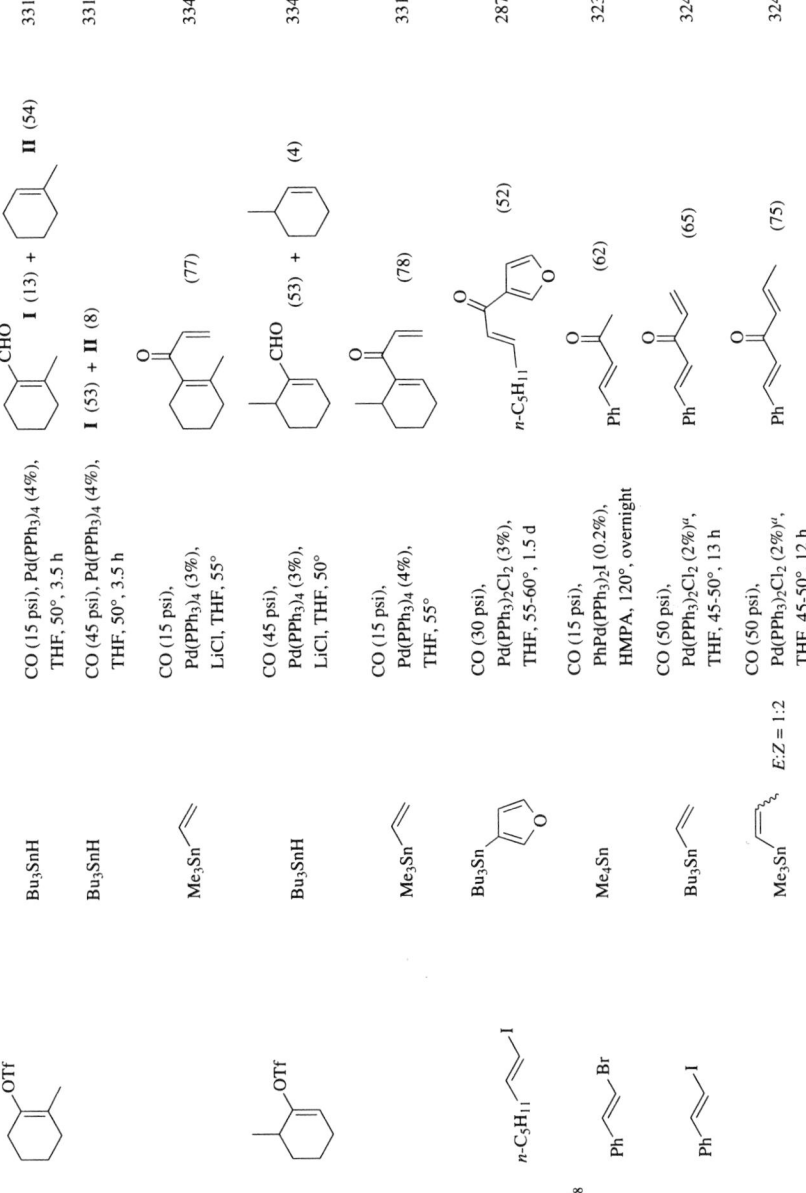

Substrate	Organotin	Conditions	Product (% yield)	Refs.
(cyclohexenyl-methyl OTf)	Bu₃SnH	CO (15 psi), Pd(PPh₃)₄ (4%), THF, 50°, 3.5 h	(CHO) **I** (13) + **II** (54)	331
	Bu₃SnH	CO (45 psi), Pd(PPh₃)₄ (4%), THF, 50°, 3.5 h	**I** (53) + **II** (8)	331
(methyl-cyclohexenyl OTf)	Me₃Sn⌒	CO (15 psi), Pd(PPh₃)₄ (3%), LiCl, THF, 55°	(vinyl ketone) (77)	334
	Bu₃SnH	CO (45 psi), Pd(PPh₃)₄ (3%), LiCl, THF, 50°	(CHO) (53) + (4)	334
	Me₃Sn⌒	CO (15 psi), Pd(PPh₃)₄ (4%), THF, 55°	(vinyl ketone) (78)	331
n-C₅H₁₁⌒I	Bu₃Sn (3-furyl)	CO (30 psi), Pd(PPh₃)₂Cl₂ (3%), THF, 55-60°, 1.5 d	(52)	287
Ph⌒Br	Me₄Sn	CO (15 psi), PhPd(PPh₃)₂I (0.2%), HMPA, 120°, overnight	(62)	323
Ph⌒I	Bu₃Sn⌒	CO (50 psi), Pd(PPh₃)₂Cl₂ (2%)ᵃ, THF, 45-50°, 13 h	(65)	324
Me₃Sn⌒		CO (50 psi), Pd(PPh₃)₂Cl₂ (2%)ᵃ, THF, 45-50°, 12 h	(75), E:Z = 1:2	324

C₈

TABLE XXV. CARBONYLATIVE CROSS-COUPLING OF ALKENYL ELECTROPHILES (Continued)

Substrate	Stannane	Conditions	Product(s) and Yield(s) (%)	Refs.
(cyclooctenyl iodide)	Bu₃Sn⁓⁓ $E:Z = 1:6$	CO (50 psi), Pd(PPh₃)₂Cl₂ (2%)[u], THF, 45-50°, 22 h	(40)	324
	Bu₃Sn-(furanyl)	CO (30 psi), Pd(PPh₃)₄ (3%), THF, 50°, 1 d	(40) + (20)	287
	Bu₃Sn⁓Ph	CO (50 psi), Pd(PPh₃)₂Cl₂ (2%)[u], THF, 45-50°, 66 h	Ph (70)	324
	Bu₃Sn⁓	CO (50 psi), Pd(PPh₃)₂Cl₂ (2%)[u], THF, 65°, 8 h	(74)	324
(5,5-dimethylcyclohexenone iodide)	Bu₃Sn⁓⁓ $E:Z = 1:6$	CO (50 psi), Pd(PPh₃)₂Cl₂ (2%)[u], THF, 45-50°, 12 h	(71)	324
(cyclopentenyl OTf)	Bu₃SnH	CO (15 psi), Pd(PPh₃)₄ (4%), LiCl, PhMe, 50°, 3.5 h	I (27) + (4)	331
	Bu₃SnH	CO (45 psi), Pd(PPh₃)₄ (4%), LiCl, PhMe, 50°, 3.5 h	I (1)	331
	Me₃Sn(OEt vinyl)	CO (15 psi), Pd(PPh₃)₄ (2%), LiCl, THF, 55°, 16 h	(63)	270

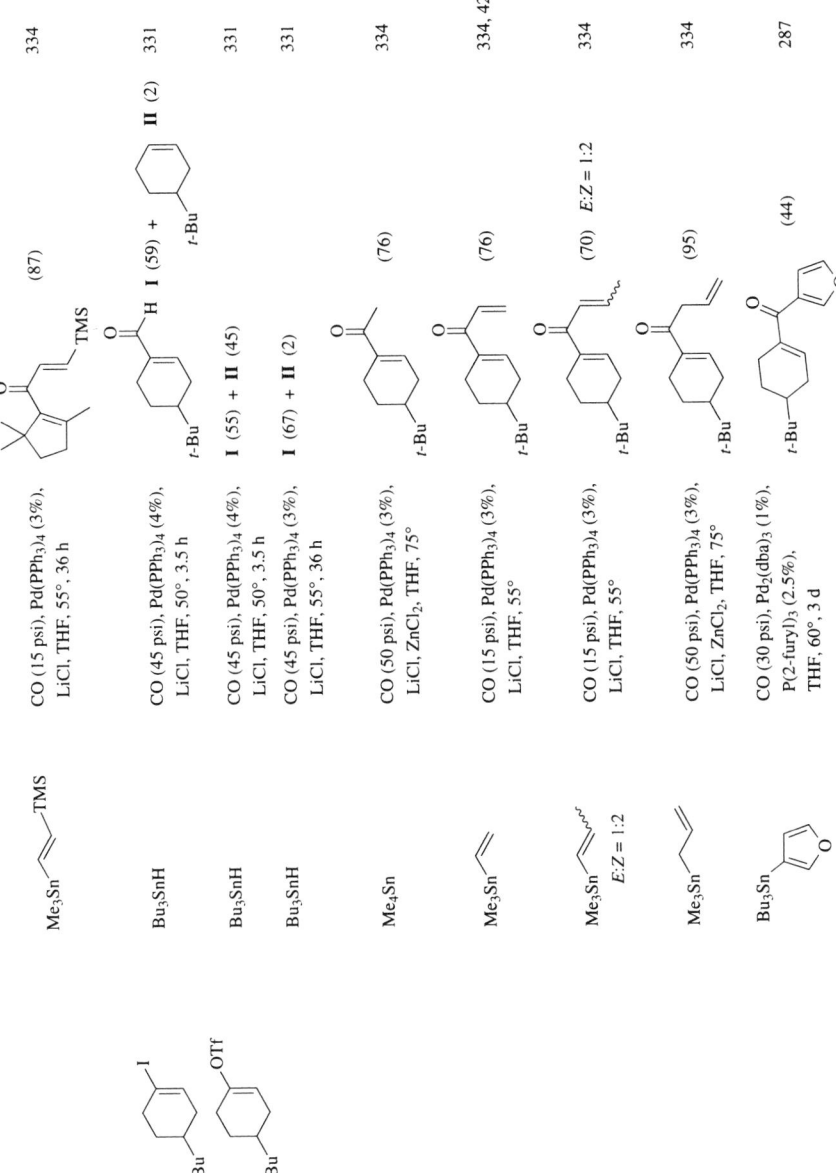

Substrate	Stannane	Conditions	Product(s) and Yield(s) (%)	Refs.
	Me₃SnC≡CTMS	CO (50 psi), Pd(PPh₃)₄ (3%), LiCl, THF, 20°	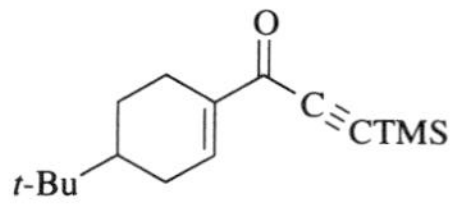 (95)	334
	Me₃SnPh	CO (50 psi), Pd(PPh₃)₄ (3%), LiCl, ZnCl₂, THF, 75°	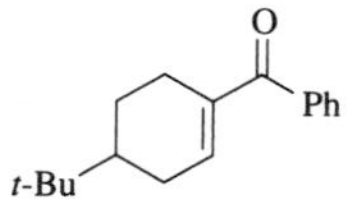 (93)	334
	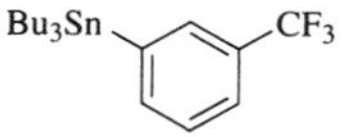 Bu₃Sn—⟨CF₃⟩	CO (50 psi), Pd(PPh₃)₄ (3%), LiCl, ZnCl₂, THF, 75°	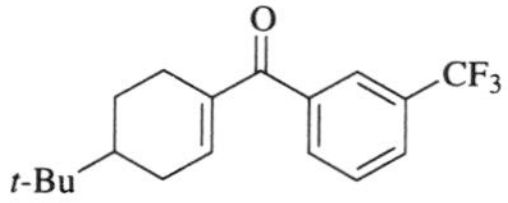 (76)	334
	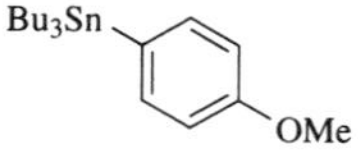 Bu₃Sn—⟨OMe⟩	CO (50 psi), Pd(PPh₃)₄ (3%), LiCl, ZnCl₂, THF, 75°	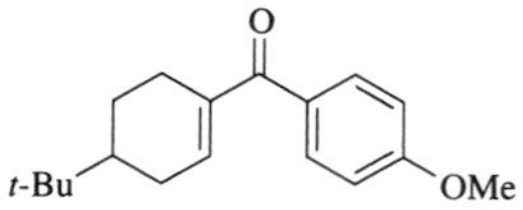 (88)	334
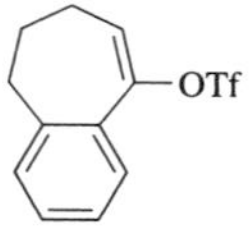 C₁₁	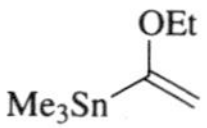 Me₃Sn—C(OEt)=CH₂	CO (15 psi), Pd(PPh₃)₄ (2%), LiCl, THF, 55°, 21 h	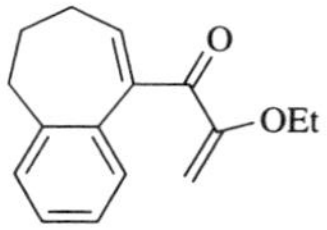 (92)	270
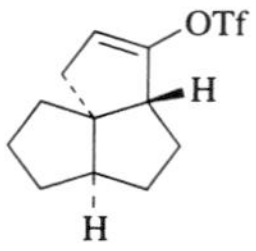	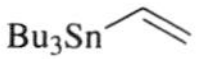 Bu₃Sn—CH=CH₂	CO (15 psi), Pd(PPh₃)₄ (4%), LiCl, THF, 55°, 48 h	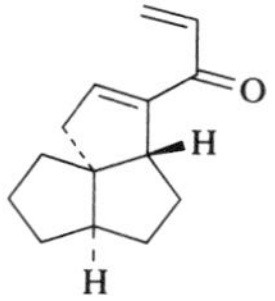 (50)	487

556

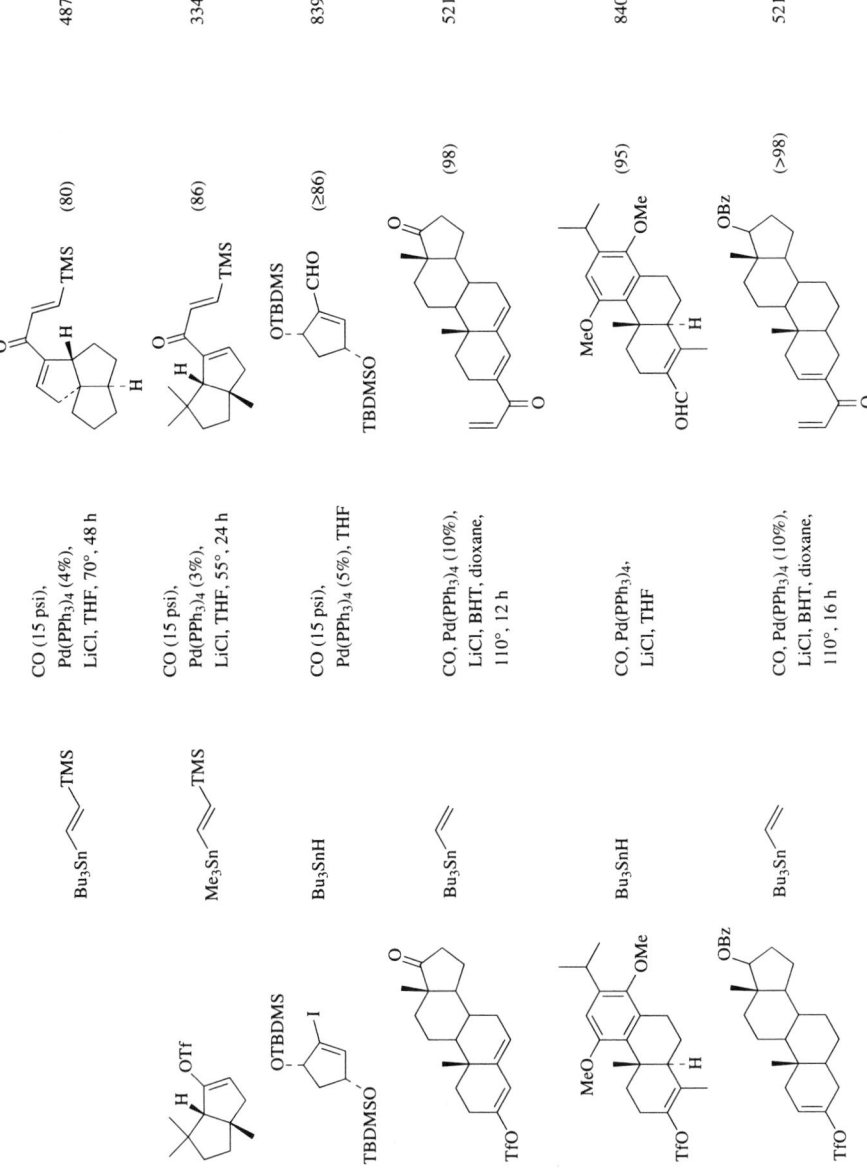

487

334

839

521

840

521

TABLE XXV. CARBONYLATIVE CROSS-COUPLING OF ALKENYL ELECTROPHILES (*Continued*)

Substrate	Stannane	Conditions	Product(s) and Yield(s) (%)	Refs.
C_{27}				
	Bu_3Sn	CO, Pd(PPh$_3$)$_4$ (10%), LiCl, BHT, dioxane, 110°, 10 h	(>98)	521

^{a} An alternative reagent was BnPd(PPh$_3$)$_2$I (2%).

TABLE XXVI. CARBONYLATIVE CROSS-COUPLING OF ARYL ELECTROPHILES

Substrate	Stannane	Conditions	Product(s) and Yield(s) (%)	Refs.
C_6 PhBr	Bu_3SnH	CO (450 psi), $PhPd(PPh_3)_2I$ (4%), PhMe, 50°, 3.5 h	$PhCHO$ (2) + PhBr (70) + C_6H_6 (28)	331
PhI	Bu_3SnH	CO (15 psi), $Pd(PPh_3)_4$ (4%), PhMe, 50°, 3.5 h	$PhCHO$ (93)	325, 331
	Me_4Sn	CO (15 psi), $PhPd(PPh_3)_2I$ (0.2%), HMPA, 120°, overnight	[structure] (85)	323
	Me_4Sn	CO (15 psi), $Ni(PPh_3)_2(CO)_2$ (3%), HMPA, 120°, 24 h	[structure] (73)	841
	Bu_4Sn	CO (15 psi), $PhPd(PPh_3)_2I$ (0.2%), HMPA, 120°, overnight	[structure] Bu-n (73)	841
	Bu_3Sn [furanyl]	CO (30 psi), $Pd(PPh_3)_2Cl_2$ (3%), THF, 55°, 3 d	[structure] (60)	287
	Me_3Sn [OEt vinyl]	CO (15 psi), $Pd(PPh_3)_4$ (2%), dioxane, 95°, 19.5 h	[structure] OEt (67)	270
	Bu_3Sn [OEt vinyl]	CO (75 psi), $Pd(PPh_3)_2Cl_2$ (3%), $CHCl_3$, 80°	[structure] OEt (60)	842
	Me_3SnPh	CO (15 psi), $[(\eta^3\text{-}C_3H_5)PdCl]_2$ (1%), HMPA, 20°, 48 h	$PhCOPh$ (42) + Ph-Ph (58)	326, 843

TABLE XXVI. CARBONYLATIVE CROSS-COUPLING OF ARYL ELECTROPHILES (*Continued*)

Substrate	Stannane	Conditions	Product(s) and Yield(s) (%)	Refs.
	Ph$_4$Sn	CO (15 psi), PhPd(PPh$_3$)$_2$I (0.2%), HMPA, 120°, overnight	PhC(O)Ph (68)	323
	(Bu$_3$Sn-substituted dioxolane-spiro cyclobutenone)	CO (32 psi), BnPd(PPh$_3$)$_2$Cl (1%), C$_6$H$_6$, 80°, 18 h	(dioxolane-spiro cyclobutenone benzoyl product) (86)	844
	(Bu$_3$Sn-substituted piperidinyl cyclobutenedione)	CO (15 psi), CuI (2%), BnPd(PPh$_3$)$_2$Cl (2%), THF, 50°, 40 h	(piperidinyl cyclobutenedione benzoyl product) (65)	844
	Bu$_3$Sn—furan—SnBu$_3$	CO (30 psi), Pd(PPh$_3$)$_4$ (10%), THF, 55°, 3 d	(furan-3,4-dibenzoyl product) (80)	287
	Me$_3$SnNEt$_2$	CO (15 psi), PhPd(PPh$_3$)$_2$I (2%), 3 h	PhCONEt$_2$ (90)	329, 330
PhN$_2^+$ PF$_6^-$	Me$_4$Sn	CO (130 psi), Pd(OAc)$_2$ (2%), CH$_3$CN, reflux, 1 h	PhC(O)CH$_3$ (55)	845
PhN$_2^+$ BF$_4^-$	Et$_4$Sn	CO (130 psi), Pd(OAc)$_2$ (2%), CH$_3$CN, reflux, 6 h	PhC(O)Et (40)	845
(4-bromoiodobenzene)	Bu$_3$SnH	CO (45 psi), Pd(PPh$_3$)$_4$ (4%), PhMe, 50°, 3.5 h	4-Br-C$_6$H$_4$-CHO (2) + bromobenzene (28)	325, 331

Substrate	Reagent	Conditions	Product(s) (% yield)	Refs.
4-Br-C6H4-N2+ BF4-	R4Sn	CO (130 psi), Pd(OAc)2 (2%), CH3CN, reflux	4-Br-C6H4-C(O)R, R = Me, 1 h (84); R = Ph, 6 h (59)	845
4-Br-C6H4-OTf	Bu3Sn-C6H4-OMe	CO (15 psi), Pd(dppf)Cl2 (4%), LiCl, BHT, DMF, 95°, 27 h	(45) + (17) + (6)	336
	Bu3Sn–CH=CH–Ph	CO (15 psi), Pd(dppf)Cl2 (4%), BHT, DMF, 70°, 7 h	(84)	336
2-Cl-C6H4-N2+ BF4-	Me4Sn	CO (130 psi), Pd(OAc)2 (4%), CH3CN, reflux, 1 h	(76)	340, 845
3-Cl-iodobenzene	Bu3SnH	CO (45 psi), Pd(PPh3)4 (4%), THF, 50°, 3.5 h	(78) + (6)	331
4-Cl-iodobenzene	Bu3SnH	CO (45 psi), Pd(PPh3)4 (4%), THF, 50°, 3.5 h	(77) + (9)	331
4-Cl-C6H4 (CH2=C(OEt)SnMe3)	Me3Sn–C(OEt)=CH2	1. CO (15 psi), Pd(PPh3)4 (2%), dioxane, 95°, 24 h 2. O3	(78)	270

561

TABLE XXVI. CARBONYLATIVE CROSS-COUPLING OF ARYL ELECTROPHILES (Continued)

Substrate	Stannane	Conditions	Product(s) and Yield(s) (%)	Refs.
X-C$_6$H$_4$-N$_2^+$ BF$_4^-$ (X = Cl)	Me$_3$SnPh	CO (15 psi), [(η³-C$_3$H$_5$)PdCl]$_2$ (1%), HMPA, 20°, 12 h	Cl-C$_6$H$_4$-CO-Ph (98)	326
1,4-diiodobenzene	Me$_4$Sn	CO (130 psi), Pd(OAc)$_2$ (2%), CH$_3$CN, reflux, 1 h	X-C$_6$H$_4$-CO-CH$_3$, X = Cl (76); X = I (79)	845
1,4-diiodobenzene	Me$_3$SnPh	CO (15 psi), [(η³-C$_3$H$_5$)PdCl]$_2$ (1%), HMPA, 20°, 10 h	Ph-CO-C$_6$H$_4$-CO-Ph (78)	326
O$_2$N-C$_6$H$_3$-N$_2^+$ BF$_4^-$	Me$_4$Sn	CO (130 psi), Pd(OAc)$_2$ (2%), CH$_3$CN, reflux, 1 h	O$_2$N-C$_6$H$_4$-CO-CH$_3$ (70)	845
O$_2$N-C$_6$H$_4$-X (X = Br, Cl, I)	Bu$_3$SnH	CO (15 psi), Pd(PPh$_3$)$_4$ (4%), PhMe, 50°, 3.5 h	**I** (O$_2$N-C$_6$H$_4$-CHO) + **II** (O$_2$N-C$_6$H$_5$); Br: **I** (7) + **II** (69); Cl: **I** (0); I: **II** (84)	331
O$_2$N-C$_6$H$_4$-I	Bu$_3$SnH	CO (45 psi), Pd(PPh$_3$)$_4$ (4%), PhMe, 50°, 3.5 h	**I** (20) + **II** (62)	331
I-C$_6$H$_4$-NO$_2$	Me$_4$Sn	CO (15 psi), [(η³-C$_3$H$_5$)PdCl]$_2$ (1%), HMPA, 20°, 30 h	O$_2$N-C$_6$H$_4$-CO-CH$_3$ (95)	326, 843

Me₃Sn—CH=CH₂

CO (15 psi),
[(η³-C₃H₅)PdCl]₂ (1%),
HMPA, 20°, 1 h

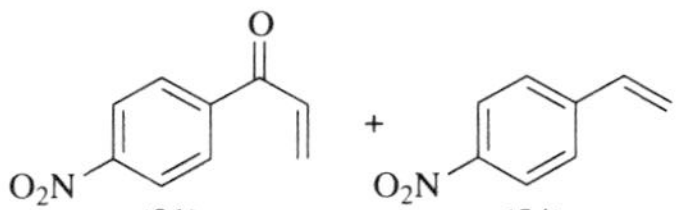

(31) + (54)

326

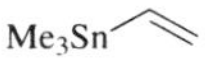

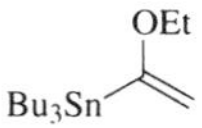

Bu₃Sn—C(OEt)=CH₂

1. CO (15 psi), Pd(PPh₃)₄ (2%),
dioxane, 95°, 25.5 h
2. O₃

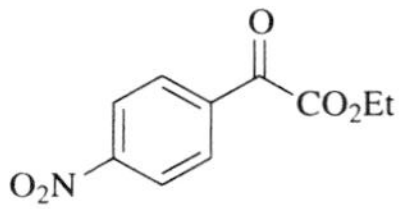

(23)

270

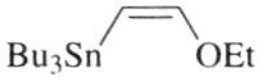

Bu₃Sn—CH=CH—OEt

CO (75 psi),
Pd(PPh₃)₂Cl₂ (3%),
CHCl₃, 80°

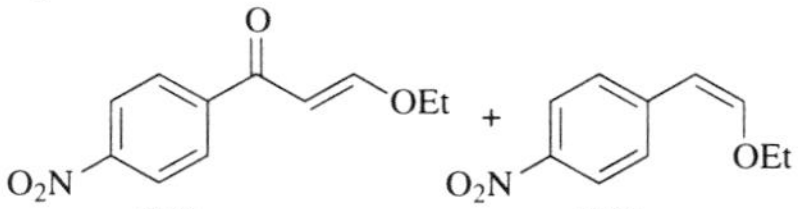

(25) + (40)

842

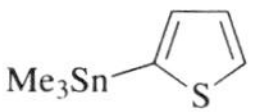

Me₃Sn—⟨thiophene⟩

CO (15 psi),
[(η³-C₃H₅)PdCl]₂ (1%),
HMPA, 20°, 0.7 h

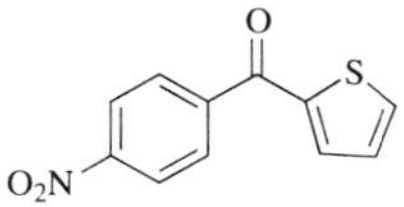

(36) + (64)

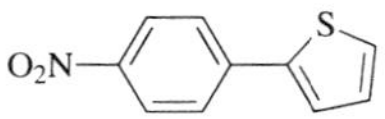

326

Me₃SnR

CO (15 psi),
[(η³-C₃H₅)PdCl]₂ (1%),
HMPA, 20°

326, 843

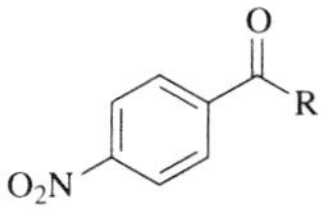

R	time	yield
Ph	4.5 h	(99)
p-ClC₆H₄	3 h	(98)
p-O₂NC₆H₄	4 h	(94)
C₆F₅	72 h	(45)
p-MeC₆H₄	5 h	(98)
p-MeOC₆H₄	3 h	(95)

TABLE XXVI. CARBONYLATIVE CROSS-COUPLING OF ARYL ELECTROPHILES (*Continued*)

Substrate	Stannane	Conditions	Product(s) and Yield(s) (%)	Refs.
O_2N-C$_6$H$_4$-N$_2^+$ PF$_6^-$	Me$_3$SnC≡CPh	CO (15 psi), [(η^3-C$_3$H$_5$)PdCl]$_2$ (1%), HMPA, 20°, 0.5 h	O_2N-C$_6$H$_4$-C(O)C≡CPh (64) + O_2N-C$_6$H$_4$-C≡CPh (36)	326
	Me$_3$SnNEt$_2$	CO (15 psi), PhPd(PPh$_3$)$_2$I (2%), 0.5 h	O_2N-C$_6$H$_4$-CONEt$_2$ (82)	329, 330
	Et$_3$SnOMe	CO (15 psi), PhPd(PPh$_3$)$_2$I (2%), 1 h	O_2N-C$_6$H$_4$-CO$_2$Me (100)	329, 330
	Et$_3$SnSPh	CO (15 psi), PhPd(PPh$_3$)$_2$I (2%), 1 h	O_2N-C$_6$H$_4$-C(O)SPh (6) + O_2N-C$_6$H$_4$-SPh-SPh (90)	329, 330
O_2N-C$_6$H$_4$-OTf	Me$_4$Sn	CO (130 psi), Pd(OAc)$_2$ (2%), CH$_3$CN, reflux, 1 h	O_2N-C$_6$H$_4$-C(O)CH$_3$ (85)	845
C$_7$ 2-iodotoluene	Bu$_3$SnC≡CPr-n	CO (15 psi), Pd(dppf)Cl$_2$ (4%), LiCl, BHT, DMF, 100°, 6 h	o-tolyl-C(O)C≡CPr-n (0)	336
	Bu$_3$SnH	CO (15 psi), Pd(PPh$_3$)$_4$ (4%), PhMe, 50°, 3.5 h	2-methylbenzaldehyde CHO (28)	325, 331
2-methyl-C$_6$H$_4$-N$_2^+$ BF$_4^-$	Me$_4$Sn	CO (130 psi), Pd(OAc)$_2$ (2%), CH$_3$CN, reflux, 1 h	2-methylacetophenone (63)	845

564

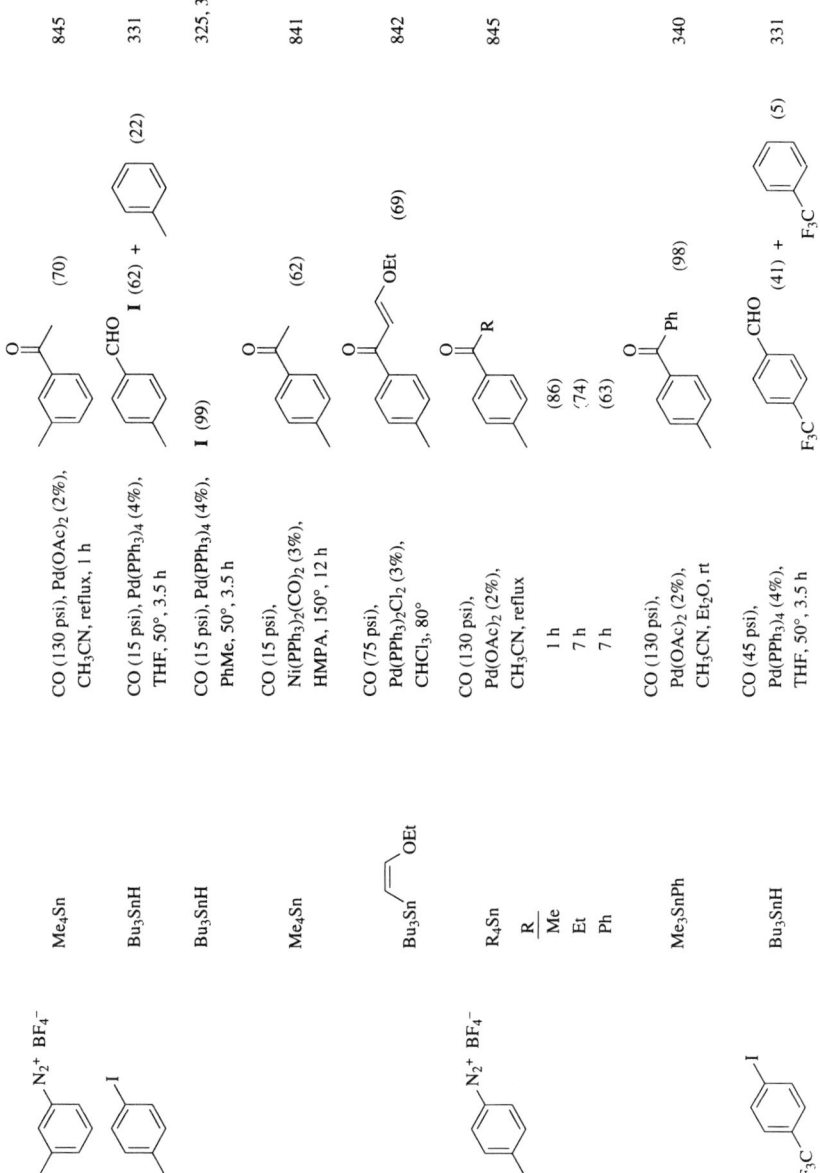

Substrate	Reagent	Conditions	Product(s) (%)	Refs.
3-methylphenyl N₂⁺ BF₄⁻	Me₄Sn	CO (130 psi), Pd(OAc)₂ (2%), CH₃CN, reflux, 1 h	methyl ketone (70)	845
4-iodotoluene	Bu₃SnH	CO (15 psi), Pd(PPh₃)₄ (4%), THF, 50°, 3.5 h	CHO **I** (62) + toluene (22)	331
4-iodotoluene	Bu₃SnH	CO (15 psi), Pd(PPh₃)₄ (4%), PhMe, 50°, 3.5 h	**I** (99)	325, 331
	Me₄Sn	CO (15 psi), Ni(PPh₃)₂(CO)₂ (3%), HMPA, 150°, 12 h	methyl ketone (62)	841
	Bu₃Sn–OEt	CO (75 psi), Pd(PPh₃)₂Cl₂ (3%), CHCl₃, 80°	enone OEt (69)	842
	R₄Sn ; R = Me / Et / Ph	CO (130 psi), Pd(OAc)₂ (2%), CH₃CN, reflux, 1 h / 7 h / 7 h	ketone (86) / (74) / (63)	845
4-methylphenyl N₂⁺ BF₄⁻	Me₃SnPh	CO (130 psi), Pd(OAc)₂ (2%), CH₃CN, Et₂O, rt	phenyl ketone (98)	340
4-F₃C-iodobenzene	Bu₃SnH	CO (45 psi), Pd(PPh₃)₄ (4%), THF, 50°, 3.5 h	CHO (41) + F₃C–C₆H₅ (5)	331

565

TABLE XXVI. CARBONYLATIVE CROSS-COUPLING OF ARYL ELECTROPHILES (*Continued*)

Substrate	Stannane	Conditions	Product(s) and Yield(s) (%)	Refs.
4-Br-C₆H₄-CN	Me₄Sn	CO (15 psi), Ni(PPh₃)₂(CO)₂ (3%), HMPA, 140°, 24 h	(41) acetyl-benzonitrile	841
4-I-C₆H₄-CN	Me₃SnPh	CO (15 psi), [(η³-C₃H₅)PdCl]₂ (1%), HMPA, 20°, 4.5 h	Ph (94) benzoyl-benzonitrile	326
2-OTf-tropone	Bu₃Sn—CH=CH₂	CO (15 psi), Pd(dppf)Cl₂ (4%), LiCl, BHT, DMF, 23°, 19 h	(50)	336
	Bu₃Sn—CH=CH—TMS	CO (15 psi), Pd(dppf)Cl₂ (4%), LiCl, BHT, DMF, 23°, 18 h	TMS (84)	336
	Bu₃Sn—C₆H₄—OMe	CO (15 psi), Pd(dppf)Cl₂ (4%), LiCl, BHT, DMF, 70°, 1 h	OMe (29) + biaryl OMe (23)	336
3-OHC-C₆H₄-OTf	Bu₃Sn—C₆H₄—OMe	CO (15 psi), Pd(dppf)Cl₂ (2%), LiCl, BHT, DMF, 90°, 8 h	OMe (98)	336

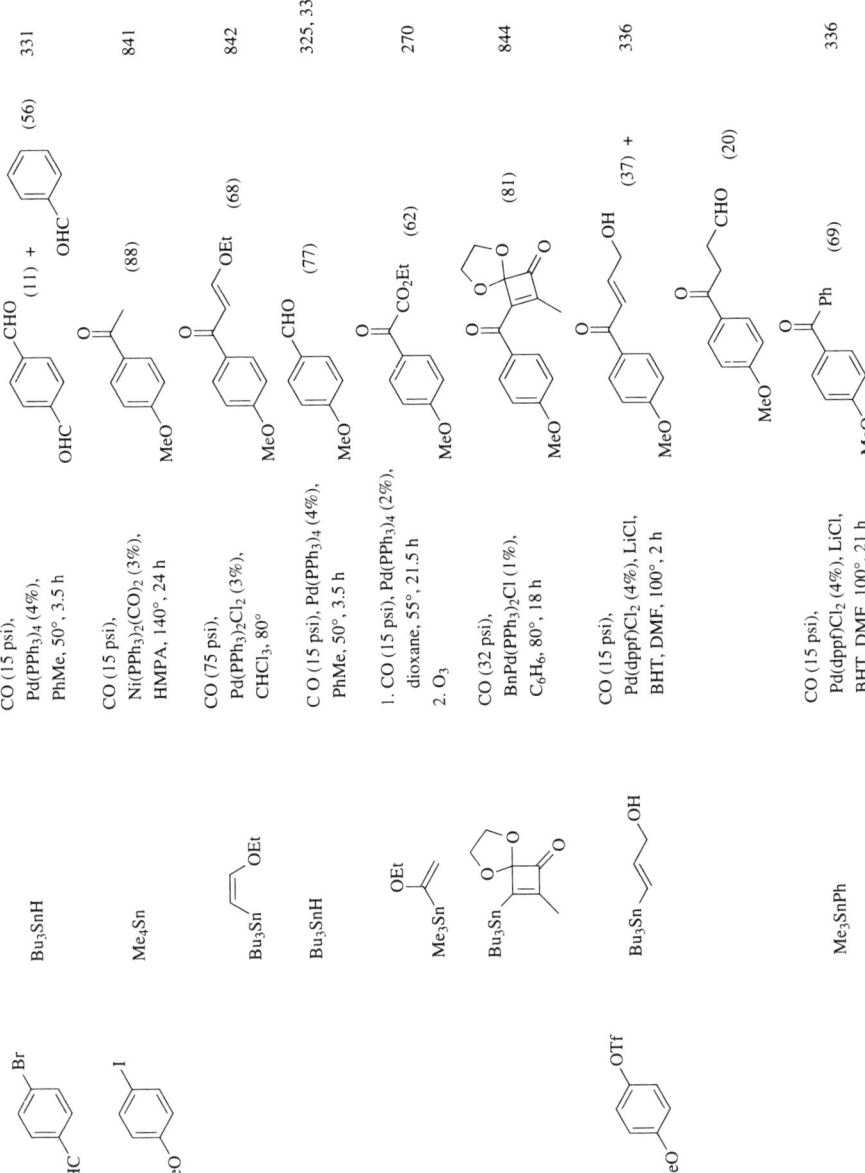

TABLE XXVI. CARBONYLATIVE CROSS-COUPLING OF ARYL ELECTROPHILES (*Continued*)

Substrate	Stannane	Conditions	Product(s) and Yield(s) (%)	Refs.
(o-iodobenzyl alcohol)	Bu₃Sn Ph	CO (15 psi), Pd(dppf)Cl₂ (2%), LiCl, BHT, DMF, 70°, 23 h	[aryl enone] Ph (68)	336
(o-OTf benzyl alcohol)	Bu₃SnH	CO (15 psi), Pd(PPh₃)₄ (4%), PhMe, 50°, 3.5 h	MeO—, CHO / OH (14) + [OH isobenzofuranol] (41) + PhCH₂OH (20)	325, 331
(3-iodobenzyl alcohol)	Bu₃Sn—OMe	CO (15 psi), Pd(dppf)Cl₂ (2%), LiCl, BHT, DMF, 90°, 13 h	[phthalide] (62)	336
(methylenedioxyphenyl OTf)	Bu₃SnH	CO (15 psi), Pd(PPh₃)₄ (4%), PhMe, 50°, 3.5 h	HO—, CHO (76) + PhCH₂OH (12)	325, 331
	Bu₄Sn	CO (15 psi), Pd(dppf)Cl₂ (4%), LiCl, BHT, DMF, 110°, 44 h	[aryl ketone] Bu-*n* (65)	336
	Bu₃SnC≡CPr-*n*	CO (15 psi), Pd(dppf)Cl₂ (4%), LiCl, BHT, DMF, 70°, 6 h	[aryl] C≡CPr-*n* (68)	336
	Bu₃Sn TMS	CO (15 psi), Pd(dppf)Cl₂ (4%), LiCl, BHT, DMF, 75°, 5 h	[aryl enone] TMS (96)	336
	Me₃SnPh	CO (15 psi), Pd(dppf)Cl₂ (4%), LiCl, BHT, DMF, 90°, 7 h	[aryl ketone] Ph (88)	336

C$_8$

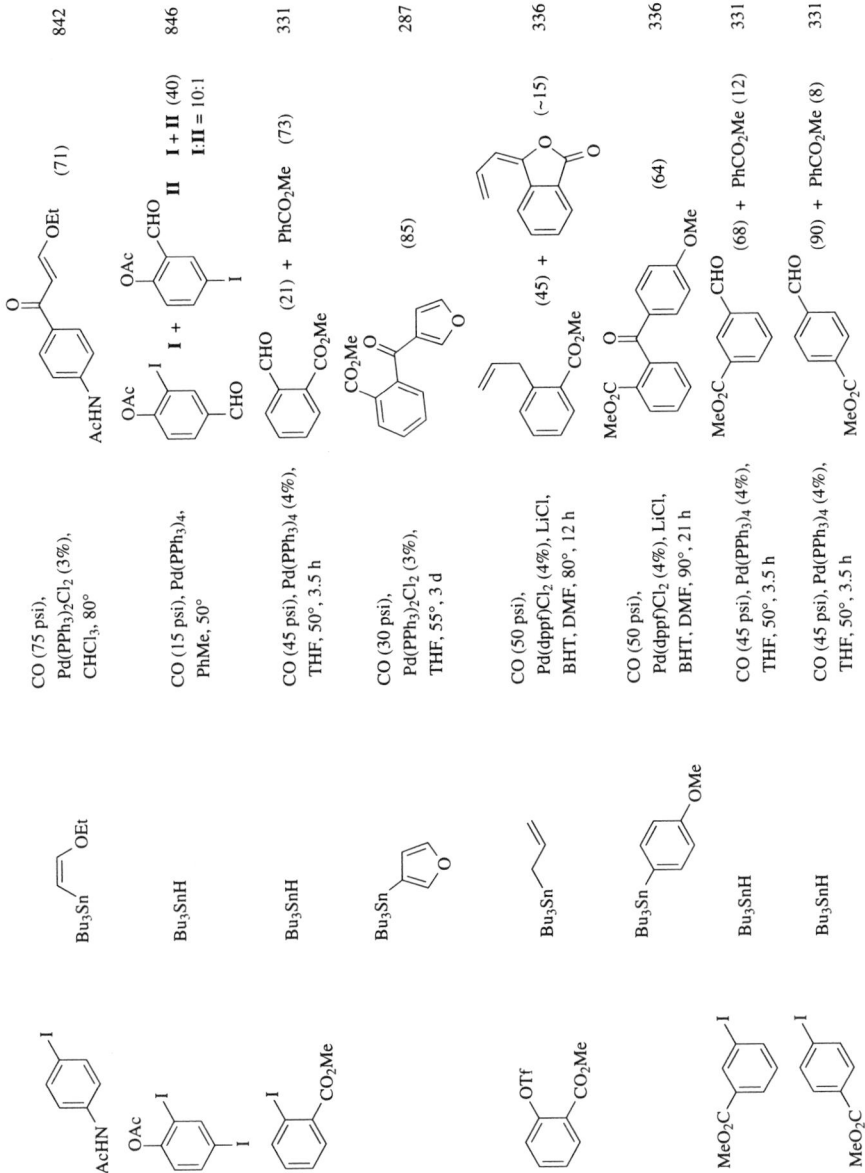

Substrate	Reagent	Conditions	Product(s) (%)	Refs.
AcHN–C$_6$H$_4$–I	Bu$_3$Sn–OEt	CO (75 psi), Pd(PPh$_3$)$_2$Cl$_2$ (3%), CHCl$_3$, 80°	AcHN–C$_6$H$_4$–C(O)–CH=CH–OEt (71)	842
OAc-diiodoarene	Bu$_3$SnH	CO (15 psi), Pd(PPh$_3$)$_4$, PhMe, 50°	I + II (40), I:II = 10:1	846
I–C$_6$H$_3$(CO$_2$Me)–I	Bu$_3$SnH	CO (45 psi), Pd(PPh$_3$)$_4$ (4%), THF, 50°, 3.5 h	(21) + PhCO$_2$Me (73)	331
—	Bu$_3$Sn–(3-furyl)	CO (30 psi), Pd(PPh$_3$)$_2$Cl$_2$ (3%), THF, 55°, 3 d	(85)	287
OTf–C$_6$H$_4$–CO$_2$Me	Bu$_3$Sn–(homoallyl)	CO (50 psi), Pd(dppf)Cl$_2$ (4%), LiCl, BHT, DMF, 80°, 12 h	(45) + (~15)	336
—	Bu$_3$Sn–C$_6$H$_4$–OMe	CO (50 psi), Pd(dppf)Cl$_2$ (4%), LiCl, BHT, DMF, 90°, 21 h	(64)	336
MeO$_2$C–C$_6$H$_4$–I (meta)	Bu$_3$SnH	CO (45 psi), Pd(PPh$_3$)$_4$ (4%), THF, 50°, 3.5 h	(68) + PhCO$_2$Me (12)	331
MeO$_2$C–C$_6$H$_4$–I (para)	Bu$_3$SnH	CO (45 psi), Pd(PPh$_3$)$_4$ (4%), THF, 50°, 3.5 h	(90) + PhCO$_2$Me (8)	331

TABLE XXVI. CARBONYLATIVE CROSS-COUPLING OF ARYL ELECTROPHILES (*Continued*)

Substrate	Stannane	Conditions	Product(s) and Yield(s) (%)	Refs.
(2-OMOM phenyl iodide)	Me$_3$SnPh	CO (15 psi), [(η^3-C$_3$H$_5$)PdCl]$_2$ (1%), HMPA, 20°, 8 h	(MeO$_2$C-C$_6$H$_4$-C(O)Ph) (98)	326
(2-allyl phenyl triflate)	Bu$_3$Sn—CH=CH—OEt	CO (75 psi), Pd(PPh$_3$)$_2$Cl$_2$ (3%), CHCl$_3$, 80°	(50)	842
	Bu$_3$Sn—CH$_2$CH=CH$_2$	CO (50 psi), Pd(dppf)Cl$_2$ (4%), LiCl, BHT, DMF, 70°, 15 h	(16) + (44)	336
			(19)	
	Bu$_3$Sn—CH=CH—TMS	CO (50 psi), Pd(dppf)Cl$_2$ (4%), LiCl, BHT, DMF, 50°, 15 h	(52)	336
(2-NHCO$_2$Et phenyl iodide)	Bu$_3$Sn—CH=CH—OEt	CO (75 psi), Pd(PPh$_3$)$_2$Cl$_2$ (3%), CHCl$_3$, 80°	(56)	842
(4-EtO$_2$C phenyl iodide)	Me$_4$Sn	CO (15 psi), Ni(PPh$_3$)$_2$(CO)$_2$ (3%), HMPA, 140°, 24 h	(EtO$_2$C-C$_6$H$_4$-C(O)CH$_3$) (87)	841

570

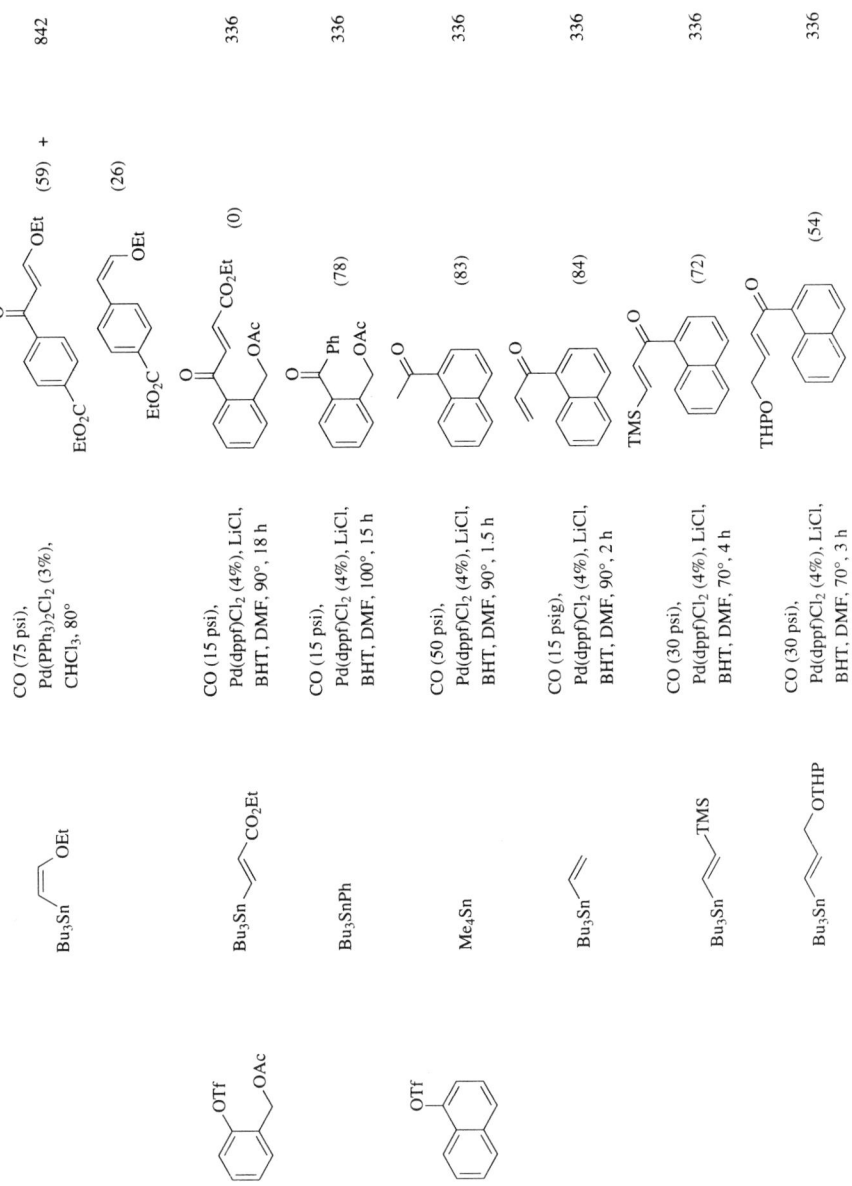

TABLE XXVI. CARBONYLATIVE CROSS-COUPLING OF ARYL ELECTROPHILES (Continued)

Substrate	Stannane	Conditions	Product(s) and Yield(s) (%)	Refs.
C11 — 4-t-Bu-phenyl OTf	Me$_3$Sn–C$_6$H$_4$–NO$_2$	CO (15 psi), Pd(dppf)Cl$_2$ (4%), LiCl, BHT, DMF, 95°, 11 h	aryl ketone (NO$_2$, t-Bu) (0)	336
4-t-Bu-cyclohexenyl OTf	Bu$_3$Sn–C(=CH$_2$)CH$_3$ (isopropenyl)	CO (15 psi), Pd(PPh$_3$)$_4$ (5.3%), LiCl, BHT, dioxane, 95°, 18 h	enone product (58)	173
2-iodophenyl dimethyl-piperazinone (Me–N, N–Me)	Bu$_3$Sn–CH$_2$...OTIPS	CO (50 psi), Pd$_2$(dba)$_3$ (2.5%), AsPh$_3$ (22%), LiCl, THF, 70°, 16 h	ketone with OTIPS (85)	511
2-iodophenyl dimethyl-piperazinone (Me–N, N–Me)	Bu$_3$Sn–...OBu-t, OTIPS	CO (50 psi), Pd$_2$(dba)$_3$ (2.5%), AsPh$_3$ (22%), LiCl, THF, 70°, 16 h	ketone with OTIPS, OBu-t (80)	328
OBoc, di-iodo arene	Bu$_3$SnH	CO (15 psi), Pd(PPh$_3$)$_4$, PhMe, 50°	**I** (OBoc, I, CHO) + **II** (OBoc, CHO, I) **I + II** (49); **I:II** = 10:1	846
C12 — 4-phenyl-phenyl iodide	Me$_3$Sn–C(OEt)=CH$_2$	1. CO (15 psi), Pd(PPh$_3$)$_4$ (2%), dioxane, 95°, 24 h 2. O$_3$	CO$_2$Et aryl ketone (Ph) (72)	270

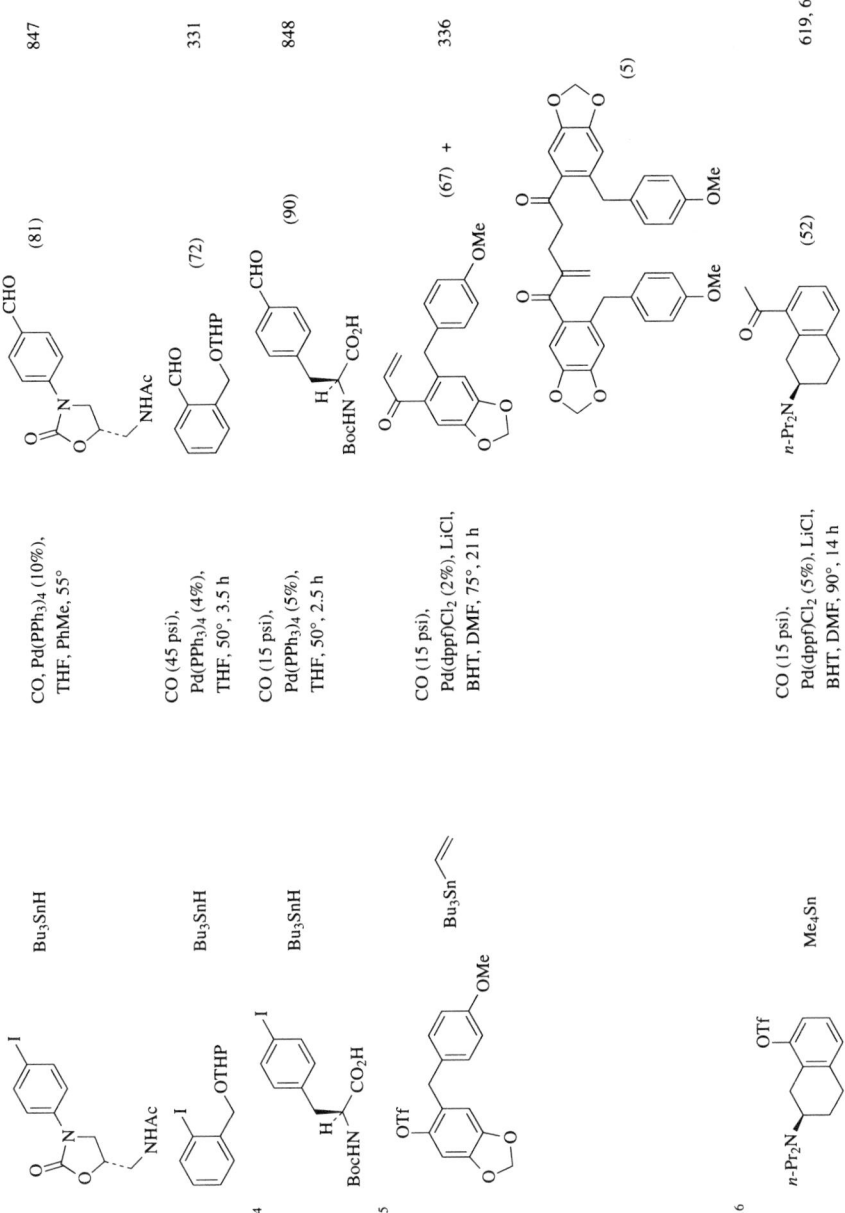

847

(81)

CO, Pd(PPh₃)₄ (10%), THF, PhMe, 55°

Bu₃SnH

331

(72)

CO (45 psi), Pd(PPh₃)₄ (4%), THF, 50°, 3.5 h

Bu₃SnH

848

(90)

CO (15 psi), Pd(PPh₃)₄ (5%), THF, 50°, 2.5 h

Bu₃SnH

336

(67) +

CO (15 psi), Pd(dppf)Cl₂ (2%), LiCl, BHT, DMF, 75°, 21 h

619, 620

(52)

CO (15 psi), Pd(dppf)Cl₂ (5%), LiCl, BHT, DMF, 90°, 14 h

Me₄Sn

C₁₄

C₁₅

C₁₆

TABLE XXVI. CARBONYLATIVE CROSS-COUPLING OF ARYL ELECTROPHILES (*Continued*)

Substrate	Stannane	Conditions	Product(s) and Yield(s) (%)	Refs.
(OTf substrate, $n\text{-}Pr_2N$)	Me_4Sn	CO (15 psi), Pd(dppf)Cl$_2$ (5%), LiCl, BHT, DMF, 90°, 14 h	(44)	619, 620
(OTf substrate, $n\text{-}Pr_2N$)	Bu_4Sn	CO (15 psi), Pd(dppf)Cl$_2$ (5%), LiCl, BHT, DMF, 90°, 14 h	(71)	620
	Me_3SnPh	CO (15 psi), Pd(dppf)Cl$_2$ (5%), LiCl, BHT, DMF, 90°, 14 h	(52)	620
C$_{18}$ (steroid TfO substrate)	Bu_3Sn (vinyl)	CO, Pd(PPh$_3$)$_4$ (5%), LiCl, BHT, dioxane, 110°, 14 h	(0)	521

TABLE XXVII. CARBONYLATIVE CROSS-COUPLING OF HETEROCYCLIC ELECTROPHILES

Substrate	Stannane	Conditions	Product(s) and Yield(s) (%)	Refs.
C_4				
(3-iodofuran)	Bu₃SnH	CO (45 psi), Pd(PPh₃)₄ (4%), THF, 50°, 3.5h	(furan-CHO) (60)	331
(2-iodothiophene)	Me₄Sn	CO (15 psi), Ni(PPh₃)₂(CO)₂ (3%), HMPA, 140°, 8.5 h	(2-acetylthiophene) (48)	841
	Bu₃Sn (cyclobutenone dioxolane) OEt	CO (32 psi), BnPd(PPh₃)₂Cl (1%), C₆H₆, 80°, 18 h	(thiophene-substituted cyclobutenone dioxolane) (53)	844
C_5	Bu₃Sn (cis) OEt	CO (75 psi), Pd(PPh₃)₂Cl₂ (3%), CHCl₃, 80°	(2-pyridyl cis-enol ether ketone) OEt X = Br (low) X = I (9)	842
(2-halopyridine, N–X)	Bu₃Sn (cis) OEt	CO (75 psi), Pd(PPh₃)₂Cl₂ (3%), CHCl₃, 80°	(3-pyridyl enone OEt) (44)	842
(3-iodopyridine)	Me₄Sn	CO (45 psi), Pd(OAc)₂ (2%), PPh₃ (8%), THF, 65°, 24 h	(6-acetyl chromanone) (64)	327
C_9 (6-bromochromanone)	Bu₃Sn (vinyl)	CO (45 psi), Pd(OAc)₂ (2%), PPh₃ (8%), THF, 40-45°, 6 h	(6-acryloyl chromanone) (64)	327
	Bu₄Sn	CO (45 psi), Pd(OAc)₂ (2%), PPh₃ (8%), THF, 85°, 6 h	(6-butanoyl chromanone, Bu-n) (74)	327

575

TABLE XXVII. CARBONYLATIVE CROSS-COUPLING OF HETEROCYCLIC ELECTROPHILES (*Continued*)

Substrate	Stannane	Conditions	Product(s) and Yield(s) (%)	Refs.
C21 (iodo-uracil nucleoside, TBDMSO / OTBDMS)	Bu₃SnPh	CO (45 psi), Pd(OAc)₂ (2%), PPh₃ (8%), THF, 65-70°, 6 h	(62)	327
	Bu₃SnC≡CPh	CO (45 psi), Pd(OAc)₂ (2%), PPh₃ (8%), THF, 65-70°, 6 h	(81)	327
	Bu₃SnH	CO, Pd(PPh₃)₄	CHO (82)	849
	Bu₃Sn–CH=CH₂	CO (50 psi), Pd(OAc)₂ (10%), PPh₃ (30%), CuI (30%), THF, 70°, 5 h	(82)	849
C30 (iodo-uracil nucleoside, MeO-trityl/OH)	Bu₃SnH	CO (50 psi), Pd(PPh₃)₄ (10%), THF, 70°, 8 h	CHO (95)	849

| Bu₃Sn reagent | Conditions | Product | Yield | Ref |

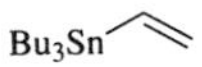

Bu_3Sn (allyl/vinyl reagent)

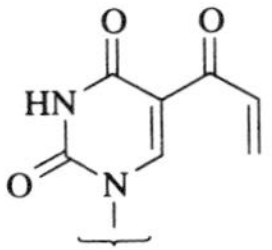

CO (50 psi), Pd(OAc)₂ (10%),
PPh₃ (30%), CuI (30%),
THF, 70°, 12 h

(77)

849

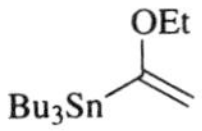

Bu_3Sn —OEt

CO (50 psi), Pd(OAc)₂ (10%),
PPh₃ (30%), CuI (30%),
THF, 70°, 12 h

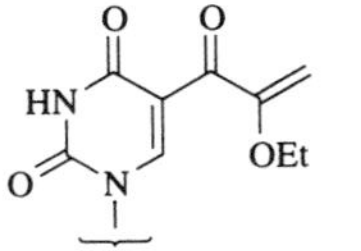

(81)

849

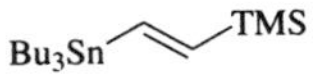

Bu_3Sn —TMS

CO (50 psi), Pd(OAc)₂ (10%),
PPh₃ (30%), CuI (30%),
THF, 70°, 8 h

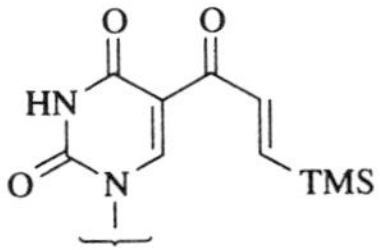

(85)

849

Bu_3SnPh

CO (50 psi), Pd(OAc)₂ (10%),
PPh₃ (30%), CuI (30%),
THF, 70°, 12 h

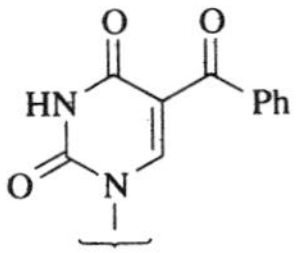

(95)

849

Substrate	Stannane	Conditions	Product(s) and Yield(s) (%)	Refs.
C₃ allyl–Cl	Me₃Sn–allyl	Pd(CH₃CN)₂Cl₂ (1.5%), PPh₃ (0.75%), CHCl₃, 25°, 48 h	divinyl ketone	333
		CO (15 psi)	(7)	
		CO (45 psi)	(62)	
		CO (90 psi)	(70)	
	Bu₃Sn–(2-methylallyl)	CO (90 psi), PPh₃ (0.75%), Pd(CH₃CN)₂Cl₂ (1.5%), CHCl₃, 25°, 48 h	(28)	333
	Me₃Sn–(2-methyl-2-butenyl)	"	(47)	333
	Me₃Sn–allyl	"	(0)	333
	Me₃Sn–(2-methyl-2-butenyl)	CO (45 psi), PPh₃ (0.75%), Pd(CH₃CN)₂Cl₂ (1.5%), CHCl₃, 25°, 48 h	(28)	333
	Me₃Sn–allyl	"	(27)	333
C₄ allyl–OMs	Me₃Sn–allyl	"	(20)	333
crotyl–Cl	Bu₃SnH	CO (45 psi), Pd(PPh₃)₄ (4%), THF, 50°, 3.5 h	NC–CH=CH–C(O)– (100)	331
2-methylallyl–Cl	Me₃Sn–allyl	CO (90 psi), PPh₃ (0.75%), Pd(CH₃CN)₂Cl₂ (1.5%), CHCl₃, 25°, 48 h	NC–ketone (16) + ketone (20)	333
3-chloro-1-butene				
NC–crotyl–Br				

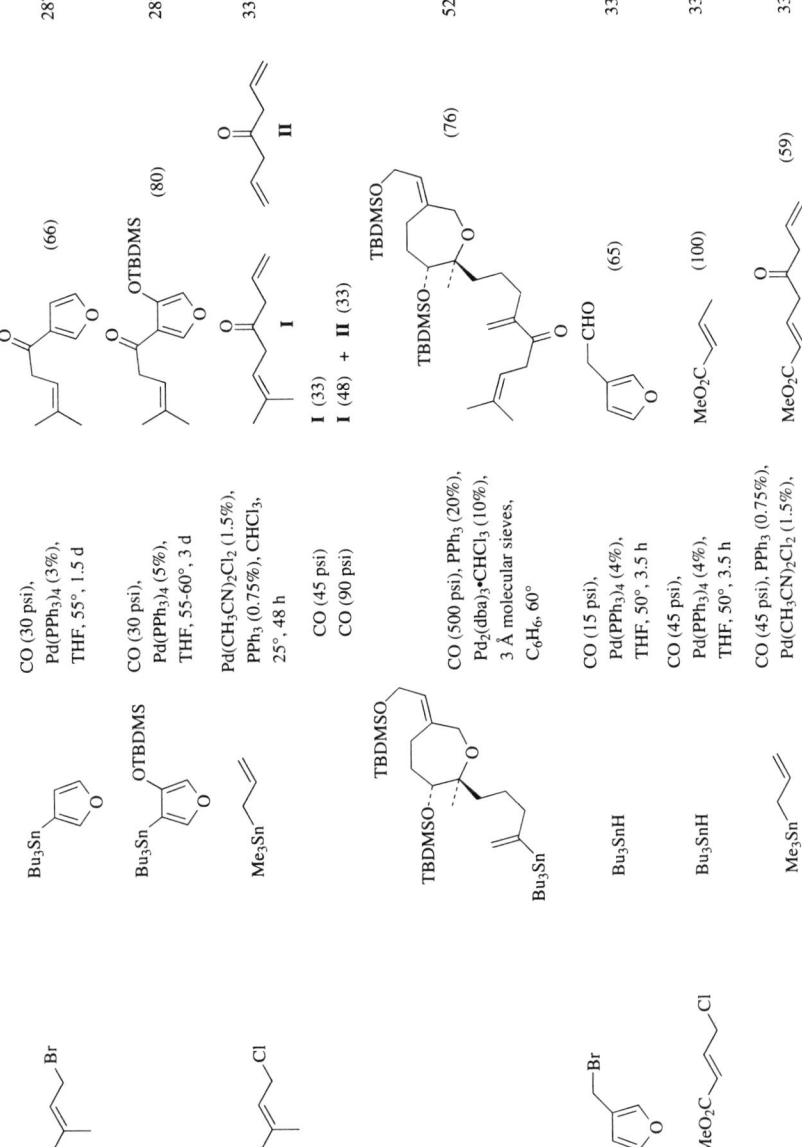

Substrate	Stannane	Conditions	Product(s) and Yield(s) (%)	Refs.
C₆				
(Br-cyclohexene)	Bu₃SnH	CO (45 psi), Pd(PPh₃)₄ (4%), PhMe, 50°, 3.5 h	I (14)	331
(Cl-cyclohexene)	Bu₃SnH	CO (15 psi), Pd(PPh₃)₄ (4%), PhMe, 50°, 3.5 h	I (65) + II (23)	325, 331
	Me₃Sn-allyl	CO (90 psi), PPh₃ (0.75%), Pd(CH₃CN)₂Cl₂ (1.5%), CHCl₃, 25°, 48 h	(21) + (13) II (22)	333
(D-Cl-cyclohexene)	Bu₃SnH	CO (45 psi), Pd(PPh₃)₄ (4%), THF, 50°, 3.5 h	I + II + III + IV I + II = (49); I:II = 1:1; III + IV = (51); III:IV = 1:1	325, 331
(D-Cl-cyclohexene)	Bu₃SnH	CO (45 psi), Pd(PPh₃)₄ (4%), THF, 50°, 3.5 h	I + II = (41); I:II = 1:1; III + IV = (59); III:IV = 1:1	325, 331
(EtO₂C-CH=CH-CH₂Br)	Bu₃Sn-furanyl	CO (30 psi), Pd(PPh₃)₄ (3%), THF, 60°, 1 d	(22) + (28)	287
C₇				
(Ph-CH₂Br)	Bu₃SnH	Pd(PPh₃)₄ (4%), THF, 50°, 3.5 h	Ph-CH₂CHO I + PhMe II	325, 331
		CO (15 psi)	I (75) + II (12)	331
		CO (45 psi)	I (94) + II (6)	

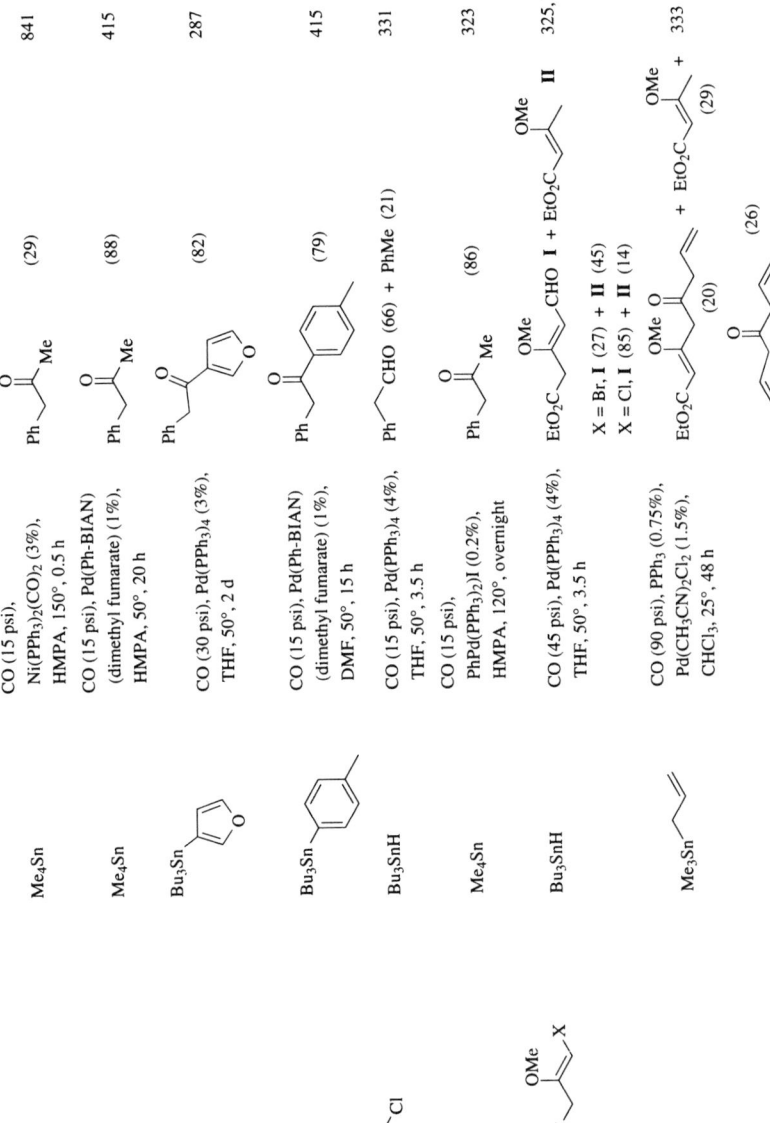

Me₄Sn	CO (15 psi), Ni(PPh₃)₂(CO)₂ (3%), HMPA, 150°, 0.5 h	Ph–C(O)–Me (29)	841
Me₄Sn	CO (15 psi), Pd(Ph-BIAN) (dimethyl fumarate) (1%), HMPA, 50°, 20 h	Ph–C(O)–Me (88)	415
Bu₃Sn	CO (30 psi), Pd(PPh₃)₄ (3%), THF, 50°, 2 d	Ph–C(O)–CH₂–(furyl) (82)	287
Bu₃Sn	CO (15 psi), Pd(Ph-BIAN) (dimethyl fumarate) (1%), DMF, 50°, 15 h	Ph–C(O)–CH₂–C₆H₄Me (79)	415
Bu₃SnH	CO (15 psi), Pd(PPh₃)₄ (4%), THF, 50°, 3.5 h	Ph–CHO (66) + PhMe (21)	331
Me₄Sn	CO (15 psi), PhPd(PPh₃)₂I (0.2%), HMPA, 120°, overnight	Ph–C(O)–Me (86)	323
Bu₃SnH	CO (45 psi), Pd(PPh₃)₄ (4%), THF, 50°, 3.5 h	EtO₂C···CHO **I** + EtO₂C···(OMe) **II** (325, 331) X = Br, **I** (27) + **II** (45) X = Cl, **I** (85) + **II** (14)	325, 331
Me₃Sn	CO (90 psi), PPh₃ (0.75%), Pd(CH₃CN)₂Cl₂ (1.5%), CHCl₃, 25°, 48 h	EtO₂C···(20) + EtO₂C···(29) + (26)	333

Substrate	Stannane	Conditions	Product(s) and Yield(s) (%)	Refs.
C_8 Ph–CH(CH₃)–Br	Me₄Sn	CO (300 psi), Pd(PPh₃)₂Cl₂ (0.3%), HMPA 120°, 1 h	Ph–C(O)–CH(CH₃)– (65) + ketone (11) + Ph-allyl (37)	337
	Ph₄Sn	CO (300 psi), Pd(PPh₃)₂Cl₂ (0.3%), HMPA 120°, 1 h	(43) + (8) + (33)	337
(chlorocyclohexene, CO₂Me)	Me₃Sn–allyl	CO (90 psi), PPh₃ (0.75%), Pd(CH₃CN)₂Cl₂ (1.5%), CHCl₃, 25°, 48 h	(23) + (27)	333
C_9 Ph–CH(CH₂CH₃)–Br	Me₄Sn	CO (300 psi), Pd(PPh₃)₂Cl₂ (0.6%), HMPA, 120°, 7 h	(50) + (1) + (36)	337
C_{10} (geranyl chloride)	Bu₃SnH	CO (45 psi), Pd(PPh₃)₄ (4%), THF, 50°, 3.5 h	CHO (54) + (46)	325, 331
	Me₃Sn–allyl	CO (90 psi), PPh₃ (0.75%), Pd(CH₃CN)₂Cl₂ (1.5%), CHCl₃, 25°, 48 h	(22)	333

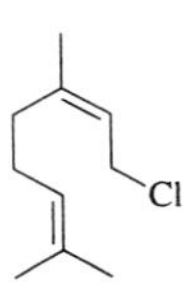

331

(41)

(59) +

Bu₃SnH

CO (45 psi), Pd(PPh₃)₄ (4%),
THF, 50°, 3.5 h

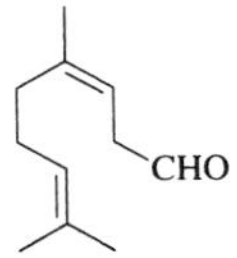

CHO

C₂₄

850

(94)

TMS

OMe
OMe

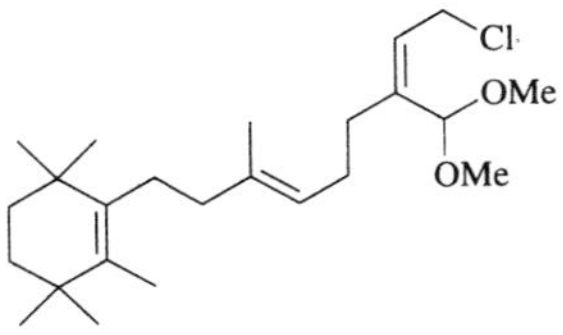

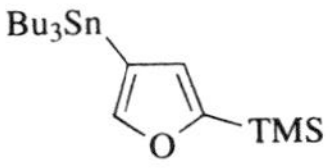

CO (45 psi),
Pd₂(dba)₃ (10%),
PPh₃ (20%), 50°, 36 h

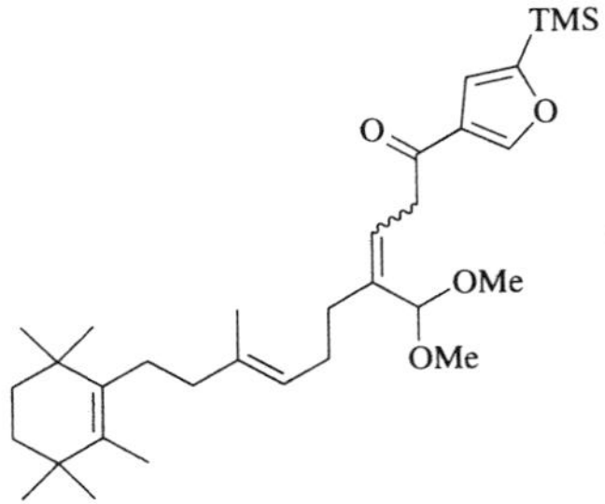

Bu₃Sn — TMS
O

Cl
OMe
OMe

TABLE XXIX. CARBONYLATIVE CROSS-COUPLING OF MISCELLANEOUS ELECTROPHILES

Substrate	Stannane	Conditions	Product(s) and Yield(s) (%)	Refs.
C₄				
(ethyl bromoacetate)	Ph₄Sn	CO (15 psi), PhPd(PPh₃)₂I (0.2%), HMPA, 120°, overnight	(67)	323
C₅				
	Me₄Sn	CO (300 psi), Pd(PPh₃)₂Cl₂ (1%), HMPA, 120°, 1 h	**I** (14) + **II** (25)	337
	Me₄Sn	CO (300 psi), Pd(AsPh₃)₂Cl₂ (1%), HMPA, 120°, 3 h	**I** (62) + **II** (25)	337
C₆				
n-BuC≡CCl	Bu₃SnH	CO (45 psi), Pd(PPh₃)₄ (4%), THF, 50°, 3.5 h	n-BuC≡CCHO (0) + CHO (28) + (38)	331

TABLE XXX. INTRAMOLECULAR CARBONYLATIVE CROSS-COUPLING REACTIONS

Substrate	Conditions	Product(s) and Yield(s) (%)	Refs.
C_{11}	CO (15 psi), Pd(dba)$_2$ (5%), LiCl, K$_2$CO$_3$, THF or DMF, 60°	**I** (38-45)	186
	CO (15 psi), polymer-supported Pd(0)(dppf) (5%), LiCl, K$_2$CO$_3$, dioxane, reflux, 3-4 h	**I** (0)	186
C_{12}	CO (15 psi), Pd(dba)$_2$ (5%), LiCl, K$_2$CO$_3$, THF or DMF, 60°	**I** (33-46)	186
	CO (15 psi), Pd(dppf)Cl$_2$ (5%), LiCl, K$_2$CO$_3$, THF, 65°	**I** (47)	186
	CO (15 psi), polymer-supported Pd(0)(dppf) (5%), LiCl, K$_2$CO$_3$, dioxane, reflux, 3-4 h	**I** (28)	186
C_{13}	CO (15 psi), Pd(dba)$_2$ (5%), LiCl, K$_2$CO$_3$, THF or DMF, 60°	**I** (35-45)	186
	CO (15 psi), Pd(dppf)Cl$_2$ (5%), LiCl, K$_2$CO$_3$, THF, 65°	**I** (60)	186
	CO (15 psi), polymer-supported Pd(0)(dppf) (5%), LiCl, K$_2$CO$_3$, dioxane, reflux, 3-4 h	**I** (78)	186
C_{14}	CO (15 psi), Pd(dba)$_2$ (5%), LiCl, K$_2$CO$_3$, THF or DMF, 60°	**I** (44-52)	186
	CO (15 psi), Pd(PPh$_3$)$_4$ (5%), LiCl, K$_2$CO$_3$, dioxane, 90°	**I** (59)	186
	CO (15 psi), Pd(dppf)Cl$_2$ (5%), LiCl, K$_2$CO$_3$, THF, 65°	**I** (55)	186

TABLE XXX. INTRAMOLECULAR CARBONYLATIVE CROSS-COUPLING REACTIONS (*Continued*)

Substrate	Conditions	Product(s) and Yield(s) (%)	Refs.
C$_{15}$	CO (15 psi), polymer-supported Pd(0)(dppf) (5%), LiCl, K$_2$CO$_3$, dioxane, reflux, 3-4 h	**I** (76)	186
	CO (15 psi), Pd(dba)$_2$ (5%), LiCl, K$_2$CO$_3$, THF or DMF, 60°	**I** (34-39)	186
	CO (15 psi), Pd(PPh$_3$)$_4$ (5%), LiCl, K$_2$CO$_3$, dioxane, 90°	**I** (60)	186
	CO (15 psi), polymer-supported Pd(0)(dppf) (5%), LiCl, K$_2$CO$_3$, dioxane, reflux, 3-4 h	**I** (70)	186
C$_{19}$	CO (50 psi), Pd(CH$_3$CN)$_2$Cl$_2$ (10%), LiCl, DMF, rt, 13 h	(24)	335
	CO (50 psi), Pd(CH$_3$CN)$_2$Cl$_2$ (10%), LiCl, DMF, rt, 13 h	(53)	335

TABLE XXXI. CROSS-COUPLING REACTIONS THAT FORM POLYMERS

Substrate	Stannane	Conditions	Product(s) and Yield(s) (%)	Refs.
C$_4$				
Br—CH=CH—CH$_2$—Br	(Bu$_3$Sn—⬡—CO)$_2$O	Pd(PPh$_3$)$_2$Cl$_2$ (2%), PPh$_3$ (4%), DMA, 165°, 4 h	n = 21.8	851, 852
	Me$_3$Sn—⬠(N-Boc)—SnMe$_3$ (2-bromo)	Pd(PPh$_3$)$_2$Cl$_2$ (2%), PPh$_3$ (4%), DMA, 165°, 4 h	n = 4	659
	(Bu$_3$Sn—⬡—CO)$_2$O	Pd(PPh$_3$)$_2$Cl$_2$ (2%), PPh$_3$ (4%), DMA, 165°, 4 h	n = 18.1	851, 852
C$_7$				
(thiophene 2,5-diBr)	SnBu$_3$	Pd(PPh$_3$)$_2$Cl$_2$ (2%), PhMe, 70°, 12 h	n = —	225
C$_8$				
(BrCH$_2$—⬡—CH$_2$Br)	Bu$_3$SnC≡C—⬡—C≡CSnBu$_3$	Pd(PPh$_3$)$_2$Cl$_2$ (2%), PPh$_3$ (4%), DMA, 120°, 4 h	n = 24.6	851, 852a
	(Bu$_3$Sn—⬡—CO)$_2$O	Pd(PPh$_3$)$_2$Cl$_2$ (2%), PPh$_3$ (4%), DMA, 165°, 4 h	n = 17.8	851, 852

587

TABLE XXXI. CROSS-COUPLING REACTIONS THAT FORM POLYMERS (Continued)

Substrate	Stannane	Conditions	Product(s) and Yield(s) (%)	Refs.
	$Bu_3SnC{\equiv}C$—⟨aryl⟩—$C{\equiv}CSnBu_3$	$Pd(PPh_3)_2Cl_2$ (2%), PPh_3 (4%), DMA, 120°, 4 h	$n = 5.3$	851, 852a
	Me_3Sn—⟨aryl, Bu-t⟩—$SnMe_3$	$Pd(AsPh_3)_2Cl_2$ (1%), neat, 70°, 24 h	$n = 49$	232
	⟨crown-thiophene $SnMe_3$⟩	$Pd(AsPh_3)_2Cl_2$ (1%), THF, reflux	$x = 2, 3$ $n = $ —	853
	⟨crown-thiophene $SnMe_3$⟩	$Pd(AsPh_3)_2Cl_2$ (1%), THF, reflux	$x = 2, 3$ $n = $ —	853

588

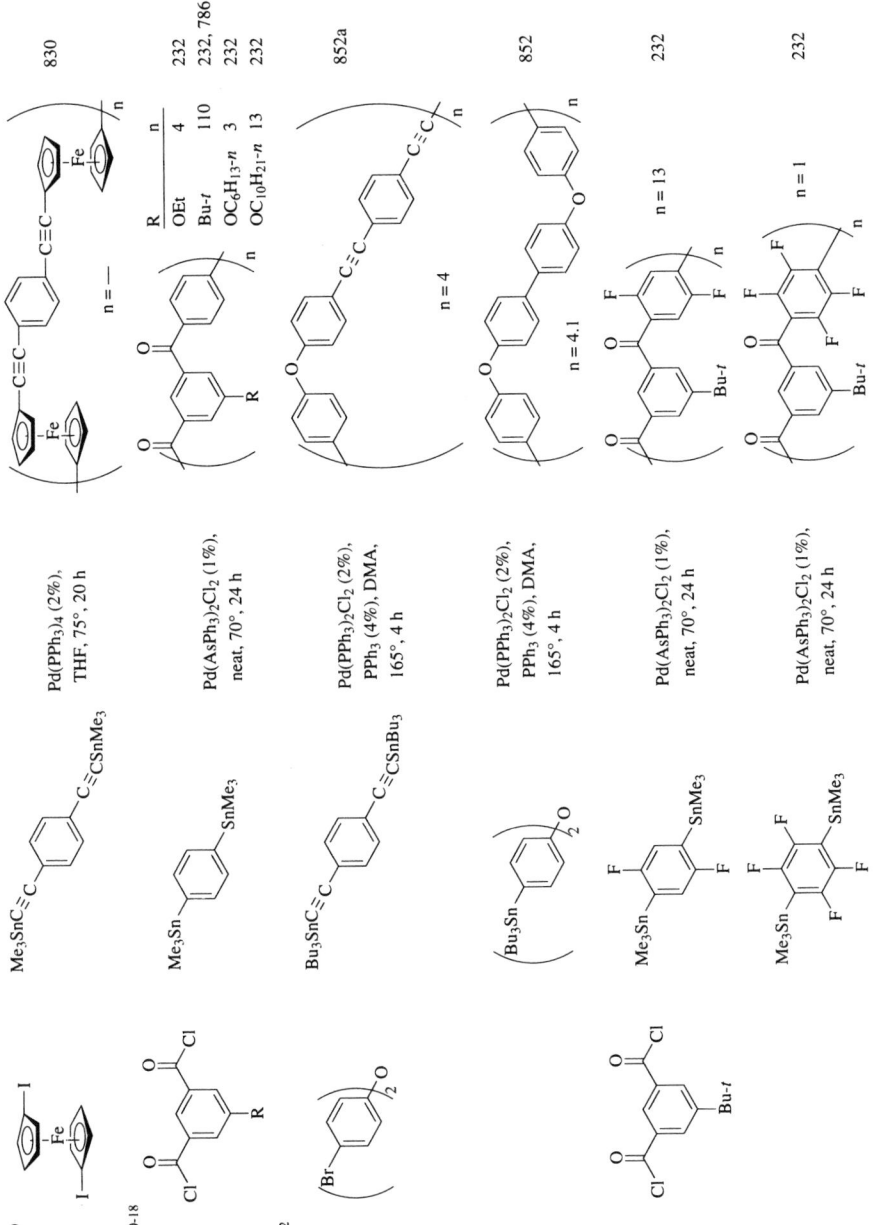

TABLE XXXI. CROSS-COUPLING REACTIONS THAT FORM POLYMERS (*Continued*)

Substrate	Stannane	Conditions	Product(s) and Yield(s) (%)	Refs.
	Me_3Sn—SnMe$_3$ (with Bu-t groups)	Pd(AsPh$_3$)$_2$Cl$_2$ (1%), neat, 70°, 24 h	(polymer with Bu-t groups, diketone), $n = —$	232
	Me$_3$Sn—biphenyl—SnMe$_3$	Pd(AsPh$_3$)$_2$Cl$_2$ (1%), neat, 70°, 24 h	(polymer with Bu-t), $n = 109$	232
	(Me$_3$Sn) ... O stannane	Pd(AsPh$_3$)$_2$Cl$_2$ (1%), neat, 70°, 24 h	(polymer with Bu-t, O), $n = 148$	232
C_{13} (Br)$_2$—SO$_2$	(Me$_3$Sn) ... O stannane	Pd(PPh$_3$)$_2$Cl$_2$ (2%), PPh$_3$ (4%), DMA, 160°, 4 h	(polymer with SO$_2$, O), $n = 21$	852
(Br)$_2$—CO	Bu$_3$SnC≡C—C≡CSnBu$_3$	Pd(PPh$_3$)$_2$Cl$_2$ (2%), PPh$_3$ (4%), DMA, 160°, 4 h	(polymer with C≡C), $n = 5.7$	851, 852a

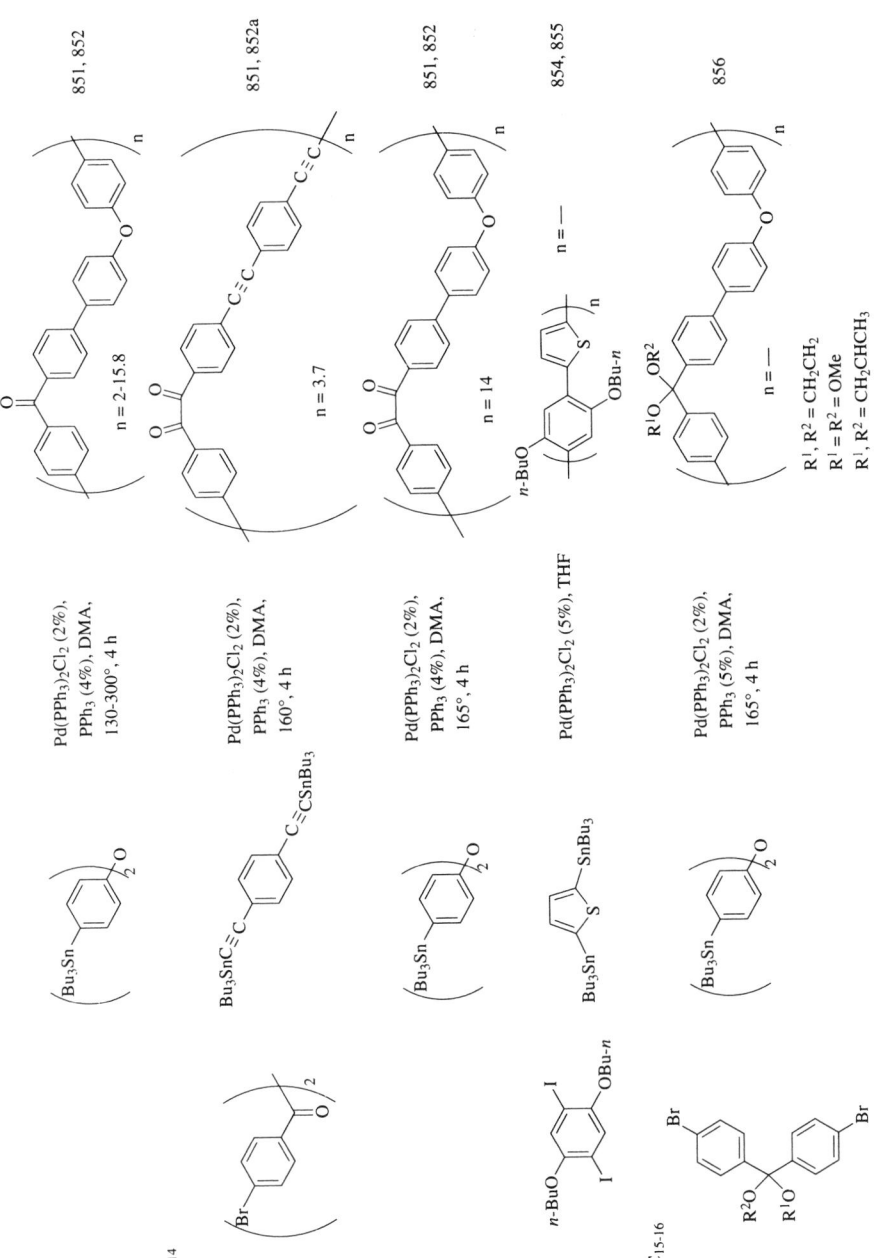

TABLE XXXI. CROSS-COUPLING REACTIONS THAT FORM POLYMERS (Continued)

Substrate	Stannane	Conditions	Product(s) and Yield(s) (%)	Refs.
C_{16} (diiodo dialkoxybenzene, RO, OR, I, I)	Bu_3Sn-thiophene-$SnBu_3$	$Pd(PPh_3)_2Cl_2$ (5%), THF	polymer (thiophene–dialkoxybenzene)$_n$, RO/OR	
			R	n
			$n\text{-}C_5H_{11}$	—	855
			$n\text{-}C_6H_{13}$	—	855
			$n\text{-}C_7H_{15}$	—	855
			$n\text{-}C_8H_{17}$	—	855
			$n\text{-}C_9H_{19}$	—	855
			$n\text{-}C_{12}H_{25}$	—	854, 855
			$n\text{-}C_{16}H_{33}$	22	854, 855
C_{20} (naphthalene diacid chloride, t-Bu, Bu-t)	Me_3Sn–C$_6$H$_4$–$SnMe_3$	$Pd(AsPh_3)_2Cl_2$ (1%), neat, 70°, 24 h	polymer, $n = —$	232
	Me_3Sn, $SnMe_3$, Bu-t (substituted benzene)	$Pd(AsPh_3)_2Cl_2$ (1%), neat, 70°, 24 h	polymer, $n = —$	232

857

SO_2Me

n

x

y

$(CH_2)_6$

Me—N

$n\text{-}C_6H_{13}$ N $n\text{-}C_6H_{13}$

O   O

S

S

n = —
x + y = 1
y = 0–0.49

Pd(PPh$_3$)$_4$,
LiCl,
dioxane,
90°, 16 h

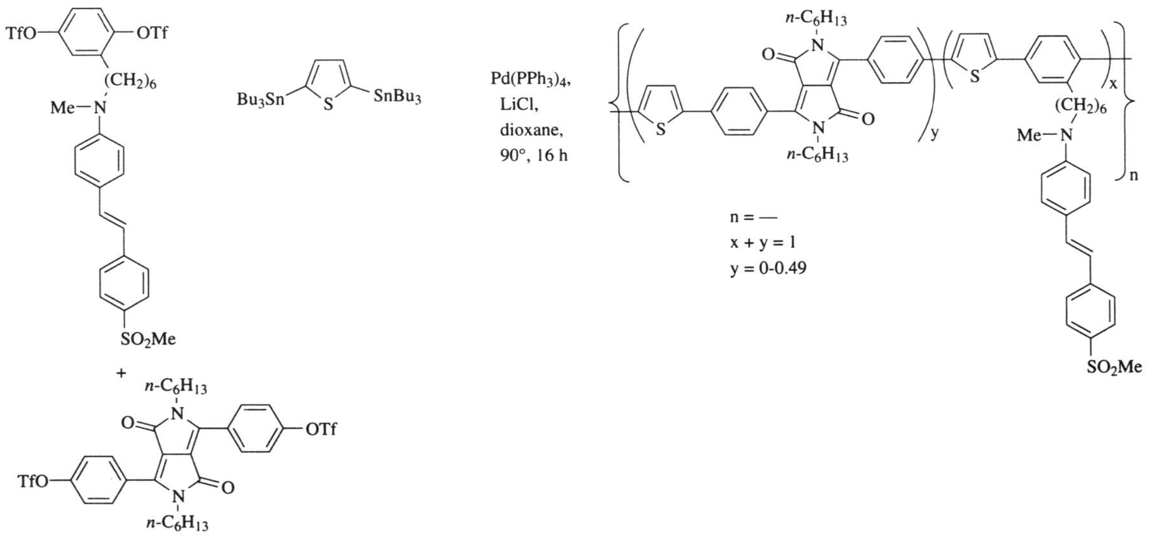

858

n = 11

OTBDMS

TBDMSO

Si
Ph   Ph

S

S

Pd(0)

C_{29}

SnBu$_3$

Bu$_3$Sn   S

OTf

$(CH_2)_6$

Me—N

SO_2Me

+

$n\text{-}C_6H_{13}$ N $n\text{-}C_6H_{13}$

O   O

OTf

TfO

TfO

C_{41}

Ph   Ph

Bu$_3$Sn   S

OTBDMS   OTBDMS

S   SnBu$_3$

Si

Br   S

OTBDMS   OTBDMS

Si
Ph   Ph

S   Br

TABLE XXXI. CROSS-COUPLING REACTIONS THAT FORM POLYMERS (*Continued*)

Substrate	Stannane	Conditions	Product(s) and Yield(s) (%)	Refs.
C_{42} (tetra-acyl chloride substrate with $C_{16}H_{33}$-n groups)	(Me$_3$Sn–C$_6$H$_4$–SnMe$_3$)	Pd(AsPh$_3$)$_2$Cl$_2$ (1%), neat, 70°, 24 h	(polymer with $C_{16}H_{33}$-n groups), $n = 9$	232
$>C_{50}$ (iodo-substituted polymer with P ring)	Me$_3$SnC≡CPh	Pd(PPh$_3$)$_2$Cl$_2$ (2%), PhMe, 70°, 12 h	(polymer, CPh alkyne), $n = —$	225
	Bu$_3$Sn(CH=CH$_2$)	Pd$_2$(dba)$_3$ (5%), AsPh$_3$ (20%), NMP, 45°, overnight	(vinyl-substituted P-ring ketone) (≥89)	145
	Bu$_3$Sn(CH=CH–CH$_3$)	Pd$_2$(dba)$_3$ (5%), AsPh$_3$ (20%), NMP, 45°, overnight	(propenyl-substituted P-ring ketone) (≥91)	145
	Bu$_3$Sn(CH=C(CH$_3$)$_2$)	Pd$_2$(dba)$_3$ (5%), AsPh$_3$ (20%), NMP, 45°, overnight	(isobutenyl-substituted P-ring ketone) (≥85)	145
	Bu$_3$Sn–(benzodioxole)	Pd$_2$(dba)$_3$ (5%), AsPh$_3$ (20%), NMP, 45°, overnight	(benzodioxole-substituted P-ring ketone) (≥90)	145

594

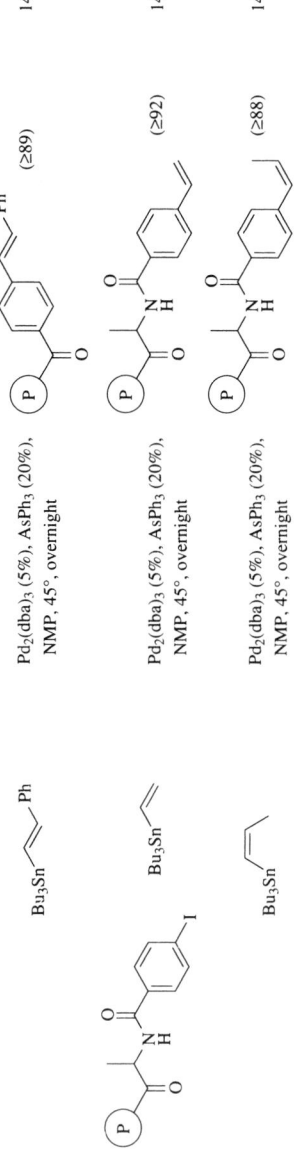

Pd$_2$(dba)$_3$ (5%), AsPh$_3$ (20%), NMP, 45°, overnight   (≥89)   145

Pd$_2$(dba)$_3$ (5%), AsPh$_3$ (20%), NMP, 45°, overnight   (≥92)   145

Pd$_2$(dba)$_3$ (5%), AsPh$_3$ (20%), NMP, 45°, overnight   (≥88)   145

TABLE XXXII. MULTI-STEP TRANSFORMATIONS INVOLVING DIRECT CROSS-COUPLING REACTIONS

Substrate	Stannane and Other Components	Conditions	Product(s) and Yield(s) (%)	Refs.
C$_2$				
Br	Bu$_3$Sn⟍ + ⬡	Pd(PPh$_3$)$_4$ (1%), C$_6$H$_6$, 120°, 20 h	(64)	342
O Cl	Bu$_3$Sn⟍ + ⬡	Pd(PPh$_3$)$_4$ (1%), C$_6$H$_6$, 80°, 20 h	(0)	342
C$_3$				
Br	Bu$_3$Sn⟍ + HC≡CCO$_2$Me	Pd(CH$_3$CN)$_2$Cl$_2$ (5%), P(2-furyl)$_3$ (2%), NMP, 80°, 24 h	CO$_2$Me (16)	356
	Bu$_3$Sn⟍ + HC≡CTMS	Pd(CH$_3$CN)$_2$Cl$_2$ (5%), P(2-furyl)$_3$ (2%), PhMe, 80°, 24 h	TMS (16)	356
	Bu$_3$Sn⟍ + HC≡CBu-n	Pd(CH$_3$CN)$_2$Cl$_2$ (5%), P(2-furyl)$_3$ (2%), HMPA, 80°, 24 h	n-Bu (20)	356
	Bu$_3$SnPh + HC≡CCO$_2$Me	Pd(CH$_3$CN)$_2$Cl$_2$ (5%), P(2-furyl)$_3$ (2%), HMPA, 80°, 24 h	CO$_2$Me Ph (36)	356
	Bu$_3$SnPh + HC≡CR	Pd(CH$_3$CN)$_2$Cl$_2$ (5%), Et$_4$NCl, HMPA, 80°, 24 h	R Ph	356

R
CH$_2$OEt (21)
TMS (23)
Bu-n (24)
Ph (26)

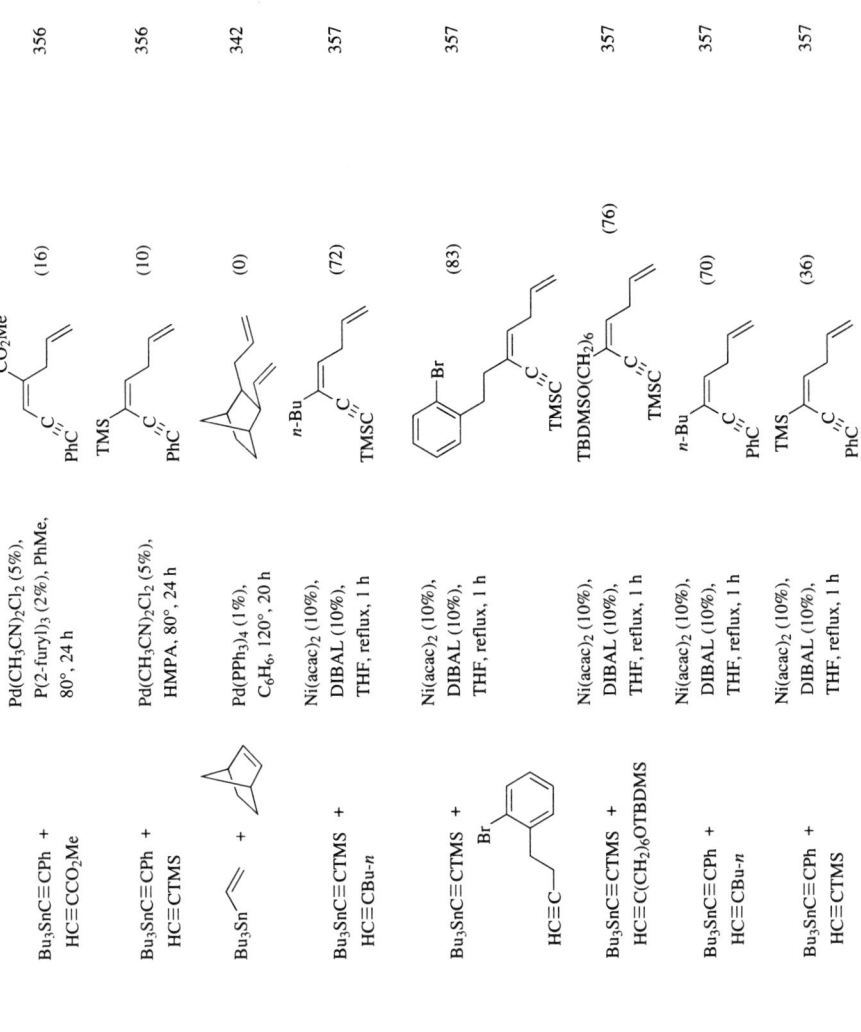

Substrate	Stannane and Other Components	Conditions	Product(s) and Yield(s) (%)	Refs.
	Bu₃SnC≡CPh + HC≡CPh	Ni(acac)₂ (10%), DIBAL (10%), THF, reflux, 1 h	(8 0)	357
	Bu₃SnC≡CPh + $HC≡C$ (cyclohexane-OH)	Ni(acac)₂ (10%), DIBAL (10%), THF, reflux, 1 h	(79)	357
(O=, CH₃, CH₂Cl ketone)	Bu₃Sn (allyl)	Pd(PPh₃)₄ (1%), CH₂Cl₂, 80°, 20 h	(63)	859
TMSI	Bu₃Sn (allyl) + HC≡CPh	Pd(PPh₃)₄ (2%), dioxane, 60°, 7 h	Ph, TMS (80)	364
	Bu₃SnC≡CMe + HC≡CBu-n	Pd(PPh₃)₄ (2%), dioxane, 60°, 3 h	n-Bu, TMS, MeC (71)	364
	Bu₃SnC≡CMe + HC≡CPh	Pd(PPh₃)₄ (2%), dioxane, 60°, 2.5 h	Ph, TMS, MeC (97)	364
	Bu₃SnC≡CMe + HC≡CC₆H₄Cl-p	Pd(PPh₃)₄ (2%), dioxane, 60°, 2.5 h	p-ClC₆H₄, TMS, MeC (70)	364
	Bu₃SnC≡CMe + HC≡C(CH₂)₂C₆H₄Br-o	Pd(PPh₃)₄ (2%), dioxane, 60°, 2 h	o-BrC₆H₄(CH₂)₂, TMS, MeC (70)	364

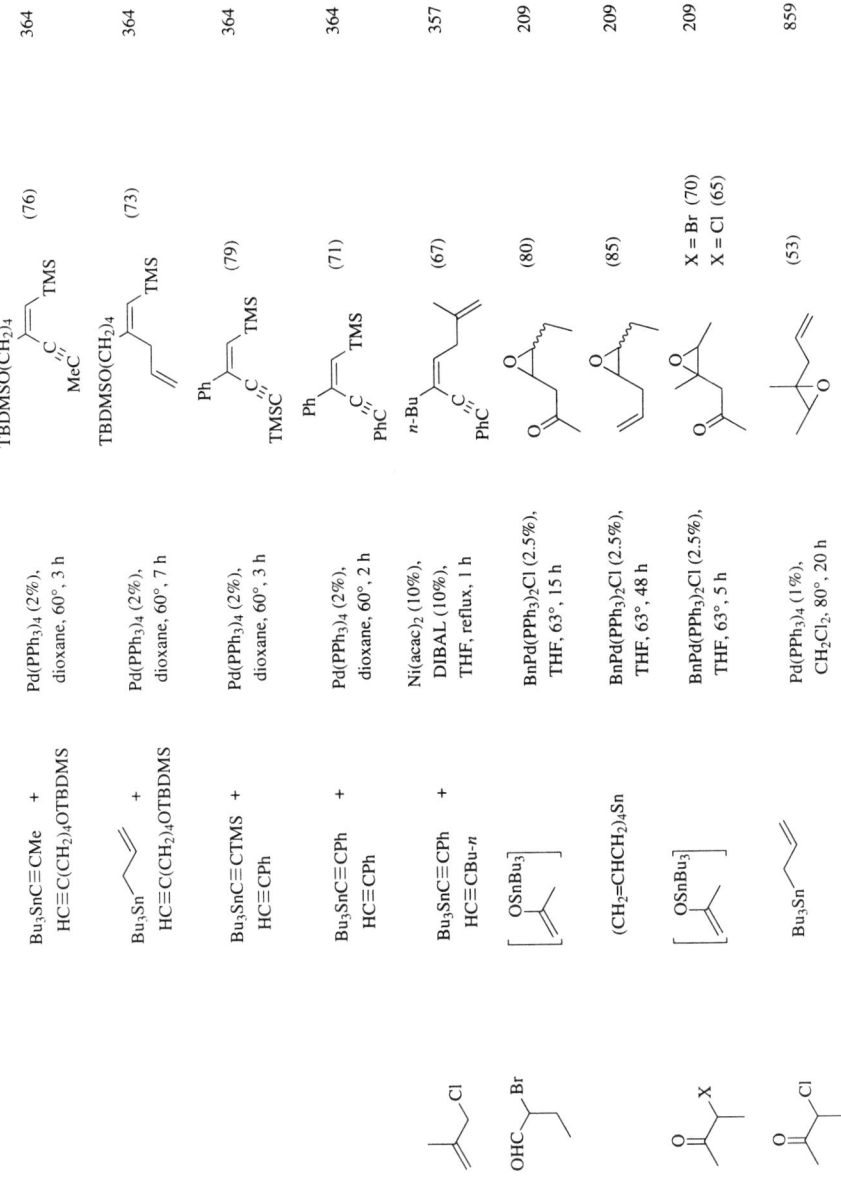

C₄

Substrate	Reagent	Conditions	Product (%)	Refs.
(2-chloromethylallyl)	Bu₃SnC≡CMe + HC≡C(CH₂)₄OTBDMS	Pd(PPh₃)₄ (2%), dioxane, 60°, 3 h	TBDMSO(CH₂)₄ ... TMS (76)	364
OHC-CH(Br)-	Bu₃Sn (allyl) + HC≡C(CH₂)₄OTBDMS	Pd(PPh₃)₄ (2%), dioxane, 60°, 7 h	TBDMSO(CH₂)₄ ... TMS (73)	364
	Bu₃SnC≡CTMS + HC≡CPh	Pd(PPh₃)₄ (2%), dioxane, 60°, 3 h	Ph ... TMSC ... TMS (79)	364
	Bu₃SnC≡CPh + HC≡CPh	Pd(PPh₃)₄ (2%), dioxane, 60°, 2 h	Ph ... PhC ... TMS (71)	364
	Bu₃SnC≡CPh + HC≡CBu-n	Ni(acac)₂ (10%), DIBAL (10%), THF, reflux, 1 h	n-Bu ... PhC (67)	357
O=C(X)...	[OSnBu₃ isopropenyl]	BnPd(PPh₃)₂Cl (2.5%), THF, 63°, 15 h	(80)	209
	(CH₂=CHCH₂)₄Sn	BnPd(PPh₃)₂Cl (2.5%), THF, 63°, 48 h	(85)	209
	[OSnBu₃ isopropenyl]	BnPd(PPh₃)₂Cl (2.5%), THF, 63°, 5 h	X = Br (70) / X = Cl (65)	209
O=C... Cl	Bu₃Sn (allyl)	Pd(PPh₃)₄ (1%), CH₂Cl₂, 80°, 20 h	(53)	859

TABLE XXXII. MULTI-STEP TRANSFORMATIONS INVOLVING DIRECT CROSS-COUPLING REACTIONS (Continued)

Substrate	Stannane and Other Components	Conditions	Product(s) and Yield(s) (%)	Refs.
		1. Pd(PPh₃)₄ (3%), Ag₂O, DMF, 100°, 15 min 2. HCl (2 N), 100°, 1 h	(71)	860
		1. Pd(PPh₃)₄ (3%), CuO, DMF, 100° 2. HCl (2 N)	(5)	861
	(CH₂=CHCH₂)₄Sn	BnPd(PPh₃)₂Cl (2.5%), THF, 63°, 48 h	(90)	209
		Pd(PPh₃)₄ (1%), CH₂Cl₂, 80°, 20 h	(30)	859
Me₃SiSiMe₂I	Bu₃SnC≡CMe + HC≡CPh	Pd(PPh₃)₄ (2%), dioxane, 60°, 2 h	(72)	364
C₆ PhBr	Bu₃SnH + *t*-BuNC	Pd(PPh₃)₄ (10%), C₆H₆, 120°	(0)	862
	+ *t*-BuNC	Pd(PPh₃)₄ (10%), C₆H₆, 120°	(tr)	862
	+	Pd(PPh₃)₄ (1%), C₆H₆, 100°	(87)	342
	+	Pd(PPh₃)₄ (1%), C₆H₆, 100°, 10 h	(87)	341, 342

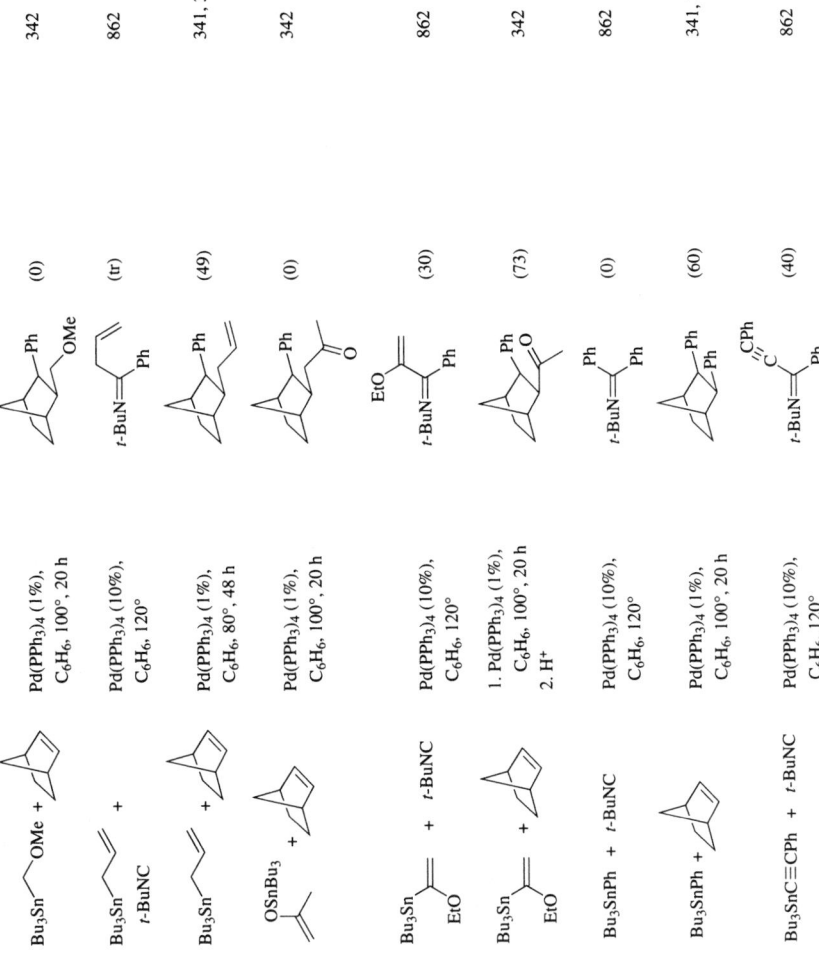

Substrate	Stannane and Other Components	Conditions	Product(s) and Yield(s) (%)	Refs.
	Bu₃SnC≡CPh +	Pd(PPh₃)₄ (1%), C₆H₆, 100°, 12 h	(28)	341, 342
	—SnMe₃	Pd(PPh₃)₄, *i*-Pr₂NEt, BHT, PhMe, reflux	**I** + **II** **I** + **II** (58) **I:II** = 9:91	373
	Bu₃SnC≡CPh +	Pd(PPh₃)₄ (1%), C₆H₆, 100°	(46)	342
	Bu₃SnR + *t*-BuNC R NEt₂ OMe OEt SPh CN	Pd(PPh₃)₄ (10%), C₆H₆, 120°	(22) (63) (48) (10) (40)	862
PhI	Bu₃Sn +	Pd(PPh₃)₄ (1%), C₆H₆	(49)	342
	Bu₃SnC≡CPh +	Pd(PPh₃)₄ (1%), C₆H₆, 100°, 10 h	(56)	341, 342

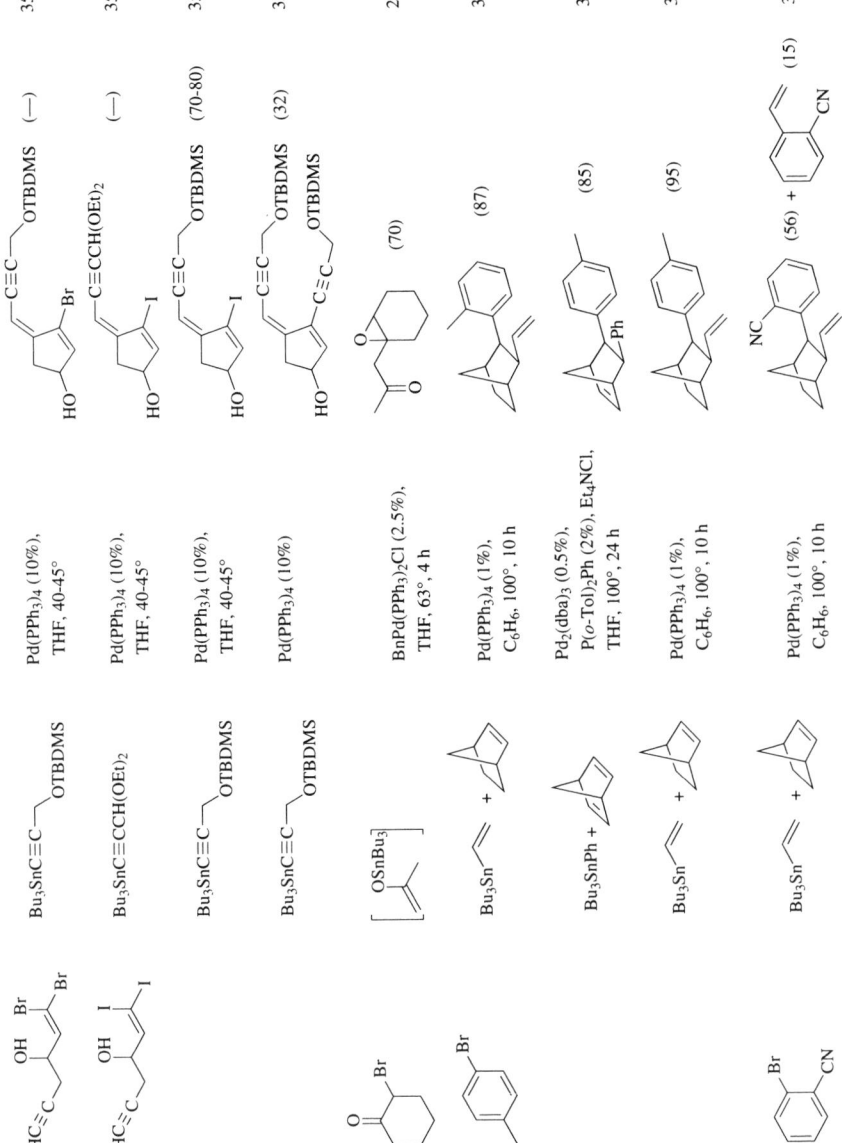

Substrate	Stannane and Other Components	Conditions	Product(s) and Yield(s) (%)	Refs.
pyridine (NHAc, Br)	OHC-thiophene, Bu₃Sn	Pd(PPh₃)₄ (3%), DMF, 100°, 8 h	(69)	860
pyridine (Br, NHAc)	OHC-thiophene, Bu₃Sn	1. Pd(PPh₃)₄ (5%), CuO, DMF, 100° 2. HCl (2 N)	(61)	861
pyridine (NHAc, Br)	OHC-thiophene, Bu₃Sn	Pd(PPh₃)₄ (3%), DMF, 100°, 8 h	(60)	860
pyridine N⁺–O⁻ (NHAc, Br)	OHC-thiophene, R₃Sn	Pd(PPh₃)₄ (5%), CuO, DMF, 100°	R = Me (54), R = Me, no CuO (46), R = Bu (33)	694
PhC(O)Cl	Bu₃Sn–vinyl + norbornene	Pd(PPh₃)₄ (1%), C₆H₆, 80°, 12 h	(36)	341, 342
2-bromobenzaldehyde (Br, CHO)	Bu₃Sn–CH=CH–CH₂–N(TMS)₂	Pd(PPh₃)₄, PhMe, reflux, 24 h	(89)	462
2-iodobenzaldehyde (I, CHO)	Bu₃Sn–squarate (O, O, H₂N)	BnPh(PPh₃)₂Cl (6.5%), CuI, C₆H₆, rt, 20 h	(92)	644

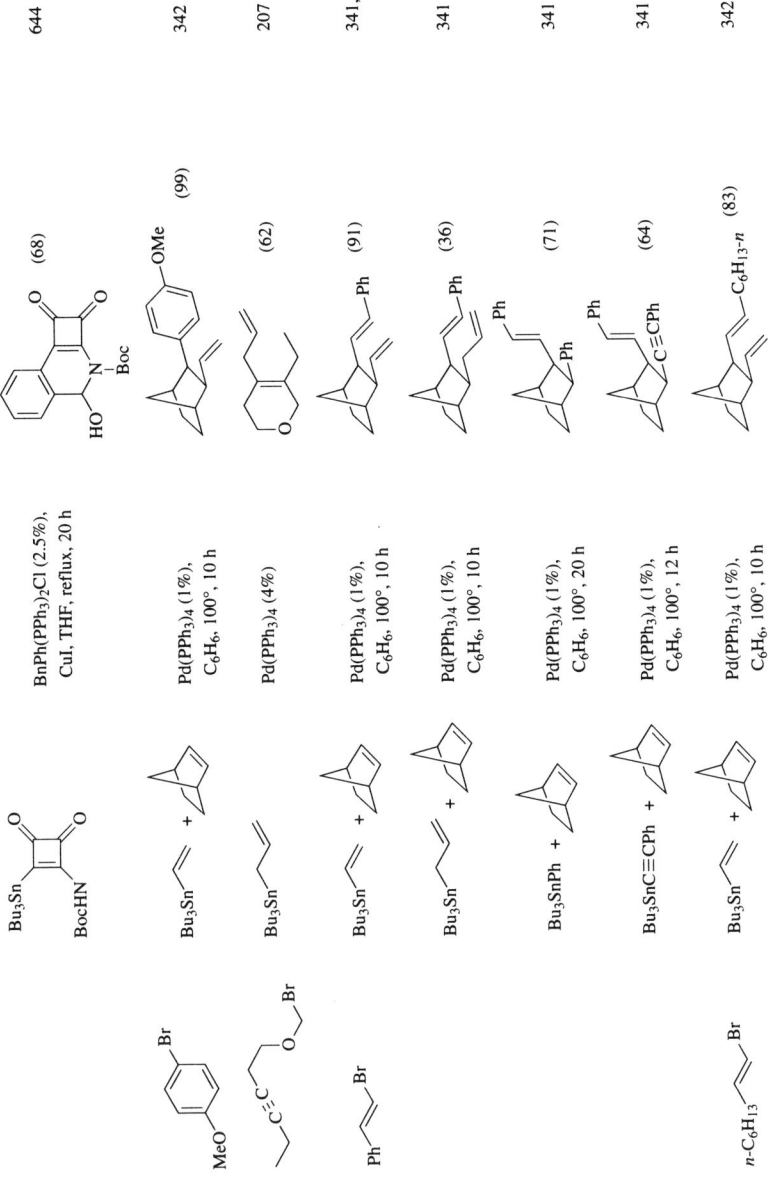

TABLE XXXII. MULTI-STEP TRANSFORMATIONS INVOLVING DIRECT CROSS-COUPLING REACTIONS (*Continued*)

Substrate	Stannane and Other Components	Conditions	Product(s) and Yield(s) (%)	Refs.
MeO$_2$C— (thiophene)—HgCl	Bu$_3$Sn— C$_5$H$_{11}$-n / OSiEt$_3$ + norbornene	1. PdCl$_2$, LiCl, norbornene, CH$_3$CN 2. stannane, PPh$_3$, THF	Et$_3$SiO— C$_5$H$_{11}$-n ... CO$_2$Me (34)	863
2-Br, NHAc aniline	SnMe$_3$ / CHO pyridine	Pd(dppb)Cl$_2$, CuO, DMF, 105–110°, 48 h	(78)	864
	Me$_3$Sn / acetyl pyridine	Pd(dppb)Cl$_2$, CuO, DMF, 105–110°, 48 h	(80)	864
	SnMe$_3$ / CHO, OMe pyridine	Pd(dppb)Cl$_2$, CuO, DMF, 105–110°, 48 h	OMe (68)	864
	Bu$_3$Sn— N(TMS)$_2$	Pd(PPh$_3$)$_4$, PhMe, reflux, 48 h	(86)	464, 540
2-Br acetophenone	Bu$_3$Sn— + norbornene	Pd(PPh$_3$)$_4$ (1%), C$_6$H$_6$, 100°, 10 h	(96)	341, 342
4-Br acetophenone	Bu$_3$Sn— allyl	BnPd(PPh$_3$)$_2$Cl (2.5%), THF, 63°, 5 h	Ph (75)	209
PhCOCH$_2$Br				

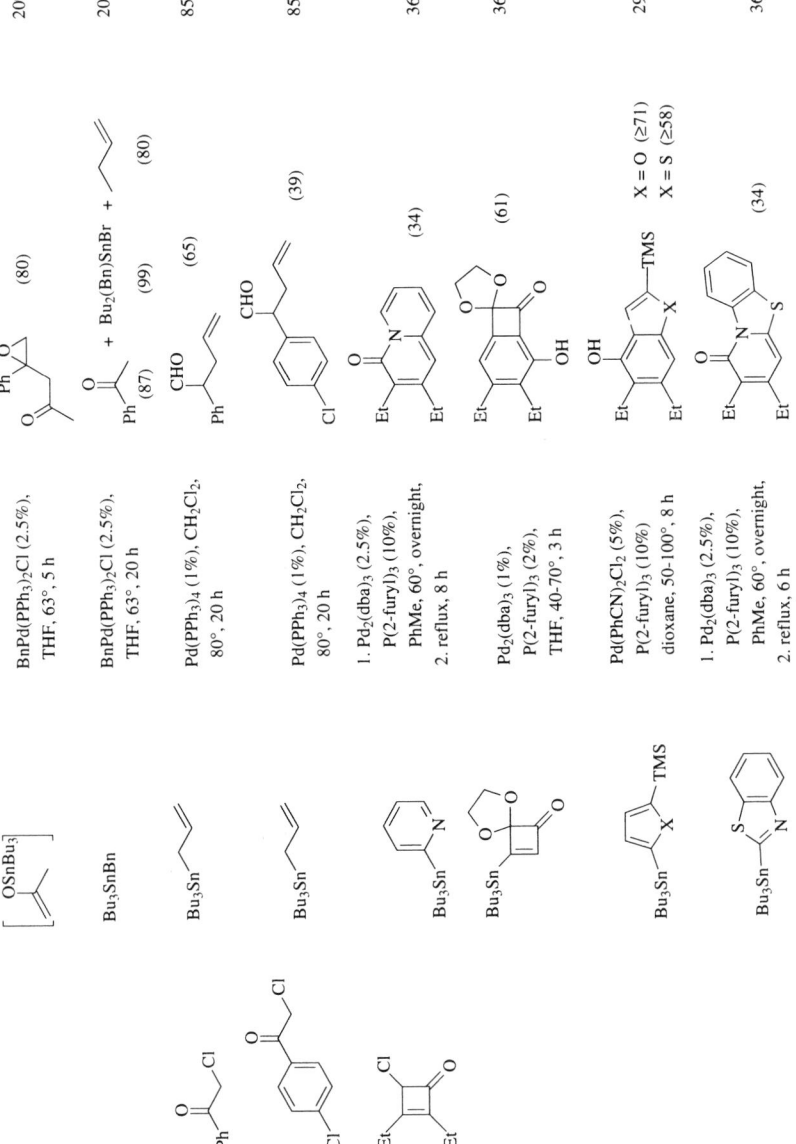

TABLE XXXII. MULTI-STEP TRANSFORMATIONS INVOLVING DIRECT CROSS-COUPLING REACTIONS (*Continued*)

Substrate	Stannane and Other Components	Conditions	Product(s) and Yield(s) (%)	Refs.
		Pd(PhCN)$_2$Cl$_2$ (5%), P(2-furyl)$_3$ (10%) dioxane, 50-100°, 16 h	X = O (≥94) X = S (≥85)	295
		Pd(PhCN)$_2$Cl$_2$ (5%), P(2-furyl)$_3$ (10%), dioxane, 50-100°, 16 h	(≥92)	295
		Pd$_2$(dba)$_3$ (1%), P(2-furyl)$_3$ (4%), dioxane, 80-100°	(95)	360
		Pd(PhCN)$_2$Cl$_2$ (5%), P(2-furyl)$_3$ (10%), dioxane, 50-100°, 16 h	X = O (≥76) X = S (≥78)	295
		Pd(dba)$_2$ (5%), P(2-furyl)$_3$ (10%), dioxane, 60°, 5 h	(67)	359
		Pd(PhCN)$_2$Cl$_2$ (5%), P(2-furyl)$_3$ (10%), dioxane, 100°, 6 h	(55)	359

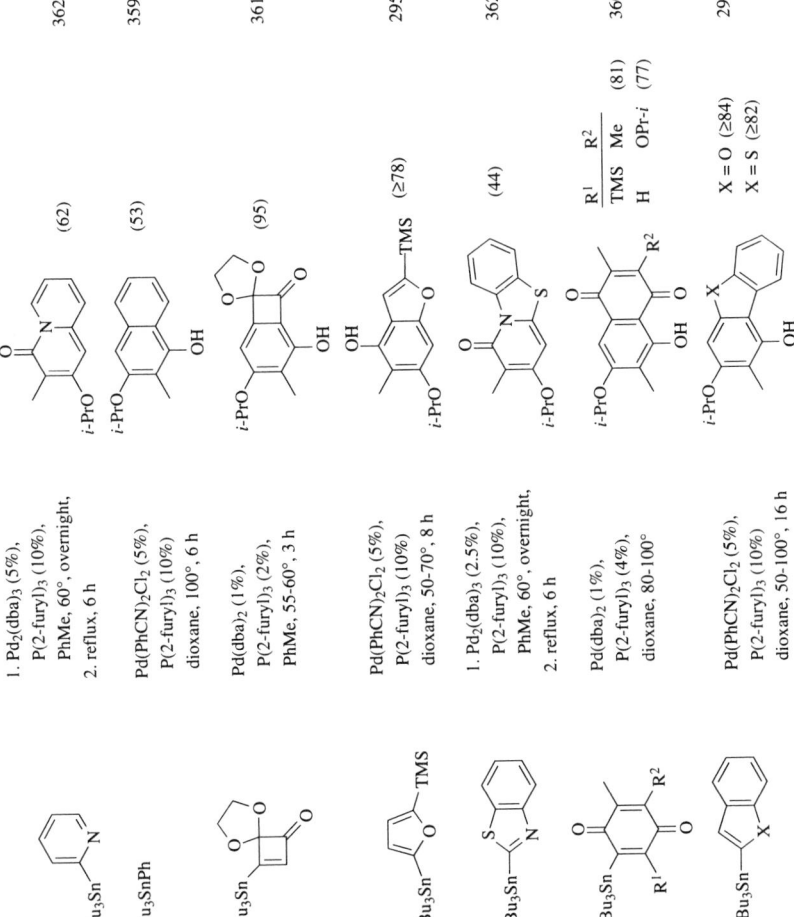

Organostannane	Conditions	Product (yield)	Refs.
Bu₃Sn–(2-pyridyl)	1. Pd₂(dba)₃ (5%), P(2-furyl)₃ (10%), PhMe, 60°, overnight, 2. reflux, 6 h	(62)	362
Bu₃SnPh	Pd(PhCN)₂Cl₂ (5%), P(2-furyl)₃ (10%) dioxane, 100°, 6 h	(53)	359
(spiro dioxolane cyclobutenone stannane)	Pd(dba)₂ (1%), P(2-furyl)₃ (2%), PhMe, 55–60°, 3 h	(95)	361
Bu₃Sn–(5-TMS-2-furyl)	Pd(PhCN)₂Cl₂ (5%), P(2-furyl)₃ (10%) dioxane, 50–70°, 8 h	(≥78)	295
Bu₃Sn–benzothiazole	1. Pd₂(dba)₃ (2.5%), P(2-furyl)₃ (10%), PhMe, 60°, overnight, 2. reflux, 6 h	(44)	362
Bu₃Sn–quinone (R¹, R²)	Pd(dba)₂ (1%), P(2-furyl)₃ (4%), dioxane, 80–100°	R¹=TMS, R²=Me (81); R¹=H, R²=OPr-i (77)	360
Bu₃Sn–(2-benzofuran/thiophene, X)	Pd(PhCN)₂Cl₂ (5%), P(2-furyl)₃ (10%) dioxane, 50–100°, 16 h	X = O (≥84); X = S (≥82)	295

TABLE XXXII. MULTI-STEP TRANSFORMATIONS INVOLVING DIRECT CROSS-COUPLING REACTIONS (*Continued*)

Substrate	Stannane and Other Components	Conditions	Product(s) and Yield(s) (%)	Refs.
C₉				
		Pd(dba)₂ (1%), P(2-furyl)₃ (4%), dioxane, 80-100°	(75)	360
		Pd(dppb)Cl₂, CuO, DMF, 105-110°, 48 h	R = Me (87) R = OMe (91)	864
		Pd(PPh₃)₄, DMF, 100°, 24 h	(59)	101
		1. Pd(PPh₃)₄ (3%), DMF, 100°, 24 h 2. HCl (2 N)	(43)	102
		1. Pd(dppb)Cl₂ (5%), CuO, DMF, 100°, 3 h 2. HCl (2 N) 3. NaOH (2 N)	(57)	96
		Pd(PPh₃)₄, DMF, 100°, 24 h	(75)	101
		1. Pd(PPh₃)₄ (3%), DMF, 100°, 24 h 2. HCl (2 N)	(63)	102

610

Substrate	Reagent	Conditions	Product (yield)	Ref.
I, NHBoc-thiophene	OHC–thiophene–Bu₃Sn	Pd(PPh₃)₄, DMF, 100°, 24 h	thieno-pyridine (42)	101
	Me₃Sn–thiophene(1,3-dioxolane)	1. Pd(PPh₃)₄ (3%), DMF, 100°, 24 h 2. HCl (2 N)	(27)	102
Ph–C(O)–CHBr–CH₃	[OSnBu₃] isopropenyl	BnPd(PPh₃)₂Cl (2.5%), THF, 63°, 5 h	Ph, O epoxide (55)	209
Cl–CH₂CH₂–C(O)–Ph	Bu₃Sn–allyl	BnPd(PPh₃)₂Cl (2.5%), THF, 63°, 48 h	Ph, O oxetane (85)	209
Br, CO₂Et benzene	Bu₃Sn N(TMS)₂	Pd(PPh₃)₄, PhMe, reflux, 96 h	benzazepinone N–H, O (90)	462
Br, CO₂Et, MeO, MeO benzene	Bu₃Sn N(TMS)₂	Pd(PPh₃)₄, PhMe, reflux, 24 h	MeO, MeO benzazepinone N–H, O (18)	462
	Bu₃Sn–pyridine	1. Pd₂(dba)₃ (2.5%), P(2-furyl)₃ (10%), PhMe, 60°, overnight. 2. reflux, 8 h	n-Bu pyridinone (31)	362
Cl, O, n-Bu cyclobutenone	Bu₃Sn cyclobutenone(1,3-dioxolane)	Pd₂(dba)₃ (1%), P(2-furyl)₃ (2%), THF, 40-70°, 3 h	n-Bu, OH (71)	361

TABLE XXXII. MULTI-STEP TRANSFORMATIONS INVOLVING DIRECT CROSS-COUPLING REACTIONS (*Continued*)

Substrate	Stannane and Other Components	Conditions	Product(s) and Yield(s) (%)	Refs.
C$_{10}$	Bu$_3$Sn—	Pd$_2$(dba)$_3$ (1%), P(2-furyl)$_3$ (2%), THF, 40-70°, 3 h	(72)	361
	Bu$_3$Sn—	Pd(PPh$_3$)$_4$ (5%), LiCl, DMF, 60°, 8 h	(43) + (5)	349
	Bu$_3$SnTMS	Pd(PPh$_3$)$_4$ (5%), LiCl, DMF, 60°, 8 h	(54)	349
	Bu$_3$Sn—	Pd(OAc)$_2$ (5%), PPh$_3$ (10%), Et$_3$N, LiCl, DMF, 60°, 8 h	(63)	349
	Bu$_3$Sn—	Pd(PPh$_3$)$_4$ (5%), LiCl, DMF, 60°, 8 h	(46) + (39)	349
	Bu$_3$SnC≡CTMS	Pd(OAc)$_2$ (5%), PPh$_3$ (10%), Et$_3$N, LiCl, DMF, 60°, 8 h	(45)	349

612

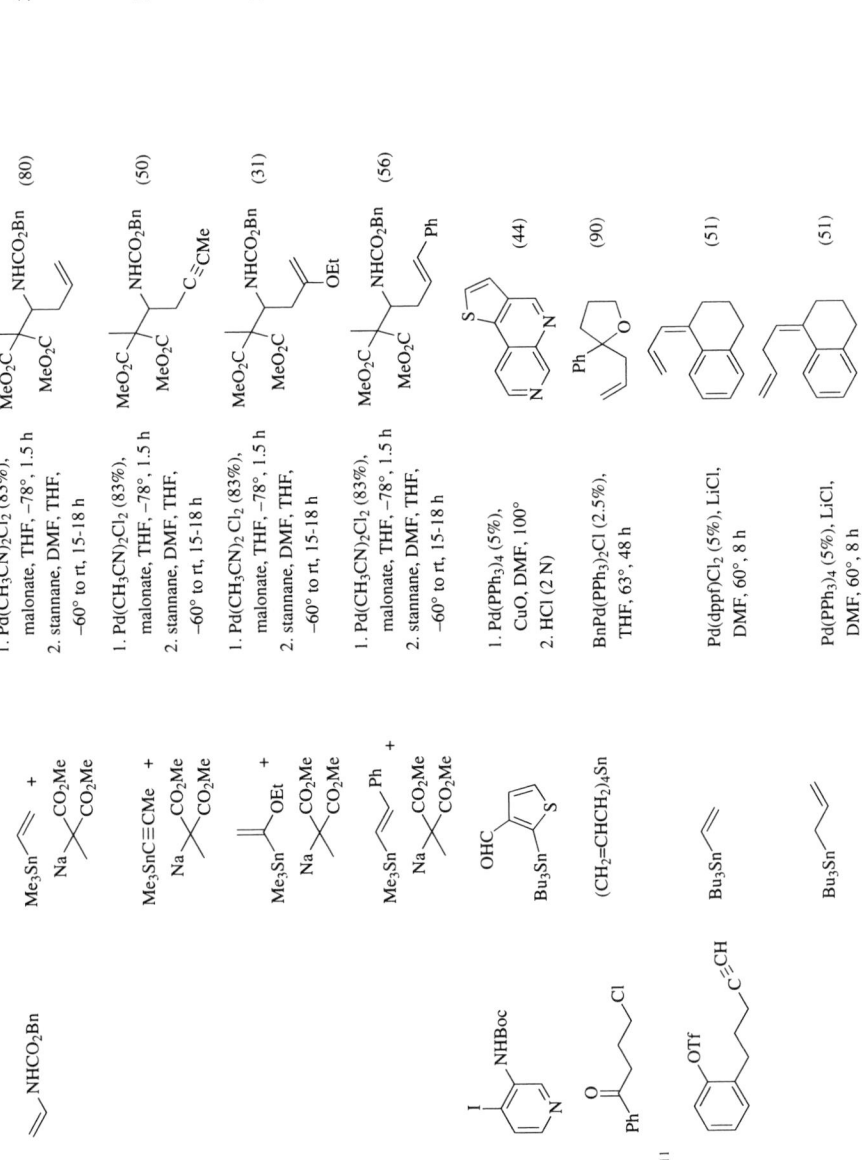

Substrate	Reagents	Conditions	Product (%)	Refs.
CH₂=CH-NHCO₂Bn	Me₃Sn-CH=CH₂ + Na·C(CO₂Me)₂	1. Pd(CH₃CN)₂Cl₂ (83%), malonate, THF, −78°, 1.5 h 2. stannane, DMF, THF, −60° to rt, 15-18 h	MeO₂C, MeO₂C —NHCO₂Bn (80)	339
	Me₃SnC≡CMe + Na·C(CO₂Me)₂	1. Pd(CH₃CN)₂Cl₂ (83%), malonate, THF, −78°, 1.5 h 2. stannane, DMF, THF, −60° to rt, 15-18 h	MeO₂C, MeO₂C —NHCO₂Bn, C≡CMe (50)	339
	Me₃Sn-C(OEt)=CH₂ + Na·C(CO₂Me)₂	1. Pd(CH₃CN)₂Cl₂ (83%), malonate, THF, −78°, 1.5 h 2. stannane, DMF, THF, −60° to rt, 15-18 h	MeO₂C, MeO₂C —NHCO₂Bn, OEt (31)	339
	Me₃Sn-CH=CH-Ph + Na·C(CO₂Me)₂	1. Pd(CH₃CN)₂Cl₂ (83%), malonate, THF, −78°, 1.5 h 2. stannane, DMF, THF, −60° to rt, 15-18 h	MeO₂C, MeO₂C —NHCO₂Bn, Ph (56)	339

C_{11}

Substrate	Reagents	Conditions	Product (%)	Refs.
(I, NHBoc-pyridine)	OHC-thiophene-Bu₃Sn	1. Pd(PPh₃)₄ (5%), CuO, DMF, 100° 2. HCl (2 N)	(thieno-quinoline) (44)	861
Cl(CH₂)₃C(O)Ph	(CH₂=CHCH₂)₄Sn	BnPd(PPh₃)₂Cl (2.5%), THF, 63°, 48 h	(Ph-tetrahydrofuran, allyl) (90)	209
(OTf, C≡CH aryl)	Bu₃Sn-CH=CH₂	Pd(dppf)Cl₂ (5%), LiCl, DMF, 60°, 8 h	(dihydronaphthalene, vinyl) (51)	349
(OTf, C≡CH aryl)	Bu₃Sn-CH₂CH=CH₂	Pd(PPh₃)₄ (5%), LiCl, DMF, 60°, 8 h	(dihydronaphthalene, allyl) (51)	349

TABLE XXXII. MULTI-STEP TRANSFORMATIONS INVOLVING DIRECT CROSS-COUPLING REACTIONS (Continued)

Substrate	Stannane and Other Components	Conditions	Product(s) and Yield(s) (%)	Refs.
(2-iodophenyl, CH≡C, N–Ac)	$Bu_3SnC{\equiv}CTMS$	Pd(PPh₃)₄ (5%), LiCl, DMF, 60°, 8 h	TMSC≡C= (tetralinylidene) (54)	349
	Bu_3Sn (allyl)	Pd(OAc)₂ (10%), PPh₃ (20%), CH₃CN, 5–25°, 2–6 h	(indoline, N–Ac) (40)	347
	Bu_3Sn (homoallyl)	Pd(OAc)₂ (10%), PPh₃ (20%), Et₄NCl, CH₃CN, 5–25°, 2–6 h	(indoline, N–Ac) (54)	347
(2-iodophenyl, C≡CH, N–Me)	Bu_3Sn (allyl)	Pd(OAc)₂ (10%), PPh₃ (20%), LiCl, CH₃CN, 60°, 2 h	(N–Me isoquinolinone + isomer) (20) 4:1	347
	Bu_3Sn (homoallyl)	Pd(OAc)₂ (10%), PPh₃ (20%), Et₄NCl, CH₃CN, 60°, 1.5 h	(N–Me isoquinolinone) (50)	347
(chlorocyclobutenone, Me, Ph)	Bu_3Sn (allyl)	Pd(PhCN)₂Cl₂ (4%), P(2-furyl)₃ (10%), dioxane, 100°, 3 h	(phenol, Me, Ph, OH) (75)	359

Common substrate:

Organostannane	Conditions	Product	Yield (%)	Ref.
Bu$_3$Sn (dioxolane spirocyclobutenone)	Pd$_2$(dba)$_3$ (1%), P(2-furyl)$_3$ (2%), THF, 40–70°, 3 h		(75)	361
Bu$_3$Sn, TMS	Pd$_2$(dba)$_3$ (1%), P(2-furyl)$_3$ (4%), dioxane, 80–100°		(94)	360
Bu$_3$Sn, X	Pd(PhCN)$_2$Cl$_2$ (5%), P(2-furyl)$_3$ (10%), dioxane, 50–100°, 16 h		X = O (≥89); X = S (≥51)	295
Bu$_3$Sn, S	Pd(PhCN)$_2$Cl$_2$ (5%), P(2-furyl)$_3$ (10%), dioxane, 50–100°, 16 h		(≥63)	295
Bu$_3$Sn (allyl)	Pd(PhCN)$_2$Cl$_2$ (5%), P(2-furyl)$_3$ (10%), dioxane, 100°, 3 h		(75)	359
Bu$_3$Sn, EtO	Pd(PhCN)$_2$Cl$_2$ (5%), P(2-furyl)$_3$ (10%), dioxane, 100°, 3 h		(50)	359
Bu$_3$Sn, N (pyridyl)	1. Pd$_2$(dba)$_3$ (2.5%), P(2-furyl)$_3$ (10%), PhMe, 60°, overnight; 2. reflux, 10 h		(42)	362

Substrate	Stannane and Other Components	Conditions	Product(s) and Yield(s) (%)	Refs.
	Bu_3Sn	$Pd_2(dba)_3$ (1%), P(2-furyl)$_3$ (2%), THF, 40-70°, 3 h	(56)	361
	Bu_3Sn	1. $Pd_2(dba)_3$ (2.5%), P(2-furyl)$_3$ (10%), PhMe, 60°, overnight, 2. reflux, 8 h	(58)	362
	Bu_3Sn	$Pd(PhCN)_2Cl_2$ (5%), P(2-furyl)$_3$ (10%), dioxane, 50-100°, 16 h	X = O (≥75) X = S (≥63)	295
	Bu_3Sn	$Pd(PhCN)_2Cl_2$ (5%), P(2-furyl)$_3$ (10%), dioxane, 50-100°, 16 h	(≥60)	295
	Bu_3Sn TMS	$Pd_2(dba)_3$ (1%), P(2-furyl)$_3$ (4%), dioxane, 80-100 1	(74)	360
	Bu_3Sn TMS	$Pd_2(dba)_3$ (1%), P(2-furyl)$_3$ (4%), dioxane, 80-100°	(82)	360

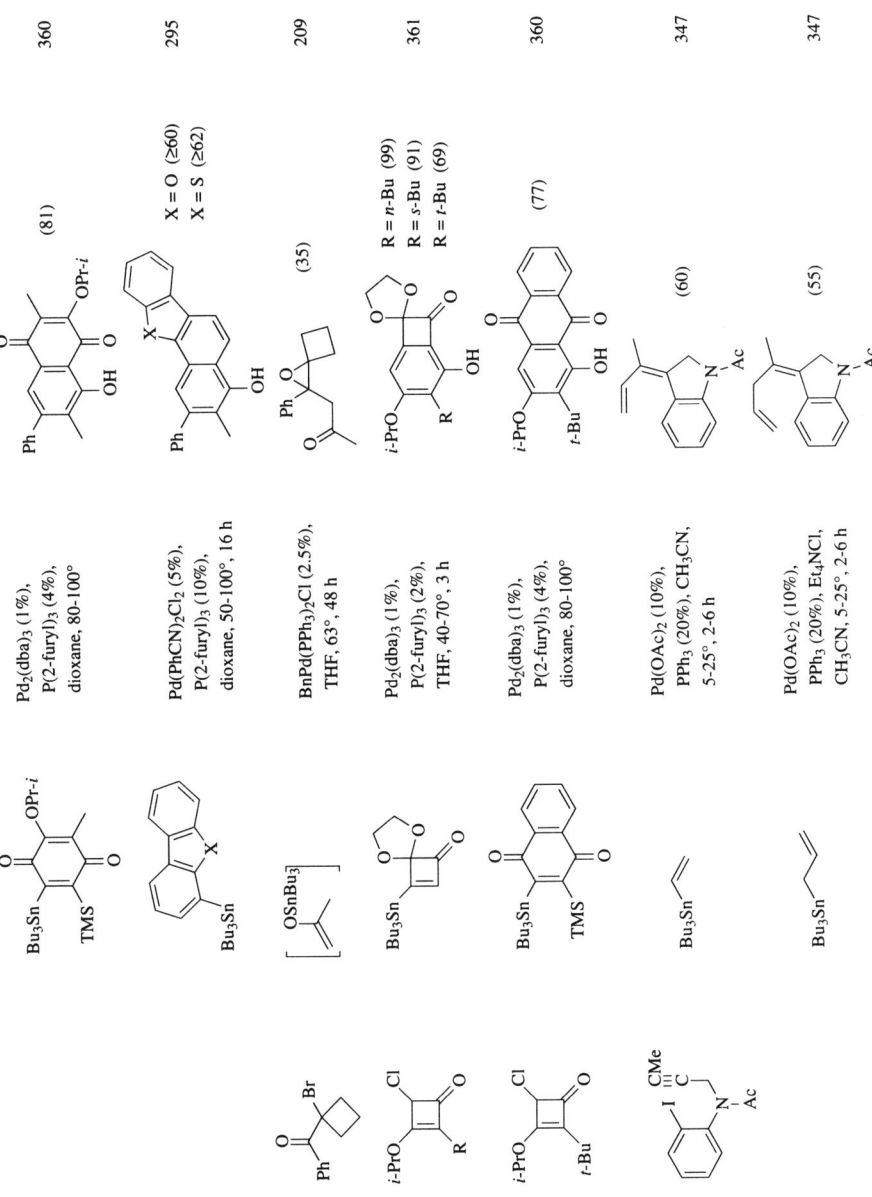

360

295

209

361

360

347

347

TABLE XXXII. MULTI-STEP TRANSFORMATIONS INVOLVING DIRECT CROSS-COUPLING REACTIONS (Continued)

Substrate	Stannane and Other Components	Conditions	Product(s) and Yield(s) (%)	Refs.
	Bu_3Sn	Pd(PhCN)$_2$Cl$_2$ (5%), P(2-furyl)$_3$ (10%), dioxane, 100°, 3 h	(74)	359
	Bu_3Sn (OEt)	Pd(PhCN)$_2$Cl$_2$ (5%), P(2-furyl)$_3$ (10%), dioxane, 100°, 4 h	(54)	359
	Me$_3$SnPh	Pd(PhCN)$_2$Cl$_2$ (5%), P(2-furyl)$_3$ (10%), dioxane, 100°, 53 h	(77)	359
	Bu_3Sn (TMS)	Pd(PhCN)$_2$Cl$_2$ (5%), P(2-furyl)$_3$ (10%), dioxane, 50–70°, 8 h	(≥65)	295
	Bu_3Sn	Pd(PhCN)$_2$Cl$_2$ (5%), P(2-furyl)$_3$ (10%), dioxane, 50–100°, 16 h	X = O (≥65), X = S (≥61)	295
	Bu_3Sn	Pd(PhCN)$_2$Cl$_2$ (5%), P(2-furyl)$_3$ (10%), dioxane, 50–100°, 16 h	(≥58)	295
	Bu_3Sn	Pd(dba)$_2$ (10%), P(2-furyl)$_3$ (30%), ZnCl$_2$, THF, reflux, 0.5 h	(72)	345

618

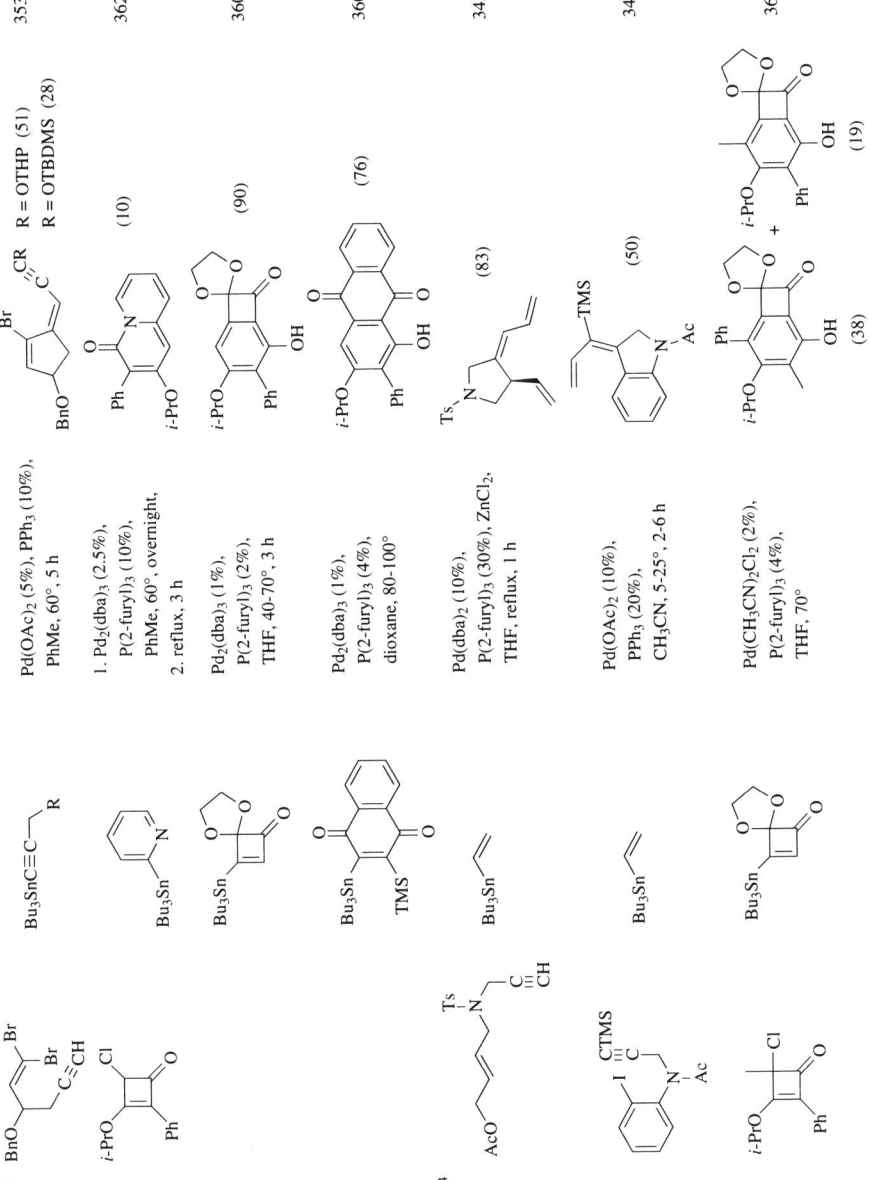

353

362

360

360

345

347

361

TABLE XXXII. MULTI-STEP TRANSFORMATIONS INVOLVING DIRECT CROSS-COUPLING REACTIONS (*Continued*)

Substrate	Stannane and Other Components	Conditions	Product(s) and Yield(s) (%)	Refs.
	SnBu3	Pd(PPh3)4 (10%), THF, 60°	I + II (72), I:II = 11:1	354
	Bu3Sn	Pd(CH3CN)2Cl2 (2%), P(2-furyl)3 (4%), THF, 70°	(79)	361
	Bu3Sn⌒⌇	Pd(OAc)2 (10%), PPh3 (20%), LiCl, CH3CN, 80°, 48 h	(60)	347
C17	Bu3SnH	Pd2(dba)3	(96)	865
	Bu3Sn⌒⌇	Pd(0) (10%), THF, 60°, 24 h	(42)	866

620

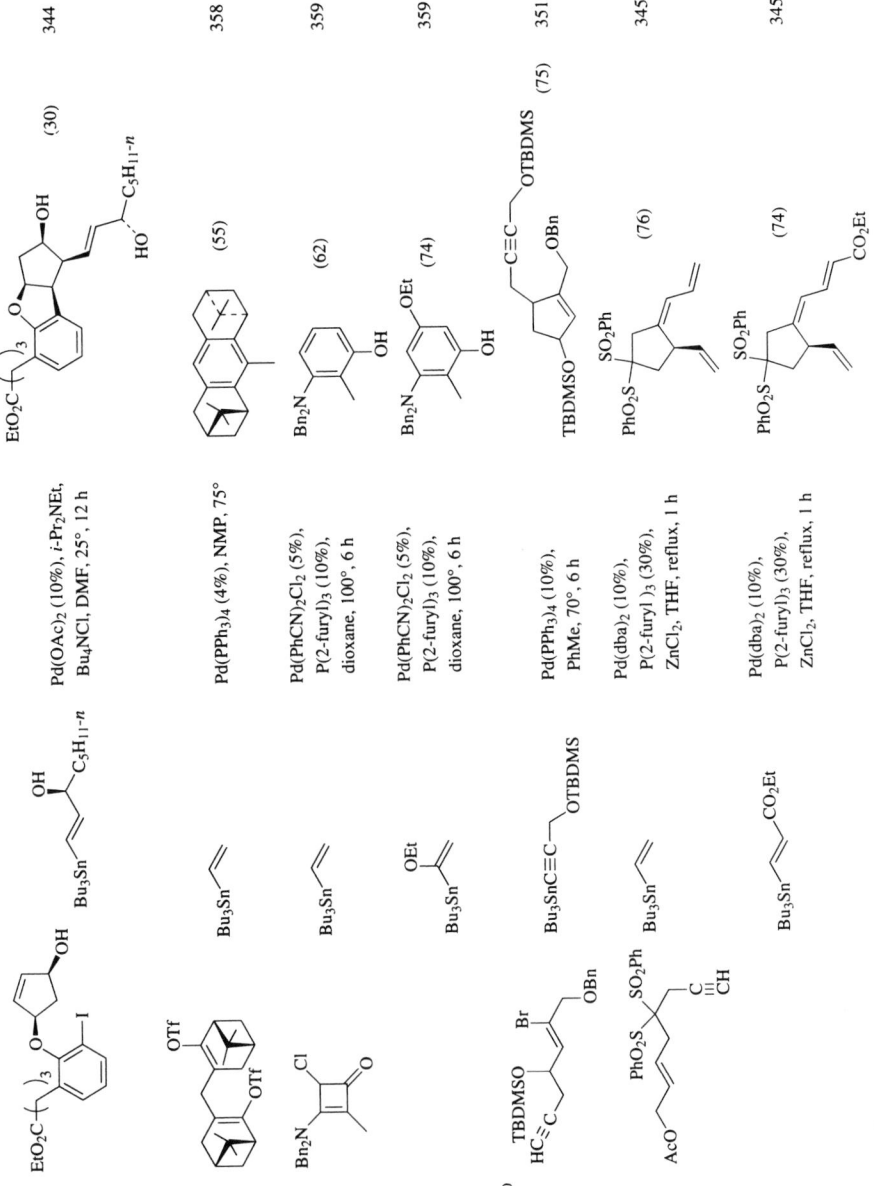

TABLE XXXII. MULTI-STEP TRANSFORMATIONS INVOLVING DIRECT CROSS-COUPLING REACTIONS (*Continued*)

Substrate	Stannane and Other Components	Conditions	Product(s) and Yield(s) (%)	Refs.
	Bu₃Sn⌕OTHP	Pd(dba)₂ (10%), P(2-furyl)₃ (30%), ZnCl₂, THF, reflux, 4 h	(61)	345
	Bu₃Sn⌕	Pd(PPh₃)₄ (15%), THF, 40°, 6 h	(69)	345
	Bu₃Sn⌕	Pd(dba)₂ (10%), P(2-furyl)₃ (30%), ZnCl₂, THF, reflux, 1 h	(76)	345
	Bu₃Sn⌕ OEt or Bu₃Sn⌕	Pd(dba)₂ (10%), P(2-furyl)₃ (30%), ZnCl₂, THF, reflux, 1 h	(—)	345
	Bu₃Sn⌕R	Pd(dba)₂ (10%), P(2-furyl)₃ (30%), ZnCl₂, THF, reflux, 2 h	R = CO₂Et (77) / R = TMS (85)	345
	Bu₃Sn⌕CO₂Et	Pd(dba)₂ (10%), P(2-furyl)₃ (30%), ZnCl₂, THF, reflux, 4 h	(67)	345

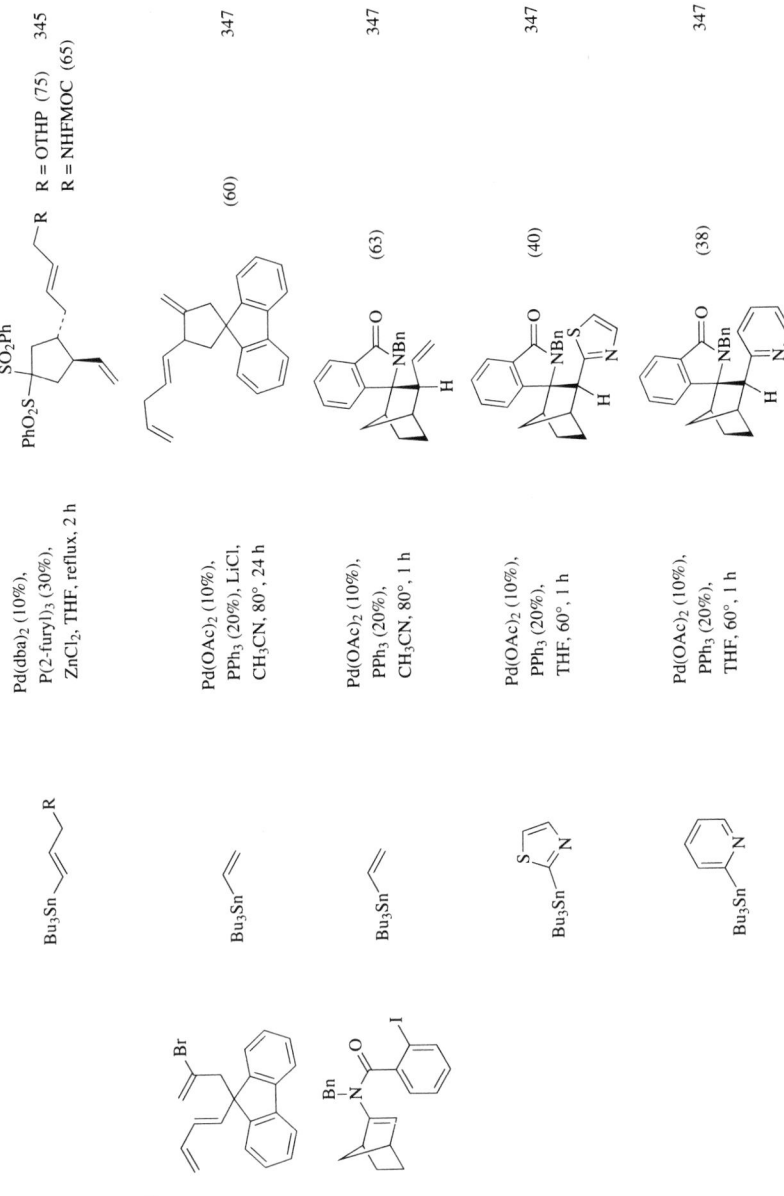

Pd(dba)$_2$ (10%),
P(2-furyl)$_3$ (30%),
ZnCl$_2$, THF, reflux, 2 h

R = OTHP (75)   345
R = NHFMOC (65)

Pd(OAc)$_2$ (10%),
PPh$_3$ (20%), LiCl,
CH$_3$CN, 80°, 24 h

(60)   347

Pd(OAc)$_2$ (10%),
PPh$_3$ (20%),
CH$_3$CN, 80°, 1 h

(63)   347

Pd(OAc)$_2$ (10%),
PPh$_3$ (20%),
THF, 60°, 1 h

(40)   347

Pd(OAc)$_2$ (10%),
PPh$_3$ (20%),
THF, 60°, 1 h

(38)   347

C$_{21}$

TABLE XXXII. MULTI-STEP TRANSFORMATIONS INVOLVING DIRECT CROSS-COUPLING REACTIONS (*Continued*)

Substrate	Stannane and Other Components	Conditions	Product(s) and Yield(s) (%)	Refs.
	Bu₃Sn⌣⌣Ph	Pd(OAc)₂ (10%), PPh₃ (20%), THF, 60°, 1 h	**I + II** (46), **I:II** = 1:8	347
	Bu₃SnC≡CPh	Pd(OAc)₂ (10%), PPh₃ (20%), THF, 60°, 1 h	**I + II** (38), **I:II** = 1:1	347
	Me₃SnSnMe₃	Pd(OAc)₂ (10%), PPh₃ (20%), CH₃CN, 80°, 1 h	(80)	347
	Bu₃Sn (pyranose, OTBDMS, Ph)	Pd(PPh₃)₄ (10%), PhMe, reflux, 12 h	(72) + (15)	297

624

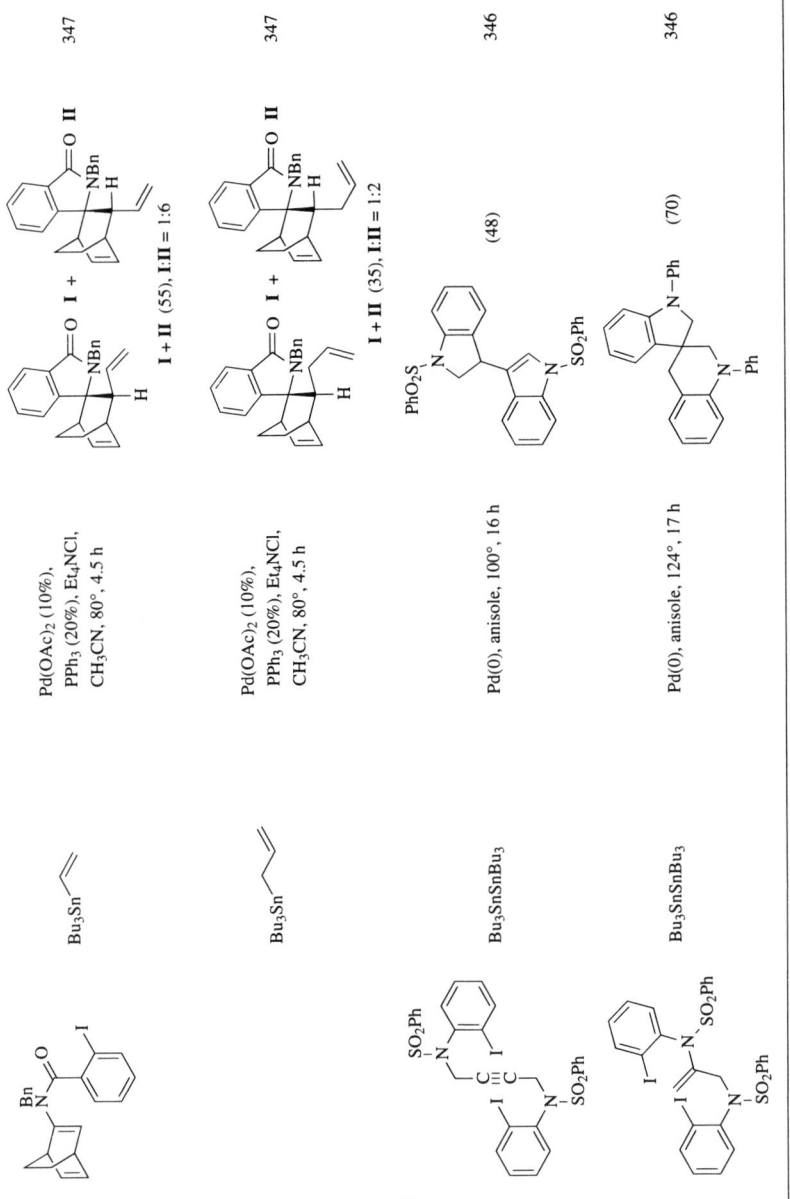

Substrate	Stannane and Other Components	Conditions	Product(s) and Yield(s) (%)	Refs.
C₃	Me₃SnR + Na—⟨CO₂Me / CO₂Me	1. Pd(CH₃CN)₂Cl₂ (83%), THF, rt 2. malonate, Et₃N, −78 to −60°, 2 h 3. CO (15 psi), −30°, 2 h 4. stannane, −30° to rt, 15-18 h	MeO₂C⟨CO₂Me ... R, O	339
	R CH=CMe₂ 2-furyl Ph C≡CPh		(73) (63) (75) (65)	
C₈	Bu₃SnR	CO (45 psi), BnPd(PPh₃)₂Cl (5%), dioxane, 50-100°, 16 h	(structure with Et, Et, R)	363
	R CH=CH₂ CH=CHCH₃ 3-furyl Ph C₆H₄Cl-p C₆H₄OMe-p		(60) (65) stereochemistry not determined (80) (63) (84) (80)	
	Bu₃Sn—⟨X	CO (45 psi), BnPd(PPh₃)₂Cl (5%), dioxane, 50-100°, 16 h	(i-PrO structure) X = O (89) X = S (74)	363
	Bu₃SnPh	CO (45 psi), BnPd(PPh₃)₂Cl (5%), dioxane, 50-100°, 16 h	(i-PrO, Ph structure) (60)	363

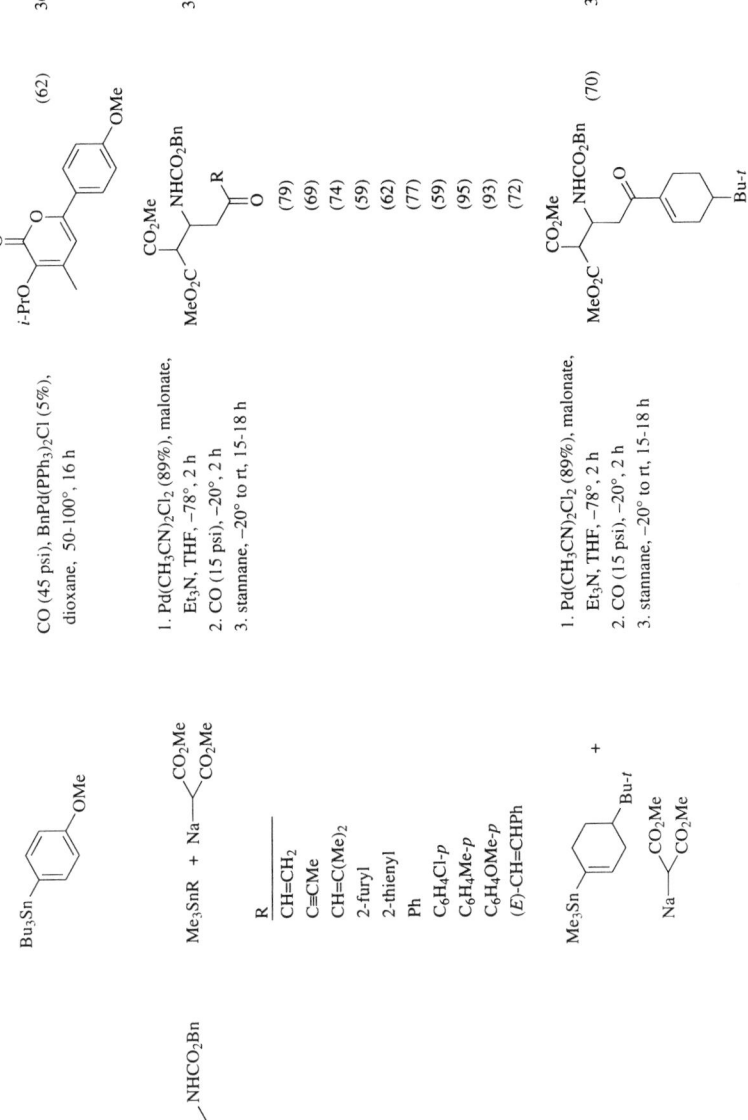

363

339

339

C_{10}

CO (45 psi), BnPd(PPh$_3$)$_2$Cl (5%), dioxane, 50-100°, 16 h

(62)

Me$_3$SnR + Na—CH(CO$_2$Me)$_2$

1. Pd(CH$_3$CN)$_2$Cl$_2$ (89%), malonate, Et$_3$N, THF, −78°, 2 h
2. CO (15 psi), −20°, 2 h
3. stannane, −20° to rt, 15-18 h

R	
CH=CH$_2$	(79)
C≡CMe	(69)
CH=C(Me)$_2$	(74)
2-furyl	(59)
2-thienyl	(62)
Ph	(77)
C$_6$H$_4$Cl-p	(59)
C$_6$H$_4$Me-p	(95)
C$_6$H$_4$OMe-p	(93)
(E)-CH=CHPh	(72)

1. Pd(CH$_3$CN)$_2$Cl$_2$ (89%), malonate, Et$_3$N, THF, −78°, 2 h
2. CO (15 psi), −20°, 2 h
3. stannane, −20° to rt, 15-18 h

(70)

Substrate	Stannane and Other Components	Conditions	Product(s) and Yield(s) (%)	Refs.
	Bu₃SnR	CO, Pd(OAc)₂ (10%), PPh₃ (20%), Et₄NCl, PhMe, 100°, 15 h		355
	R			
	2-furyl		(88)	
	2-pyridyl		(83)	
	(*E*)-CH=CHPh		(61)	
	Bu₃Sn	CO, Pd(OAc)₂ (10%), PPh₃ (20%), Et₄NCl, PhMe, 100°, 15 h	(52)	355
C₁₁	Me₃Sn + Na malonate (CO₂Me, CO₂Me)	1. Pd(PhCN)₂Cl₂ (71%), rt 2. malonate, Et₃N, 2.5 h 3. CO (15 psi), −78 to −20° 4. stannane, −20° to rt, 15-18 h	(65)	339
	Me₃Sn + Na (CO₂Me, CO₂Me)	1. Pd(PhCN)₂Cl₂ (71%), rt 2. malonate, Et₃N, 2.5 h 3. CO (15 psi), −78 to −20° 4. stannane, −20° to rt, 15-18 h	(68)	339

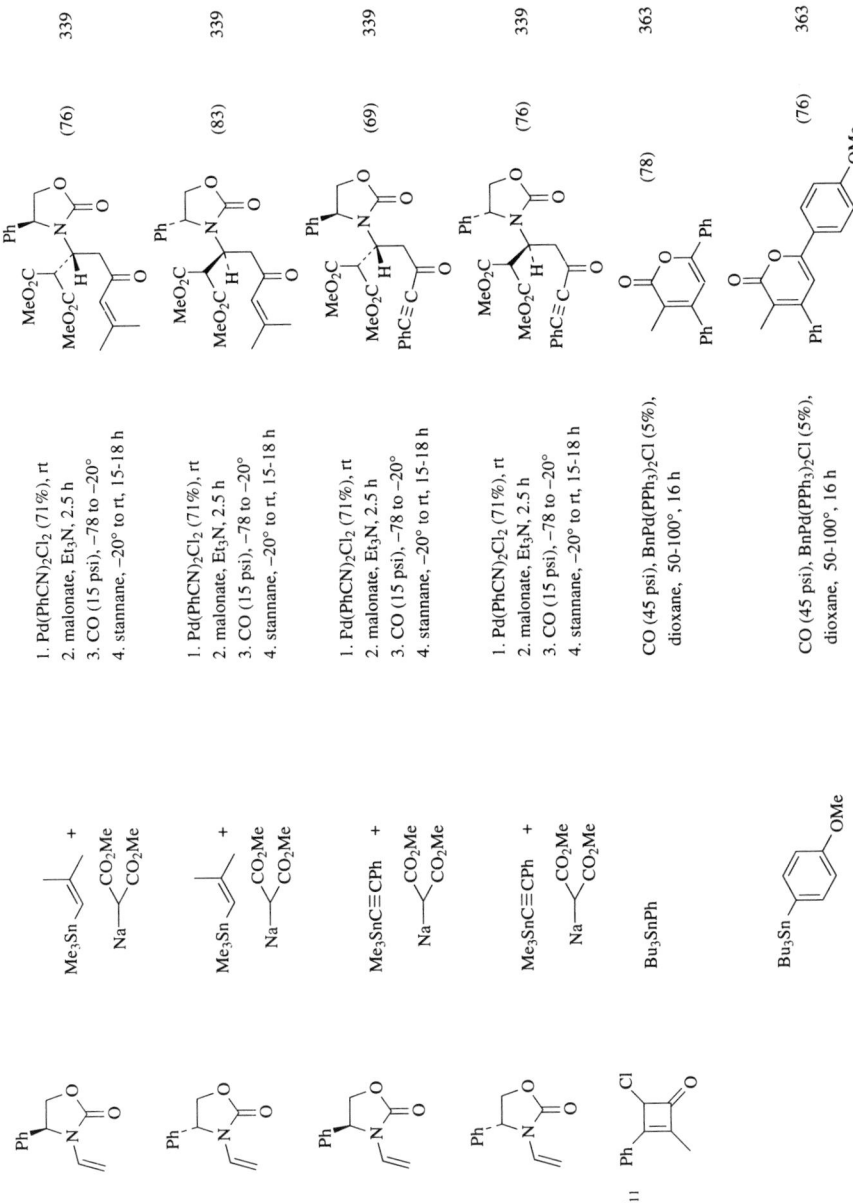

		(76)	339
		(83)	339
		(69)	339
		(76)	339
		(78)	363
		(76)	363

1. Pd(PhCN)₂Cl₂ (71%), rt
2. malonate, Et₃N, 2.5 h
3. CO (15 psi), −78 to −20°
4. stannane, −20° to rt, 15-18 h

1. Pd(PhCN)₂Cl₂ (71%), rt
2. malonate, Et₃N, 2.5 h
3. CO (15 psi), −78 to −20°
4. stannane, −20° to rt, 15-18 h

1. Pd(PhCN)₂Cl₂ (71%), rt
2. malonate, Et₃N, 2.5 h
3. CO (15 psi), −78 to −20°
4. stannane, −20° to rt, 15-18 h

1. Pd(PhCN)₂Cl₂ (71%), rt
2. malonate, Et₃N, 2.5 h
3. CO (15 psi), −78 to −20°
4. stannane, −20° to rt, 15-18 h

CO (45 psi), BnPd(PPh₃)₂Cl (5%),
dioxane, 50-100°, 16 h

CO (45 psi), BnPd(PPh₃)₂Cl (5%),
dioxane, 50-100°, 16 h

Substrate	Stannane and Other Components	Conditions	Product(s) and Yield(s) (%)	Refs.
C12				
(n-Bu pyranone with Cl)	Bu₃SnR	CO (45 psi), BnPd(PPh₃)₂Cl (5%), dioxane, 50-100°, 16 h	(n-Bu pyranone with R)	363
	R			
	C(OEt)=CH₂		(60)	
	3-furyl		(72)	
	2-thienyl		(72)	
	2-N-methylpyrryl		(60)	
	Ph		(86)	
	C₆H₄Cl-p		(60)	
	C₆H₄OMe-p		(60)	
C16				
(iodo aniline PhO₂S substrate)	Bu₃Sn-pyridine	CO, Pd(OAc)₂ (10%), PPh₃ (20%), Et₄NCl, PhMe, 100°, 15 h	(indoline product with pyridyl ketone)	(71)	355
(iodo aniline PhO₂S substrate)	Bu₃Sn-indole-SEM	CO, Pd(OAc)₂ (10%), PPh₃ (20%), Et₄NCl, PhMe, 100°, 15 h	(indoline product with indolyl ketone SEM)	(71)	355
C17				
(iodo aniline Bn substrate)	Bu₃Sn-furyl	CO, Pd(OAc)₂ (10%), PPh₃ (20%), Et₄NCl, PhMe, 100°, 15 h	(oxindole product with furyl ketone Bn)	(87)	355

630

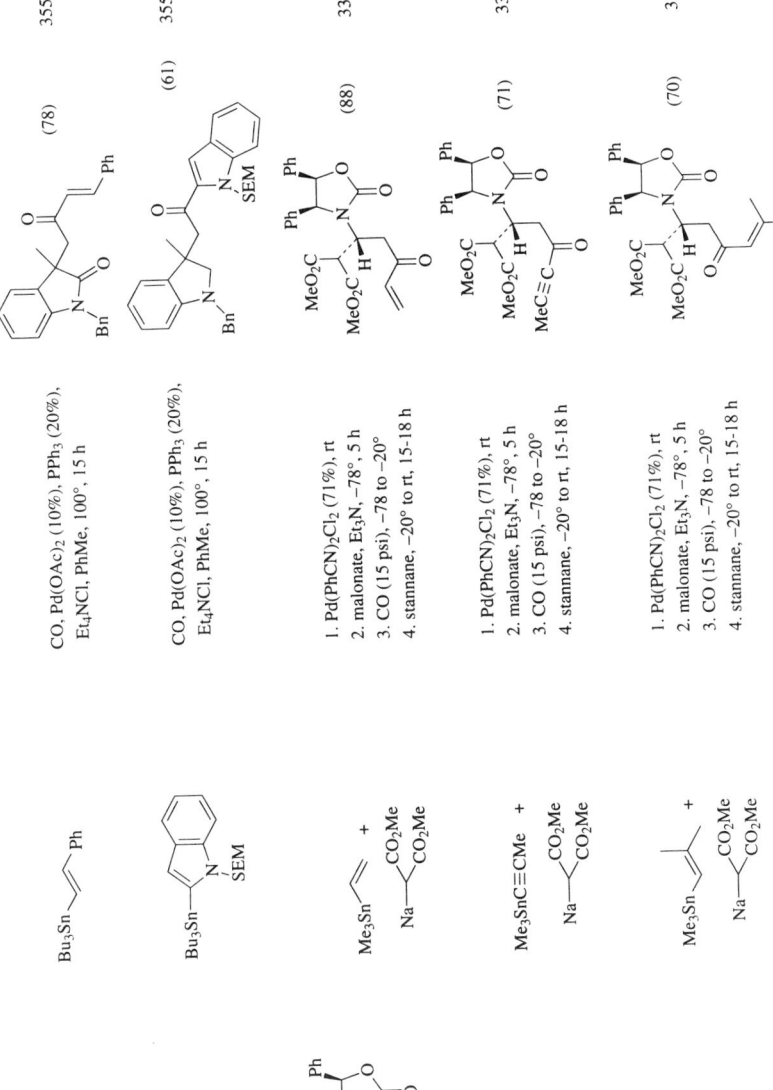

TABLE XXXIII. MULTI-STEP TRANSFORMATIONS INVOLVING CARBONYLATIVE CROSS-COUPLING (*Continued*)

Substrate	Stannane and Other Components	Conditions	Product(s) and Yield(s) (%)	Refs.
	Me$_3$Sn⟍⟍ + Na⟍(CO$_2$Et)(CO$_2$Bu-t)	1. Pd(PhCN)$_2$Cl$_2$ (71%), rt 2. malonate, Et$_3$N, −78°, 5 h 3. CO (15 psi), −78 to −20° 4. stannane, −20° to rt, 15-18 h	(68-77) 1:1 diastereomers	339
	Me$_3$SnR + Na⟍(CO$_2$Me)(CO$_2$Me) R ― 2-furyl 2-thienyl Ph C$_6$H$_4$Cl-p C$_6$H$_4$Me-p	1. Pd(PhCN)$_2$Cl$_2$ (71%), rt 2. malonate, Et$_3$N, −78°, 5 h 3. CO (15 psi), −78 to −20° 4. stannane, −20° to rt, 15-18 h	(65) (68) (37) (25) (69)	339
	Me$_3$SnC≡CPh + Na⟍(CO$_2$Me)(CO$_2$Me)	1. Pd(PhCN)$_2$Cl$_2$ (71%), rt 2. malonate, Et$_3$N, −78°, 5 h 3. CO (15 psi), −78 to −20° 4. stannane, −20° to rt, 15-18 h	(90)	339

632

REFERENCES

[1] Kosugi, M.; Shimizu, Y.; Migita, T. *Chem. Lett.* **1977**, 1423.

[2] Kosugi, M.; Shimizu, Y.; Migita, T. *J. Organomet. Chem.* **1977**, *129*, C36.

[3] Kosugi, M.; Sasazawa, K.; Shimizu, Y.; Migita, T. *Chem. Lett.* **1977**, 301.

[4] Milstein, D.; Stille, J. K. *J. Am. Chem. Soc.* **1978**, *100*, 3636.

[5] Beletskaya, I. P. *J. Organomet. Chem.* **1983**, *250*, 551.

[6] Stille, J. K. *Angew. Chem., Int. Ed. Engl.* **1986**, *25*, 508.

[7] Kumada, M. *Pure Appl. Chem.* **1980**, *52*, 669.

[8] Erdik, E. *Tetrahedron* **1992**, *48*, 9577.

[9] Miyaura, N.; Ishiyama, T.; Sasaki, H.; Ishikawa, M.; Satoh, M.; Suzuki, A. *J. Am. Chem. Soc.* **1989**, *111*, 314.

[10] Hatanaka, Y.; Hiyama, T. *Synlett* **1991**, 845.

[11] Farina, V.; Krishnan, B. *J. Am. Chem. Soc.* **1991**, *113*, 9585.

[12] Liebeskind, L. S.; Fengl, R. W. *J. Org. Chem.* **1990**, *55*, 5359.

[13] Mitchell, T. N. *Synthesis* **1992**, 803.

[14] Farina, V.; Roth, G. P. in *Advances in Metal-Organic Chemistry*, Liebeskind, L. S. Ed., Vol. 5, JAI Press, Greenwich, CT, **1995**.

[15] Stille, J. K. in *The Chemistry of the Metal-Carbon Bond*, Hartley, F. R., Patai S., Eds., Vol. 2, John Wiley, New York, **1985**; p. 625.

[16] Amatore, C.; Azzabi, M.; Jutand, A. *J. Organomet. Chem.* **1989**, *363*, C41.

[17] Fauvarque, J. F.; Pflüger, F.; Troupel, M. *J. Organomet. Chem.* **1981**, *208*, 419.

[18] Ugo, R.; Pasini, A.; Fusi, A.; Cenini, S. *J. Am. Chem. Soc.* **1972**, *94*, 7364.

[19] Milstein, D.; Stille, J. K. *J. Am. Chem. Soc.* **1979**, *101*, 4992.

[20] Amatore, C.; Azzabi, M.; Jutand, A. *J. Am. Chem. Soc.* **1991**, *113*, 1670.

[21] Lau, K. S. Y.; Wong, P. K.; Stille, J. K. *J. Am. Chem. Soc.* **1976**, *98*, 5832.

[22] Becker, Y.; Stille, J. K. *J. Am. Chem. Soc.* **1978**, *100*, 838.

[23] Kramer, A. V.; Osborn, J. A. *J. Am. Chem. Soc.* **1974**, *96*, 7832.

[24] Sheffy, F. K.; Godschalx, J. P.; Stille, J. K. *J. Am. Chem. Soc.* **1984**, *106*, 4833.

[25] Kurosawa, H.; Ogoshi, S.; Kawasaki, Y.; Murai, S.; Miyoshi, M.; Ikeda, I. *J. Am. Chem. Soc.* **1990**, *112*, 2813.

[26] Kurosawa, H.; Kajimaru, H.; Ogoshi, S.; Yoneda, H.; Miki, K.; Kasai, N.; Murai, S.; Ikeda, I. *J. Am. Chem. Soc.* **1992**, *114*, 8417.

[27] Labadie, J. W.; Stille, J. K. *J. Am. Chem. Soc.* **1983**, *105*, 6129.

[28] Scott, W. J.; Stille, J. K. *J. Am. Chem. Soc.* **1986**, *108*, 3033.

[29] Chen, Q.-Y.; He, Y.-B. *Chin. J. Chem.* **1990**, 451.

[30] Farina, V.; Krishnan, B.; Marshall, D. R.; Roth, G. P. *J. Org. Chem.* **1993**, *58*, 5434.

[31] Piers, E.; Friesen, R. W.; Keay, B. A. *J. Chem. Soc., Chem. Commun.* **1985**, 809.

[32] Gronowitz, S.; Messmer, A.; Timari, G. *J. Heterocycl. Chem.* **1992**, *29*, 1049.

[33] Farina, V.; Kapadia, S.; Krishnan, B.; Wang, C.; Liebeskind, L. S. *J. Org. Chem.* **1994**, *59*, 5905.

[34] Milstein, D.; Stille, J. K. *J. Am. Chem. Soc.* **1979**, *101*, 4981.

[35] Godschalx, J.; Stille, J. K. *Tetrahedron Lett.* **1980**, *21*, 2599.

[36] Trost, B. M.; Keinan, E. *Tetrahedron Lett.* **1980**, *21*, 2595.

[37] Keinan, E.; Roth, Z. *J. Org. Chem.* **1983**, *48*, 1769.

[38] Goliaszewski, A.; Schwartz, J. *Organometallics* **1985**, *4*, 417.

[39] Goliaszewski, A.; Schwartz, J. *Tetrahedron* **1985**, *41*, 5779.

[40] Farina, V.; Baker, S. R.; Benigni, D. A.; Hauck, S. I.; Sapino, C., Jr. *J. Org. Chem.* **1990**, *55*, 5833.

[41] Vedejs, E.; Haight, A. R.; Moss, W. O. *J. Am. Chem. Soc.* **1992**, *114*, 6556.

[42] Brown, J. M.; Pearson, M.; Jastrzebski, J. T. B. H.; van Koten, G. *J. Chem. Soc., Chem. Commun.* **1992**, 1440.

[43] Peet, W. G.; Tam, W. *J. Chem. Soc., Chem. Commun.* **1983**, 853.

[44] Kobayashi, Y.; Kato, N.; Shimazaki, T.; Sato, F. *Tetrahedron Lett.* **1988**, *29*, 6297.

[45] Angara, G. J.; Bovonsombat, P.; McNelis, E. *Tetrahedron Lett.* **1992**, *33*, 2285.

[46] Stille, J. K.; Groh, B. L. *J. Am. Chem. Soc.* **1987**, *109*, 813.

[47] Stille, J. K.; Simpson, J. H. *J. Am. Chem. Soc.* **1987**, *109*, 2138.

[48] Taniguchi, M.; Takeyama, Y.; Fugami, K.; Oshima, K.; Utimoto, K. *Bull. Chem. Soc. Jpn.* **1991**, *64*, 2593.

[49] Murakami, M.; Amii, H.; Takizawa, N.; Ito, Y. *Organometallics* **1993**, *12*, 4223.

[50] Potter, G. A.; McCague, R. *J. Org. Chem.* **1990**, *55*, 6184.

[51] Pearson, W. H.; Postich, M. J. *J. Org. Chem.* **1994**, *59*, 5662.

[52] Stille, J. K.; Sweet, M. P. *Tetrahedron Lett.* **1989**, *30*, 3645.

[53] Stille, J. K.; Sweet, M. P. *Organometallics* **1990**, *9*, 3189.

[54] Eisley, D. A.; MacLeod, D.; Miller, J. A.; Quayle, P. *Tetrahedron Lett.* **1992**, *33*, 409.

[55] Tamayo, N.; Echavarren, A. M.; Paredes, M. C. *J. Org. Chem.* **1991**, *56*, 6488.

[56] Echavarren, A. M.; Tamayo, N.; Paredes, M. C. *Tetrahedron Lett.* **1993**, *34*, 4713.

[57] Echavarren, A. M.; Tamayo, N.; Cárdenas, D. J. *J. Org. Chem.* **1994**, *59*, 6075.

[58] Chan, K. S.; Mak, C. C. *Tetrahedron* **1994**, *50*, 2003.

[59] Paley, R. S.; de Dios, A.; Fernández de la Pradilla, R. *Tetrahedron Lett.* **1993**, *34*, 2429.

[60] Paley, R. S.; Lafontaine, J. A.; Ventura, M. P. *Tetrahedron Lett.* **1993**, *34*, 3663.

[61] Johnson, C. R.; Adams, J. P.; Braun, M. P.; Senanayake, C. B. W. *Tetrahedron Lett.* **1992**, *33*, 919.

[62] Nishikawa, T.; Isobe, M. *Tetrahedron* **1994**, *50*, 5621.

[63] Liebeskind, L. S.; Wang, J. *Tetrahedron Lett.* **1990**, *31*, 4293.

[64] Liebeskind, L. S.; Yu, M. S.; Yu, R. H.; Wang, J.; Hagen, K. S. *J. Am. Chem. Soc.* **1993**, *115*, 9048.

[65] Siesel, D. A.; Staley, S. W. *Tetrahedron Lett.* **1993**, *34*, 3679.

[66] Siesel, D. A.; Staley, S. W. *J. Org. Chem.* **1993**, *58*, 7870.

[67] Banwell, M. G.; Cameron, J. M.; Collis, M. P.; Crisp, G. T.; Gable, R. W.; Hamel, E.; Lambert, J. N.; Mackay, M. F.; Reum, M. E.; Scoble, J. A. *Aust. J. Chem.* **1991**, *44*, 705.

[68] Piers, E.; Lu, Y.-F. *J. Org. Chem.* **1988**, *53*, 926.

[69] Piers, E.; Skerlj, R. T. *J. Chem. Soc., Chem. Commun.* **1987**, 1025.

[70] Fujiwara, K.; Kurisaki, A.; Hirama, M. *Tetrahedron Lett.* **1990**, *31*, 4329.

[71] Hirama, M.; Fujiwara, K.; Shigematu, K.; Fukazawa, Y. *J. Am. Chem. Soc.* **1989**, *111*, 4120.

[72] Hirama, M.; Gomibuchi, T.; Fujiwara, K.; Sugiura, Y.; Uesugi, M. *J. Am. Chem. Soc.* **1991**, *113*, 9851.

[73] Tokuda, M.; Fujiwara, K.; Gomibuchi, T.; Hirama, M.; Uesugi, M.; Sugiura, Y. *Tetrahedron Lett.* **1993**, *34*, 669.

[74] Palmisano, G.; Santagostino, M. *Synlett* **1993**, 771.

[75] Barrett, A. G. M.; Boys, M. L.; Boehm, T. L. *J. Chem. Soc., Chem. Commun.* **1994**, *16*, 1881.

[76] Pattenden, G.; Thom, S. M. *Synlett* **1993**, 215.

[77] Kende, A. S.; Kawamura, K.; DeVita, R. J. *J. Am. Chem. Soc.* **1990**, *112*, 4070.

[78] Hong, C. Y.; Kishi, Y. *J. Am. Chem. Soc.* **1991**, *113*, 9693.

[79] Férézou, J. P.; Julia, M.; Liu, L. W.; Pancrazi, A. *Synlett* **1991**, 614.

[80] Evans, D. A.; Gage, J. R.; Leighton, J. L. *J. Am. Chem. Soc.* **1992**, *114*, 9434.

[81] Barrett, A. G. M.; Edmunds, J. J.; Hendrix, J. A.; Malecha, J. W.; Parkinson, C. J. *J. Chem. Soc., Chem. Commun.* **1992**, 1238.

[82] Tanimoto, N.; Gerritz, S. W.; Sawabe, A.; Noda, T.; Filla, S. A.; Masamune, S. *Angew. Chem., Int. Ed. Engl.* **1994**, *33*, 673.

[83] Kende, A. S.; Koch, K.; Dorey, G.; Kaldor, I.; Liu, K. *J. Am. Chem. Soc.* **1993**, *115*, 9842.

[84] Evans, D. A.; Black, W. C. *J. Am. Chem. Soc.* **1993**, *115*, 4497.

[85] Smith, A. B., III; Maleczka, R. E., Jr.; Leazer, J. L., Jr.; Leahy, J. W.; McCauley, J. A.; Condon, S. M. *Tetrahedron Lett.* **1994**, *35*, 4911.

[86] Nicolaou, K. C.; Chakraborty, T. K.; Piscopio, A. D.; Minowa, N.; Bertino, P. *J. Am. Chem. Soc.* **1993**, *115*, 4419.

[87] Kashin, A. N.; Bumagina, I. G.; Bumagin, N. A.; Beletskaya, I. P. *J. Org. Chem. USSR* **1981**, *17*, 18; *Chem. Abstr.* **1981**, *95*, 43254.

[88] McKean, D. R.; Parrinello, G.; Renaldo, A. F.; Stille, J. K. *J. Org. Chem.* **1987**, *52*, 422.

[89] Krolski, M. E.; Renaldo, A. F.; Rudisill, D. E.; Stille, J. K. *J. Org. Chem.* **1988**, *53*, 1170.

[90] Kosugi, M.; Kameyama, M.; Migita, T. *Chem. Lett.* **1983**, 927.

[91] Guram, A. S.; Buchwald, S. L. *J. Am. Chem. Soc.* **1994**, *116*, 7901.

[92] Paul, F.; Patt, J.; Hartwig, J. F. *J. Am. Chem. Soc.* **1994**, *116*, 5969.

[93] Yamamoto, Y.; Azuma, Y.; Mitoh, H. *Synthesis* **1986**, 564.

[94] Alves, T.; B., d. O. A.; Snieckus, V. *Tetrahedron Lett.* **1988**, *29*, 2135.

[95] Sakamoto, T.; Satoh, C.; Kondo, Y.; Yamanaka, H. *Heterocycles* **1992**, *34*, 2379.

[96] Gronowitz, S.; Björk, P.; Malm, J.; Hörnfeldt, A. B. *J. Organomet. Chem.* **1993**, *460*, 127.

[97] Laborde, E.; Kiely, J. S.; Lesheski, L. E.; Schroeder, M. C. *J. Heterocycl. Chem.* **1991**, *28*, 191.

[98] Porco, J. A., Jr.; Schoenen, F. J.; Stout, T. J.; Clardy, J.; Schreiber, S. L. *J. Am. Chem. Soc.* **1990**, *112*, 7410.

[99] Rocca, P.; Marsais, F.; Godard, A.; Quéguiner, G. *Tetrahedron Lett.* **1993**, *34*, 2937.

[100] Ishida, H.; Yui, K.; Aso, Y.; Otsubo, T.; Ogura, F. *Bull. Chem. Soc. Jpn.* **1990**, *63*, 2828.

[101] Gronowitz, S.; Hörnfeldt, A.-B.; Yang, Y. *Chem. Scr.* **1988**, *28*, 281.

[102] Yang, Y.; Hörnfeldt, A.-B.; Gronowitz, S. *Synthesis* **1989**, *2*, 130.

[103] Crisp, G. T. *Synth. Commun.* **1989**, *19*, 307.

[104] Rossi, R.; Carpita, A.; Ciofalo, M.; Houben, J. L. *Gazz. Chim. Ital.* **1990**, *120*, 793.

[105] Rossi, R.; Carpita, A.; Messeri, T. *Synth. Commun.* **1991**, *12*, 1875.

[106] Barbarella, G.; Zambianchi, M. *Tetrahedron* **1994**, *50*, 1249.

[107] Kelly, T. R.; Jagoe, C. T.; Gu, Z. *Tetrahedron Lett.* **1991**, *32*, 4263.

[108] Dondoni, A.; Fogagnolo, M.; Medici, A.; Negrini, E. *Synthesis* **1987**, 185.

[109] Somei, M.; Sayama, S.; Naka, K.; Yamada, F. *Heterocycles* **1988**, *27*, 1585.

[110] Yang, Y.; Martin, A. R. *Synth. Commun.* **1992**, *22*, 1757.

[111] Wang, D.; Haseltine, J. *J. Heterocycl. Chem.* **1994**, *31*, 1637.

[112] Minakawa, N.; Sasaki, T.; Matsuda, A. *Bioorg. Med. Chem. Lett.* **1993**, *3*, 183.

[113] Matsuda, A.; Minakawa, N.; Sasaki, T.; Ueda, T. *Chem. Pharm. Bull.* **1988**, *36*, 2730.

[114] Labadie, S. S. *Synth. Commun.* **1994**, *24*, 709.

[115] Tamao, K.; Yamaguchi, S.; Shiro, M. *J. Am. Chem. Soc.* **1994**, *116*, 11715.

[116] Le Floch, P.; Carmichael, D.; Ricard, L.; Mathey, F. *J. Am. Chem. Soc.* **1993**, *115*, 10665.

[117] Nair, V.; Turner, G. A.; Chamberlain, S. D. *J. Am. Chem. Soc.* **1987**, *109*, 7223.

[118] Nair, V.; Turner, G. A.; Buenger, G. S.; Chamberlain, S. D. *J. Org. Chem.* **1988**, *53*, 3051.

[119] Nair, V.; Purdy, D. F.; Sells, T. B. *J. Chem. Soc., Chem. Commun.* **1989**, 878.

[120] Nair, V.; Lyons, A. G. *Tetrahedron* **1989**, *45*, 3653.

[121] Crisp, G. T. *Synth. Commun.* **1989**, *19*, 2117.

[122] Crisp, G. T.; Macolino, V. *Synth. Commun.* **1990**, *20*, 413.

[123] Hassan, M. E. *Collect. Czech. Chem. Commun.* **1991**, *56*, 1944.

[124] Wigerinck, P.; Pannecouque, C.; Snoeck, R.; Claes, P.; De Clercq, E.; Herdewijn, P. *J. Med. Chem.* **1991**, *34*, 2383.

[125] Herdewijn, P.; Kerremans, L.; Wigerinck, P.; Vandendriessche, F.; Van Aerschot, A. *Tetrahedron Lett.* **1991**, *32*, 4397.

[126] Peters, D.; Hörnfeld, A.-B.; Gronowitz, S. *J. Heterocycl. Chem.* **1991**, *28*, 1629.

[127] Yamamoto, Y.; Seko, T.; Nemoto, H. *J. Org. Chem.* **1989**, *54*, 4734.

[128] Farina, V.; Hauck, S. I. *Synlett* **1991**, 157.

[129] Peters, D.; Hörnfeldt, A.-B.; Gronowitz, S. *J. Heterocycl. Chem.* **1991**, *28*, 1613.

[130] Mamos, P.; Van Aerschot, A. A.; Weyns, N. J.; Herdewijn, P. A. *Tetrahedron Lett.* **1992**, *33*, 2413.

[131] Moriarty, R. M.; Epa, W. R.; Awasthi, A. K. *Tetrahedron Lett.* **1990**, *31*, 5877.

[132] Tanaka, H.; Hayakawa, H.; Shibata, S.; Haraguchi, K.; Miyasaka, T. *Nucleosides Nucleotides* **1992**, *11*, 319.

[133] Gundersen, L.-L. *Tetrahedron Lett.* **1994**, *35*, 3155.

[134] Gundersen, L.-L.; Bakkestuen, A. K.; Aasen, A. J.; Øveråa, H.; Rise, F. *Tetrahedron* **1994**, *50*, 9743.

[135] Solberg, J.; Undheim, K. *Acta Chem. Scand., Ser. B* **1987**, *B41*, 712.

[136] Brakta, M.; Daves, G. D., Jr. *J. Chem. Soc., Perkin Trans. 1* **1992**, 1883.

[137] Benneche, T. *Acta Chem. Scand.* **1990**, *44*, 927.

[138] Kondo, Y.; Watanabe, R.; Sakamoto, T.; Yamanaka, H. *Chem. Pharm. Bull.* **1989**, *37*, 2814.

[139] Kondo, Y.; Watanabe, R.; Sakamoto, T.; Yamanaka, H. *Chem. Pharm. Bull.* **1989**, *37*, 2933.

[140] Majeed, A. J.; Antonsen, O.; Benneche, T.; Undheim, K. *Tetrahedron* **1989**, *45*, 993.

[141] Solberg, J.; Undheim, K. *Acta Chem. Scand.* **1989**, *43*, 62.

[142] Watanabe, T.; Hayashi, K.; Sakurada, J.; Ohki, M.; Takamatsu, N.; Hirohata, H.; Takeuchi, K.; Yuasa, K.; Ohta, A. *Heterocycles* **1989**, *29*, 123.

[143] DiMagno, S. G.; Lin, V. S.-Y.; Therien, M. J. *J. Org. Chem.* **1993**, *58*, 5983.

[144] DiMagno, S. G.; Lin, V. S.-Y.; Therien, M. J. *J. Am. Chem. Soc.* **1993**, *115*, 2513.

[145] Deshpande, M. S. *Tetrahedron Lett.* **1994**, *35*, 5613.

[146] Labadie, J. W.; Tueting, D.; Stille, J. K. *J. Org. Chem.* **1983**, *48*, 4634.

[147] Milstein, D.; Stille, J. K. *J. Org. Chem.* **1979**, *44*, 1613.

[148] Echavarren, A. M.; Pérez, M.; Castaño, A. N.; Cuerva, J. M. *J. Org. Chem.* **1994**, *59*, 4179.

[149] Logue, M. W.; Teng, K. *J. Org. Chem.* **1982**, *47*, 2549.

[150] Andrianome, M.; Delmond, B. *J. Org. Chem.* **1988**, *53*, 542.

[151] Gaare, K.; Repstad, T.; Benneche, T.; Undheim, K. *Acta Chem. Scand.* **1993**, *47*, 57.

[152] Degl'Innocenti, A.; Dembech, P.; Mordini, A.; Ricci, A.; Seconi, G. *Synthesis* **1991**, 267.

[153] Baldwin, J. E.; Adlington, R. M.; Ramcharitar, S. H. *J. Chem. Soc., Chem. Commun.* **1991**, 940.

[154] Baldwin, J. E.; Adlington, R. M.; Ramcharitar, S. H. *Tetrahedron* **1992**, *48*, 2957.

[155] Linderman, R. J.; Graves, D. M.; Kwochka, W. R.; Ghannam, A. F.; Anklekar, T. V. *J. Am. Chem. Soc.* **1990**, *112*, 7438.

[156] Four, P.; Guibé, F. *J. Org. Chem.* **1981**, *46*, 4439.

[157] Balas, L.; Jousseaume, B.; Shin, H.; Verlhac, J.-B.; Wallian, F. *Organometallics* **1991**, *10*, 366.

[158] Jousseaume, B.; Kwon, H.; Verlhac, J.-B.; Denat, F.; Dubac, J. *Synlett* **1993**, 117.

[159] Adlington, R. M.; Baldwin, J. E.; Gansaeuer, A.; McCoull, W.; Russell, A. T. *J. Chem. Soc., Perkin Trans. 1* **1994**, 1697.

[160] Trost, B. M.; Keinan, E. *Tetrahedron Lett.* **1980**, *21*, 2591.

[161] Bumagin, N. A.; Kasatkin, A. N.; Beletskaya, I. P. *Dokl. Akad. Nauk SSSR* **1982**, *266*, 862; *Chem. Abstr.* **1982**, *98*, 143554.

[162] Kosugi, M.; Ohashi, K.; Akuzawa, K.; Kawazoe, T.; Sano, H.; Migita, T. *Chem. Lett.* **1987**, 1237.

[163] Del Valle, L.; Stille, J. K.; Hegedus, L. S. *J. Org. Chem.* **1990**, *55*, 3019.

[164] Tueting, D. R.; Echavarren, A. M.; Stille, J. K. *Tetrahedron* **1989**, *45*, 979.

[165] Keinan, E.; Peretz, M. *J. Org. Chem.* **1983**, *48*, 5302.

[166] Ni, Z.; Padwa, A. *Synlett* **1992**, 869.

[167] Boden, C.; Pattenden, G. *Synlett* **1994**, 181.

[168] Dangles, O.; Guibé, F.; Balavoine, G.; Lavielle, S.; Marquet, A. *J. Org. Chem.* **1987**, *52*, 4984.

[169] Azizian, H.; Eaborn, C.; Pidcock, A. *J. Organomet. Chem.* **1981**, *215*, 49.

[170] Palmisano, G.; Santagostino, M. *Tetrahedron* **1993**, *49*, 2533.

[171] Sustmann, R.; Lau, J.; Zipp, M. *Tetrahedron Lett.* **1986**, *27*, 5207.

[172] Rayner, C. M.; Astles, P. C.; Paquette, L. A. *J. Am. Chem. Soc.* **1992**, *114*, 3926.

[173] Hettrick, C. M.; Kling, J. K.; Scott, W. J. *J. Org. Chem.* **1991**, *56*, 1489.

[174] Marino, J. P.; Long, J. K. *J. Am. Chem. Soc.* **1988**, *110*, 7916.

[175] Scott, W. J.; McMurry, J. E. *Acc. Chem. Res.* **1988**, *21*, 47.

[176] Wulff, W. D.; Peterson, G. A.; Bauta, W. E.; Chan, K. S.; Faron, K. L.; Gilbertson, S. R.; Kaesler, R. W.; Yang, D. C.; Murray, C. K. *J. Org. Chem.* **1986**, *51*, 277.

[177] Gibbs, R. A.; Krishnan, U. *Tetrahedron Lett.* **1994**, *35*, 2509.

[178] Piers, E.; Friesen, R. W. *J. Chem. Soc., Chem. Commun.* **1988**, 125.

[179] Piers, E.; Friesen, R. W. *Can. J. Chem.* **1987**, *65*, 1681.

[180] Piers, E.; Llinas-Brunet, M. *J. Org. Chem.* **1989**, *54*, 1483.

[181] Piers, E.; Friesen, R. W.; Keay, B. A. *Tetrahedron* **1991**, *47*, 4555.

[182] Piers, E.; Friesen, R. W. *Can. J. Chem.* **1992**, *70*, 1204.

[183] Piers, E.; Llinas-Brunet, M.; Oballa, R. M. *Can. J. Chem.* **1993**, *71*, 1484.

[184] Piers, E.; Friesen, R. W. *J. Org. Chem.* **1986**, *51*, 3405.

[185] Stille, J. K.; Tanaka, M. *J. Am. Chem. Soc.* **1987**, *109*, 3785.

[186] Stille, J. K.; Su, H.; Hill, D. H.; Schneider, P.; Tanaka, M.; Morrison, D. L.; Hegedus, L. S. *Organometallics* **1991**, *10*, 1993.

[187] Moriarty, R. M.; Epa, W. R. *Tetrahedron Lett.* **1992**, *33*, 4095.

[188] Hinkle, R. J.; Poulter, G. T.; Stang, P. J. *J. Am. Chem. Soc.* **1993**, *115*, 11626.

[189] Echavarren, A. M.; Stille, J. K. *J. Am. Chem. Soc.* **1987**, *109*, 5478.

[190] Saá, J. M.; Martorell, G.; García-Raso, A. *J. Org. Chem.* **1992**, *57*, 678.

[191] Saá, J. M.; Martorell, G. *J. Org. Chem.* **1993**, *58*, 1963.

[192] Crisp, G. T.; Papadopoulos, S. *Aust. J. Chem.* **1988**, *41*, 1711.

[193] Edstrom, E. D.; Wei, Y. *J. Org. Chem.* **1994**, *59*, 6902.

[194] Crisp, G. T.; Papadopoulos, S. *Aust. J. Chem.* **1989**, *42*, 279.

[195] Robl, J. A. *Synthesis* **1991**, 56.

[196] Sandosham, J.; Undheim, K. *Heterocycles* **1994**, *37*, 501.

[197] Cook, G. K.; Hornback, W. J.; Jordan, C. L.; McDonald, J. H., III; Munroe, J. E. *J. Org. Chem.* **1989**, *54*, 5828.

[198] Rano, T. A.; Greenlee, M. L.; DiNinno, F. P. *Tetrahedron Lett.* **1990**, *31*, 2853.

[199] Crisp, G. T.; Flynn, B. L. *Tetrahedron Lett.* **1990**, *31*, 1347.

[200] Peña, M. R.; Stille, J. K. *J. Am. Chem. Soc.* **1989**, *111*, 5417.

[201] Chen, Q.-Y.; He, Y.-B.; Yang, Z.-Y. *Youji Huaxue* **1987**, 474; *Chem. Abstr.* **1987**, *109*, 109940.

[202] Badone, D.; Cecchi, R.; Guzzi, U. *J. Org. Chem.* **1992**, *57*, 6321.

[203] Roth, G. P.; Fuller, C. E. *J. Org. Chem.* **1991**, *56*, 3493.

[204] Kikukawa, K.; Kono, K.; Wada, F.; Matsuda, T. *J. Org. Chem.* **1983**, *48*, 1333.

[205] Brigas, A. F.; Johnstone, R. A. *J. Chem. Soc., Chem. Commun.* **1994**, 1923.

[206] Bumagin, N. A.; Sukhomlinova, A. N.; Igushkina, S. O.; Banchikov, A. N.; Tolstaya, T. P.; Beletskaya, I. P. *Bull. Acad. Sci. USSR, Div. Chem. Sci.* **1992**, *42*, 2128; not in *Chem. Abstr.*

[207] Bhatt, R. K.; Shin, D. S.; Falck, J. R.; Mioskowski, C. *Tetrahedron Lett.* **1992**, *33*, 4885.

[208] Simpson, J. H.; Stille, J. K. *J. Org. Chem.* **1985**, *50*, 1759.

[209] Pri-Bar, I.; Pearlman, P. S.; Stille, J. K. *J. Org. Chem.* **1983**, *48*, 4629.

[210] Matsubara, S.; Mitani, M.; Utimoto, K. *Tetrahedron Lett.* **1987**, *28*, 5857.

[211] Kobayashi, T.; Sakakura, T.; Tanaka, M. *Tetrahedron Lett.* **1985**, *26*, 3463.

[212] Ito, Y.; Inouye, M.; Yokota, H.; Murakami, M. *J. Org. Chem.* **1990**, *55*, 2567.

[213] Bhatt, R. K.; Chauhan, K.; Wheelan, P.; Murphy, R. C.; Falck, J. R. *J. Am. Chem. Soc.* **1994**, *116*, 5050.

[214] Hollingworth, G. J.; Sweeney, J. B. *Synlett* **1993**, 463.

[215] Beaudet, I.; Parrain, J. L.; Quintard, J. P. *Tetrahedron Lett.* **1992**, *33*, 3647.

[216] Shair, M. D.; Yoon, T.; Danishefsky, S. J. *J. Org. Chem.* **1994**, *59*, 3755.

[217] Gilbert, A. M.; Wulff, W. D. *J. Am. Chem. Soc.* **1994**, *116*, 7449.

[218] Bunz, U. H. F.; Enkelmann, V.; Räder, J. *Organometallics* **1993**, *12*, 4745.

[219] Bunz, U. H. F.; Enkelmann, V. *Angew. Chem., Int. Ed. Engl.* **1993**, *32*, 1653.

[220] Bunz, U. H. F.; Enkelmann, V. *Organometallics* **1994**, *13*, 3823.

[221] Jevnaker, N.; Benneche, T.; Undheim, K. *Acta Chem. Scand.* **1993**, *47*, 406.

[222] Uemura, M.; Nishimura, H.; Hayashi, T. *Tetrahedron Lett.* **1993**, *34*, 107.

[223] Wiegelmann, J. E. C.; Bunz, U. H. F. *Organometallics* **1993**, *12*, 3792.

[224] Wiegelmann, J. E. C.; Bunz, U. H. F.; Schiel, P. *Organometallics* **1994**, *13*, 4649.

[225] Wright, M. E.; Pulley, S. R. *Macromolecules* **1989**, *22*, 2542.

[226] Lo Sterzo, C.; Miller, M. M.; Stille, J. K. *Organometallics* **1989**, *8*, 2331.

[227] Scott, W. J. *J. Chem. Soc., Chem. Commun.* **1987**, *23*, 1755.

[228] Rolland, H.; Potin, P.; Majoral, J.-P.; Bertrand, G. *Tetrahedron Lett.* **1992**, *33*, 8095.

[229] Labadie, S. S. *J. Org. Chem.* **1989**, *54*, 2496.

[230] Lo Sterzo, C. *J. Chem. Soc., Dalton Trans.* **1992**, 1989.

[231] Crescenzi, R.; Lo Sterzo, C. *Organometallics* **1992**, *11*, 4301.

[232] Deeter, G. A.; Moore, J. S. *Organometallics* **1993**, *26*, 2535.

[233] Kosugi, M.; Sumiya, T.; Ohhashi, K.; Sano, H.; Migita, T. *Chem. Lett.* **1985**, 997.

[234] Kosugi, M.; Sumiya, T.; Ogata, T.; Sano, H.; Migita, T. *Chem. Lett.* **1984**, 1225.

[235] Kosugi, M.; Ishiguro, M.; Negishi, Y.; Sano, H.; Migita, T. *Chem. Lett.* **1984**, 1511.

[236] Kosugi, M.; Negishi, Y.; Kameyama, M.; Migita, T. *Bull. Chem. Soc. Jpn.* **1985**, *58*, 3383.

[237] Kosugi, M.; Suzuki, M.; Hagiwara, I.; Goto, K.; Saitoh, K.; Migita, T. *Chem. Lett.* **1982**, 939.

[238] Kosugi, M.; Hagawara, I.; Sumiya, T.; Migita, T. *J. Chem. Soc., Chem. Commun.* **1983**, 344.

[239] Kosugi, M.; Hagiwara, I.; Migita, T. *Chem. Lett.* **1983**, 839.

[240] Kosugi, M.; Hagiwara, I.; Sumiya, T.; Migita, T. *Bull. Chem. Soc. Jpn.* **1984**, *57*, 242.

[241] Kuwajima, I.; Urabe, H. *J. Am. Chem. Soc.* **1982**, *104*, 6831.

[242] Ye, J.; Bhatt, R. K.; Falck, J. R. *Tetrahedron Lett.* **1993**, *34*, 8007.

[243] Ye, J.; Bhatt, R. K.; Falck, J. R. *J. Am. Chem. Soc.* **1994**, *116*, 1.

[244] Nativi, C.; Ricci, A.; Taddei, M. *Tetrahedron Lett.* **1990**, *31*, 2637.

[244a] Crisp, G. T.; Glink, P. T. *Tetrahedron* **1994**, *50*, 2623.

[245] Busacca, C. A.; Swestock, J.; Johnson, R. E.; Bailey, T. R.; Musza, L.; Roger, C. A. *J. Org. Chem.* **1994**, *59*, 7553.

[246] Levin, J. I. *Tetrahedron Lett.* **1993**, *34*, 6211.

[247] Acuña, A. C.; Zapata, A. *Synth. Commun.* **1988**, *18*, 1133.

[248] Acuña, A. C.; Zapata, A. *Synth. Commun.* **1988**, *18*, 1125.

[249] Kikukawa, K.; Umekawa, H.; Matsuda, T. *J. Organomet. Chem.* **1986**, *311*, C44.

[250] Renaldo, A. F.; Ito, H. *Synth. Commun.* **1987**, *17*, 1823.

[251] Cummins, C. H.; Gordon, E. J. *Tetrahedron Lett.* **1994**, *35*, 8133.

[252] Takle, A.; Kocienski, P. *Tetrahedron* **1990**, *46*, 4503.

[253] Pimm, A.; Kocienski, P.; Street, S. D. A. *Synlett* **1992**, 886.

[254] Mitchell, T. N.; Reimann, W. *Organometallics* **1986**, *5*, 1991.

[255] Chenard, B. L.; Van Zyl, C. M.; Sanderson, D. R. *Tetrahedron Lett.* **1986**, *27*, 2801.

[256] Mitchell, T. N.; Wickenkamp, R.; Amamria, A.; Dicke, R.; Schneider, U. *J. Org. Chem.* **1987**, *52*, 4868.

[257] Kiely, J. S.; Laborde, E.; Lesheski, L. E.; Bucsh, R. A. *J. Heterocycl. Chem.* **1991**, *28*, 1581.

[258] Laborde, E.; Lesheski, L. E.; Kiely, J. S. *Tetrahedron Lett.* **1990**, *31*, 1837.

[259] Houpis, I. N.; DiMichele, L.; Molina, A. *Synlett* **1993**, 365.

[260] Farina, V.; Hauck, S. I. *J. Org. Chem.* **1991**, *56*, 4317.

[261] Hollingworth, G. J.; Sweeney, J. B. *Tetrahedron Lett.* **1992**, *33*, 7049.

[262] Xu, Y.; Jin, F.; Huang, W. *J. Org. Chem.* **1994**, *59*, 2638.

[263] Matthews, D. P.; Gross, R. S.; McCarthy, J. R. *Tetrahedron Lett.* **1994**, *35*, 1027.

[264] Matthews, D. P.; Wadi, P. P.; Sabol, J. S.; McCarthy, J. R. *Tetrahedron Lett.* **1994**, *35*, 5177.

[265] Sorokina, R. S.; Rybakova, L. F.; Kalinovskii, I. O.; Chernoplekova, V. A.; Beletskaya, I. P. *J. Org. Chem. USSR* **1982**, *18*, 2180.

[266] Sorokina, R. S.; Rybakova, L. F.; Kalinovskii, I. O.; Beletskaya, I. P. *Bull. Acad. Sci. USSR, Div. Chem. Sci.* **1985**, *34*, 1506; not in *Chem. Abstr.*

[267] Liebeskind, L. S.; Stone, G. B.; Zhang, S. *J. Org. Chem.* **1994**, *59*, 7917.

[268] Liebeskind, L. S.; Yu, M. S.; Fengl, R. W. *J. Org. Chem.* **1993**, *58*, 3543.

[269] Kosugi, M.; Sumiya, T.; Obara, Y.; Suzuki, M.; Sano, H.; Migita, T. *Bull. Chem. Soc. Jpn.* **1987**, *60*, 767.

[270] Kwon, H. B.; McKee, B. H.; Stille, J. K. *J. Org. Chem.* **1990**, *55*, 3114.

[271] Blanchot, V.; Fétizon, M.; Hanna, I. *Synthesis* **1990**, 755.

[272] Sakamoto, T.; Kondo, Y.; Yasuhara, A.; Yamanaka, H. *Heterocycles* **1990**, *31*, 219.

[273] Sakamoto, T.; Kondo, Y.; Yasuhara, A.; Yamanaka, H. *Tetrahedron* **1991**, *47*, 1877.

[274] Sakamoto, T.; Satoh, C.; Kondo, Y.; Yamanaka, H. *Chem. Pharm. Bull.* **1993**, *41*, 81.

[275] Aidhen, I. S.; Braslau, R. *Synth. Commun.* **1994**, *24*, 789.

[276] Badone, D.; Cardamone, R.; Guzzi, U. *Tetrahedron Lett.* **1994**, *35*, 5477.

[277] Nativi, C.; Taddei, M.; Mann, A. *Tetrahedron* **1989**, *45*, 1131.

[278] Lipshutz, B. H.; Alami, M. *Tetrahedron Lett.* **1993**, *34*, 1433.

[279] Haack, R. A.; Penning, T. D.; Djuric, S. W.; Dziuba, J. A. *Tetrahedron Lett.* **1988**, *29*, 2783.

[280] Gómez-Bengoa, E.; Echavarren, A. M. *J. Org. Chem.* **1991**, *56*, 3497.

[281] Rai, R.; Aubrecht, K. B.; Collum, D. B. *Tetrahedron Lett.* **1995**, *36*, 3111.

[282] Roshchin, A. I.; Bumagin, N. A.; Beletskaya, I. P. *Tetrahedron Lett.* **1995**, *36*, 125.

[283] Garcia Martínez, A.; J., O. B.; de Fresno Cerezo, A.; Subramanian, L. R. *Synlett* **1994**, 1047.

[284] Yamamoto, Y.; Yanagi, A. *Chem. Pharm. Bull.* **1982**, *30*, 2003.

[285] Yamamoto, Y.; Yanagi, A. *Heterocycles* **1982**, *19*, 41.

[286] Bailey, T. R. *Tetrahedron Lett.* **1986**, *27*, 4407.

[287] Yang, Y.; Wong, H. N. C. *Tetrahedron* **1994**, *50*, 9583.

[288] Bailey, T. R. *Synthesis* **1991**, 242.

[289] Palmisano, G.; Santagostino, M. *Helv. Chim. Acta* **1993**, *76*, 2356.

[290] Fukuyama, T.; Chen, X.; Peng, G. *J. Am. Chem. Soc.* **1994**, *116*, 3127.

[291] Ciattini, P. G.; Morera, E.; Ortar, G. *Tetrahedron Lett.* **1994**, *35*, 2405.

[292] Kondo, Y.; Uchiyama, D.; Sakamoto, T.; Yamanaka, H. *Tetrahedron Lett.* **1989**, *30*, 4249.

[293] Gothelf, K.; Thomsen, I. B.; Torssell, K. B. G. *Acta Chem. Scand.* **1992**, *46*, 494.

[294] Aoyagi, Y.; Inoue, A.; Koizumi, I.; Hashimoto, R.; Tokunaga, K.; Gohma, K.; Komatsu, J.; Sekine, K.; Miyafuji, A.; Kunoh, J.; Honma, R.; Akita, Y.; Ohta, A. *Heterocycles* **1992**, *33*, 257.

[295] Liebeskind, L. S.; Wang, J. *J. Org. Chem.* **1993**, *58*, 3550.

[296] Pearce, B. C. *Synth. Commun.* **1992**, *22*, 1627.

[297] Dubois, E.; Beau, J.-M. *Tetrahedron Lett.* **1990**, *31*, 5165.

[298] Friesen, R. W.; Sturino, C. F. *J. Org. Chem.* **1990**, *55*, 5808.

[299] Friesen, R. W.; Sturino, C. F. *J. Org. Chem.* **1990**, *55*, 2572.

[300] Friesen, R. W.; Loo, R. W.; Sturino, C. F. *Can. J. Chem.* **1994**, *72*, 1262.

[301] Zhang, H.-C.; Brakta, M.; Daves, G. D., Jr. *Tetrahedron Lett.* **1993**, *34*, 1571.

[302] Sakamoto, T.; Yasuhara, A.; Kondo, Y.; Yamanaka, H. *Synlett* **1992**, 502.

[303] Farina, V. *Comprehensive Organometallic Chemistry* **1995**, *12*, 161.

[304] Andrianome, M.; Häberle, K.; Delmond, B. *Tetrahedron* **1989**, *45*, 1079.

[305] Verlhac, J.-B.; Pereyre, M.; Quintard, J.-P. *Tetrahedron* **1990**, *46*, 6399.

[306] Yamamoto, Y.; Hatsuya, S.; Yamada, J.-i. *J. Org. Chem.* **1990**, *55*, 3118.

[307] Verlhac, J.-B.; Chanson, E.; Jousseaume, B.; Quintard, J.-P. *Tetrahedron Lett.* **1985**, *26*, 6075.

[308] Bumagin, N. A.; Gulevich, Y. V.; Beletskaya, I. P. *J. Organomet. Chem.* **1985**, *282*, 421.

[309] Mitchell, T. N.; Kwetkat, K. *Synthesis* **1990**, 1001.

[310] Kosugi, M.; Shimizu, K.; Ohtani, A.; Migita, T. *Chem. Lett.* **1981**, 829.

[311] Kosugi, M.; Ohya, T.; Migita, T. *Bull. Chem. Soc. Jpn.* **1983**, *56*, 3855.

[312] Bumagin, N. A.; Bumagina, I. G.; Beletskaya, I. P. *Dokl. Akad. Nauk SSSR* **1984**, *274*, 1103; *Chem. Abstr.* **1984**, *101*, 72854.

[313] Bumagin, N. A.; Gulevich, Y. V.; Beletskaya, I. P. *Bull. Acad. Sci. USSR, Div. Chem. Sci.* **1984**, *33*, 1044; not in *Chem. Abstr.*

[314] Bumagin, N. A.; Kasatkin, A. N.; Beletskaya, I. P. *Bull. Acad. Sci. USSR, Div. Chem. Sci.* **1984**, *33*, 588; not in *Chem. Abstr.*

[315] Bumagin, N. A.; Gulevich, Y. V.; Artamkina, G. A.; Beletskaya, I. P. *Bull. Acad. Sci. USSR, Div. Chem. Sci.* **1984**, *33*, 1098; not in *Chem. Abstr.*

[316] Kosugi, M.; Kameyama, M.; Sano, H.; Migita, T. *Nippon Kagaku Kaishi* **1985**, *3*, 547; *Chem. Abstr.* **1985**, *104*, 129990.

[317] Carpita, A.; Rossi, R.; Scamuzzi, B. *Tetrahedron Lett.* **1989**, *30*, 2699.

[318] Kosugi, M.; Ogata, T.; Terada, M.; Sano, H.; Migita, T. *Bull. Chem. Soc. Jpn.* **1985**, *58*, 3657.

[319] Jixiang, C.; Crisp, G. T. *Synth. Commun.* **1992**, *22*, 683.

[320] Lebedev, S. A.; Starosel'skaya, L. F.; Shifrina, R. R.; Beletskaya, I. P. *Bull. Acad. Sci. USSR, Div. Chem. Sci.* **1983**, *32*, 597; not in *Chem. Abstr.*

[321] Tunney, S. E.; Stille, J. K. *J. Org. Chem.* **1987**, *52*, 748.

[322] Keinan, E.; Sahai, M.; Roth, Z.; Nudelman, A.; Herzig, J. *J. Org. Chem.* **1985**, *50*, 3558.

[323] Tanaka, M. *Tetrahedron Lett.* **1979**, *28*, 2601.

[324] Goure, W. F.; Wright, M. E.; Davis, P. D.; Labadie, S. S.; Stille, J. K. *J. Am. Chem. Soc.* **1984**, *106*, 6417.

[325] Baillargeon, V. P.; Stille, J. K. *J. Am. Chem. Soc.* **1983**, *105*, 7175.

[326] Bumagin, N. A.; Bumagina, I. G.; Kashin, A. N.; Beletskaya, I. P. *Dokl. Akad. Nauk SSSR* **1981**, *261*, 1141; *Chem. Abstr.* **1981**, *96*, 104426.

[327] Davies, S. G.; Pyatt, D.; Thomson, C. *J. Organomet. Chem.* **1990**, *387*, 381.

[328] Knight, S. D.; Overman, L. E.; Pairaudeau, G. *J. Am. Chem. Soc.* **1993**, *115*, 9293.

[329] Bumagin, N. A.; Gulevich, Y. V.; Beletskaya, I. P. *Bull. Acad. Sci. USSR, Div. Chem. Sci.* **1984**, *33*, 879; not in *Chem. Abstr.*

[330] Bumagin, N. A.; Gulevich, Y. V.; Beletskaya, I. P. *J. Organomet. Chem.* **1985**, *285*, 415.

[331] Baillargeon, V. P.; Stille, J. K. *J. Am. Chem. Soc.* **1986**, *108*, 452.

[332] Cowell, A.; Stille, J. K. *J. Am. Chem. Soc.* **1980**, *102*, 4193.

[333] Merrifield, J. H.; Godschalx, J. P.; Stille, J. K. *Organometallics* **1984**, *3*, 1108.

[334] Crisp, G. T.; Scott, W. J.; Stille, J. K. *J. Am. Chem. Soc.* **1984**, *106*, 7500.

[335] Gyorkos, A. C.; Stille, J. K.; Hegedus, L. S. *J. Am. Chem. Soc.* **1990**, *112*, 8465.

[336] Echavarren, A. M.; Stille, J. K. *J. Am. Chem. Soc.* **1988**, *110*, 1557.

[337] Kobayashi, T.; Tanaka, M. *J. Organomet. Chem.* **1981**, *205*, C27.

[338] Masters, J. J.; Hegedus, L. S. *J. Org. Chem.* **1993**, *58*, 4547.

[339] Masters, J. J.; Hegedus, L. S.; Tamariz, J. *J. Org. Chem.* **1991**, *56*, 5666.

[340] Kikukawa, K.; Idemoto, T.; Katayama, A.; Kono, K.; Wada, F.; Matsuda, T. *J. Chem. Soc., Perkin Trans. 1* **1987**, 1511.

[341] Kosugi, M.; Tamura, H.; Sano, H.; Migita, T. *Chem. Lett.* **1987**, 193.

[342] Kosugi, M.; Tamura, H.; Sano, H.; Migita, T. *Tetrahedron* **1989**, *45*, 961.

[343] Oda, H.; Ito, K.; Kosugi, M.; Migita, T. *Chem. Lett.* **1994**, *8*, 1443.

[344] Larock, R. C.; Lee, N. H. *J. Org. Chem.* **1991**, *56*, 6253.

[345] Oppolzer, W.; Ruiz-Montes, J. *Helv. Chim. Acta* **1993**, *76*, 1266.

[346] Grigg, R.; Sukirthalingam, S.; Sridharan, V. *Tetrahedron Lett.* **1991**, *32*, 2545.

[347] Burns, B.; Grigg, R.; Ratananukul, P.; Sridharan, V.; Stevenson, P.; Sukirthalingam, S.; Worakun, T. *Tetrahedron Lett.* **1988**, *29*, 5565.

[348] Wang, R.-T.; Chou, F.-L.; Luo, F.-T. *J. Org. Chem.* **1990**, *55*, 4846.

[349] Luo, F.-T.; Wang, R.-T. *Tetrahedron Lett.* **1991**, *32*, 7703.

[350] Negishi, E.-i.; Noda, Y.; Lamaty, F.; Vawter, E. J. *Tetrahedron Lett.* **1990**, *31*, 4393.

[351] Nuss, J. M.; Levine, B. H.; Rennels, R. A.; Heravi, M. M. *Tetrahedron Lett.* **1991**, *32*, 5243.

[352] Nuss, J. M.; Rennels, R. A.; Levine, B. H. *J. Am. Chem. Soc.* **1993**, *115*, 6991.

[353] Torii, S.; Okumoto, H.; Tadokoro, T.; Nishimura, A.; Rashid, M. A. *Tetrahedron Lett.* **1993**, *34*, 2139.

[354] Nuss, J. M.; Murphy, M. M.; Rennels, R. A.; Heravi, M. H.; Mohr, B. J. *Tetrahedron Lett.* **1993**, *34*, 3079.

[355] Grigg, R.; Redpath, J.; Sridharan, V.; Wilson, D. *Tetrahedron Lett.* **1994**, *35*, 4429.

[356] Kosugi, M.; Sakaya, T.; Ogawa, S.; Migita, T. *Bull. Chem. Soc. Jpn.* **1993**, *66*, 3058.

[357] Ikeda, S.-i.; Cui, D.-M.; Sato, Y. *J. Org. Chem.* **1994**, *59*, 6877.

[358] Barry, J.; Kodadek, T. *Tetrahedron Lett.* **1994**, *35*, 2465.

[359] Krysan, D. J.; Gurski, A.; Liebeskind, L. S. *J. Am. Chem. Soc.* **1992**, *114*, 1412.

[360] Edwards, J. P.; Krysan, D. J.; Liebeskind, L. S. *J. Am. Chem. Soc.* **1993**, *115*, 9868.

[361] Edwards, J. P.; Krysan, D. J.; Liebeskind, L. S. *J. Org. Chem.* **1993**, *58*, 3942.

[362] Birchler, A. G.; Liu, F.; Liebeskind, L. S. *J. Org. Chem.* **1994**, *59*, 7737.

[363] Liebeskind, L. S.; Wang, J. *Tetrahedron* **1993**, *49*, 5461.

[364] Chatani, N.; Amishiro, N.; Murai, S. *J. Am. Chem. Soc.* **1991**, *113*, 7778.

[365] Tolstikov, G. A.; Miftakhov, M. S.; Danilova, N. A.; Vel'der, Y. L.; Spirikhin, L. V. *Synthesis* **1989**, 625.

[366] Tolstikov, G. A.; Miftakhov, M. S.; Danilova, N. A.; Vel'der, Y. L.; Spirikhin, L. V. *Synthesis* **1989**, 633.

[367] Bumagin, N. A.; Ponomarev, A. B.; Beletskaya, I. P. *J. Org. Chem. USSR* **1988**, *23*, 1222.

[368] van Asselt, R.; Elsevier, C. J. *Organometallics* **1994**, *13*, 1972.

[369] Brehm, E. C.; Stille, J. K.; Meyers, A. I. *Organometallics* **1992**, *11*, 938.

[370] Tamayo, N.; Echavarren, A. M.; Paredes, M. C.; Fariña, F.; Noheda, P. *Tetrahedron Lett.* **1990**, *31*, 5189.

[371] Keay, B. A.; Bontront, J. L. *J. Can. J. Chem.* **1991**, *69*, 1326.

[372] Tius, M. A.; Gu, X.; Gomez-Galeno, J. *J. Am. Chem. Soc.* **1990**, *112*, 8188.

[373] Stork, G.; Isaacs, R. C. A. *J. Am. Chem. Soc.* **1990**, *112*, 7399.

[374] Flynn, B. L.; Macolino, V.; Crisp, G. T. *Nucleosides Nucleotides* **1991**, *10*, 763.

[375] Crisp, G. T.; Glink, P. T. *Tetrahedron* **1994**, *50*, 3213.

[376] Kuhn, H.; Neumann, W. *Synlett* **1994**, 123.

[377] Kong, K.-C.; Cheng, C.-H. *J. Am. Chem. Soc.* **1991**, *113*, 6313.

[378] Sagelstein, B. E.; Butler, T. W.; Chenard, B. L. *J. Org. Chem.* **1995**, *60*, 12.

[379] Martorell, G.; Garcia-Raso, A.; Saá, J. M. *Tetrahedron Lett.* **1990**, *31*, 2357.

[380] Renaldo, A. F.; Labadie, J. W.; Stille, J. K. *Org. Synth.* **1989**, *67*, 86.

[381] Crisp, G. T.; Bubner, T. P. *Synth. Commun.* **1990**, *20*, 1665.

[382] Lee, E.; Hur, C. U.; Jeong, Y. C.; Rhee, Y. H.; Chang, M. H. *J. Chem. Soc., Chem. Commun.* **1991**, 1314.

[383] Tilley, J. W.; Sarabu, R.; Wagner, R.; Mulkerins, K. *J. Org. Chem.* **1990**, *55*, 906.

[384] Gothelf, K. V.; Torssell, K. B. G. *Acta Chem. Scand.* **1994**, *48*, 165.

[385] Zapata, A. J.; Ruíz, J. *J. Organomet. Chem.* **1994**, *479*, C6.

[386] Negishi, E.-i.; Owczarczyk, Z. *Tetrahedron Lett.* **1991**, *32*, 6683.

[387] Stracker, E. C.; Zweifel, G. *Tetrahedron Lett.* **1991**, *32*, 3329.

[388] Friesen, R. W.; Loo, R. W. *J. Org. Chem.* **1991**, *56*, 4821.

[389] Tius, M.; Gomez-Galeno, J.; Gu, X.-Q.; Zaidi, J. H. *J. Am. Chem. Soc.* **1991**, *113*, 5775.

[390] Lamba, J. J. S.; Tour, J. M. *J. Am. Chem. Soc.* **1994**, *116*, 11723.

[391] Clough, J. M.; Mann, I. S.; Widdowson, D. A. *Tetrahedron Lett.* **1987**, *28*, 2645.

[392] Mann, I. S.; Widdowson, D. A.; Clough, J. M. *Tetrahedron* **1991**, *47*, 7981.

[393] Ishiyama, T.; Miyaura, N.; Suzuki, A. *Synlett* **1991**, 687.

[394] Krigman, M. R.; Silverman, A. P. *Neurotoxicology* **1984**, *5*, 129.

[395] Pereyre, M.; Quintard, J.-P.; Rahm, A. *Tin in Organic Synthesis*; Butterworths: London, 1987.

[396] Jones, K.; Lappert, M. F. *J. Organomet. Chem.* **1965**, *3*, 295.

[397] Farina, V. *J. Org. Chem.* **1991**, *56*, 4895.

[398] Stang, P. J.; Treptow, W. *Synthesis* **1980**, 283.

[399] Stang, P. J.; Fox, T. E. *Synthesis* **1979**, 438.

[400] Scott, W. J.; McMurry, J. E. *Tetrahedron Lett.* **1983**, *24*, 979.

[401] Crisp, G. T.; Scott, W. J. *Synthesis* **1985**, 335.

[402] Stang, P. J.; Summerville, R. *J. Am. Chem. Soc.* **1969**, *91*, 4600.

[403] Summerville, R. H.; Senkler, C. A.; Schleyer, P. v. R.; Dueber, T. E.; Stang, P. J. *J. Am. Chem. Soc.* **1974**, *96*, 1100.

[404] Hendrickson, J. B.; Bergeron, R. *Tetrahedron Lett.* **1973**, *14*, 4607.

[405] Stang, P. J.; Hanack, M.; Subramanian, L. R. *Synthesis* **1982**, 85.

[406] Ritter, K. *Synthesis* **1993**, 735.

[407] Coulson, D. R. *Inorg. Synth.* **1972**, *13*, 121.

[408] Takahashi, I.; Ito, T.; Sakai, S.; Ishii, Y. *J. Chem. Soc., Chem. Commun.* **1970**, 1065.

[409] Kharash, M. S.; Seyler, R. C.; Mayo, F. R. *J. Am. Chem. Soc.* **1938**, *60*, 882.

[410] Schoenberg, A.; Bartoletti, I.; Heck, R. F. *J. Org. Chem.* **1974**, *39*, 3318.

[411] Feltham, R. D.; Elbaze, G.; Ortega, R.; Eck, C.; Dubrawski, J. *Inorg. Chem.* **1985**, *24*, 1503.

[412] Fitton, P.; McKeon, J. E.; Ream, B. C. *J. Chem. Soc., Chem. Commun.* **1969**, 370.

[413] Hayashi, T.; Konishi, M.; Kobori, Y.; Kumada, M.; Higuchi, T.; Hirotsu, K. *J. Am. Chem. Soc.* **1984**, *106*, 158.

[414] Dent, W. T.; Long, R.; Wilkinson, A. J. *J. Chem. Soc.* **1964**, 1585.

[415] van Asselt, R.; Elsevier, C. J. *Tetrahedron* **1994**, *50*, 323.

[416] Wright, S. W.; Harris, R. R.; Collins, R. J.; Corbett, R. L.; Green, A. M.; Wadman, E. A.; Batt, D. G. *J. Med. Chem.* **1992**, *35*, 3148.

[417] Mori, M.; Kaneta, N.; Shibasaki, M. *J. Org. Chem.* **1991**, *56*, 3486.

[418] Patel, H. K.; Kilburn, J. D.; Langley, G. J.; Edwards, P. D.; Mitchell, T.; Southgate, R. *Tetrahedron Lett.* **1994**, *35*, 481.

[419] Schwede, W.; Cleve, A.; Neef, G.; Ottow, E.; Stöckemann, K.; Wiechert, R. *Steroids* **1994**, *59*, 176.

[420] Stille, J. K.; Echavarren, A. M.; Williams, R. M.; Hendrix, J. A. *Org. Synth.* **1993**, *71*, 97.

[420a] Curran, D. P.; Chang, C.-T. *J. Org. Chem.* **1989**, *54*, 3140.

[421] Scott, W. J.; Crisp, G. T.; Stille, J. K. *Org. Synth.* **1989**, *68*, 116.

[422] Untiedt, S.; de Meijere, A. *Chem. Ber.* **1954**, *127*, 1511.

[423] Dubois, E.; Beau, J.-M. *J. Chem. Soc., Chem. Commun.* **1990**, *17*, 1191.

[424] Dubois, E.; Beau, J.-M. *Carbohydr. Res.* **1992**, *228*, 103.

[425] Labadie, S. S.; Teng, E. *J. Org. Chem.* **1994**, *59*, 4250.

[426] Kosugi, M.; Fukiage, A.; Takayanagi, M.; Sano, H.; Migita, T.; Satoh, M. *Chem. Lett.* **1988**, 1351.

[427] Yamamoto, Y.; Hatsuya, S.; Yamada, J.-i. *J. Chem. Soc., Chem. Commun.* **1988**, 86.

[428] Rubin, Y.; Knobler, C. B.; Diederich, F. *J. Am. Chem. Soc.* **1990**, *112*, 1607.

[429] MacLeod, D.; Moorcroft, D.; Quayle, P.; Dorrity, M. R. J.; Malone, J. F.; Davies, G. M. *Tetrahedron Lett.* **1990**, *31*, 6077.

[430] Duchene, A.; Abarbri, M.; Parrain, J.-L.; Kitamura, M.; Noyori, R. *Synlett* **1994**, *7*, 524.

[431] Hatanaka, Y.; Matsui, K.; Hiyama, T. *Tetrahedron Lett.* **1989**, *30*, 2403.

[432] Yang, Y.; Wong, H. N. C. *J. Chem. Soc., Chem. Commun.* **1992**, 1723.

[433] Keenan, R. M.; Weinstock, J.; Finkelstein, J. A.; Franz, R. G.; Gaitanopoulos, D. E.; Girard, G. R.; Hill, D. T.; Morgan, T. M.; Samanen, J. M.; Hempel, J.; Eggleston, D. S.; Aiyar, N.; Griffin, E.; Olhstein, E. H.; Stack, E. J.; Weidley, E. F.; Edwards, R. *J. Med. Chem.* **1992**, *35*, 3858.

[434] Rossi, R.; Carpita, A.; Ciofalo, M.; Lippolis, V. *Tetrahedron* **1991**, *47*, 8443.

[435] Bellina, F.; Carpita, A.; De Santis, M.; Rossi, R. *Tetrahedron* **1994**, *50*, 12029.

[436] Houpis, I. N. *Tetrahedron Lett.* **1991**, *32*, 6675.

[437] Lindsay, C. M.; Widdowson, D. A. *J. Chem. Soc., Perkin Trans. 1* **1988**, 569.

[438] Takayama, H.; Suzuki, T. *J. Chem. Soc., Chem. Commun.* **1988**, 1044.

[439] Casson, S.; Kocienski, P. *J. Chem. Soc., Perkin Trans. 1* **1994**, 1187.

[440] Adam, W.; Klug, P. *J. Org. Chem.* **1994**, *59*, 2695.

[441] Férézou, J. P.; Julia, M.; Li, Y.; Liu, L. W.; Pancrazi, A. *Synlett* **1991**, 53.

[442] Sharma, S.; Oehlschlager, A. C. *J. Org. Chem.* **1989**, *54*, 5064.

[443] Kiehl, A.; Eberhardt, A.; Adam, M.; Enkelmann, V.; Müllen, K. *Angew. Chem., Int. Ed. Engl.* **1992**, *31*, 1588.

[444] Scott, W. J.; Crisp, G. T.; Stille, J. K. *J. Am. Chem. Soc.* **1984**, *106*, 4630.

[445] Lin, H.-S.; Rampersaud, A. A.; Zimmerman, K.; Steinberg, M. I.; Boyd, D. B. *J. Med. Chem.* **1992**, *35*, 2658.

[446] Bellina, F.; Carpita, A.; Ciucci, D.; De Santis, M.; Rossi, R. *Tetrahedron* **1993**, *49*, 4677.

[447] Ostwald, R.; Chavant, P.-Y.; Stadtmüller, H.; Knochel, P. *J. Org. Chem.* **1994**, *59*, 4143.

[448] Farina, V.; Roth, G. P. *Tetrahedron Lett.* **1991**, *32*, 4243.

[449] Wender, P. A.; Tebbe, M. J. *Synthesis* **1991**, 1089.

[450] Boyd, D. R.; Hand, M. V.; Sharma, N. D.; Chima, J.; Dalton, H.; Sheldrake, G. N. *J. Chem. Soc., Chem. Commun.* **1991**, 1630.

[451] Pearson, A. J.; Holden, M. S. *J. Organomet. Chem.* **1990**, *383*, 307.

[452] Lee, J.; Snyder, J. K. *J. Org. Chem.* **1990**, *55*, 4995.

[453] Baker, S. R.; Roth, G. P.; Sapino, C. *Synth. Commun.* **1990**, *20*, 2185.

[454] Niwa, H.; Watanabe, M.; Inagaki, H.; Yamada, K. *Tetrahedron* **1994**, *50*, 7385.

[455] Paterson, I.; Gardner, M.; Banks, B. J. *Tetrahedron* **1989**, *45*, 5283.

[456] Banwell, M. G.; Collis, M. P.; Crisp, G. T.; Lambert, J. N.; Reum, M. E.; Scoble, J. A. *J. Chem. Soc., Chem. Commun.* **1989**, 616.

[457] Verlhac, J.-B.; Pereyre, M.; Shin, H. *Organometallics* **1991**, *10*, 3007.

[458] Sandosham, J.; Undheim, K. *Tetrahedron* **1994**, *50*, 275.

[459] Arukwe, J.; Benneche, T.; Undheim, K. *J. Chem. Soc., Perkin Trans. 1* **1989**, 255.

[460] Kosugi, M.; Ogata, T.; Terada, M.; Sano, H.; Migita, T. *Bull. Chem. Soc. Jpn.* **1985**, *58*, 3657.

[461] Roth, G. P.; Farina, V.; Liebeskind, L. S.; Pena-Cabrera, E. *Tetrahedron Letters* **1995**, *36*, 2191.

[462] Corriu, R. J. P.; Geng, B.; Moreau, J. J. E. *J. Org. Chem.* **1993**, *58*, 1443.

[463] Bumagin, N. A.; Bumagina, I. G.; Beletskaya, I. P. *Dokl. Chem.* **1983**, 333; not in *Chem. Abstr.*

[464] Corriu, R. J. P.; Bolin, G.; Moreau, J. J. E. *Bull. Soc. Chim. Fr.* **1993**, *130*, 273.

[465] Capella, L.; Degl'Innocenti, A.; Mordini, A.; Reginato, G.; Ricci, A.; Seconi, G. *Synthesis* **1991**, 1201.

[466] Kende, A. S.; DeVita, R. J. *Tetrahedron Lett.* **1990**, *31*, 307.

[467] Degl'Innocenti, A.; Stucchi, E.; Capperucci, A.; Mordini, A.; Reginato, G.; Ricci, A. *Synlett* **1992**, 332.

[468] Naruse, Y.; Esaki, T.; Yamamoto, H. *Tetrahedron Lett.* **1988**, *29*, 1417.

[469] Naruse, Y.; Esaki, T.; Yamamoto, H. *Tetrahedron* **1988**, *44*, 4747.

[470] Becicka, B. T.; Koerwitz, F. L.; Drtina, G. J.; Baenziger, N. C.; Wiemer, D. F. *J. Org. Chem.* **1990**, *55*, 5613.

[471] Gothelf, K. V.; Torssell, K. G. *Acta Chem. Scand.* **1994**, *48*, 61.

[472] Bovonsombat, P.; McNelis, E. *Tetrahedron Lett.* **1992**, *33*, 7705.

[473] Crisp, G. T.; Glink, P. T. *Tetrahedron* **1994**, *50*, 2623.

[474] Hettrick, C. M.; Scott, W. J. *J. Am. Chem. Soc.* **1991**, *113*, 4903.

[475] Ley, S. V.; Redgrave, A. J.; Taylor, S. C.; Ahmed, S.; Ribbons, D. W. *Synlett* **1991**, 741.

[476] Haiza, M.; Lee, J.; Snyder, J. K. *J. Org. Chem.* **1990**, *55*, 5008.

[477] Bestmann, H. J.; Attygalle, A. B.; Schwarz, J.; Garbe, W.; Vostrowsky, O.; Tomida, I. *Tetrahedron Lett.* **1989**, *30*, 2911.

[478] McLaughlin, M. L.; McKinney, J. A.; Paquette, L. A. *Tetrahedron Lett.* **1986**, *27*, 5595.

[479] Paquette, L. A.; Moriarty, K. J.; McKinney, J. A.; Rogers, R. D. *Organometallics* **1989**, *8*, 1707.

[480] Paquette, L. A.; Ra, C. S.; Edmonson, S. D. *J. Org. Chem.* **1990**, *55*, 2443.

[481] Paquette, L. A.; Shi, Y. J. *J. Org. Chem.* **1989**, *54*, 5205.

[482] Paquette, L. A.; Shi, Y.-J. *J. Am. Chem. Soc.* **1990**, *112*, 8478.

[483] Paquette, L. A.; Ross, R. J.; Shi, Y. J. *J. Org. Chem.* **1990**, *55*, 1589.

[484] Lee, J.; Li, J.-H.; Oya, S.; Snyder, J. K. *J. Org. Chem.* **1992**, *57*, 5301.

[485] Forsyth, C. J.; Clardy, J. *J. Am. Chem. Soc.* **1988**, *110*, 5911.

[486] Forsyth, C. J.; Clardy, J. *J. Am. Chem. Soc.* **1990**, *112*, 3497.

[487] Cheney, D. L.; Paquette, L. A. *J. Org. Chem.* **1989**, *54*, 3334.

[488] Paquette, L. A.; Sivik, M. R. *Organometallics* **1992**, *11*, 3503.

[489] Leanna, M. R.; Morton, H. E. *Tetrahedron Lett.* **1993**, *34*, 4485.

[490] Papageorgiou, C.; Florineth, A.; Mikol, V. *J. Med. Chem.* **1994**, *37*, 3674.

[491] Queneau, Y.; Krol, W. J.; Bornmann, W. G.; Danishefsky, S. J. *J. Org. Chem.* **1992**, *57*, 4043.

[492] Chan, C.; Cox, P. B.; Roberts, S. M. *J. Chem. Soc., Chem. Commun.* **1988**, 971.

[493] Desmaele, D.; d'Angelo, J. *J. Org. Chem.* **1994**, *59*, 2292.

[494] Nicolaou, K. C.; Nadin, A.; Leresche, J. E.; La Greca, S.; Tsuri, T.; Yue, E. W.; Yang, Z. *Angew. Chem., Int. Ed. Engl.* **1994**, *33*, 2187.

[495] Johnson, C. R.; Adams, J. P.; Collins, M. A. *J. Chem. Soc., Perkins Trans. 1* **1993**, 1.

[496] Braisted, A. C.; Schultz, P. G. *J. Am. Chem. Soc.* **1994**, *116*, 2211.

[497] Tamura, R.; Kohno, M.; Utsunomiya, S.; Yamawaki, K.; Azuma, N.; Matsumoto, A.; Ishii, Y. *J. Org. Chem.* **1993**, *58*, 3953.

[498] Oh, J.; Cha, J. K. *Synlett* **1994**, 967.

[499] Burke, S. D.; Piscopio, A. D.; Kort, M. E.; Matulenko, M. A.; Parker, M. H.; Armistead, D. M.; Shankaran, K. *J. Org. Chem.* **1994**, *59*, 332.

[500] Djuric, S. W.; Haack, R. A.; Yu, S. S. *J. Chem. Soc., Perkin Trans. 1* **1989**, 2133.

[501] Butera, J.; Bagli, J.; Doubleday, W.; Humber, L.; Treasurywala, A.; Loughney, D.; Sestanj, K.; Millen, J.; Sredy, J. *J. Med. Chem.* **1989**, *32*, 757.

[502] Mascareñas, J. L.; Garcia, A. M.; Castedo, L.; Mouriño, A. *Tetrahedron Lett.* **1992**, *33*, 7589.

[503] Han, Q.; Wiemer, D. F. *J. Am. Chem. Soc.* **1992**, *114*, 7692.

[504] Niwa, H.; Ieda, S.; Inagaki, H.; Yamada, K. *Tetrahedron Lett.* **1990**, *31*, 7157.

[505] Rudisill, D. E.; Castonguay, L. A.; Stille, J. K. *Tetrahedron Lett.* **1988**, *29*, 1509.

[506] Corey, E. J.; Houpis, I. N. *J. Am. Chem. Soc.* **1990**, *112*, 8997.

[507] Yokokawa, F.; Hamada, Y.; Shioiri, T. *Tetrahedron Lett.* **1993**, *34*, 6559.

[508] Myers, A. G.; Dragovich, P. S. *J. Am. Chem. Soc.* **1993**, *115*, 7021.

[509] Castedo, L.; Mouriño, A.; Sarandeses, L. A. *Tetrahedron Lett.* **1986**, *27*, 1523.

[510] Takeyama, Y.; Ichinose, Y.; Oshima, K.; Utimoto, K. *Tetrahedron Lett.* **1989**, *30*, 3159.

511 Angle, S. R.; Fevig, J. M.; Knight, S. D.; Marquis, R. W. J.; Overman, L. E. *J. Am. Chem. Soc.* **1993**, *115*, 3966.

512 Skoda-Földes, R.; Kollár, L.; Heil, B.; Gálik, G.; Tuba, Z.; Arcadi, A. *Tetrahedron: Asymmetry.* **1991**, *2*, 633.

513 Schweder, B.; Uhlig, E.; Döring, M.; Kosemund, D. *J. Prakt. Chem.* **1993**, *335*, 439.

514 Chu-Moyer, M. Y.; Danishefsky, S. J.; Schulte, G. K. *J. Am. Chem. Soc.* **1994**, *116*, 11213.

515 Tius, M. A.; Kannangara, G. S. K.; Kerr, M. A.; Grace, K. J. S. *Tetrahedron* **1993**, *49*, 3291.

516 Corey, E. J.; Wu, L. I. *J. Am. Chem. Soc.* **1993**, *115*, 9327.

517 Ciattini, P. G.; Morera, E.; Ortar, G. *Tetrahedron Lett.* **1990**, *31*, 1889.

518 Degl'Innocenti, A.; Capperucci, A.; Bartoletti, L.; Mordini, A.; Reginato, G. *Tetrahedron Lett.* **1994**, *35*, 2081.

519 Evans, D. A.; Black, W. C. *J. Am. Chem. Soc.* **1992**, *114*, 2260.

520 Tanaka, H.; Kameyama, Y.; Sumida, S.-i.; Shiroi, T.; Sasaoka, M.; Taniguchi, M.; Torii, S. *Synlett* **1992**, 351.

521 Skoda-Földes, R.; Kollár, L.; Marinelli, F.; Arcadi, A. *Steroids* **1994**, *59*, 691.

522 Trost, B. M.; Greenspan, P. D.; Geisser, H.; Kim, J. H.; Greeves, N. *Angew. Chem., Int. Ed. Engl.* **1994**, *33*, 2182.

523 Frye, S. V.; Haffner, C. D.; Maloney, P. R.; Mook, R. A., Jr.; Dorsey, G. F., Jr.; Hiner, R. N.; Cribbs, C. M.; Wheeler, T. N.; Ray, J. A.; Andrews, R. C.; Batchelor, K. W.; Bramson, H. N.; Stuart, J. D.; Schweiker, S. L.; van Arnold, J.; Croom, S.; Bickett, D. M.; Moss, M. L.; Tian, G.; Unwalla, R. J.; Lee, F. W.; Tippin, T. K.; James, M. K.; Grizzle, M. K.; Long, J. E.; Schuster, S. V. *J. Med. Chem.* **1994**, *37*, 2352.

524 Congreve, M. S.; Holmes, A. B.; Looney, M. G. *J. Am. Chem. Soc.* **1993**, *115*, 5815.

525 Hashimoto, S.-i.; Suzuki, A.; Shinoda, T.; Miyazaki, Y.; Ikegami, S. *Chem. Lett.* **1992**, 1835.

526 Zhang, H. X.; Guibé, F.; Balavoine, G. *J. Org. Chem.* **1990**, *55*, 1857.

527 Piers, E.; Ellis, K. A. *Tetrahedron Lett.* **1993**, *34*, 1875.

528 Piers, E.; Brunet, M.-L.; Oballa, R. M. *Can. J. Chem.* **1993**, *71*, 1484.

529 Kosugi, M.; Naka, H.; Harada, S.; Sano, H.; Migita, T. *Chem. Lett.* **1987**, 1371.

530 Sakamoto, T.; Kondo, Y.; Uchiyama, D.; Yamanaka, H. *Tetrahedron* **1991**, *47*, 5111.

531 Kosugi, M.; Koshiba, M.; Atoh, A.; Sano, H.; Migita, T. *Bull. Chem. Soc. Jpn.* **1986**, *59*, 677.

532 Verlhac, J.-B.; Quintard, J.-P.; Pereyre, M. *J. Chem. Soc., Chem. Commun.* **1988**, 503.

533 Galarini, R.; Musco, A.; Pontellini, R.; Santi, R. *J. Mol. Catal.* **1992**, *72*, L11.

534 Quintard, J. P.; Dumartin, G.; Elissondo, B.; Rahm, A.; Pereyre, M. *Tetrahedron* **1989**, *45*, 1017.

535 Liu, B.; Zhu, D.; Pan, H.; Zhang, A. *Cuihua Xuebao* **1994**, *15*, 85; *Chem. Abstr.* **1994**, *121*, 133462.

536 Iyoda, M.; Kuwatani, Y.; Ueno, N.; Oda, M. *J. Chem. Soc., Chem. Commun.* **1992**, 158.

537 Kang, K.-T.; Kim, S. S.; Lee, J. C. *Tetrahedron Lett.* **1991**, *32*, 4341.

538 Kosugi, M.; Ishikawa, T.; Nogami, T.; Migita, T. *Nippon Kagaku Kaishi* **1985**, 520; *Chem. Abstr.* **1985**, *104*, 68496.

539 Parrain, J.-L.; Duchene, A.; Quintard, J.-P. *Tetrahedron Lett.* **1990**, *31*, 1857.

540 Corriu, R. J. P.; Bolin, G.; Moreau, J. J. E. *Tetrahedron Lett.* **1991**, *32*, 4121.

541 Donnelly, D. M. X.; Finet, J.-P.; Stenson, P. H. *Heterocycles* **1989**, *28*, 15.

542 Uemura, M.; Nishimura, H.; Kamikawa, K.; Nakayama, K.; Hayashi, Y. *Tetrahedron Lett.* **1994**, *35*, 1909.

543 Schreiber, S. L.; Porco, J. A., Jr. *J. Org. Chem.* **1989**, *54*, 4721.

544 Urabe, H.; Matsuka, T.; Sato, F. *Tetrahedron Lett.* **1992**, *33*, 4183.

545 Sakamoto, T.; Funami, N.; Kondo, Y.; Yamanaka, H. *Heterocycles* **1991**, *32*, 1387.

546 Yang, Y.; Wong, H. N. C. *J. Chem. Soc., Chem. Commun.* **1992**, 656.

547 Azizian, H.; Eaborn, C.; Pidcock, A. *J. Organomet. Chem.* **1981**, *215*, 49.

548 Azarian, D.; Dua, S. S.; Eaborn, C.; Walton, D. R. M. *J. Organomet. Chem.* **1976**, *117*, C55.

549 Kosugi, M.; Kato, Y.; Kiuchi, K.; Migita, T. *Chem. Lett.* **1981**, 69.

550 Dondoni, A.; Fantin, G.; Fogagnolo, M.; Medici, A.; Pedrini, P. *Synthesis* **1987**, 693.

551 Dondoni, A.; Fogagnolo, M.; Fantin, G.; Medici, A.; Pedrini, P. *Tetrahedron Lett.* **1986**, *27*, 5269.

552 Sakamoto, T.; Shiga, F.; Yasuhara, A.; Uchiyama, D.; Kondo, Y.; Yamanaka, H. *Synthesis* **1992**, 746.

553 Bumagin, N. A.; Bumagina, I. G.; Beletskaya, I. P. *Dokl. Akad. Nauk SSSR* **1984**, *274*, 818; *Chem. Abstr.* **1984**, *101*, 111062.

554 Liebeskind, L. S.; Riesinger, S. W. *J. Org. Chem.* **1993**, *58*, 408.

555 Bellina, F.; Carpita, A.; De Santis, M.; Rossi, R. *Tetrahedron Lett.* **1994**, *35*, 6913.

556 Rocca, P.; Marsais, F.; Godard, A.; Queguiner, G. *Tetrahedron* **1993**, *49*, 3325.

557 Kashin, A. N.; Bumagina, I. G.; Bumagin, N. A.; Bakunin, V. N.; Beletskaya, I. P. *J. Org. Chem. USSR* **1981**, *17*, 789; *Chem. Abstr.* **1981**, *95*, 133056.

558 Somei, M.; Yamada, F.; Naka, K. *Chem. Pharm. Bull.* **1987**, *35*, 1322.

559 Weller, P. E.; Hanzlik, R. P. *J. Labelled Compd. Radiopharm.* **1988**, *25*, 991.

560 Takahashi, K.; Nihira, T. *Bull. Chem. Soc. Jpn.* **1992**, *65*, 1855.

561 Takahashi, K.; Nihira, T.; Akiyama, K.; Ikegami, Y.; Fukuyo, E. *J. Chem. Soc., Chem. Commun.* **1992**, 620.

562 Beley, M.; Chodorowski, S.; Collin, J.-P.; Sauvage, J.-P. *Tetrahedron Lett.* **1993**, *34*, 2933.

563 Sakamoto, T.; Yasuhara, A.; Kondo, Y.; Yamanaka, H. *Heterocycles* **1993**, *36*, 2597.

564 Iwao, M.; Takehara, H.; Furukawa, S.; Watanabe, M. *Heterocycles* **1993**, *36*, 1483.

565 Alvarez, A.; Guzman, A.; Ruiz, A.; Velarde, E.; Muchowski, J. M. *J. Org. Chem.* **1992**, *57*, 1653.

566 Bumagin, N. A.; Gulevich, Y. V.; Artamkina, G. A.; Beletskaya, I. P. *Bull. Acad. Sci. USSR, Div. Chem. Sci.* **1984**, *33*, 1098; not in *Chem. Abstr.*

567 Wentland, M. P.; Lesher, G. Y.; Reuman, M.; Gruett, M. D.; Singh, B.; Aldous, S. C.; Dorff, P. H.; Rake, J. B.; Coughlin, S. A. *J. Med. Chem.* **1993**, *36*, 2801.

568 Turner, W. R.; Suto, M. J. *Tetrahedron Lett.* **1993**, *34*, 281.

569 Gothelf, K. V.; Torssell, K. B. G. *Acta Chem. Scand.* **1994**, *48*, 165.

570 Gronowitz, S.; Timari, G. *J. Heterocycl. Chem.* **1990**, *27*, 1159.

571 Gronowitz, S.; Timari, G. *J. Heterocycl. Chem.* **1990**, *27*, 1127.

572 Walsh, T. F.; Fitch, K. J.; MacCoss, M.; Chang, R. S. L.; Kivlighn, S. D.; Lotti, V. J.; Siegl, P. K. S.; Patchett, A. A.; Greenlee, W. J. *Bioorg. Med. Chem. Lett.* **1994**, *4*, 219.

573 Kashin, A. N.; Bumagina, I. G.; Bumagin, N. A.; Bakunin, V. N.; Beletskaya, I. P. *Izv. Akad. Nauk SSSR, Ser. Khim.* **1980**, 2185; *Chem. Abstr.* **1980**, *94*, 30858.

574 Wang, S.; Yan, S.; Hu, X.; Guo, H. *Huaxue Xuebao* **1993**, *51*, 393; *Chem. Abstr.* **1993**, *119*, 139027.

575 Olszewski, J. D.; Marshalla, M.; Sabat, M.; Sundberg, R. J. *J. Org. Chem.* **1994**, *59*, 4285.

576 Kashin, A. N.; Bumagina, I. G.; Bumagin, N. A.; Beletskaya, I. P.; Reutov, O. A. *Izv. Akad. Nauk SSSR, Ser. Khim.* **1980**, 479; *Chem. Abstr.* **1980**, *93*, 26019.

577 Nikanorov, V. A.; Rozenberg, V. I.; Kharitonov, V. G.; Yatsenko, E. V.; Mikul'shina, V. V.; Bumagin, N. A.; Beletskaya, I. P.; Guryshev, V. N.; Yur'ev, V. V.; Reutov, O. A. *Metalloorg. Khim.* **1991**, *4*, 689; *Chem. Abstr.* **1991**, *115*, 92458.

578 Sun, Q.; Gatto, B.; Yu, C.; Liu, A.; Liu, L. F.; LaVoie, E. J. *Bioorg. Med. Chem. Lett.* **1994**, *4*, 2871.

579 Gothelf, K.; Thomsen, I. B.; Torssell, K. B. G. *Acta Chem. Scand.* **1992**, *46*, 494.

580 Booth, C.; Imanieh, H.; Quayle, P.; Lu, S. Y. *Tetrahedron Lett.* **1992**, *33*, 413.

581 Duchene, A.; Quintard, J.-P. *Synth. Commun.* **1985**, *15*, 873.

582 Achab, S.; Guyot, M.; Potier, P. *Tetrahedron Lett.* **1993**, *34*, 2127.

583 Carpino, P. A.; Sneddon, S. F.; da Silva Jardine, P.; Magnus-Ayritey, G. T.; Rauch, A. L.; Burkard, M. R. *Bioorg. Med. Chem. Lett.* **1994**, *4*, 93.

584 Rivero, R.; Kevin, N. J.; Allen, E. E. *Bioorg. Med. Chem. Lett.* **1993**, *3*, 1119.

585 Ellingboe, J. W.; Antane, M.; Nguyen, T. T.; Collini, M. D.; Antane, S.; Bender, R.; Hartupee, D.; White, V.; McCallum, J.; Park, C. H.; Russo, A.; Osler, M. B.; Wojdan, A.; Dinsih, J.; Ho, D. M.; Bagli, J. F. *J. Med. Chem.* **1994**, *37*, 542.

[586] Perrier, H.; Prasit, P.; Wang, Z. *Tetrahedron Lett.* **1994**, *35*, 1501.

[587] Cuevas, J.-C.; Patil, P.; Snieckus, V. *Tetrahedron Lett.* **1989**, *30*, 5841.

[588] DuMartin, G.; Pereyre, M.; Quintard, J.-P. *Tetrahedron Lett.* **1987**, *28*, 3935.

[589] Cummins, C. H. *Tetrahedron Lett.* **1994**, *35*, 857.

[590] Bumagin, N. A.; Ponomarev, A. B.; Beletskaya, I. P. *J. Organomet. Chem.* **1985**, *291*, 129.

[591] Rudisill, D. E.; Stille, J. K. *J. Org. Chem.* **1989**, *54*, 5856.

[592] Kurth, M.; Pèlegrin, A.; Rose, K.; Offord, R. E.; Pochon, S.; Mach, J.-P.; Buchegger, F. *J. Med. Chem.* **1993**, *36*, 1255.

[593] Arano, Y.; Wakisaka, K.; Ohmomo, Y.; Uezono, T.; Mukai, T.; Motonari, H.; Shiono, H.; Sakahara, H.; Konishi, J.; Tanaka, C.; Yokoyama, A. *J. Med. Chem.* **1994**, *37*, 2609.

[594] Müller, G.; Dürner, G.; Bats, J. W.; Göbel, M. W. *Liebigs Ann. Chem.* **1994**, 1075.

[595] Schreiber, S. L.; Desmaele, D.; Porco, J. A., Jr. *Tetrahedron Lett.* **1988**, *29*, 6689.

[596] Iwao, M.; Takehara, H.; Obata, S.; Watanabe, M. *Heterocycles* **1994**, *38*, 1717.

[597] Cooper, C. B.; McFarland, J. W.; Blair, K. T.; Fontaine, E. H.; Jones, C. S.; Muzzi, M. L. *Bioorg. Med. Chem. Lett.* **1994**, *4*, 835.

[598] Takle, A.; Kocienski, P. *Tetrahedron Lett.* **1989**, *30*, 1675.

[599] Takeuchi, M.; Tuihiji, T.; Nishimura, J. *J. Org. Chem.* **1993**, *58*, 7388.

[600] Chang, L. L.; Ashton, W. T.; Flanagan, K. L.; Naylor, E. M.; Chakravarty, P. K.; Patchett, A. A.; Greenlee, W. J.; Bendesky, R. J.; Chen, T.-B.; Faust, K. A.; Kling, P. J.; Schaffer, L. W.; Schorn, T. W.; Zingaro, G. J.; Chang, R. S. L.; Lotti, V. J.; Kivlighn, S. D.; Siegl, P. K. S. *Bioorg. Med. Chem. Lett.* **1994**, *4*, 115.

[601] Negishi, E.-i.; Noda, Y.; Lamaty, F.; Vawter, E. J. *Tetrahedron Lett.* **1990**, *31*, 4393.

[602] Wentland, M. P.; Lesher, G. Y.; Reuman, M.; Pilling, G. M.; Saindane, M. T.; Perni, R. B.; Eissenstat, M. A.; Weaver, J. D., III; Singh, B.; Rake, J.; Coughlin, S. A. *Bioorg. Med. Chem. Lett.* **1993**, *3*, 1711.

[603] Salituro, F. G.; Tomlinson, R. C.; Baron, B. M.; Palfreyman, M. G.; McDonald, I. A. *J. Med. Chem.* **1994**, *37*, 334.

[604] Hark, R. R.; Hauze, D. B.; Petrovskaia, O.; Joullie, M. M.; Jaouhari, R.; McComiskey, P. *Tetrahedron Lett.* **1994**, *35*, 7719.

[605] Stafford, J. A.; Valvano, N. L. *J. Org. Chem.* **1994**, *59*, 4346.

[606] Kelly, T. R.; Bridger, G. J.; Zhao, C. *J. Am. Chem. Soc.* **1990**, *112*, 8024.

[607] Smyth, M. S.; Stefanova, I.; Horak, I. D.; Burke, T. R., Jr. *J. Med. Chem.* **1993**, *36*, 3015.

[608] Azzena, U.; Melloni, G.; Pisano, L. *Tetrahedron Lett.* **1993**, *34*, 5635.

[609] John, C. S.; Saga, T.; Kinuya, S.; Le, N.; Jeong, J. M.; Paik, C. H.; Reba, R. C.; Varma, V. M.; McAfee, J. G. *Nucl. Med. Biol.* **1993**, *20*, 75.

[610] Robl, J. A. *Tetrahedron Lett.* **1990**, *31*, 3421.

[611] Fu, J.-m.; Sharp, M. J.; Snieckus, V. *Tetrahedron Lett.* **1988**, *29*, 5459.

[612] Sonesson, C.; Waters, N.; Svensson, K.; Carlsson, A.; Smith, M. W.; Piercey, M. F.; Meier, E.; Wikström, H. *J. Med. Chem.* **1993**, *36*, 3188.

[613] Rybakova, L. F.; Sorokina, R. S.; Petrov, E. S.; Val'kova, G. A.; Shifrina, R. R.; Beletskaya, I. P. *Bull. Acad. Sci. USSR, Div. Chem. Sci.* **1985**, *34*, 1108; not in *Chem. Abstr.*

[614] Mori, M.; Kaneta, N.; Shibasaki, M. *J. Org. Chem.* **1991**, *56*, 3486.

[615] Bailey, T. R.; Diana, G. D.; Kowalczyk, P. J.; Akullian, V.; Eissenstat, M. A.; Cutcliffe, D.; Mallamo, J. P.; Carabateas, P. M.; Pevear, D. C. *J. Med. Chem.* **1992**, *35*, 4628.

[616] Zimmermann, E. K.; Stille, J. K. *Macromolecules* **1985**, *18*, 321.

[617] Namavari, M.; Satyamurthy, N.; Phelps, M. E.; Barrio, J. R. *Appl. Radiat. Isot.* **1993**, *44*, 527.

[618] Matsumoto, T.; Hosoya, T.; Suzuki, K. *Synlett* **1991**, 709.

[619] Liu, Y.; Svensson, B. E.; Yu, H.; Cortizo, L.; Ross, S. B.; Lewander, T.; Hacksell, U. *Bioorg. Med. Chem. Lett.* **1991**, *1*, 257.

[620] Liu, Y.; Yu, H.; Svensson, B. E.; Cortizo, L.; Lewander, T.; Hacksell, U. *J. Med. Chem.* **1993**, *36*, 4221.

[621] de Paulis, T.; Smith, H. E. *Synth. Commun.* **1991**, *21*, 1091.

[622] Tilley, J. W.; Clader, J. W.; Zawoiski, S.; Wirkus, M.; LeMahieu, R. A.; O'Donnell, M.; Crowley, H.; Welton, A. F. *J. Med. Chem.* **1989**, *32*, 1814.

[623] Hanefeld, W.; Jung, M. *Liebigs Ann. Chem.* **1994**, 59.

[624] Tilley, J. W.; Danho, W.; Lovey, K.; Wagner, R.; Swistok, J.; Makofske, R.; Michalewsky, J.; Triscari, J.; Nelson, D.; Weatherford, S. *J. Med. Chem.* **1991**, *34*, 1125.

[625] Kollár, L.; Skoda-Földes, R.; Mahó, S.; Tuba, Z. *J. Organomet. Chem.* **1993**, *453*, 159.

[626] Huang, F.-C.; Chan, W.-K.; Warus, J. D.; Morrissette, M. M.; Moriarty, K. J.; Chang, M. N.; Travis, J. J.; Mitchell, L. S.; Nuss, G. W.; Sutherland, C. A. *J. Med. Chem.* **1992**, *35*, 4253.

[627] Hanefeld, W.; Jung, M. *Pharmazie* **1994**, *49*, 18.

[628] Hanefeld, W.; Jung, M. *Tetrahedron* **1994**, *50*, 2459.

[629] Patel, H. K.; Kilburn, J. D.; Langley, G. J.; Edwards, P. D.; Mitchell, T.; Southgate, R. *Tetrahedron Lett.* **1994**, *35*, 481.

[630] Urones, J. G.; Marcos, I. S.; Basabe, P.; Garrido, N. M.; Jorge, A.; Moro, R. F.; Lithgow, A. M. *Tetrahedron* **1993**, *49*, 6079.

[631] Blaszczak, L. C.; Halligan, N. G.; Seitz, D. E. *J. Labelled Compd. Radiopharm.* **1989**, *27*, 401.

[632] Soll, R. M.; Kinney, W. A.; Primeau, J.; Garrick, L.; McCaully, R. J.; Colatsky, T.; Oshiro, G.; Park, C. H.; Hartupee, D.; White, V.; McCallum, J.; Russo, A.; Dinish, J.; Wojdan, A. *Bioorg. Med. Chem. Lett.* **1993**, *3*, 757.

[633] Rychnovsky, S. D.; Hwang, K. *J. Org. Chem.* **1994**, *59*, 5414.

[634] Holt, D. A.; Oh, H.-J.; Rozamus, L. W.; Yen, H.-K.; Brandt, M.; Levy, M. A.; Metcalf, B. W. *Bioorg. Med. Chem. Lett.* **1993**, *3*, 1735.

[635] Zhuang, Z.-P.; Kung, M.-P.; Kung, H. F. *J. Med. Chem.* **1994**, *37*, 1406.

[636] Liljebris, C.; Resul, B.; Hacksell, U. *Bioorg. Med. Chem. Lett.* **1993**, *3*, 241.

[637] Rama Rao, A. V.; Gurjar, M. K.; Bhaskar Reddy, A.; Khare, V. B. *Tetrahedron Lett.* **1993**, *34*, 1657.

[638] Kelly, T. R.; Xu, W.; Ma, Z.; Li, Q.; Bhushan, V. *J. Am. Chem. Soc.* **1993**, *115*, 5843.

[639] Saulnier, M. G.; LeBoulluec, K. L.; Vyas, D. M.; Crosswell, A. R.; Doyle, T. W. *Bioorg. Med. Chem. Lett.* **1992**, *2*, 1213.

[640] Takeuchi, M.; Nishimura, J. *Tetrahedron Lett.* **1992**, *33*, 5563.

[641] Rama Rao, A. V.; Gurjar, M. K.; Kaiwar, V.; Khare, V. B. *Tetrahedron Lett.* **1993**, *34*, 1661.

[642] Chan, K. S.; Chan, C. S. *Synth. Commun.* **1993**, *23*, 1489.

[643] Rama Rao, A. V.; Laxma Reddy, K.; Srinivasa Rao, A. *Tetrahedron Lett.* **1994**, *35*, 5047.

[644] Liebeskind, L. S.; Zhang, J. *J. Org. Chem.* **1991**, *56*, 6379.

[645] Grigg, R.; Teasdale, A.; Sridharan, V. *Tetrahedron Lett.* **1991**, *32*, 3859.

[646] Kalivretenos, A.; Stille, J. K.; Hegedus, L. S. *J. Org. Chem.* **1991**, *56*, 2883.

[647] Kelly, T. R.; Li, Q.; Bhushan, V. *Tetrahedron Lett.* **1990**, *31*, 161.

[648] Bradley, J. C.; Durst, T. *J. Org. Chem.* **1991**, *56*, 5459.

[649] Magnus, P.; Witty, D.; Stamford, A. *Tetrahedron Lett.* **1993**, *34*, 23.

[650] Finch, H.; Pegg, N. A.; Evans, B. *Tetrahedron Lett.* **1993**, *34*, 8353.

[651] Sandosham, J.; Undheim, K. *Acta Chem. Scand.* **1989**, *43*, 684.

[652] Djuric, S. W.; Huff, R. M.; Penning, T. D.; Clare, M.; Swenton, L.; Kachur, J. F.; Villani-Price, D.; Krivi, G. G.; Pyla, E. Y.; Warren, T. G. *Bioorg. Med. Chem. Lett.* **1992**, *2*, 1367.

[653] Sasaki, S.; Takao, F.; Watanabe, K.; Obana, N.; Maeda, M.; Fukumura, T.; Takehara, S. *Chem. Pharm. Bull.* **1993**, *41*, 296.

[654] Birkett, M. A.; Knight, D. W.; Mitchell, M. B. *Synlett* **1994**, 253.

[655] Engler, T. A.; Reddy, J. P.; Combrink, K. D.; Vander Velde, D. *J. Org. Chem.* **1990**, *55*, 1248.

[656] Engler, T. A.; Combrink, K. D.; Letavic, M. A.; Lynch, K. O., Jr.; Ray, J. E. *J. Org. Chem.* **1994**, *59*, 6567.

[657] Haraguchi, K.; Itoh, Y.; Tanaka, H.; Miyasaka, T. *Tetrahedron Lett.* **1991**, *32*, 3391.

[658] Haraguchi, K.; Itoh, Y.; Tanaka, H.; Akita, M.; Miyasaka, T. *Tetrahedron* **1993**, *49*, 1371.

[659] Martina, S.; Enkelmann, V.; Wegener, G.; Schlüter, A.-D. *Synth. Metals* **1992**, *51*, 299.

[660] Dupré, B.; Meyers, A. I. *J. Org. Chem.* **1991**, *56*, 3197.

[661] Hegedus, L. S.; Holden, M. S. *J. Org. Chem.* **1986**, *51*, 1171.

[662] Tidwell, J. H.; Peat, A. J.; Buchwald, S. L. *J. Org. Chem.* **1994**, *59*, 7164.

[663] Vaillancourt, V.; Albizati, K. F. *J. Am. Chem. Soc.* **1993**, *115*, 3499.

[664] Yokoyama, Y.; Ikeda, M.; Saito, M.; Yoda, T.; Suzuki, H.; Murakami, Y. *Heterocycles* **1990**, *31*, 1505.

[665] Danheiser, R. L.; Brisbois, R. G.; Kowalczyk, J. J.; Miller, R. F. *J. Am. Chem. Soc.* **1990**, *112*, 3093.

[666] Sheppard, G. S.; Pireh, D.; Carrera, G. M., Jr.; Bures, M. G.; Heyman, H. R.; Steinman, D. H.; Davidsen, S. K.; Phillips, J. G.; Guinn, D. E.; May, P. D.; Conway, R. G.; Rhein, D. A.; Calhoun, W. C.; Albert, D. H.; Magoc, T. J.; Carter, G. W.; Summers, J. B. *J. Med. Chem.* **1994**, *37*, 2011.

[667] Gronowitz, S.; Peters, D. *Heterocycles* **1990**, *30*, 645.

[668] Catellani, M.; Luzzati, S.; Musco, A.; Speroni, F. *Synth. Metals* **1994**, *62*, 223.

[669] Malm, J.; Björk, P.; Gronowitz, S.; Hörnfeldt, A.-B. *Tetrahedron Lett.* **1992**, *33*, 2199.

[670] Wigerinck, P.; Kerremans, L.; Claes, P.; Snoeck, R.; Maudgal, P.; De Clercq, E.; Herdewijn, P. *J. Med. Chem.* **1993**, *36*, 538.

[671] Kitimura, C.; Tanaka, S.; Yamashita, Y. *J. Chem. Soc., Chem. Commun.* **1994**, 1585.

[672] Nordvall, G.; Sundquist, S.; Nilvebrant, L.; Hacksell, U. *Bioorg. Med. Chem. Lett.* **1994**, *4*, 2837.

[673] Otsubo, T.; Kono, Y.; Hozo, N.; Miyamoto, H.; Aso, Y.; Ogura, F.; Tanaka, T.; Sawada, M. *Bull. Chem. Soc. Jpn.* **1993**, *66*, 2033.

[674] Bridges, A. J.; Lee, A.; Schwartz, C. E.; Towle, M. J.; Littlefield, B. A. *Bioorg. Med. Chem. Lett.* **1993**, *1*, 403.

[675] Kevin, N. J.; Rivero, R. A.; Greenlee, W. J.; Chang, R. S. L.; Chen, T. B. *Bioorg. Med. Chem. Lett.* **1994**, *4*, 189.

[676] Sanfilippo, P. J.; McNally, J. J.; Press, J. B.; Fitzpatrick, L. J.; Urbanski, M. J.; Katz, L. B.; Giardino, E.; Falotico, R.; Salata, J.; Moore, J. B., Jr.; Miller, W. *J. Med. Chem.* **1992**, *35*, 4425.

[677] Tamao, K.; Yamaguchi, S.; Shiozaki, M.; Nakagawa, Y.; Ito, Y. *J. Am. Chem. Soc.* **1992**, *114*, 5867.

[678] Barber, C.; Jarowicki, K.; Kocienski, P. *Synlett* **1991**, 197.

[679] Wattanasin, S. *Synth. Commun.* **1988**, *18*, 1919.

[680] Koch, K.; Biggers, M. S. *J. Org. Chem.* **1994**, *59*, 1216.

[681] Taka, N.; Koga, H.; Sato, H.; Ishizawa, T.; Takahashi, T.; Imagawa, J.-i. *Bioorg. Med. Chem. Lett.* **1994**, *4*, 2893.

[682] Takahashi, T.; Koga, H.; Sato, H.; Ishizawa, T.; Taka, N.; Imagawa, J.-i. *Bioorg. Med. Chem. Lett.* **1994**, *4*, 2899.

[683] Yoo, S.-e.; Suh, J. H.; Joeng, N. *Bioorg. Med. Chem. Lett.* **1992**, *2*, 381.

[684] Al-Abed, Y.; Al-Tel, T. H.; Schröder, C.; Voelter, W. *Angew. Chem., Int. Ed. Engl.* **1994**, *33*, 1499.

[685] Jarowicki, K.; Kocienski, P.; Marczak, S.; Willson, T. *Tetrahedron Lett.* **1990**, *31*, 3433.

[686] Morris, J.; Wishka, D. G.; Lin, A. H.; Humphrey, W. R.; Wiltse, A. L.; Gammill, R. B.; Judge, T. M.; Bisaha, S. N.; Olds, N. L.; Jacob, C. S.; Bergh, C. L.; Cudahy, M. M.; Williams, D. J.; Nishizawa, E. E.; Thomas, E. W.; Gorman, R. R.; Benjamin, C. W.; Shebuski, R. J. *J. Med. Chem.* **1993**, *36*, 2026.

[687] Kelly, T. R.; Kim, M. H. *J. Org. Chem.* **1992**, *57*, 1593.

[688] Tius, M.; Gomez-Galeno, J.; Gu, X.-Q.; Zaidi, J. H. *J. Am. Chem. Soc.* **1991**, *113*, 5775.

[689] Macdonald, S. J. F.; McKenzie, T. C.; Hassen, W. D. *J. Chem. Soc., Chem. Commun.* **1987**, 1528.

[690] Paquette, L. A.; Wang, T.-Z.; Sivik, M. R. *J. Am. Chem. Soc.* **1994**, *116*, 11323.

[691] Paquette, L. A.; Wang, T.-Z.; Sivik, M. R. *J. Am. Chem. Soc.* **1994**, *116*, 2665.

[692] Bumagin, N. A.; Kalinovskii, I. O.; Beletskaya, I. P. *Khim. Geterotsikl. Soedin.* **1983**, 1467; *Chem. Abstr.* **1983**, *100*, 156465.

[693] Godard, A.; Rovera, J.-C.; Marsais, F.; Plé, N.; Quéguiner, G. *Tetrahedron* **1992**, *48*, 4123.

[694] Malm, J.; Hörnfeldt, A. B.; Gronowitz, S. *Heterocycles* **1993**, *35*, 245.

[695] Bumagin, N. A.; Andryukhova, N. P.; Beletskaya, I. P. *Dokl. Akad. Nauk SSSR* **1989**, *307*, 375; *Chem. Abstr.* **1989**, *112*, 138656.

[696] Long, G. V.; Boyd, S. E.; Harding, M. M.; Buys, I. E.; Hambley, T. W. *J. Chem. Soc., Dalton Trans.* **1993**, 3175.

[697] Dehmlow, E. V.; Sleegers, A. *Liebigs Ann. Chem.* **1992**, 953.

[698] Kelly, T. R.; Bowyer, M. C.; Bhaskar, K. V.; Bebbington, D.; Garcia, A.; Lang, F.; Kim, M. H.; Jette, M. P. *J. Am. Chem. Soc.* **1994**, *116*, 3657.

[699] Maguire, M. P.; Sheets, K. R.; McVety, K.; Spada, A. P.; Zilberstein, A. *J. Med. Chem.* **1994**, *37*, 2129.

[700] Ghadiri, M. R.; Soares, C.; Choi, C. *J. Am. Chem. Soc.* **1992**, *114*, 825.

[701] Marsais, F.; Pineau, P.; Nivolliers, F.; Mallet, M.; Turck, A.; Godard, A.; Queguiner, G. *J. Org. Chem.* **1992**, *57*, 565.

[702] Odobel, F.; Sauvage, J.-P.; Harriman, A. *Tetrahedron Lett.* **1993**, *34*, 8113.

[703] Collin, J.-P.; Harriman, A.; Heitz, V.; Odobel, F.; Sauvage, J.-P. *J. Am. Chem. Soc.* **1994**, *116*, 5679.

[704] Potts, K. T.; Konwar, D. *J. Org. Chem.* **1991**, *56*, 4815.

[705] Bracher, F.; Hildebrand, D. *Tetrahedron* **1994**, *50*, 12329.

[706] Bantick, J. R.; Beaton, H. G.; Cooper, S. L.; Hill, S.; Hirst, S. C.; McInally, T.; Spencer, J.; Tinker, A. C.; Willis, P. A. *Bioorg. Med. Chem. Lett.* **1994**, *4*, 121.

[707] Zhang, H. C.; Daves, G. D., Jr. *Organometallics* **1993**, *12*, 1499.

[708] Sandosham, J.; Undheim, K. *Acta Chem. Scand.* **1994**, *48*, 279.

[709] Sandosham, J.; Undheim, K.; Rise, F. *Heterocycles* **1993**, *35*, 235.

[710] Blough, B. E.; Mascarella, S. W.; Rothman, R. B.; Carroll, F. I. *J. Chem. Soc., Chem. Commun.* **1993**, 758.

[711] Echavarren, A. M.; Stille, J. K. *J. Am. Chem. Soc.* **1988**, *110*, 4051.

[712] Godard, A.; Fourquez, J. M.; Tamion, R.; Marsais, F.; Quéguiner, G. *Synlett* **1994**, *4*, 235.

[713] Wentland, M. P.; Perni, R. B.; Dorff, P. H.; Brundage, R. P.; Castaldi, M. J.; Bailey, T. R.; Carabateas, P. M.; Bacon, E. R.; Young, D. C.; Woods, M. G.; Rosi, D.; Drozd, M. L.; Kullnig, R. K.; Dutko, F. J. *J. Med. Chem.* **1993**, *36*, 1580.

[714] VanAtten, M. K.; Ensinger, C. L.; Chiu, A. T.; McCall, D. E.; Nguyen, T. T.; Wexler, R. R.; Timmermans, P. B. M. W. M. *J. Med. Chem.* **1993**, *36*, 3985.

[715] Farina, V.; Firestone, R. A. *Tetrahedron* **1993**, *49*, 803.

[716] Van Aken, K. J.; Lux, G. M.; Deroover, G. G.; Mererpoel, L.; Hoornaertt, G. J. *Tetrahedron* **1994**, *50*, 5211.

[717] Wigerinck, P.; Pannecouque, C.; Snoeck, R.; Claes, P.; De Clercq, E.; Herdewijn, P. *J. Med. Chem.* **1991**, *34*, 2383.

[718] Gutierrez, A. J.; Terhorst, T. J.; Matteucci, M. D.; Froehler, B. C. *J. Am. Chem. Soc.* **1994**, *116*, 5540.

[719] Chou, W.-N.; White, J. B. *Tetrahedron Lett.* **1991**, *32*, 157.

[720] Wang, L. R. R.; Benneche, T.; Undheim, K. *Acta Chem. Scand.* **1990**, *44*, 726.

[721] Street, L. J.; Baker, R.; Book, T.; Reeve, A. J.; Saunders, J.; Willson, T.; Marwood, R. S.; Patel, S.; Freedman, S. B. *J. Med. Chem.* **1992**, *35*, 295.

[722] Van Aerschot, A. A.; Mamos, P.; Weyns, N. J.; Ikeda, S.; De Clercq, E.; Herdewijn, P. A. *J. Med. Chem.* **1993**, *36*, 2938.

[723] Nair, V.; Purdy, D. F. *Tetrahedron* **1991**, *47*, 365.

[724] Bell, A. S.; Fishwick, C. W. G.; Reed, J. E. *Tetrahedron Lett.* **1994**, *35*, 6551.

[725] Bracher, F.; Hildebrand, D. *Liebigs Ann. Chem.* **1992**, 1315.

[726] Bracher, F.; Hildebrand, D. *Liebigs Ann. Chem.* **1993**, 837.

[727] Newhouse, B. J.; Meyers, A. I.; Sirisoma, N. S.; Braun, M. P.; Johnson, C. R. *Synlett* **1993**, 573.

[728] Sjögren, M.; Hansson, S.; Norrby, P.-O.; Åkermark, B.; Cucciolito, M. E.; Vitagliano, A. *Organometallics* **1992**, *11*, 3954.

[729] Laborde, E.; Kiely, J.; Lesheski, L. E.; Schroeder, M. C. *J. Heterocyclic Chem.* **1991**, *28*, 191.

[730] Peña, M. R.; Stille, J. K. *Tetrahedron Lett.* **1987**, *28*, 6573.

[731] Peters, D.; Hoernfeldt, A. B.; Gronowitz, S.; Johansson, N. G. *J. Heterocycl. Chem.* **1991**, *28*, 529.

[732] Verlinde, C. L. M. J.; Callens, M.; Van Calenbergh, S.; Van Aerschot, A.; Herdewijn, P.; Hannaert, V.; Michels, P. A. M.; Opperdoes, F. R.; Hol, W. G. J. *J. Med. Chem.* **1994**, *37*, 3605.

[733] Hedberg, M. H.; Johansson, A. M.; Hacksell, U. *J. Chem. Soc., Chem. Commun.* **1992**, 845.

[734] Hedberg, M. H.; Johansson, A. M.; Fowler, C. J.; Terenius, L.; Hacksell, U. *Bioorg Med. Chem. Lett.* **1994**, *4*, 2527.

[735] Harmata, M.; Barnes, C. L.; Karra, S. R.; Elahmad, S. *J. Am. Chem. Soc.* **1994**, *116*, 8392.

[736] Venkatesan, A. M.; Levin, J. I.; Baker, J. S.; Chan, P. S.; Bailey, T.; Couplet, J. *Bioorg. Med. Chem. Lett.* **1994**, *4*, 183.

[737] Davies, S. G.; Pyatt, D. *Heterocycles* **1989**, *28*, 163.

[738] Levin, J. I.; Chan, P. S.; Couplet, J.; Thibault, L.; Venkatesan, A. M.; Bailey, T. K.; Vice, G.; Cobuzzi, A.; Lai, F.; Mellish, N. *Bioorg. Med. Chem. Lett.* **1994**, *4*, 1709.

[739] Sessler, J. L.; Wang, B.; Harriman, A. *J. Am. Chem. Soc.* **1993**, *115*, 10418.

[740] Edstrom, E. D.; Wei, Y. *J. Org. Chem.* **1993**, *58*, 403.

[741] Nair, V.; Buenger, G. S. *Synthesis* **1988**, 848.

[742] Farina, V.; Baker, S. R.; Sapino, C., Jr. *Tetrahedron Lett.* **1988**, *29*, 6043.

[743] Bateson, J. H.; Burton, G.; Elsmere, S. A.; Elliott, R. L. *Synlett* **1994**, 152.

[744] Roth, G. P.; Sapino, C. *Tetrahedron Lett.* **1991**, *32*, 4073.

[745] Minnetian, O. M.; Morris, I. K.; Snow, K. M.; Smith, K. M. *J. Org. Chem.* **1989**, *54*, 5567.

[746] Herdewijn, P.; Kerremans, L.; Snoeck, R.; Van Aerschot, A.; Esmans, E.; De Clercq, E. *Bioorg. Med. Chem. Lett.* **1992**, *2*, 1057.

[747] Levin, J. I.; Chan, P. S.; Couplet, J.; Bailey, T. K.; Vice, G.; Thibault, L.; Lai, F.; Venkatesan, A. M.; Cobuzzi, A. *Bioorg. Med. Chem. Lett.* **1994**, *4*, 1703.

[748] de Laszlo, S. E.; Allen, E. E.; Quagliato, C. S.; Greenlee, W. J.; Patchett, A. A.; Nachbar, R. B.; Siegl, P. K.; Chang, R. S.; Kivlighn, S. D.; Schorn, T. S.; Faust, K. A.; Chen, T.-B.; Zingaro, G. J.; Lotti, V. J. *Bioorg. Med. Chem. Lett.* **1993**, *3*, 1299.

[749] Soderquist, J. A.; Leong, W. W.-H. *Tetrahedron Lett.* **1983**, *24*, 2361.

[750] Bumagin, N. A.; Bumagina, I. G.; Kashin, A. N.; Beletskaya, I. P. *J. Org. Chem. USSR* **1982**, *8*, 977; *Chem. Abstr.* **1982**, *97*, 216343.

[751] Kashin, A. N.; Bumagina, I. G.; Bumagin, N. A.; Beletskaya, I. P. *Izv. Akad. Nauk SSSR, Ser. Khim.* **1981**, 1433; *Chem. Abstr.* **1981**, *95*, 114976.

[752] Ando, K.; Hatano, C.; Akadegawa, N.; Shigihara, A.; Takayama, H. *J. Chem. Soc., Chem. Commun.* **1992**, 870.

[753] Pérez, M.; Castaño, A. M.; Echavarren, A. M. *J. Org. Chem.* **1992**, *57*, 5047.

[754] Colson, P.-J.; Franck-Neumann, M.; Sedrati, M. *Tetrahedron Lett.* **1989**, *30*, 2393.

[755] Kashin, A. N.; Bumagin, N. A.; Kalinovskii, I. O.; Beletskaya, I. P.; Reutov, O. A. *J. Org. Chem. USSR* **1980**, *16*, 1329; *Chem. Abstr.* **1980**, *94*, 14747.

[756] Sewald, N.; Gaa, K.; Burger, K. *Heteroatom Chemistry* **1993**, *4*, 253.

[757] Lander, P. A.; Hegedus, L. S. *J. Am. Chem. Soc.* **1994**, *116*, 8126.

[758] Comins, D. L.; Mantlo, N. B. *Tetrahedron Lett.* **1987**, *28*, 759.

[759] Mitchell, T. N.; Kwetkat, K. *J. Organomet. Chem.* **1992**, *439*, 127.

[760] Barbry, D.; Couturier, D. *J. Labelled Compd. Radiopharm.* **1987**, *24*, 603.

[761] Ley, S. V.; Trudell, M. L.; Wadsworth, D. J. *Tetrahedron* **1991**, *47*, 8285.

[762] Ley, S. V.; Wadsworth, D. J. *Tetrahedron Lett.* **1989**, *30*, 1001.

[763] Parrain, J.-L.; Beaudet, I.; DuchɅne, A.; Watrelot, S.; Quintard, J.-P. *Tetrahedron Lett.* **1993**, *34*, 5445.

[764] Huffman, J. W.; Potnis, S. M.; Satish, A. V. *J. Org. Chem.* **1985**, *50*, 4266.

[765] Norley, M. C.; Kocienski, P. J.; Faller, A. *Synlett* **1994**, 77.

[766] Bonnaffé, D.; Simon, H. *Tetrahedron* **1992**, *48*, 9695.

[767] Ackroyd, J.; Karpf, M.; Dreiding, A. S. *Helv. Chim. Acta* **1985**, *68*, 338.

[768] Brieden, W.; Ostwald, R.; Knochel, P. *Angew. Chem., Int. Ed. Engl.* **1993**, *32*, 582.

[769] Guibé, F.; Four, P.; Riviere, H. *J. Chem. Soc., Chem. Commun.* **1980**, 432.

[770] Baldwin, J. E.; Adlington, R. M.; Ramcharitar, S. H. *Synlett* **1992**, 875.

[771] Castaño, A. M.; Cuerva, J. M.; Echavarren, A. M. *Tetrahedron Lett.* **1994**, *35*, 7435.

[772] Eicher, T.; Massonne, K.; Herrmann, M. *Synthesis* **1991**, 1173.

[773] Ireland, R. E.; Obrecht, D. M. *Helv. Chim. Acta* **1986**, *69*, 1273.

[774] Pellicciari, R.; Gallo-Mezo, M. A.; Natalini, B.; Amer, A. M. *Tetrahedron Lett.* **1992**, *33*, 3003.

[775] Salituro, F. G.; McDonald, I. A. *J. Org. Chem.* **1988**, *53*, 6138.

[776] Ornstein, P. L.; Melikian, A.; Martinelli, M. *J. Tetrahedron Lett.* **1994**, *35*, 5759.

[777] Kende, A. S.; Roth, B.; Sanfilippo, P. J.; Blacklock, T. J. *J. Am. Chem. Soc.* **1982**, *104*, 5808.

[778] Darwish, I. S.; Patel, C.; Miller, M. J. *J. Org. Chem.* **1993**, *58*, 6072.

[779] Ho, T. L.; Gopalan, B.; Nestor, J. J., Jr. *J. Org. Chem.* **1986**, *51*, 2405.

[780] Darwish, I. S.; Miller, M. J. *J. Org. Chem.* **1994**, *59*, 451.

[781] Burke, S. D.; Piscopio, A. D.; Kort, M. E.; Matulenko, M. A.; Parker, M. H.; Armistead, D. M.; Shankaran, K. *J. Org. Chem.* **1994**, *59*, 332.

[782] Salvi, J.-P.; Walchshofer, N.; Paris, J. *Tetrahedron Lett.* **1994**, *35*, 1181.

[783] Labadie, J. W.; Stille, J. K. *Tetrahedron Lett.* **1983**, *24*, 4283.

[784] Mazur, P.; Nakanishi, K. *J. Org. Chem.* **1992**, *57*, 1047.

[785] Wright, M. E.; Lowe-Ma, C. K. *Organometallics* **1990**, *9*, 347.

[786] Moore, J. S. *Makromol. Chem., Rapid Commun.* **1992**, *13*, 91.

[787] Jousseaume, B.; Villeneuve, P. *Tetrahedron* **1989**, *45*, 1145.

[788] Sakamoto, T.; Shiga, F.; Uchiyama, D.; Kondo, Y.; Yamanaka, H. *Heterocycles* **1992**, *33*, 813.

[789] Hibino, J.-i.; Matsubara, S.; Morizawa, Y.; Oshima, K.; Nozaki, H. *Tetrahedron Lett.* **1984**, *25*, 2151.

[790] Baxter, A. J. G.; Dixon, J.; Ince, F.; Manners, C. N.; Teague, S. J. *J. Med. Chem.* **1993**, *36*, 2739.

[791] Hodgson, D. M.; Boulton, L. T.; Maw, G. N. *Tetrahedron Lett.* **1994**, *35*, 2231.

[792] Yu, K.-L.; Mansuri, M. M.; Starrett, J. E., Jr. *Tetrahedron Lett.* **1994**, *35*, 8955.

[793] Torok, D. S.; Scott, W. J. *Tetrahedron Lett.* **1993**, *34*, 3067.

[794] Crisp, G. T.; O'Donoghue, A. I. *Synth. Commun.* **1989**, *19*, 1745.

[795] Kende, A. S.; Mendoza, J. S.; Fujii, Y. *Tetrahedron* **1993**, *49*, 8015.

[796] Claesson, A.; Swahn, B. M.; Edvinsson, K. M.; Molin, H.; Sandberg, M. *Bioorg. Med. Chem. Lett.* **1992**, *2*, 1247.

[797] Monclus, M.; Luxen, A. *Org. Prep. Proced. Int.* **1992**, *24*, 692.

[798] Adlington, R. M.; Baldwin, J. E.; Gansaeuer, A.; McCoull, W.; Russell, A. T. *J. Chem. Soc., Perkin Trans. 1* **1994**, 1697.

[799] Miftakhov, M. S.; Lesnikova, E. T.; Tolstikov, G. A. *J. Org. Chem. USSR* **1986**, *22*, 2007; *Chem. Abstr.* **1986**, *107*, 115397.

[800] Liebeskind, L. S.; Foster, B. S. *J. Am. Chem. Soc.* **1990**, *112*, 8612.

[801] Guibé, F.; Zigna, A.-M.; Balavoine, G. *J. Organomet. Chem.* **1986**, *306*, 257.

[802] Echavarren, A. M.; Tueting, D. R.; Stille, J. K. *J. Am. Chem. Soc.* **1988**, *110*, 4039.

[803] White, J. D.; Jensen, M. S. *J. Am. Chem. Soc.* **1993**, *115*, 2970.

[804] Guibé, F.; Xian, Y. T.; Balavoine, G. *J. Organomet. Chem.* **1986**, *306*, 267.

[805] Kosugi, M.; Miyajima, Y.; Nakanishi, H.; Sano, H.; Migita, T. *Bull. Chem. Soc. Jpn.* **1989**, *62*, 3383.

[806] Owton, W. M.; Brunavs, M. *Synth. Commun.* **1991**, *21*, 981.

[807] Sheffy, F. K.; Stille, J. K. *J. Am. Chem. Soc.* **1983**, *105*, 7173.

[808] Yoshida, J.; Funahashi, H.; Iwasaki, H.; Kawabata, N. *Tetrahedron Lett.* **1986**, *27*, 4469.

[809] Kurosawa, H.; Kajimaru, H.; Miyoshi, M.-A.; Ohnishi, H.; Ikeda, I. *J. Mol. Catal.* **1992**, *74*, 481.

[810] Keinan, E.; Greenspoon, N. *Tetrahedron Lett.* **1982**, *23*, 241.

[811] Sano, H.; Okawara, M.; Ueno, Y. *Synthesis* **1984**, *11*, 933.

[812] Keinan, E.; Bosch, E. *J. Org. Chem.* **1986**, *51*, 4006.

[813] Lampilas, M.; Lett, R. *Tetrahedron Lett.* **1992**, *33*, 773.

[814] Katsumura, S.; Fujiwara, S.; Isoe, S. *Tetrahedron Lett.* **1987**, *28*, 1191.

[815] Lampilas, M.; Lett, R. *Tetrahedron Lett.* **1992**, *33*, 773.

[816] Nagano, N.; Itahana, H.; Hisamichi, H.; Sakamoto, K.; Hara, R. *Tetrahedron Lett.* **1994**, *35*, 4577.

[817] Mori, K.; Koga, Y. *Bioorg. Med. Chem. Lett.* **1992**, *2*, 391.

[818] Farina, V.; Baker, S. R.; Benigni, D.; Sapino, C., Jr. *Tetrahedron Lett.* **1988**, *29*, 5739.

[819] van Asselt, R.; Elsevier, C. J. *Organometallics* **1992**, *11*, 1999.

[820] Kraus, G. A.; Ridgeway, J. *J. Org. Chem.* **1994**, *59*, 4735.

[821] Paquette, L. A.; Rayner, C. M.; Doherty, A. M. *J. Am. Chem. Soc.* **1990**, *112*, 4078.

[822] Astles, P. C.; Paquette, L. A. *Synlett* **1992**, 444.

[823] Paquette, L. A.; Astles, P. C. *J. Org. Chem.* **1993**, *58*, 165.

[824] Trost, B. M.; Pietrusiewicz, K. M. *Tetrahedron Lett.* **1985**, *26*, 4039.

[825] Lo Sterzo, C.; Stille, J. K. *Organometallics* **1990**, *9*, 687.

[826] Saha, A. K.; Hossain, M. M. *J. Organomet. Chem.* **1993**, *445*, 137.

[827] Uemura, M.; Nishimura, H.; Hayashi, T. *J. Organomet. Chem.* **1994**, *473*, 129.

[828] Wright, M. E. *J. Organomet. Chem.* **1989**, *376*, 353.

[829] Mitchell, T. N.; Kwetkat, K.; Rutschow, D.; Schneider, U. *Tetrahedron* **1989**, *45*, 969.

[830] Ingham, S. L.; Khan, M. S.; Lewis, J.; Long, N. J.; Raithby, P. R. *J. Organomet. Chem.* **1994**, *470*, 153.

[831] Jevnaker, N.; Benneche, T.; Undheim, K. *Acta Chem. Scand.* **1993**, *47*, 406.

[832] Kosugi, M.; Koshiba, M.; Sano, H.; Migita, T. *Bull. Chem. Soc. Jpn.* **1985**, *58*, 1075.

[833] Kosugi, M.; Ohya, T.; Migita, T. *Bull. Chem. Soc. Jpn.* **1983**, *56*, 3539.

[834] Kosugi, M.; Takano, I.; Sakurai, M.; Sano, H.; Migita, T. *Chem. Lett.* **1984**, 1221.

[835] Ito, Y.; Inouye, M.; Murakami, M. *Tetrahedron Lett.* **1988**, *29*, 5379.

[836] Ito, Y.; Inouye, M.; Murakami, M. *Chem. Lett.* **1989**, 1261.

[837] Kuniyasu, H.; Ogawa, A.; Sonoda, N. *Tetrahedron Lett.* **1993**, *34*, 2491.

[838] Shair, M. D.; Yoon, T.-y.; Danishefsky, S. J. *Angew. Chem., Int. Ed. Engl.* **1995**, *34*, 1721.

[839] Johnson, C. R.; Golebiowski, A.; Braun, M. P.; Sundram, H. *Tetrahedron Lett.* **1994**, *35*, 1833.

[840] Shishido, K.; Goto, K.; Miyoshi, S.; Takaisi, Y.; Shibuya, M. *J. Org. Chem.* **1994**, *59*, 406.

[841] Tanaka, M. *Synthesis* **1981**, 47.

[842] Sakamoto, T.; Yasuhara, A.; Kondo, Y.; Yamanaka, H. *Chem. Pharm. Bull.* **1992**, *40*, 1137.

[843] Bumagin, N. A.; Bumagina, I. G.; Kashin, A. N.; Beletskaya, I. P. *Izv. Akad. Nauk SSSR, Ser. Khim.* **1981**, *7*, 1675; *Chem. Abstr.* **1981**, *95*, 114980.

[844] Liebeskind, L. S.; Yu, M. S.; Fengl, R. W. *J. Org. Chem.* **1993**, *58*, 3543.

[845] Kikukawa, K.; Kono, K.; Wada, F.; Matsuda, T. *Chem. Lett.* **1982**, 35.

[846] Bates, R. W.; Gabel, C. J.; Ji, J. *Tetrahedron Lett.* **1994**, *35*, 6993.

[847] Gregory, W. A.; Brittelli, D. R.; Wang, C. L. J.; Kezar, I., Hollis S.; Carlson, R. K.; Park, C.-H.; Corless, P. F.; Miller, S. J.; Rajagopalan, P.; Wounola, M. A.; McRipley, R. J.; Eberly, V. S.; Slee, A. M.; Forbes, M. *J. Med. Chem.* **1990**, *33*, 2569.

[848] Hartman, G. D.; Halczenko, W. *Synth. Commun.* **1991**, *21*, 2103.

[849] Crouch, G. J.; Eaton, B. E. *Nucleosides & Nucleotides* **1994**, *13*, 939.

[850] Katsumura, S.; Fujiwara, S.; Isoe, S. *Tetrahedron Lett.* **1988**, *29*, 1173.

[851] Bochmann, M.; Kelly, K. *J. Chem. Soc., Chem. Commun.* **1989**, 532.

[852] Bochmann, M.; Kelly, K.; Lu, J. *J. Polym. Sci., Polym. Chem.* **1992**, *30A*, 2503.

[852a] Bochmann, M.; Kelly, K. *J. Polym. Sci., Polym. Chem.* **1992**, *30A*, 2511.

[853] Marsella, M. J.; Swager, T. M. *J. Am. Chem. Soc.* **1993**, *115*, 12214.

[854] Bao, Z.; Chan, W.; Yu, L. *Chem. Mater.* **1993**, *5*, 2; *Chem. Abstr.* **1993**, *118*, 192407.

[855] Yu, L.; Bao, Z.; Cai, R. *Angew. Chem., Int. Ed. Engl.* **1993**, *32*, 1345.

[856] Bochmann, M.; Lu, J. *J. Polym. Sci.: Pt. A. Polym. Chem.* **1994**, *32*, 2493.

[857] Chan, W.-K.; Chen, Y.; Peng, Z.; Yu, L. *J. Am. Chem. Soc.* **1993**, *115*, 11735.

[858] Tamao, K.; Yamaguchi, S.; Shiozaki, M.; Nakagawa, Y.; Ito, Y. *J. Am. Chem. Soc.* **1992**, *114*, 5867.

[859] Kosugi, M.; Arai, H.; Yoshino, A.; Migita, T. *Chem. Lett.* **1978**, 795.

[860] Gronowitz, S.; Malm, J.; Hörnfeldt, A.-B. *Collect. Czech. Chem. Commun.* **1991**, *56*, 2340.

[861] Malm, J.; Rehn, B.; Hörnfeldt, A.-B.; Gronowitz, S. *J. Heterocycl. Chem.* **1994**, *31*, 11.

[862] Kosugi, M.; Ogata, T.; Tamura, H.; Sano, H.; Migita, T. *Chem. Lett.* **1986**, 1197.

[863] Larock, R. C.; Leach, D. R.; Bjorge, S. M. *J. Org. Chem.* **1986**, *51*, 5221.

[864] Malm, J.; Björk, P.; Gronowitz, S.; Hörnfeldt, A.-B. *Tetrahedron Lett.* **1994**, *35*, 3195.

[865] Takacs, J. M.; Chandramouli, S. *Organometallics* **1990**, *9*, 2877.

INDEX

Acetonyltributylstannane, coupling of, 26
Acyl chlorides, coupling of, 16
Acylstannanes, coupling of, 34
Ag(I) salts, and coupling improvement, 7
Air, effect on coupling reaction, 54
Aldehyde synthesis, 17
Alkenyl bromides
  coupling of, 9
  isomerization of, 9
Alkenyl epoxides
  in coupling reaction, 18
  stereochemistry of, 18
Alkenyl halides, reactions of, 9
Alkenyl iodides, coupling of, 9
Alkenyl phenyliodonium salts, coupling of, 21
Alkenyl sulfonates, coupling of, 19–21
Alkenyl triflates, carbonylative coupling of, 40
Alkenylstannanes, 27–30
  coupling of, 27
  formation of, 20
  stereospecificity, 27
  substitution effects, 27
α-Alkoxyalkenylstannanes, coupling of, 30
β-Alkoxyalkenylstannanes, coupling of, 30
α-Alkoxylallylstannanes, coupling of, 34
α-Alkoxystannanes, coupling of, 27
Alkyl halides, coupling of, 23
Alkylstannanes, 25–27
  activation of, 26
  limitations of, 25
  "nontransferable" ligands, 25
  triphenyl arsine as ligand, 25
Alkynyl halides, coupling of, 23
Alkynylstannanes, 32
  coupling of, 32
Allenyl acetates, coupling of, 18
Allenylstannanes, coupling of, 30
Allyl chlorides, carbonylative coupling of, 39
π-Allyl complex, as postulated intermediate, 8
Allylic acetates, cross-coupling of, 17
Allylic chlorides, in oxidative addition, 5
Allylic electrophiles

coupling of, 17
  mechanistic considerations, 17
Allylic halides, coupling with organostannanes, 7
Allylstannanes, 32–34
  isomerization of, 32
  regiochemistry of, 33
Aminostannanes, coupling of, 13
α-Aminostannanes, coupling of, 27
Aminostannanes, coupling of, 35
Ammonia complexes, 55
Anthramycin, coupling of. 22
anti Oxidative addition, 5
Antioxidants, in cross coupling reaction, 54
Arene metallocarbonyls, coupling of, 24
Aryl bromides, coupling of, 12
Aryl chlorides
  coupling of, 12
  effect of activating groups, 12
Aryl diazonium salts
  carbonylative coupling of, 42
  coupling of, 22
Aryl halides
  carbonylative coupling of, 37–39
  coupling of, 12
Aryl ketones, formation of, 40
Aryl stannanes, 30–32
  effect of Cu(I) salts, 30
  electronic effects, 30
  substitution effects, 30
Aryl trichlorostannanes, coupling of, 31
Aryl triflates
  comparison with iodides, 21
  compatible reagents, 22
  coupling of, 21, 25
  and double bond migration, 22
  effect of Cu(I) cocatalyst, 22
  and isomerization, 22
  with LiCl, 21
  optimizing conditions, 22
Arylboron compound, coupling of, 51
5-Arylcytosines, preparation of, 15
Aryldiazonium compounds, coupling of, 28

N-Aryltriflimides, 52
Arylzinc compounds, coupling of, 51
AsPh, as "soft" ligand, 10
AsPh₃, as ligand, 16

9-BBN, coupling of, 51
Benzoprostacyclins, formation of, 43
Benzyl bromide, coupling of, 19
Benzyl chlorides, carbonylative coupling of, 39
Benzyl halides, coupling of, 19
Benzyl trialkylstannanes, 25
Benzylic bromides, oxidative addition, 5
Benzylic stannanes, formation of, 19
Biaryls, formation of, 46
Bifunctional electrophiles, coupling products of, 24
(2,2′-Bipyridine)fumaronitrile palladium(0), as coupling catalyst, 19
1,2-Bis(stannyl)ethylenes, 30
8-Bromoadenosines, coupling of, 15
2-Bromophosphinines, coupling of, 15
Bromopyridines, coupling of, 13
3-Bromoquinolines, coupling of, 13
Bromotropolones, coupling of, 11
5-Bromouracil, coupling of, 15
Butyltin fluoride, 55

Carbacephem triflates, coupling of, 22
γ-Carbalkoxyallylstannanes
   coupling of, 34, 35
   regiochemistry of, 35
Carbamoyl chlorides, coupling of, 17
Carbapenem triflates, coupling of, 22
Carbonylative couplings, 36–46
   with alkenyl triflates, 40
   of aryl diazonium salts, 42
   of aryl halides, 37–39
   with aryl sulfonates, 40–41
   background, 36
   and bond migration, 40
   conditions, 36
   and cross-coupling, 41
   formylation, 38
   of heterocyclic halides, 37–39
   with heterocyclic sulfonates, 40–41
   and lactone formation, 39
   with LiCl additive, 40
   Pd catalyzed, 36
   role of additives, 38
   stereochemistry of, 39
   with α,β-unsaturated aldehydes, 40
   with α,β-unsaturated ketones, 40
   using polymer-supported Pd catalyst, 40

and *Z/E* isomerization, 37
   with zinc chloride additive, 40
Catalyst activation, 53
Cephalosporins, semisynthetic, synthesis of, 17
Cephem triflates, coupling of, 22
α-Chlorocyclobutenones, coupling of, 31
Chloroformates, coupling of, 17
6-Chloropurines, coupling of, 15
2-Chloropyrazines, coupling of, 15
4-Chloropyridine, coupling of, 13
Cine substitution, 48
   and silver carbonate, 49
Cine-substitution process, 28
CO atmosphere, 46
Cocatalyst, 7
Colchicine, 11
Combinatorial libraries, by coupling, 16
Coordinatively unsaturated Pd(0), 5
"Copper effect," 54
Cross-coupling, 17
   effect of LiCl, 17, 40
Crotyltrimethylstannane, coupling of, 33
Cu(I)
   as additive, 11
   as cocatalyst, 7, 10, 11, 20, 27, 28, 29, 30, 22, 47, 48, 56
CuI, as ligand scavenger, 7
Cyclohexenylstannanes, coupling of, 29
Cyclooctatetraenyl bromide, coupling of, 11
Cyclopropyltributylstannane, as cyclopropyl group transfer agent, 27

DBU, to remove tin residue, 55
Decarbonylation, in coupling of acyl halides, 16
Destannylation, 48
Diaryliodonium salts, coupling of, 23
Diazonium salts, coupling of, 22
DIBAL, for catalyst activation, 53
1,1-Dibromoolefins, coupling of, 50
Dibromosiloles, coupling of, 15
2,5-Dibromothiazole, coupling of, 14
Dichloro[1,1′-bis(diphenylphosphino)ferrocene]-palladium, as catalyst, 40
1,4-Diketones, as coupling product, 16
α-Diketones, formation of, 34
Distannane derivatives, as reagents, 34
1,1-Distannylalkenes, coupling of, 30
3,4-Distannylfurans, coupling of, 31
Double bond migration, 22, 49
Dynemicin, synthesis of, 24

E/Z isomerizations, 9, 10, 20, 37, 49, 50
β-Elimination, of Pd hydride, 42, 44, 48

Flash chromatography, for purification of stannanes, 52
p-Fluorophenyl sulfonates, coupling of, 22
Fluorosulfonates, coupling of, 22
Formylation, of aryl iodides, 38
Furanocembranolides, synthesis of, 19
Furanones, formation of, 31
Furyl halides, coupling of, 14
2-Furylstannanes, coupling of, 31

Glycals, α-substituted, 32
Grignard reagent, for catalyst activation, 53

α-Halo ethers, coupling of, 23
α-Halo thioethers, 23
β-Halo-α,β-unsaturated esters, coupling of, 10
β-Halo-α,β-unsaturated ketones, coupling of, 10, 11
β-Halo-α,β-unsaturated sulfoxides, coupling of, 10
Halocyclobutenediones, 11
α-Halolactones, 23
Halopyrimidines, coupling of, 15
Heteroaryl halides, coupling of, 13
Heteroatom-halogen bonds, activation of, 24
Heterocyclic halides, carbonylative couplings of, 37–39
Heterocyclic stannanes, coupling of, 32
Hexaalkyldistannanes, to form benzylic stannanes, 19
Hexabutyldistannane, 35
Hexamethyldistannane, 35
    in experimental procedure, 58
Homocoupling reactions, 46, 51
    of electrophiles, 46
    to form biaryls, 46
    to form symmetrical dienes, 46
    of stannanes, 46

Imidazolyl bromides, coupling of, 14
Imidazolyl dibromides, coupling of, 14
Imidazolyl iodides, coupling of, 14
Imidoyl chlorides, coupling of, 23
Indole derivatives, synthesis of, 13
Indolyl halides, coupling of, 14
Indolyl triflates, coupling of, 22
Insertion of aryl-Pd intermediate, 48
Intramolecular coupling, 7, 16, 18, 20
    and formation of large rings, 18
Intramolecular cross-coupling, 11
8-Iodoadenosines, coupling of, 15
Iodoaromatics, coupling of, 36
Iodobenzene, coupling of, 28

5-Iodocytosines, coupling of, 15
4-Iodoisoxazoles, coupling of, 15
2-Iodopurines, coupling of, 15
5-Iodouracil, coupling of, 15
5-Iodouridines, coupling of, 15
6-Iodouridines, coupling of, 15
Isomerizations, 22
Isoquinolyl triflates, coupling of, 22
Isoquinolylstannanes, coupling of, 31
5-Isoxazolylstannanes, coupling of, 31

Lacrimin A, synthesis of, 28
β-Lactam antibacterials, synthesis of, 22
Lactones, 39
Lavendamycin analog, synthesis of, 13
Leinamycin, synthesis of, 11
LiCl
    as yield enhancer, 20
    influence on transmetallation, 6
LiCl effect, 20, 54
Ligand dissociation, in transmetallation, 6
Ligand scavenger, effect of Cu(I), 7, 11
Ligand transfer, 47–48
"Ligandless" catalysts, in cross-coupling reactions, 3, 53
"Ligandless" conditions, 35, 37
"Ligandless" Pd species, 10
Ligands, proper choice of, 52, 53

Macrocycles, synthesis of, 11, 20
Maleic anhydride, effect of on regiochemistry, 8
Mechanism
    catalytic cycle, 4
    oxidative addition, 5
Mechanistic considerations, 4–9
Mesylates, coupling of, 19
Metallocarbonyls, coupling of, 24
Moisture, effect on coupling reaction, 54

Naphthyl triflates, coupling of, 22
(+)-Negamycin, synthesis of, 41
(−)-5-epi-Negamycin, synthesis of, 41
Ni(0) catalysis, as coupling catalyst, 23, 44
Nickel catalysis, 35
Nitrogen-based ligands, 53
"Nontransferable" ligands, 3, 25
Norbornadiene, coupling of, 43
Nucleophilic assistance at Sn(IV), 8
Nucleosides, substituted, 51

One-electron transfer, during oxidative addition, 5
Organoboron, coupling of, 51

Organoboron reagents, in cross-coupling reactions, 3
Organocopper species, 7, 27
Organomagnesium reagents, in cross-coupling reactions, 3
Organosilicon reagents, in cross-coupling reactions, 3
Organostannanes
  allylic, 8
  aromatic, 9
  cyclic, 9
  reaction with allylic halides, 7
Organotin sulfides, to form C-S bonds, 35
Organozinc, coupling of, 51
Organozinc reagents
  as alternatives, 51
  in cross-coupling reactions, 3
Oxalyl chloride, in coupling reaction, 16
Oxidative addition, 5
  of allylic chlorides, 5
  anti, 5
  CIDNP studies of, 5
  electronic considerations, 5
  one-electron transfer in, 5
  role of oxygen, 5
  stereochemistry of, 5
  syn, 5
Oxygen
  effect on coupling reaction, 54
  in oxidative addition, 5

Pd(0)-carbene species, 48
Pentacoordinate tin, 27
Perfluorinated alkyl iodides, coupling of, 23
Polyene metallocarbonyls, coupling of, 24
Polymer-bound aryl iodides, coupling of, 16
Polymer-supported Pd catalyst, 40
Polymeric materials, formation of, 24
Polypheylenes, synthesis of, 51
Porphyrins, coupling of, 15
Propargyl halides
  coupling of, 19
  limitation of, 19
Propargylic acetates, coupling of, 18
Propargylic derivatives, formation of, 30
Protecting groups, use of, 4
Protodestannylation, 48
Pyridylstannanes, coupling of, 31
Pyrimidines, polyhalogenated, 15
Pyrimidyl triflates, coupling of, 22
2-Pyrrolylstannanes, coupling of, 31

Quinolyl, coupling of, 22

Quinolylstannanes, coupling of, 31
Quinone halides, coupling of, 10

Rapamycin, synthesis of, 12
Reductive elimination, 7
  of π-allyl complexes, 8
  T-shaped intermediate in, 7
Regiochemistry, 4–9
  and maleic anhydride, 8
Removal of tin halides, 54–55
Rhodium catalysis, for coupling of acyl chlorides, 16

pseudo-Saccharyl *O*-ethers, coupling of, 23
S_E2 mechanism, of transmetallation, 6
Selective transfer, 27
Selectivity, of group transfer, 7
Selectride, for catalyst activation, 53
Side reactions, 46–50
  cine substitution, 48
  destannylation, 48
  double bond insertion, 50
  double bond migration, 49
  electrophile reduction, 49
  homocoupling reactions, 46
  hydrolysis, 50
  ligand transfer, 47
  product isomerization, 49
  reduction, 50
Silver carbonate, to prevent cine substitution, 49
"Soft" ligands, 7
  $AsPh_3$, 20
Solvents, for coupling reaction, 53
Sonogashira coupling, 32
β-Stannyl enones, coupling of, 29
3-Stannylfurans, coupling of, 31
Stannylindoles, coupling of, 31
Stereochemistry, 4–9
Steric hindrance, 47
α-Styrylstannanes
  coupling of, 28
  substitution effects, 28
Sulfonates, polyfluorinated, coupling of, 22
β-Sulfonyl alkenylstannanes, coupling of, 29
Symmetrical dienes, formation of, 46
syn Oxidative addition, 5

T-shaped intermediate, in reductive eliminations, 7
Tamoxifen analogs, formation of, 51
Tandem Heck-Stille sequence, 42
  intramolecular, 43
  limitations of, 43
  reverse strategy, 45
Tandem Stille coupling, 12

Terpenic allylstannanes, coupling of, 33
2-Thiazolylstannanes, coupling of, 31
Thienyl halides, coupling of, 14
2-Thienylstannanes, coupling of, 31
Three-component condensation, 46
Tin alkoxides, coupling of, 36
Tosylates, coupling of, 19
Toxicity, of tin compounds, 52
Transmetallation, 5, 7, 48
    LiCl effects in, 6
    ligand dissociation in, 6
    mechanism in, 6
    rate enhancements in, 6
Tri(2-furyl)phosphine
    and aryl transfer, 49
    and *E/Z* isomerization, 50
    as ligand, 6, 16, 33, 43, 53
    experimental procedure for, 57, 58
Tri-*ortho*-tolyl phosphine, as ligand, 11, 43, 53
Tribromophosphinines, coupling of, 15
4-(Tributylstannyl)-2-azetidinones, coupling of, 27
α-(Tributylstannyl)acrylates, coupling of, 28
(Tributylstannyl)acrylates, coupling of, 28
Tributylstannyl-2-(5*H*)-furanones, coupling of, 29
Tributyltin hydride, for reduction of vinyl iodides, 10
Trichlorostannates, 53
Triflates
    coupling mechanism of, 6
    coupling of, 6, 19
    intramolecular coupling of, 7
Triflic anhydride, to form triflates, 52

(Trimethylstannyl)diphenylphosphine, coupling of, 36
Triphenylarsine, as Pd ligand, 6, 53
Triphenylarsine ligand, 25
Triphenylarsine ligand, and aryl transfer, 49

α,β-Unsaturated aldehydes, carbonylative coupling of, 40
α,β-Unsaturated ketones, carbonylative coupling of, 40
Uridine triflates, coupling of, 22

Vinyl iodides
    experimental procedure for coupling, 55
    reduction of, with tributyltin hydride, 10
    tetrasubstituted, 10
Vinyl triflates
    coupling of, 19
    effect of LiCl, 20
    experimental for coupling, 55, 56
Vitamin D, synthesis of, 44

Workup, 54

Zinc acetylides, coupling of, 51
Zinc chloride, as additive in carbonylative coupling reaction, 40
Zn(II) salts
    as yield enhancer, 26
    effect of, 7, 54

DATE DUE